POROUS POLYMERS

About the Cover

Top: Courtesy of Younan Xia, Department of Biomedical Engineering, Washington University in St. Louis, St. Louis, Missouri, USA.

2nd from top: Courtesy of Marc A. Hillmyer, Department of Chemistry, University of Minnesota, Minneapolis, Minnesota, USA. From Mao, H.; Hillmyer, M. A. *Soft Matter* 2006, **2**, 57–59, with permission from the Royal Society of Chemistry.

3rd from top: Courtesy of Michael S. Silverstein, Department of Materials Engineering, Technion – Israel Institute of Technology, Haifa, Israel. From Sergienko, A. Y.; Tai, H. W.; Narkis, M.; Silverstein, M. S. *J Appl Polym Sci* 2004, **94**, 2233–2239, with permission from John Wiley & Sons, Inc.

Bottom: Courtesy of Haifei Zhang, Department of Chemistry, University of Liverpool, Liverpool, UK.

POROUS POLYMERS

Edited by

Michael S. Silverstein
Department of Materials Engineering
Technion – Israel Institute of Technology
Haifa, Israel

Neil R. Cameron
Department of Chemistry
University of Durham
Durham, UK

Marc A. Hillmyer
Department of Chemistry
University of Minnesota
Minneapolis, Minnesota, USA

WILEY

A JOHN WILEY & SONS, INC., PUBLICATION

Published by John Wiley & Sons, Inc., Hoboken, New Jersey.
Published simultaneously in Canada.

Library of Congress Cataloging-in-Publication Data

Porous polymers / edited by Michael S. Silverstein, Neil R. Cameron, Marc A. Hillmyer.
p. cm.
Includes bibliographical references and index.
ISBN 978-0-470-39084-9 (cloth)
1. Porous materials. 2. Plastic foams. I. Silverstein, Michael S. II. Cameron, Neil R. (Neil Ronald), 1969- III. Hillmyer, Marc A.
TA418.9.P6P684 2011
620.1′92–dc22

2010025450

10 9 8 7 6 5 4 3 2 1

CONTENTS

PREFACE

For a very long time there were only a limited number of porous polymeric systems. On one side of the size spectrum, copolymerizations with rigid, multifunctional crosslinking comonomers were employed for the synthesis of glassy polymers with inherent microporosity (we have tried to maintain a certain sense of uniformity and consistency by encouraging the use of International Union of Pure and Applied Chemistry nomenclature: microporous for pore sizes less than 2 nm, mesoporous for pore sizes between 2 and 50 nm, and macroporous for pore sizes greater than 50 nm). In addition, macroporous, crosslinked polymers were synthesized through the use of sacrificial porogens. Both of these porous polymer systems were used in ion-exchange and liquid chromatography applications. On the other side of the size spectrum, polymer foams based on thermosets (e.g., polyurethanes) or thermoplastics (e.g., polystyrene) contained millimeter-sized pores. These lightweight foams were of interest for their mechanical properties per unit mass and for their heat- and sound-insulating properties.

Recently, a large number of innovative routes to myriad porous polymeric systems have been explored to generate new materials with wide-ranging technological applicability. As the need for porous polymers with more complex structures and functions has increased, so has the ability to synthesize such systems with tunable mechanical properties, well-defined pore sizes, specified pore wall functionality, and controlled pore size distributions and interconnectivities. Moreover, such polymers can be generated with inherent microporosity (subnanometer pores) or with templated macroporosity (micrometer-scale pores). Recent advances in the field of porous polymers extend beyond the ability to synthesize novel materials. These new porous polymers challenged the capabilities of the standard characterization methodologies. In response, innovative characterization methodologies were developed to support the synthetic efforts. These novel characterization methodologies provided insight into the porous structures and their effects on the material's properties. These new porous polymeric materials are now being adapted for applications in the fields of microelectronics, biomedical devices, membrane processes, and catalysis. In addition, porous polymers are being used as templates for the production of porous ceramics and other materials. As novel materials, innovative characterization methodologies, and new applications have been developing rapidly, we saw a need to collect and organize the information under one compendium.

This book gathers the various aspects of the porous polymer field into one volume. The book not only presents a fundamental description of the field, but also describes the state of the art for such materials and provides a glimpse into the future. Emphasizing a different aspect of the ongoing research and development in porous polymers, the book is divided into three sections: Synthesis, Characterization, and Applications. The Synthesis section explores the different methods developed to synthesize porous polymers, from systems with inherent porosity to templating approaches. The Characterization section collects different approaches to describing the pore size, pore size distribution, and porous structure, as well as the effects that porosity has on the properties. The Applications section gathers the various ways in which contemporary porous polymers are incorporated into a diverse and growing number of practical applications. The first part of each chapter presents the basic scientific and engineering principles underlying the topic, while the second part presents state of the art results based on those principles. In this fashion, the book connects and integrates topics from seemingly disparate fields, each of which embodies different aspects inherent in the diverse field of porous polymeric materials.

ACKNOWLEDGMENTS

The editors would like to thank Edmund Immergut for suggesting that a comprehensive book on porous polymers would make a useful addition to the polymer literature. The editors would also like to thank all of the chapter authors for contributing their valuable time and for their cooperation throughout this project. Michael Silverstein would like to thank his extremely patient wife, Efrat, and their relatively patient children, Dana, Asaf, Gilad, and Eitan. Neil Cameron would like to dedicate this book to his friend and mentor, Professor David C. Sherrington FRS, an inspirational figure in the field of porous polymers.

Michael S. Silverstein
Zichron Yaakov, Israel

Neil R. Cameron
Durham, UK

Marc Hillmyer
Minneapolis, Minnesota, USA

January 2011

CONTRIBUTORS

Markus Antonietti, Max-Planck-Institute of Colloids and Interfaces, Potsdam, Germany

Anand Badami, The Dow Chemical Company, Midland, Michigan, USA

Mikhail R. Baklanov, IMEC, Leuven, Belgium

Jonathan Behrendt, Aston University, Birmingham, United Kingdom

Charles T. Black, Brookhaven National Laboratory, Upton, New York, USA

Peter M. Budd, University of Manchester, Manchester, United Kingdom

Neil R. Cameron, University of Durham, Durham, United Kingdom

Bob Cieslinski, The Dow Chemical Company, Midland, Michigan, USA

William Heeschen, The Dow Chemical Company, Midland, Michigan, USA

Marc A. Hillmyer, University of Minnesota, Minneapolis, Minnesota, USA

Steven Howdle, University of Nottingham, Nottingham, United Kingdom

Peter Krajnc, University of Maribor, Maribor, Slovenia

Neil B. McKeown, Cardiff University, Cardiff, United Kingdom

Gary Mitchell, The Dow Chemical Company, Midland, Michigan, USA

Gregory Meyers, The Dow Chemical Company, Midland, Michigan, USA

Yvonne Reinwald, University of Nottingham, Nottingham, United Kingdom

Rolando M. A. Roque-Malherbe, University of Turabo, Gurabo, Puerto Rico, USA

Deborah Rothe, The Dow Chemical Company, Midland, Michigan, USA

Steve Rozeveld, The Dow Chemical Company, Midland, Michigan, USA

Lei Qian, University of Liverpool, Liverpool, United Kingdom

Denis Shamiryan, IMEC, Leuven, Belgium

Kevin Shakesheff, University of Nottingham, Nottingham, United Kingdom

Michael S. Silverstein, Technion – Israel Institute of Technology, Haifa, Israel

Andrew Sutherland, Aston University, Birmingham, United Kingdom

Arne Thomas, Max-Planck-Institute of Colloids and Interfaces, Potsdam, Germany

Clifford Todd, The Dow Chemical Company, Midland, Michigan, USA

Eric M. Todd, Monmouth College, Monmouth, Illinois, USA

Mathias Ulbricht, Universität Duisburg-Essen, Essen, Germany

Jens Weber, Max-Planck-Institute of Colloids and Interfaces, Potsdam, Germany

Charlie Wood, The Dow Chemical Company, Midland, Michigan, USA

Haifei Zhang, University of Liverpool, Liverpool, United Kingdom

SECTION I

SYNTHESIS

CHAPTER 1

Polymers with Inherent Microporosity

NEIL B. McKEOWN
Cardiff University, Cardiff, United Kingdom

PETER M. BUDD
University of Manchester, Manchester, United Kingdom

1.1 INTRODUCTION

Micropores are defined by the International Union of Pure and Applied Chemistry (IUPAC) [1] as pores with widths not exceeding 2 nm. Solids containing accessible, interconnected micropores behave as molecular sieves and possess large internal surface areas—typically 300–3000 m^2 g^{-1}, as measured by techniques based on the analysis of gas adsorption isotherms, such as the well-established Brunauer, Emmet, and Teller (BET) model [2]. Conventional microporous materials, such as zeolites and activated carbons, are widely used as adsorbents and heterogeneous catalysts and for molecular separations on the basis of size and shape [3]. Recently developed crystalline materials, such as the metal–organic frameworks (MOFs) [4] and the related covalent–organic frameworks (COFs) [5], which mimic the ordered micropore structure of zeolites, have gained much attention [6]. Similarly, work on the preparation of polymer-based organic microporous materials is of growing importance and forms the subject of this chapter.

In general, polymers have sufficient conformational and rotational flexibility to allow them to maximize intermolecular and intramolecular cohesive interactions and pack space efficiently in the solid state. Nevertheless, the packing efficiency of a polymer depends upon many factors, including its physical state (e.g., crystalline or glass), its molecular structure (e.g., flexibility, shape), and its recent history (heat treatment, swelling by solvent). Hence its "free volume"—that is, the space within the material not occupied by the polymer—can vary considerably, and if the material possesses a sufficiently large free volume, interconnectivity will occur, and it will possess significant microporosity. There are three distinct strategies for the synthesis

Porous Polymers, Edited by Michael S. Silverstein, Neil R. Cameron, and Marc A. Hillmyer

of polymers with inherent microporosity. First, excess free volume can be trapped by the formation of an amorphous hypercrosslinked polymer network, which on removal of the included solvent provides a predominantly microporous material. Second, polymers (e.g., polymers of intrinsic microporosity) can be designed to possess macromolecular structures that are both rigid and contorted so as to pack space very inefficiently, resulting in a large amount of interconnected free volume. Such polymers with "intrinsic" microporosity may be a network or a nonnetwork polymer; the former may also trap additional excess free volume, whereas the latter can be soluble and, therefore, solvent processable. Finally, the COFs are a logical extension of the MOF concept in which reversible boronic ester linkages connect rigid organic units in a crystalline network. The synthesis, properties, and applications of each of these three types of microporous polymer are fully reviewed here. First, to place each material into its appropriate context, the fundamental principles of polymer free volume will be discussed.

1.1.1 Unoccupied Space in a Polymeric Material: From Free Volume or "Excess" Free Volume to Microporosity

The specific volume of a polymer, V (the reciprocal of polymer density), is made up of both the volume occupied by the polymer molecules and the free volume of the material. As a first approximation, the occupied volume can be calculated as the specific van der Waals volume (V_w) by group contribution methods [7, 8]. However, even for perfectly ordered crystals at absolute zero, molecules cannot completely fill space. An average correction factor of 1.3 is widely used in estimates of the occupied volume, despite the fact that different polymers are sterically very different. Fractional free volume is thus often defined as $f_v = (V - 1.3V_w)/V$, which generally falls within the range of 0.10–0.23 for commonly encountered polymers [9]. However, there are other definitions of free volume that give smaller values of fractional free volume at normal temperatures because they include the effects of thermal vibrations within the occupied volume. Furthermore, in experiments involving probe molecules (e.g., nitrogen adsorption), the effective free volume depends on the size of the probe and may be affected by the presence of the probe.

For a polymer, V generally increases linearly with temperature (T), but a change of slope is encountered at the glass transition temperature (T_g), so that the expansion coefficient is higher in the rubbery state (above T_g) than in the glassy state (below T_g) (Fig. 1.1). This transition is strongly influenced by kinetics, so that, if a rubber is cooled, a 10-fold increase in the rate of cooling typically leads to a 3 K increase in T_g. If a rapidly cooled sample of polymer is then held at a constant temperature a little below T_g, it is found that V decreases over time, a process called isothermal volume recovery or physical aging. This can be understood as *excess* free volume (V_{ex}) trapped on rapid cooling and slowly lost on physical ageing. Similarly, excess free volume can be trapped by the rapid removal of a solvent from a solvent-swollen polymer. Hence, if sufficient excess free volume can be trapped within an amorphous polymer, a microporous material will result, although the microporosity may subsequently be lost through physical aging.

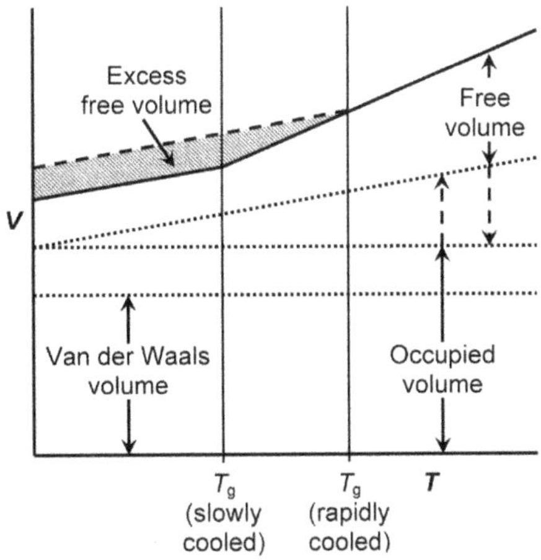

FIGURE 1.1 Schematic illustration of the dependence on temperature of the specific volume of a polymer, indicating the distinctions among molecular volume, occupied volume, free volume, and excess free volume. Note that there are various definitions of free volume, which imply different boundaries between occupied volume and free volume. Furthermore, different types of experiment may access different levels of free volume.

1.2 HYPERCROSSLINKED POLYMERS

A large amount of excess free volume can be captured by the formation of a so-called hypercrosslinked network polymer within a solvent-swollen polymer gel, which provides a microporous material on removal of the solvent. The most-investigated class of microporous polymers prepared using this concept is based on hypercrosslinked polystyrene. These polymers were first prepared and patented by Davankov and coworkers in the late 1960s and early 1970s [10–13]. Hypercrosslinked polystyrenes are now commercially available materials, obtainable with a range of microporosity and surface functionality, and are (or have been recently) marketed under the names Styrosorb and Hypersol-Macronet (Purolite); Optipore (Dow); Lewatit OC 1163 and S 7768 (AG); NG-99 and NG-100 (Jiangsu N&G Environmental Technology Co. Ltd); Amberchrom GC-161m (Toso-Haas); Chromabond HR-P (Macheney-Nagel); HySphere-SH (Spark Holland); Envi-Chrom P (Supelco); and Lichrolut EN (Merck).

1.2.1 The Synthesis of Hypercrosslinked Polymers

By far the most-studied and most-utilized hypercrosslinked polymers are derived from polystyrene. The general route to such materials involves extensive crosslinking of solvent-swollen, lightly crosslinked polystyrene beads prepared previously by suspension polymerization [12,14,15]. The crosslinking is achieved by using an efficient Friedel–Craft alkylation reaction mediated by a Lewis acid catalyst such as $FeCl_3$ or $SnCl_4$ (Fig. 1.2). The crosslinks are typically derived from reactive dichloromethyl

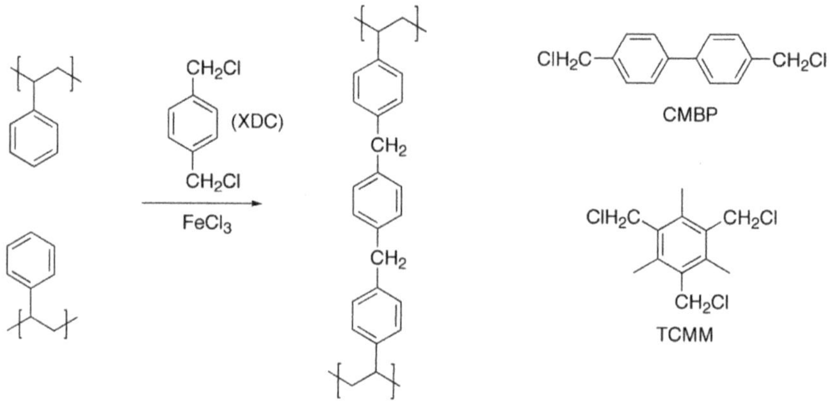

FIGURE 1.2 The synthesis of hypercrosslinked polystyrene via Friedel–Craft alkylation using a di- or trichloromethylaromatic compound.

or trichloromethyl aromatic compounds such as *p*-xylylenedichloride (XDC), 4,4′-bis-chloromethylbiphenyl (CMBP), or tri-(chloromethyl)mesitylene (TCMM) [12]. In addition, simple methylene crosslinks can be formed directly by the use of monochlorodimethylether (MCDE), or alternatively, they can be introduced by using vinylbenzyl chloride (VBC) as the original styrenic monomer (e.g., Merrifield resins) and then subjecting the resulting resin to a Lewis acid–mediated Friedel–Craft reaction (Fig. 1.3) [14]. The degree of crosslinking is best measured by a combination of elemental analysis to determine the reduction in chlorine [16] and ^{13}C solid-state nuclear magnetic resonance to evaluate the relative amount of methylene crosslinking [12, 17]. It has been shown that the subsequent properties of the hypercrosslinked

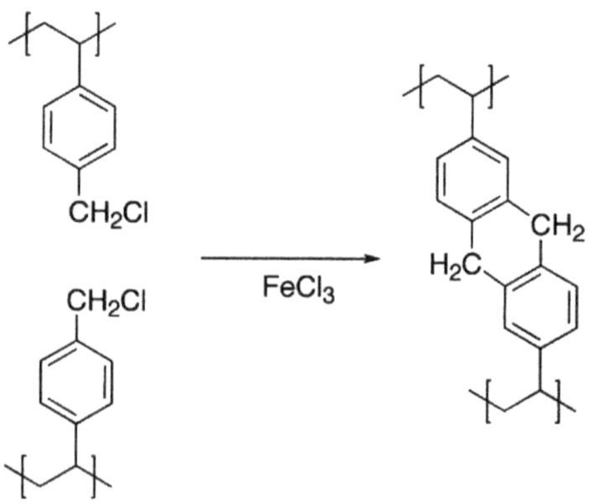

FIGURE 1.3 The synthesis of hypercrosslinked polystyrene via Friedel–Craft alkylation using chloromethylated polymer precursor derived from vinylbenzyl chloride. The fused-ring crosslinker is idealized but is consistent with the maximum number of crosslinks per repeat unit of 2 estimated by elemental analysis.

polymer (e.g., porosity) are heavily influenced by the reactions conditions used to form the crosslinkages. In particular, the average number of crosslinks per repeat unit strongly influences the mechanical properties of the material [18] and appears crucial to the attainment of a large degree of microporosity [12]. For example, it has been reported that the $FeCl_3$-mediated crosslinking of poly(vinylbenzyl chloride) beads swollen with 1,2-dichloroethane at 80°C for 12 h provides hypercrosslinked polystyrene with the greatest amount of microporosity (e.g., BET surface area >2000 $m^2\ g^{-1}$). It was postulated that the likely formation of rigid cyclic crosslinks (Fig. 1.3), which provide two crosslinks per repeat unit, helps to enhance the introduction and stability of the resulting porosity [14]. Davankov and coworkers also described the hypercrosslinking of soluble polystyrene of uniform mass distribution, which gives remarkably uniform spherical nanoparticles that remain as a suspension in solvent. Light scattering and electron microscopy suggest that the diameter of the particles is 12 nm in the unswollen state but increases to 17 nm when swollen with solvent. These particles self-assemble in tetrahydrofuran to give discrete clusters containing 13 nanoparticles [19–21].

Once a hypercrosslinked polystyrene is prepared it can be modified by the many well-established reactions of polystyrene (e.g., sulfonation) to obtain materials that are more hydrophilic and can perform efficient ion exchange [22]. Some of these ion-exchange materials are also commercially available (e.g., Hypersol-Macronet).

In addition to hypercrosslinked polystyrenes, it has been established that many other polymers exhibit significant microporosity when extensively crosslinked. Of particular note are the polymers derived from xylylenedichloride (XDC) and 4,4′-bis-chloromethylbiphenyl (CMBP) using a $FeCl_3$-mediated Friedel–Craft alkylation reaction (Fig. 1.4a), with the hypercrosslinked polymer derived from CMBP providing a degree of microporosity similar to that of the most porous hypercrosslinked polystyrenes [23]. In principle many different types of reaction can be used to prepare hypercrosslinked polymers, with successful examples including the addition of

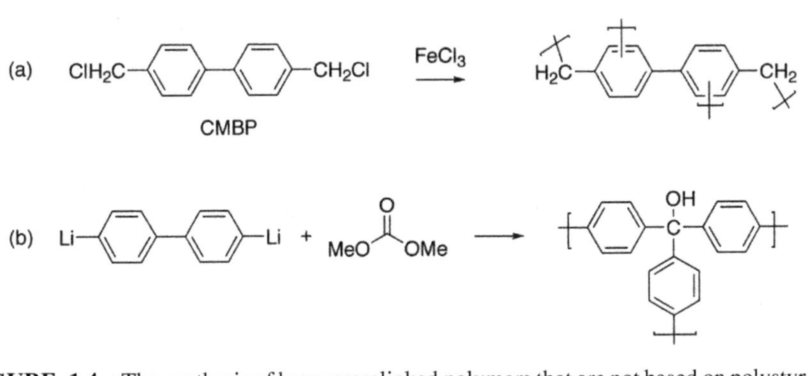

FIGURE 1.4 The synthesis of hypercrosslinked polymers that are not based on polystyrene. **(a)** The polymerization of 4,4′-bis-chloromethylbiphenyl (CMBP) using an $FeCl_3$-mediated Friedel-Craft alkylation reaction. **(b)** The reaction of an aryl dilithium with dimethylcarbonate to give a hypercrosslinked polyarylcarbinol.

aryl dilithium salts to dimethylcarbonate [24, 25] (Fig. 1.4b) or to tetraethylorthosilicate [26] and the nucleophilic substitution reaction between diiodomethane and polyaniline [27]. Recent work has also established the merit of using metal-mediated cross-coupling reactions such as the Sonigashura–Hagihari [28, 29] and Suzuki reactions [30] . With the appropriate choice of monomer, these reactions can result in the preparation of fully conjugated covalent networks with inherent microporosity, which may ultimately have applications as organic electronic materials. It is now clear that many other efficient reactions that involve the formation of rigid bonds could be used to prepare hypercrosslinked microporous polymers.

1.2.2 The Properties of Hypercrosslinked Polymers

1.2.2.1 Porosity. A number of techniques, including nitrogen adsorption [11], mercury intrusion porosimetry [31], inverse size exclusion chromatography [32], and positronium annihilation lifetime (PAL) spectroscopy [33–35], have been used to determine the amount and size distribution of the porosity obtained during hypercrosslinked polystyrene formation, and the information obtained has been reviewed comprehensively [31]. For all hypercrosslinked materials, nitrogen adsorption isotherms measured after the complete removal of solvent and adsorbed gas under vacuum show a significant degree of adsorption at low partial pressure ($p/p_o < 0.1$), consistent with a predominantly microporous structure [12, 31]. The nitrogen adsorption data can be used to calculate an apparent surface area (S) of the material using the classic BET analysis. Similarly, the apparent micropore volume can be calculated from the amount of nitrogen adsorbed at low relative pressure. Depending upon the form of the initial polymer (e.g., macroporous beads) and the solvent using during the formation of the hypercrosslinked structure, the isotherm may also demonstrate the characteristic features, at higher relative pressures, associated with nitrogen adsorption within mesoporosity and macroporosity. Mesoporosity is apparent from the appearance of a distinct hysteretic loop between the adsorption and desorption isotherms at mid values of relative pressure ($p/p_o = 0.5$–0.9) [31]. This effect is due to pore filling occurring by the condensation of liquid nitrogen within the mesopores. Some degree of mesoporosity is usually found within hypercrosslinked polymers, but the amount can be enhanced by the use of a poor solvent (e.g., hexane) to encourage microphase separation during formation of the hypercrosslinked network. Macroporosity, simply obtained from the use of macroporous beads as the polymeric precursor to the hypercrosslinked material, is evident from adsorption at high values of relative pressure ($p/p_o = 0.9$–1.0) [14].

It is clear that a significant amount of microporosity (e.g., $S > 500\ m^2\ g^{-1}$; micropore volume $>0.3\ cm^3\ g^{-1}$) is attained for a hypercrosslinked polymer only for highly rigid networks at a level of crosslinking greater than 40% (e.g., for polystyrene at least 40% of benzene units contain a crosslink). As noted, the highest amount of microporosity (e.g., $S > 2000\ m^2\ g^{-1}$; micropore volume $>0.7\ cm^3\ g^{-1}$) is attained for a degree of crosslinking approaching 200% (i.e., for polystyrene the average number of crosslinks attached to each benzene ring is two; Fig. 1.3) [14].

1.2.2.2 Swelling in Solvent. One of the most remarkable properties of hypercrosslinked polystyrenes is their tendency to swell greatly when in contact with solvent, even if the solvent is not one that is normally considered compatible with polystyrene, such as hexane, methanol, or even water [36, 37]. This effect is undoubtedly advantageous for the performance of these polymers as adsorbents, and its origin has been the subject of much debate in the literature [16, 18, 36, 38]. The most plausible explanation is related to the introduction of stress to the covalent network during the removal of solvent. During the network formation the polymer is fully solvated by a solvent that is compatible with polystyrene (e.g., dichloromethane or tetrachloroethane). Removal of the solvent results in a shrinking of the material and the introduction of internal stress due to the interconnectivity of the cyclic structures that dominate the highly rigid, covalent microporous network. The small difference in the solvation energy between a good (i.e., compatible) solvent for polystyrene (e.g., toluene) as compared with a poor (i.e., incompatible) solvent (e.g., methanol or water) is insignificant as compared to the release of internal stress that accompanies the adsorption of solvent within an evacuated hypercrosslinked polymer. It is notable that the amount of swelling on contact with poor solvents is, like the degree of microporosity, related to the average number of cross-links formed per repeat unit of the polymer [15, 18, 36–38].

1.2.3 Applications of Hypercrosslinked Polymers

1.2.3.1 Adsorption. The potential of hypercrosslinked polystyrenes as adsorbents was soon recognized and investigated by Davankov and coworkers [39, 40]. Since these early studies hypercrosslinked polymers have been used for the adsorption of organic vapors (e.g., hexane, tetrachloromethane, pyridine) [41, 42] and organic contaminants from water [43, 44] (e.g., various phenols [22, 45 49], naphthols and naphthamines [50], phenylhydrazines [51], polycyclic aromatic hydrocarbons [52,53], methomyl [54], acid red dye [55], and phthalate esters [56]). Amine-modified hypercrosslinked polystyrene has been used for the adsorption of phenols, aromatic acids, and aromatic sulfonates from water [48, 57–60], and hydroxy-modified polymers have been used for the removal of polar organic compounds (oxamyl, methomyl, and desisopropylatrazine) [61].

1.2.3.2 Substrates for Chromatography. Due to their hydrophobic nature coupled with their unusual swelling properties in highly polar solvents such as water, hypercrosslinked polystyrene beads show promise as the stationary phase for reverse-phase high-performance liquid chromatography (RPHPLC) [32, 62]. For example, they have been used for the analysis of phenols, catechols, and resorcinols in water down to a concentration of only 2 mg L^{-1} and for aqueous size-exclusion chromatography for the analysis and isolation of metal salts [63, 64]. Hypercrosslinked polystyrene beads modified by sulfonation have been used as substrates for the effective separation of organic acids [65].

1.2.3.3 Supports for Catalytic Metal Nanoparticles. The high surface area, solvent accessibility, and uniform size of hypercrosslinked polystyrene beads suggest great promise as supports for catalysts. In particular, they have been studied as substrates for the synthesis and support of metal nanoparticles for use in heterogeneous catalysis [66]. Supported platinum nanoparticles have be used successfully for the oxidation of L-sorbose [67] and phenol [68], and ruthenium nanoparticles have demonstrated high efficiency in D-glucose oxidation [69].

1.2.3.4 Hydrogen and Methane Storage. A major technical obstacle to the widespread use of hydrogen (H_2) as a nonpolluting fuel for cars is the lack of a safe and efficient system for on-board storage. Of the many potential solutions being investigated, an attractive possibility is a system based on the reversible adsorption of H_2 on the internal surface of a microporous material [70]. At present the quantity of H_2 that can be adsorbed onto any type of microporous material at a technologically reasonable range of temperature and pressure falls below the requirements of a practical H_2 storage system. Hence, there is an urgency to develop materials that can be tailored to provide a structure and chemical composition suitable for the specific demands of H_2 physisorption. Moderate values of hydrogen (up to 1.2% by mass [71,72]) can be adsorbed onto hypercrosslinked polystyrenes at 1 bar and 77 K, with the amount increasing to 3.0% at 15 bar [72]. Slightly higher values of H_2 loading (3.7% by mass at 15 bar and 77 K) can be obtained for hypercrosslinked polymers derived from the Friedel–Craft alkylation polymerization of 4,4′-bis(chloromethyl)-1,1′-biphenyl [23]. Hypercrosslinked polyanilines have also been investigated for this application and, despite exhibiting relatively modest H_2 uptake due to low apparent BET surface areas, are promising materials for hydrogen storage due to their remarkably high enthalpy of adsorption ($\sim$15 KJ mol^{-1}), which may allow practical storage at higher temperatures than 77 K [27,73].

A hypercrosslinked polymer derived from $\alpha\alpha'$-dichloro-*p*-xylene has been studied as a material for methane storage with loadings (e.g., 8.3% by mass at 20 bar and 77 K) that are competitive with other microporous materials such as activated carbons and metal–organic frameworks [74].

1.3 POLYMERS OF INTRINSIC MICROPOROSITY

1.3.1 The Concept of Polymer "Intrinsic" Microporosity

Ilinitch and coworkers first defined *intrinsic microporosity* in polymers as "a continuous network of interconnected intermolecular microcavities" [75,76]. However, the polymer used to illustrate the concept, poly(phenylene oxide), showed significant microporosity by nitrogen adsorption (e.g., BET surface area = 400 m^2 g^{-1}) only immediately after swelling by propene at high pressure, which is then lost on annealing. Therefore, this is perhaps best regarded as an example of induced excess free volume. A more complete definition of "intrinsic microporosity" is "a continuous network of interconnected intermolecular voids that forms as a direct consequence

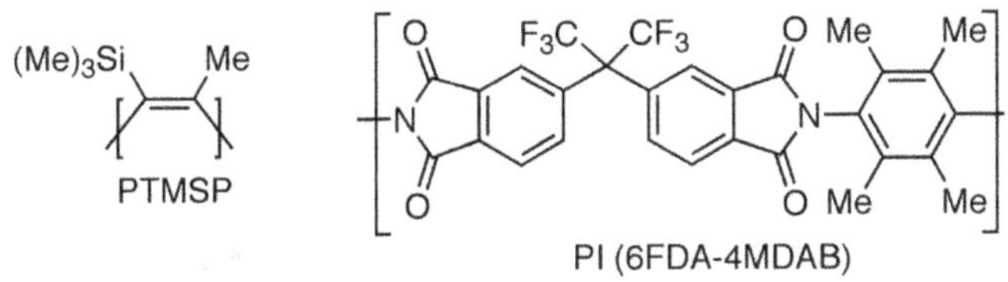

FIGURE 1.5 Examples of polymers that possess intrinsic microporosity.

of the shape and rigidity of the component macromolecules." Examples of polymers *with* intrinsic microporosity (PIMs) are encountered in the polymer membrane field, where they are more commonly referred to as high free volume or ultrapermeable polymers. In particular, poly(1-trimethysilylprop-1-yne) (PTMSP; Fig. 1.5), which is rigid and contorted due to the combination of alternating single and double bonds along with severe steric crowding caused by the bulky trimethylsilyl group, has been described as nanoporous on the basis of its very high gas permeability [77–79]. Analysis by nitrogen adsorption confirms its essentially microporous nature, with a BET surface area in excess of 900 $m^2\ g^{-1}$, although this is reduced to less than 700 $m^2\ g^{-1}$ on aging, which suggests that some of the microporosity is due to excess free volume rather than intrinsic microporosity [9]. Other examples of polymers with significant intrinsic microporosity, such as a number of fluorinated polymers (e.g., Teflon AF2400) [80] and some aromatic polyimides (e.g., 6FDA-4MAB; Fig. 1.5) [81] that were specifically designed to possess enhanced rigidity, can also be cited as polymers with intrinsic microporosity based upon their high gas permeabilities (see Section 1.3.4.1). It should be emphasized that microporosity here refers to voids with a dimension of less than 2 nm, and we are concerned with polymers that lie on the boundary between dense polymer systems and more conventional porous materials.

1.3.2 Synthesis of PIMs

PIMs are a highly versatile class of materials that possess microporosity, as demonstrated by high BET surface areas revealed by nitrogen adsorption (500–1730 $m^2\ g^{-1}$) [9, 82–84]. The microporosity is attributable directly to their highly rigid and contorted molecular structures (Fig. 1.6), which cannot pack space efficiently. In particular, the lack of rotational freedom along the polymer backbone ensures that the macromolecules cannot rearrange their conformation to cause the collapse of the micropore structure. The rigidity and lack of rotational freedom are enforced by the polymer backbone being composed solely of fused rings. The necessary sites of contortion are generally provided by spiro centers (i.e., a single tetrahedral carbon atom shared by two rings; e.g., PIM-1; Fig. 1.6b) or other rigid nonplanar structural units (e.g., triptycene; Fig. 1.6c). PIMs could be made using any number of different bond-forming reactions (e.g., aromatic imide formation) [30, 85–88], but a convenient and highly efficient method is to use the double aromatic nucleophilic substitution (S_NAr) reaction [89, 90] between monomers that incorporate catechol units

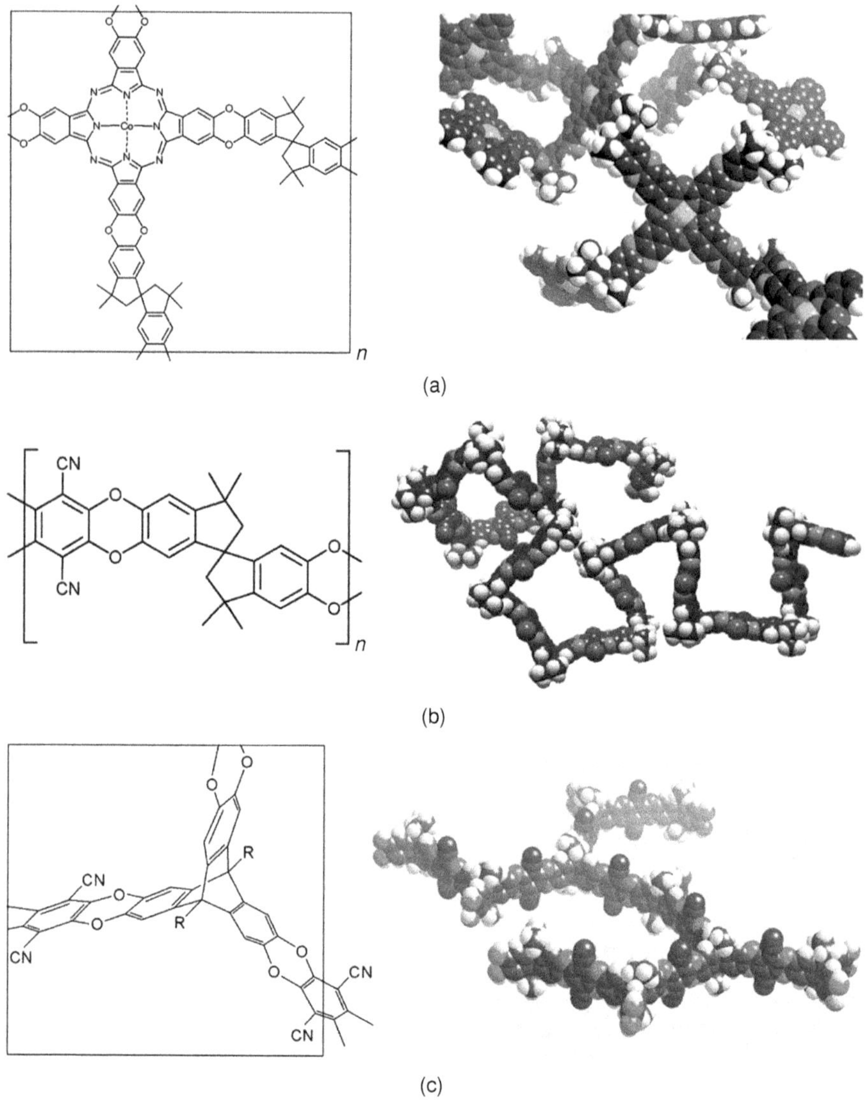

FIGURE 1.6 Molecular structures and molecular models of **(a)** a phthalocyanine-based network PIM, **(b)** the soluble PIM-1, and **(c)** a triptycene-based PIM (in the model, R = Et), showing types of structure that lead to intrinsic microporosity.

(i.e.; 1,2-dihydroxybenzene; e.g., Fig. 1.7, monomers **A1–A6**) and 1,2-difluoro- or 1,2-dichlorobenzene units (e.g., Fig. 1.7, monomers **B1–B7**) [84,91]. The key to the success of this reaction is that the second intermolecular aryl-*O*-aryl bond formation occurs rapidly to give the fused dioxane ring. Depending upon the average functionality (f_{av}) of the monomers used in their preparation (i.e., number of catechol or

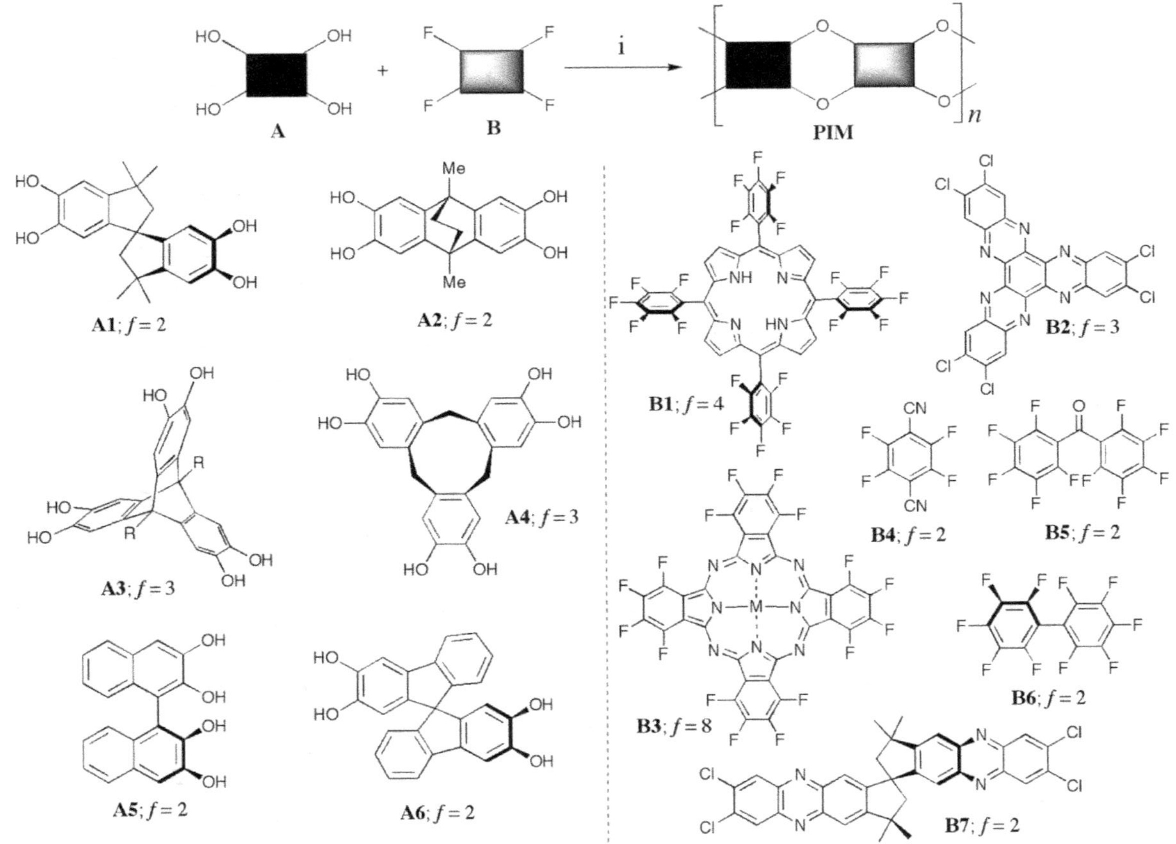

FIGURE 1.7 PIMs are prepared via a polymerization reaction using a combination of appropriate hydroxylated aromatic monomers (e.g. **A1–A6**) and fluorinated (or chlorinated) aromatic monomers (e.g. **B1–B7**). For microporosity, at least one of the monomers must contain a site of contortion, which may be a spiro center (e.g. **A1, A6,** or **B7**), a single covalent bond around which rotation is hindered (e.g. **A5, B1,** or **B6**), or a rigid, nonplanar skeleton (e.g. **A2, A3,** or **A4**). For insoluble networks the average functionality of the monomers f must be greater than 2, whereas for soluble polymers it must be equal to 2.

1,2-dihalogenated benzenes per monomer), PIMs can be prepared either as highly insoluble network polymers ($f_{av} > 2$) or as soluble polymers ($f_{av} = 2$) that are suitable for solution-based processing, which represents a unique advantage over conventional microporous materials [92]. Analysis of the soluble PIM (PIM-1) formed by the reaction between the two commercial monomers **A1** and **B4** using gel permeation chromatography shows that the weight-average molar mass (M_w) is in excess of 200,000 g mol^{-1}. This illustrates the remarkable efficiency of the double S_NAr polymerization reaction, which has been studied in depth [93], together with the unexpected predominance of cyclic oligomers that it produces [94–96]. The structural diversity of PIMs, which is provided by the wide choice of monomer precursors (Fig. 1.7), means that both the structure and properties can be tailored to fit the intended application.

1.3.3 Properties of PIMs

1.3.3.1 Microporosity. The key property of the PIMs is their microporosity, with a BET analysis of nitrogen adsorption data at 77 K being the standard technique used for the determination of apparent accessible surface area. Using this technique, it is found that PIMs display surface areas in the range of 500–1730 m^2 g^{-1}, with the triptycene-based network PIM (structure shown in Fig. 1.6c), prepared from monomers **A3** (R = Me) and **B4** (Fig. 1.7), currently holding the record for the highest-surface-area PIM. It is thought that the ribbon-like structure of this PIM afforded by the rigid triptycene unit helps to thwart space-efficient packing of the macromolecules by preventing the planar aromatic components from coalescing via cofacial π-π interactions [97]. For triptycene-based PIMs (Trip-PIMs), the microporosity can be modulated by the alkyl chain attached to the bridgehead positions of the triptycene unit, with longer alkyl chains providing materials with lower microporosity (e.g., R = octyl; BET surface area = 600 m^2 g^{-1}), presumably due to pore blocking by the flexible chains. The use of branched alkyl chains (e.g., R = *iso*-propyl; BET surface area = 1600 m^2 g^{-1}) favors greater microporosity relative to unbranched chains (R = *n*-propyl; BET surface area = 1200 m^2 g^{-1}).

Atomistic computer simulation may be used to visualize the microporosity of a PIM [98]. Four slices through a model of packed PIM-1 (structure shown in Fig. 1.6b) can be seen in Fig. 1.8, which shows both the amorphous structure of the polymer and the empty spaces that constitute free volume elements. Analysis of this type of model indicates that a small probe particle such as positronium can access extended regions of free volume, that is, intrinsic microporosity. A larger probe, such as a nitrogen molecule, cannot penetrate all the free volume elements in a static model. However, in a real experiment such as nitrogen adsorption the dynamics of the process is also important. While the polymer is rigid in the sense that there is little freedom of rotation about backbone bonds, some flexing of ribbon-like sequences is possible, and indeed some bent units can be seen in Fig. 1.8. In a nitrogen adsorption experiment, redistribution of free volume and swelling can occur, which leads to additional nitrogen uptake and gives rise to a distinct hysteresis, with the desorption curve lying above the adsorption curve down to very low relative pressures [87].

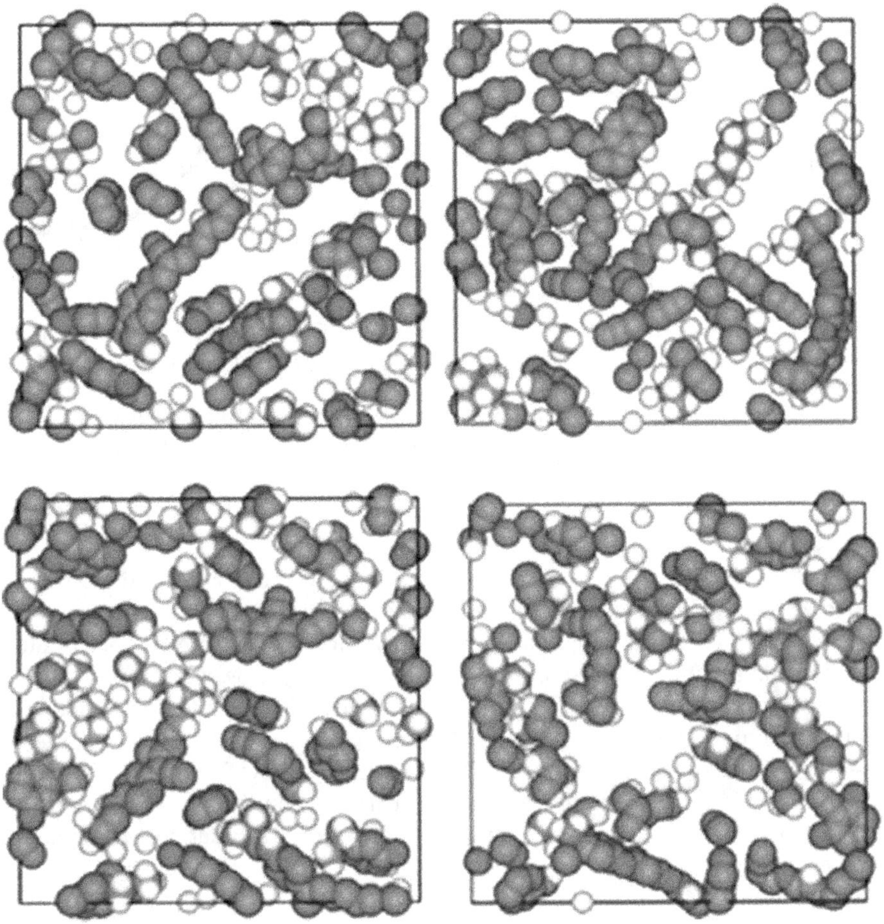

FIGURE 1.8 Four slices (each 0.31 nm thick) through a computer model of packed PIM-1 [98].

This is illustrated by the experimental data in Fig. 1.9a, which also shows simulated nitrogen adsorption data for a static model of packed PIM-1. The simulated isotherm is typical for a microporous material, reaching a plateau at low relative pressure, and gives an apparent surface area of 435 $m^2\ g^{-1}$. This may be thought of as representing the instantaneous microporosity. The experimental isotherm shows a higher BET surface area (780 $m^2\ g^{-1}$ for the data shown), and uptake continues to increase up to atmospheric pressure.

Assumptions have to be made to convert a complex arrangement of free volume into a pore size distribution. In Fig. 1.9b, the bars represent a pore size distribution from modeling, using an approach that divides highly elongated regions of free volume into smaller elements. The circles show a pore size distribution calculated from experimental data by the Horvath–Kawazoe method [99], assuming slit-shaped

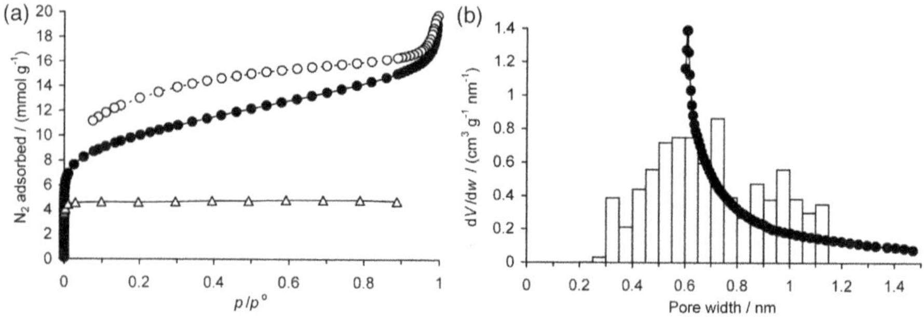

FIGURE 1.9 **(a)** The N_2 adsorption isotherm at 77 K calculated for a static computer model of packed PIM-1 (△) [98] and experimental N_2 adsorption (•) and desorption (○) curves for a sample of PIM-1. **(b)** Micropore distributions calculated by computer simulation (grand canonical Monte Carlo method) for a static model of packed PIM-1 (bars) and from the low-pressure experimental N_2 adsorption data (Horvath–Kawazoe method, assuming slit pores) (•).

pores. Both simulation and experimental data suggest a large number of pores with effective widths in the nanometer and subnanometer range, that is, microporosity as defined by IUPAC. Similar conclusions are drawn from studies by positronium annihilation lifetime spectroscopy (PALS) [35, 100, 101]. However, every technique is limited by the available analytical models, and there is currently no unambiguous way to determine a definitive pore size distribution.

Nitrogen adsorption isotherms of PIMs only rarely show the classical hysteretic loop at the mid-range of relative pressure, which can be attributable to mesoporosity caused by phase separation during formation of the network or during reprecipitation from solution.

1.3.3.2 Processability. Soluble polymers such as PIM-1 (from monomers **A1** and **B4**, Fig. 1.7; structure shown in Fig. 1.6b) and PIM-7 (from monomers **A1** and **B7**, Fig. 1.7) are readily processed to give self-standing films and coatings (see Fig. 1.10). In each form, PIM-1 exhibits a high surface area (typically, powder = 760 m^2 g^{-1}; thin film = 680 m^2 g^{-1}) as shown by nitrogen adsorption. This property is crucial to many of the potential applications of these materials as separation membranes or as functional coatings [9].

1.3.3.3 Thermal and Chemical Stability. The PIMs share many of the structural features of high-performance polymers and as such possess good thermal stability with negligible mass loss below 450°C in nitrogen (350°C in air), as indicated by thermogravimetric analysis (TGA). Solvent-cast films of PIM-1 display no thermal transitions (i.e., melting point or glass transitions) or loss in mechanical resilience as shown by dynamic mechanical thermal analysis. For example, the tensile storage modulus E' of PIM-1 is about 1 GPa, which is in the range expected for a glassy polymer, and this value is maintained until above 350°C in air [92]. In addition, the

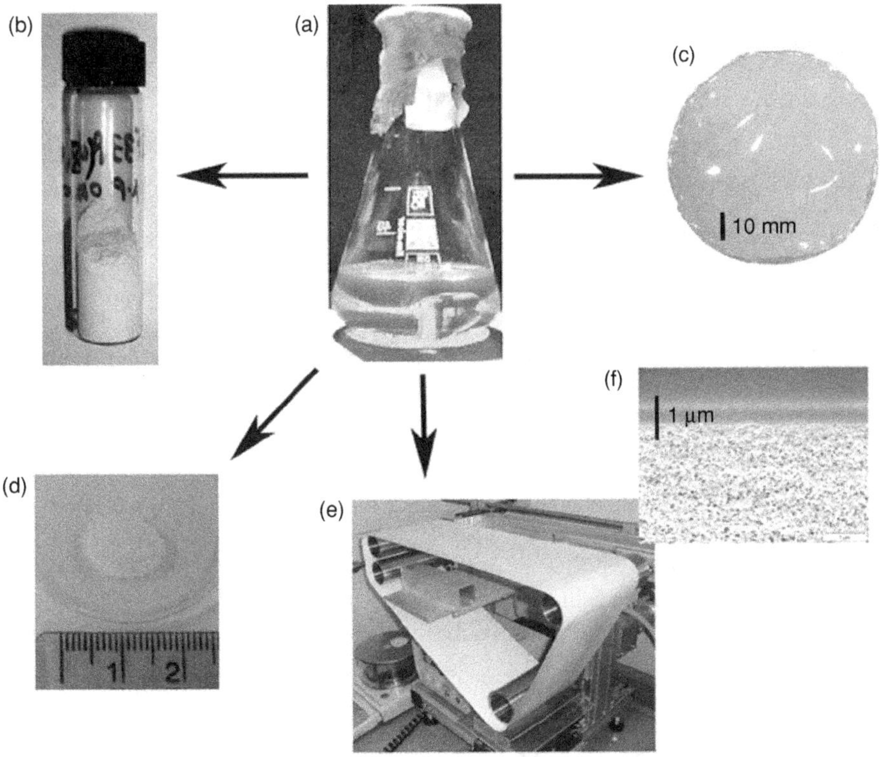

FIGURE 1.10 **(a)** PIM-1 solution in THF; **(b)** precipitated PIM-1 powder; **(c)** a solvent-cast, free-standing film; **(d)** coated alumina beads; **(e)** PIM-1 being coated on a macroporous polyacrylonitrile sheet for use as a gas separation membrane; and **(f)** an electron micrograph showing a 0.5-μm-thick coating of PIM-1 on a macroporous support.

microporosity of PIMs is retained following prolonged annealing at elevated temperature (150°C). Powdered and film samples of PIMs have been stored under ambient conditions for several years without any detectable deterioration in either chemical or microporous structure. Generally, PIMs display good stability toward acids, bases, and oxidizing reagents.

1.3.3.4 Structural and Chemical Diversity. The wide choice of monomer precursors suitable for making PIMs means that their structure and chemical nature can be tailored to the specific requirements of an application. For example, pore-size distribution can be modified by the use of bowl-shaped monomers such as cyclotricatechylene (CTC; monomer **A4**, Fig. 1.7), which provides tight binding sites for the adsorption of small molecules [102]. Metal-containing PIMs can be prepared by using fluorine-substituted, metal-containing macrocycles such as porphyrin (**B1**) [103, 104] or phthalocyanine (**B3**) as monomers [86, 103]. Alternatively, metals can

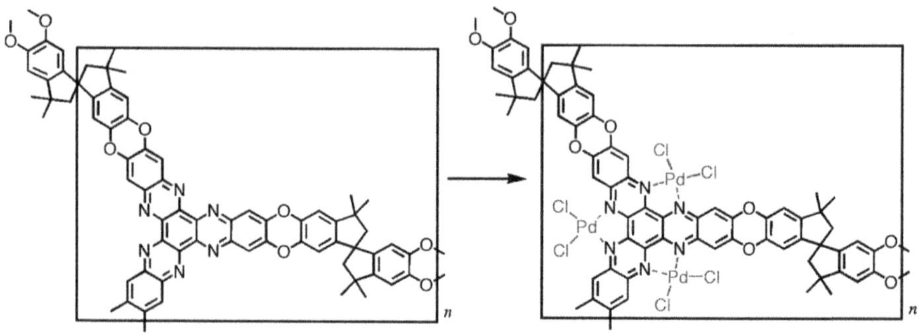

FIGURE 1.11 The incorporation of palladium dichloride into the structure of the hexaazatrinaphthylene (HATN) network PIM.

be introduced into preformed PIMs by providing suitable metal-binding sites such as the bidentate ligands embedded within the hexachlorohexaazatrinapthylene monomer (HATN; monomer **B2,** Fig. 1.7) [105]. PIMs derived from this monomer readily adsorb metal cations (e.g., 20% by mass Pd^{2+}) from solution (Fig. 1.11). Similarly, metal introduction can be achieved for PIM-7 (from monomers **A1** and **B7**, Fig. 1.7) [106] either by coprecipitation from solution or by adsorption using the powder or film form of the polymer (Fig. 1.12) [83].

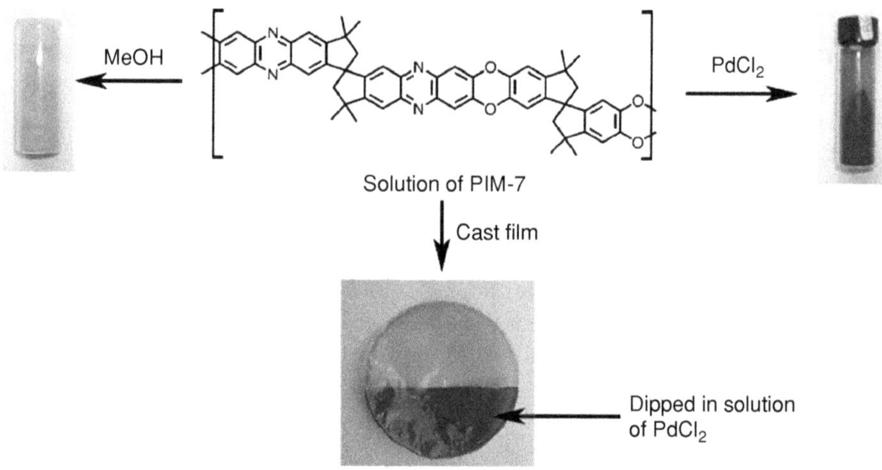

FIGURE 1.12 Methods for metal cation incorporation within PIM-7. Like PIM-1, a solution of PIM-7 can be precipitated into a microporous yellow powder (left) or cast into a self-standing film. Due to its nitrogen-binding sites, metal-containing PIM-7 can be prepared by precipitation of the polymer by metal cations (red powder on right) or by treating a preformed film with a solution of the metal salt. For both methods the metal cations act as crosslinker to render the polymer insoluble. This concept is illustrated for palladium(II) chloride, for which there is a mass loading of greater than 10%.

1.3.4 Applications of PIMs

1.3.4.1 Gas Separation Membranes. The study of the gas permeability of polymers is a well-established technological field due to the extensive commercial interest in using polymer-derived membranes for gas separations [107]. Over the last four decades an enormous volume of data has been compiled on the two main performance indicators of a polymer: the permeability coefficient $P(\mathrm{X})$ (units: Barrer $= 10^{-10}\ \mathrm{cm^3\ cm\ cm^{-2}\ s^{-1}\ cm\ Hg^{-1}} = 3.35\times10^{-16}\ \mathrm{mol\ m\ m^{-2}\ s^{-1}\ Pa^{-1}}$) for a particular gas (X), and the selectivity of one gas (X) over another (Y), which in most cases is ideal selectivity, $\alpha(\mathrm{X/Y}) = P(\mathrm{X})/P(\mathrm{Y})$, derived from single gas permeability measurements. For a useful polymer membrane it is desirable to have both high permeability and high selectivity for real gas mixtures. The selectivity of a polymer toward a gas mixture may be influenced by polymer swelling or pore-blocking by a strongly adsorbed gas, which reduces the free volume available for diffusion through the polymer.

For gas separations, most attention has been paid to membranes derived from glassy polymers. Such polymers are generally of low permeability but high selectivity. However, there are a few examples of "ultrapermeable" glassy polymers, best represented by the polyacetylene derivative poly(1-trimethylsilyl-1-propyne) (PTMSP, Fig. 1.5), which has been the focus of considerable fundamental and applied interest [77, 78, 108]. Unfortunately, such highly permeable polymers are generally of low selectivity, whereas high selectivities are obtained only for polymers with low gas permeability. Robeson quantified this trade-off by developing the idea of an "upper bound" in double-logarithmic plots of selectivity against permeability [109]. Robeson's 1991 upper bound for O_2/N_2 is shown in Fig. 1.13, together with more recent data for polymers that perform close to, or exceed, this upper bound. The data for PIM-1 and PIM-7 are also plotted [110]. As might be expected for microporous materials, films of PIM-1 [$P(O_2) = 380$ Barrer] and PIM-7 [$P(O_2) = 180$ Barrer] are highly gas permeable, with only the "ultrapermeable" polymers such as PTMSP demonstrating higher overall gas permeabilities [e.g., $P(O_2) = 6100$ Barrer but $\alpha(O_2/N_2) = 1.8$]. In addition, PIM-1 and PIM-7 [106] show substantially higher selectivities ([$\alpha(O_2/N_2)$ > 3.8] than other polymers of similar permeability and represent a significant advance on Robeson's original upper bound for O_2/N_2. PIMs also lie near or above the upper bound line for several other commercially important gas combinations, including CO_2/CH_4, H_2/N_2, and H_2/CH_4 and therefore define new upper bounds, as recently discussed by Robeson [111]. This behavior indicates that PIMs are different from the many hundreds of polymers that have been investigated for gas permeability. It has been suggested that to obtain the best permeability/selectivity properties one needs to create a polymer structure with a stiff backbone (which enhances mobility selectivity at the expense of diffusivity) while also disrupting interchain packing (to improve permeability) [112, 113]. This design principle is taken to the extreme with PIMs, for which rigidity and the prohibition of rotation are ensured by their fused ring structures, while the spirocyclic or other sites of contortion disrupt interchain packing. For the separation of O_2 and N_2 the most important factor is the mobility selectivity, which favors the smaller oxygen molecule (diameter $= 0.346$ nm) rather than the larger nitrogen molecule (diameter $= 0.364$ nm) and, for a microporous

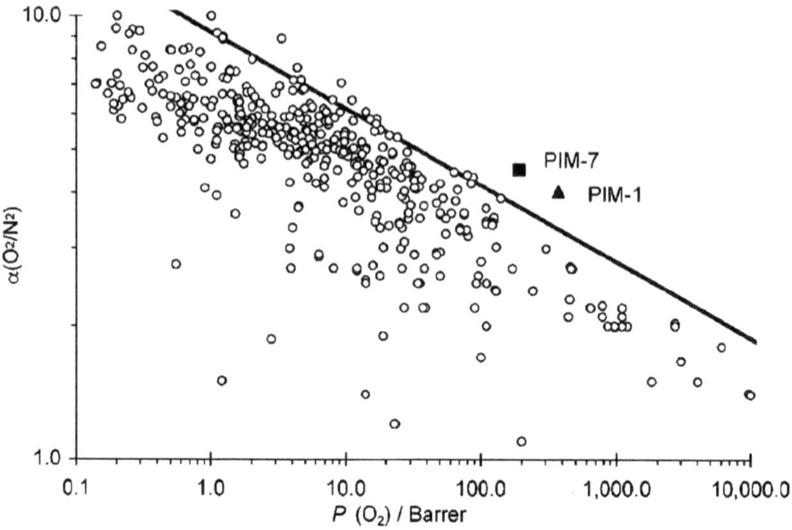

FIGURE 1.13 The Robeson plot showing the trade-off between gas permeability and ideal selectivity for oxygen versus nitrogen. The empirical upper bound line is show and is based upon the polymers demonstrating the best selectivity for a given permeability in 1991. For separation membranes it is desirable to obtain polymers whose data points lie above the upper bound line and toward the top right-hand side of the plot.

polymer, is primarily dependent on the size distribution of the micropores. For PIM-1 and PIM-7 it is evident that the pore size is smaller than that found in the microporous ultrapermeable polymers such as PTMSP, for which the high permeability arises from very large diffusion coefficients.

Of significance, the permeability of PIM-1 can be greatly enhanced [$P(O_2) = 1600$ Barrer], with only a small loss in selectivity [$\alpha(O_2/N_2) = 3.5$], by soaking the polymer in methanol [114]. Although this enhanced permeability is similar to that observed for PTMSP and appears to be based upon the introduction of excess free volume, the relaxation to a less permeable state is much slower [e.g., $P(O_2) = 1250$ Barrer after 45 days, with $\alpha(O_2/N_2) = 3.7$]. The combination of high permeabilities and good selectivities shown by PIM-1, together with its excellent processability, which allows the fabrication of very thin films (<1 μm) supported on macroporous substrates (Fig. 1.10e, f), suggests a promising future for PIMs as components of gas separation membranes. Polyimides derived from the biscatechol monomer **A1** also show exceptional permeability and selectivity due to their intrinsic microporosity [85].

1.3.4.2 Adsorption. Because they are organic microporous materials with amorphous structures related to that of activated carbon, it was anticipated that the PIMs should be suitable for the adsorption and separation of organic compounds. This was confirmed by measuring the adsorption of phenol from aqueous solution. This process is of environmental relevance, as phenols are common contaminants of wastewater

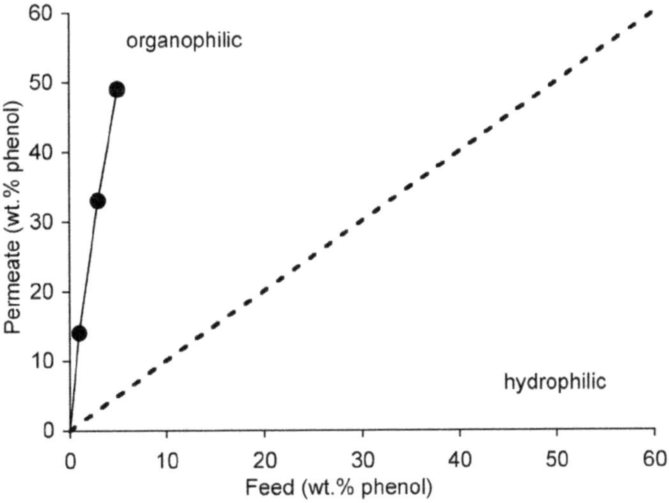

FIGURE 1.14 Pervaporation-based separation of phenol from aqueous solution using a membrane derived from PIM-1. Generally, the efficiency of separation may be expressed as a separation factor $\alpha = (Y_o/Y_w)/(X_o/X_w)$, where (Y_o/Y_w) is the weight ratio of organic compound to water in the permeate, and (X_o/X_w) is the weight ratio of organic compound to water in the feed. Values of α of 16–18 were obtained at temperatures in the range 50°C–80°C and feed compositions in the range 1–5 wt% phenol.

streams from industrial processes. The network PIM, of surface area 830 $m^2\ g^{-1}$, derived from the spiromonomer **A1** and HATN, **B2** (Figs.1.7 and 1.11), adsorbs up to 5 mmol g^{-1} of phenol from solutions of initial concentration of 0.2 mol L^{-1} (i.e., 0.5 g of phenol for 1 g of PIM) [105]. Similar performances were obtained for both PIM-1 as a powder and a phthalocyanine-based PIM, which also showed negligible uptake for a large organic dye (naphthol green B) attributable to a small micropore size (<1.0 nm) [115]. In addition, the removal of phenol from aqueous solution has also been achieved by pervaporation using a solvent-cast film derived from PIM-1 as a membrane [92]. Pervaporation is a separation process in which the feed is a liquid mixture, and a vacuum is applied to the opposite side of the membrane to remove permeate as a vapor, which is then condensed and collected. In Fig. 1.14 it can be seen that, with the PIM-1 membrane, the permeate was enriched in phenol up to 10-fold, which demonstrates that the membrane is strongly organophilic (i.e., selective for organic compounds over water), which is unusual for a polymer membrane derived from a glassy polymer. For most glassy polymers selectivity is governed predominately by size, and therefore the smaller water molecules are transported preferentially to larger organic molecules despite the organic nature of the polymer. For PIMs the intrinsic microporosity allows selectivity based upon the stronger interactions between the organic adsorbate and the organic polymer. PIM-1 is also efficient for the separation of methanol, ethanol, and butanol from water [116].

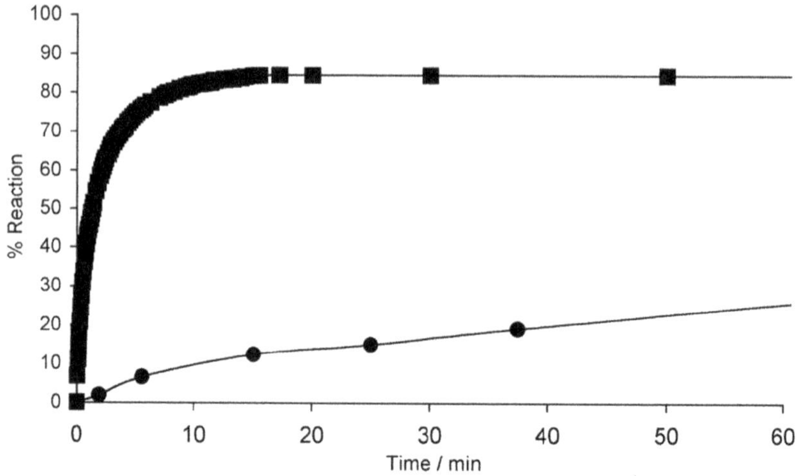

FIGURE 1.15 Dependence of the extent of reaction on time for the degradation of H_2O_2 (0.74 mol dm^{-3}, T = 30°C) with (●) low-molar-mass cobalt phthalocyanine (CoPc) and CoPc-network-PIM (■) as catalyst. Oxygen evolution was measured with a gas burette.

1.3.4.3 Heterogeneous Catalysis. As noted, for network PIMs containing either phthalocyanine, porphyrin, or hexaazatrinaphthylene subunits, it is possible to introduce appropriate transition metal ions for catalytic activity. For example, metal-containing porphyrins and phthalocyanines can display similar activity to that of the cytochrome P450 enzymes such as alkene epoxidations and hydrocarbon hydroxylations [117]. These synthetic transformations are achieved using environmentally benign oxidants such as oxygen or hydrogen peroxide. Therefore the possibility of useful heterogeneous catalysis makes these macrocycles desirable components of a microporous material. Preliminary studies on the degradation of hydrogen peroxide using the phthalocyanine-based network PIM show a greatly enhanced rate as compared to microcrystalline cobalt phthalocyanine (Fig. 1.15). In addition, this network PIM is both efficient and selective for the catalysis of the oxidation of cyclohexene to 2-cyclohexene-1-one (78% yield after 48 h), even when compared to the activity of cobalt phthalocyanine homogeneous catalysts [103]. The PIM formed by a phthalocyanine-forming reaction (of surface area = 620 m^2 g^{-1}) [86] proved to be a more efficient catalyst than that prepared via a dioxane formation between the preformed fluorinated phthalocyanine B3 and A1 (Fig. 1.7). In addition, palladium-containing PIMs (see Figs. 1.8 and 1.9) are being investigated for the catalysis of Suzuki coupling reactions [82].

1.3.4.4 Hydrogen Storage. PIMs were the first class of organic polymers that were reported to adsorb significant quantities of hydrogen at low temperature (77 K) [102]. Initially, the adsorption of H_2 was measured on three PIMs (PIM-1, HATN network, and CTC network) using both volumetric and gravimetric techniques, with consistent results for each sample. The adsorption isotherms show that the three PIMs each adsorb significant quantities of H_2 (maximum = 1.7% by mass) at relatively low pressures, with saturation being reached at less than 10-bar pressure and with most of the adsorption taking place below 1 bar. Subsequent analysis of the highly microporous triptycene-based network PIM derived from monomers **A3** (R = Et) and **B4** (Figs. 1.6c and 1.7) gave loading of around 3.0% at 20 bar and 77 K, which is competitive with hypercrosslinked polystyrenes of higher surface area (2000 m^2 g^{-1}) [97, 118, 119]. In particular, the narrower pore-size distribution, comprised of predominantly subnanometer-diameter pores, is well suited to adsorb hydrogen at relatively low pressures (e.g., 1.8% by mass at 1 bar and 77 K). Although this performance falls short of some very high surface area carbons, they adsorb larger amounts of H_2 compared to carbons of similar surface area at low pressures. However, to attain practical hydrogen storage materials from PIMs, it will be necessary to engineer examples with larger *accessible* surface areas (>2000 m^2 g^{-1}) while maintaining the predominately ultramicroporous structure necessary to retain the beneficial multiwall interactions with H_2 molecules [118, 119].

1.4 COVALENT ORGANIC FRAMEWORKS

Crystalline metal–organic frameworks (MOFs) are formed by exploiting the formation of rapidly reversible metal–organic bonds between rigid organic struts and metal ions [4, 120–122]. In an extrapolation of this successful concept, Yaghi and coworkers built purely organic frameworks by using the rapidly reversible bonding associated with the facile formation of boronic esters from monomers containing boronic acids and catechol (1,2-dihydroxybenzene) units [5, 123–125]. For example, the reaction between 1,3,5-phenylenetriboronic acid and 2,3,6,7,10,11-hexahydroxytriphenylene (Fig. 1.16), under conditions whereby the water byproduct is removed, gives a white solid. Powder X-ray diffraction analysis indicates that the material is both microcrystalline and highly porous. Heating under vacuum removes included solvent but maintains the crystallinity of the solid. Depending upon the two monomers used, the surface area of the resulting covalent organic framework (COF) can range from 500 to 4200 m^2 g^{-1}. In particular, the use of the tetrahedral monomer tetra(4-dihydroxylborylphenyl)methane or tetra(4-dihydroxylborylphenyl)silane provides crystalline organic frameworks of exceptionally high porosity and low density [123, 126]. In addition, it has been found that using monomers containing an alkyl group can tailor the porosity of the COF [127, 128]. Application of these materials as hydrogen storage materials is anticipated, and very high values of hydrogen adsorption (up to 10% by mass) have been predicted at low temperature (77 K) and high pressure (100 bar) [129, 130]. However, compared to MOFs or amorphous microporous polymers such as the PIMs, the COFs are relatively fragile, and complete removal of the included solvent has proved difficult for some of the more porous COFs for which the large values of hydrogen uptake are predicted. Problems with the hydrolytic stability of the materials might also be expected due to the reversible formation of the boronic ester.

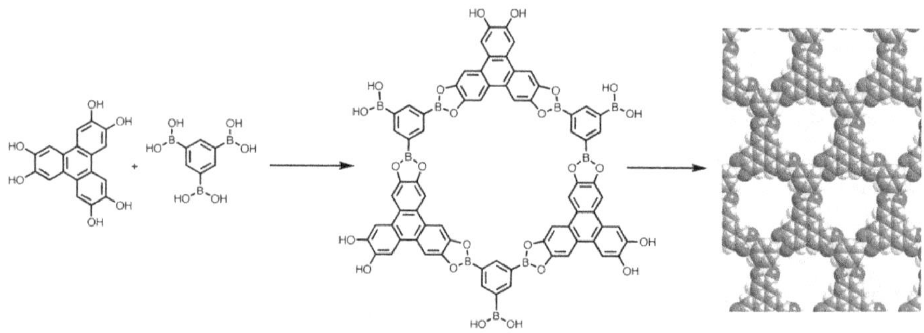

FIGURE 1.16 An illustrative example of the synthesis of a COF (COF-12) from the boronic ester–forming reaction between 1,3,5-phenylenetriboronic acid and 2,3,6,7,10,11-hexahydroxytriphenylene. The initially formed cyclic hexamers (internal pore diameter = 1.2 nm) are components of a crystalline organic framework.

1.5 CONCLUSIONS

The three basic concepts for preparing polymers with "inherent" microporosity described in this chapter—hypercrosslinked polymers, PIMs, and COFs—are especially useful for the design of multifunctional organic materials. In particular, PIMs effectively bridge the gap between conventional microporous materials and polymers because they share properties (e.g., processability and gas adsorption) that are associated with both classes of material. The potential offered by the structural diversity of PIMs and also COFs, which can be controlled simply by the choice of monomer precursors, is only just starting to be explored. Nevertheless, the few examples that have been studied suggest an enticing prospect of readily processed, bespoke organic microporous materials designed to adsorb, purify, or react with target molecules. Moreover, industrial applications (e.g., membrane separation) are likely to result from materials such as PIM-1 that can be prepared relatively cheaply on a large scale, as has been demonstrated by the commercial success of hypercrosslinked polystyrenes. With the immediate need for energy-efficient and environmentally relevant chemical processes, we can expect that polymers with inherent microporosity will be of growing importance to future research and technology.

REFERENCES

1. Everett, D. H. *Pure Appl Chem* 1972, **31**, 577.
2. Rouquerol, J.; Rouquerol, F.; Sing, K. W. *Adsorption by Powders and Porous Solids: Principles, Methodology and Applications*, Academic Press, London, 1999.
3. Schüth, F.; Sing, K.; Weitkamp, J., Eds. *Handbook of Porous Solids*, Wiley-VCH, Berlin, 2002.
4. Yaghi, O. M.; O'Keefe, M.; Ockwig, N. W.; Chae, H. K.; Eddaoudi, M.; Kim, J. *Nature* 2003, **423**, 705.

5. Coté, A. P.; Benin, A. I.; Ockwig, N. W.; O'Keeffe, M.; Matzger, A. J.; Yaghi, O. M. *Science* 2005, **310**, 1166.
6. Barton, T. J.; Bull, L. M.; Klemperer, W. G.; Loy, D. A.; McEnaney, B.; Misono, M.; Monson, P. A.; Pez, G.; Scherer, G. W.; Vartuli, J. C.; Yaghi, O. M. *Chem Mater* 1999, **11**, 2633.
7. Bondi, A. *Physical Properties of Molecular Crystals, Liquids and Glasses*, Wiley, New York, 1968.
8. van Krevelen, D. W. *Properties of Polymers*, Elsevier, Amsterdam, 1997.
9. Budd, P. M.; McKeown, N. B.; Fritsch, D., *J Mater Chem* 2005, **15**, 1977.
10. Davankov, V. A.; Rogozhin, S. V.; Tsyurupa, M. P. U.S. Patent 3,729,457, 1971.
11. Davankov, V. A.; Tsyurupa, M. P. *Pure Appl Chem* 1989, **61**, 1881.
12. Davankov, V. A.; Tsyurupa, M. P. *React Polym* 1990, **13**, 27.
13. Tsyurupa, M. P.; Davankov, V. A. *React Funct Polym* 2002, **53**, 193.
14. Ahn, J. H.; Jang, J. E.; Oh, C. G.; Ihm, S. K.; Cortez, J.; Sherrington, D. C. *Macromolecules* 2006, **39**, 627.
15. Zhang, X.; Shen, S. H.; Fan, L. Y. *J Mater Sci* 2007, **42**, 7621.
16. Veverka, P.; Jerabek, K. *React Funct Polym* 1999, **41**, 21.
17. Law, R. V.; Sherrington, D. C.; Snape, C. E.; Ando, I.; Kurosu, H. *Macromolecules* 1996, **29**, 6284.
18. Pastukhov, A. V.; Tsyurupa, M. P.; Davankov, V. A. *J Polym Sci B Polym Phys* 1999, **37**, 2324.
19. Davankov, V. A.; Ilyin, M. M.; Tsyurupa, M. P.; Timofeeva, G. I.; Dubrovina, L. V. *Macromolecules* 1996, **29**, 8398.
20. Tsyurupa, M. P.; Mrachkovskaya, T. A.; Maslova, L. A.; Timofeeva, G. I.; Dubrovina, L. V.; Titova, E. F.; Davankov, V. A.; Menshov, V. M. *React Polym* 1993, **19**, 55.
21. Arnautov, S. A.; Davankov, V. A. *Mendeleev Commun* 2006, **16**, 79.
22. Streat, M.; Sweetland, L. A. *React Funct Polym* 1997, **35**, 99.
23. Wood, C. D.; Tan, B.; Trewin, A.; Niu, H. J.; Bradshaw, D.; Rosseinsky, M. J.; Khimyak, Y. Z.; Campbell, N. L.; Kirk, R.; Stockel, E.; Cooper, A. I. *Chem Mater* 2007, **19**, 2034.
24. Webster, O. W.; Gentry, F. P.; Farlee, R. D.; Smart, B. E. *Makromol Chem Macromol Symp* 1992, **54/55**, 477.
25. Urban, C.; McCord, E. F.; Webster, O. W.; Abrams, L.; Long, H. W.; Gaede, H.; Tang, P.; Pines, A. *Chem Mater* 1995, **7**, 1325.
26. Rose, M.; Bohlmann, W.; Sabo, M.; Kaskel, S. *Chem Commun* 2008, 2462.
27. Germain, J.; Fréchet, J. M. J.; Svec, F. *J Mater Chem* 2007, **17**, 4989.
28. Jiang, J. X.; Su, F.; Trewin, A.; Wood, C. D.; Campbell, N. L.; Niu, H.; Dickinson, C.; Ganin, A. Y.; Rosseinsky, M. J.; Khimyak, Y. Z.; Cooper, A. I., *Angew Chem Int Ed* 2007, **46**, 8574.
29. Jiang, J. X.; Su, F.; Trewin, A.; Wood, C. D.; Niu, H.; Jones, J. T. A.; Khimyak, Y. Z.; Cooper, A. I. *J Am Chem Soc* 2008, **130**, 7710.
30. Weber, J.; Thomas, A. *J Am Chem Soc* 2008, **130**, 6334.
31. Tsyurupa, M. P.; Davankov, V. A. *React Funct Polym* 2006, **66**, 768.
32. Penner, N. A.; Nesterenko, P. N.; Ilyin, M. M.; Tsyurupa, M. P.; Davankov, V. A. *Chromatographia* 1999, **50**, 611.

33. Shantarovich, V. P.; Suzuki, T.; He, C.; Davankov, V. A.; Pastukhov, A. V.; Tsyurupa, M. P.; Kondo, K.; Ito, Y. *Macromolecules* 2002, **35**, 9723.
34. He, C. Q.; Suzuki, T.; Shantarovich, V. P.; Djourelov, N.; Kondo, K.; Ito, Y., *Chem Phys* 2004, **303**, 219.
35. Shantarovich, V. P.; Suzuki, T.; Yampol'skii, Y. P.; Budd, P.; Gustov, V. V.; Kevdina, I. B.; Pastukhov, A. V.; Berdonosov, S. S.; Bozhevol'nov, V. E. *High Energy Chem* 2007, **41**, 370.
36. Davankov, V. A.; Pastukhov, A. V.; Tsyurupa, M. P. J. *Polym Sci B Polym Phys* 2000, **38**, 1553.
37. Pastukhov, A. V.; Davankov, V. A.; Sidorova, E. V.; Shkol'nikov, E. I.; Volkov, V. V. *Russ Chem Bull* 2007, **56**, 484.
38. Simpson, E. J.; Koros, W. J.; Schechter, R. S., *Ind Eng Chem* 1996, **35**, 1195.
39. Belyakova, L. D.; Schevchenko, T. I.; Davankov, V. A.; Tsyurupa, M. P.; Nesmeyanov, A. N. *Adv Colloid Interface Sci* 1986, **25**, 249.
40. Tsyurupa, M. P.; Maslova, L. A.; Andreeva, A. I.; Mrachkovskaya, T. A.; Davankov, V. A. *React Polym* 1995, **25**, 69.
41. Podlesnyuk, V. V.; Hradil, J.; Kralova, E. *React Funct Polym* 1999, **42**, 181.
42. Veverka, P.; Jerabek, K., *React Funct Polym* 2004, **59**, 71.
43. Filippov, O. A.; Tikhomirova, T. I.; Tsizin, G. I.; Zolotov, Y. A. *J Anal Chem* 2003, **58**, 398.
44. Fontanals, N.; Marce, R. M.; Cormack, P. A. G.; Sherrington, D. C.; Borrull, F. *J Chromatogr A* 2008, **1191**, 118.
45. Tsyurupa, M. P.; Ilyin, M. M.; Andreeva, A. I.; Davankov, V. A., Fresenius J. *Anal Chem* 1995, **352**, 672.
46. Li, A. M.; Zhang, Q. X.; Zhang, G. C.; Chen, J. L.; Fei, Z. H.; Liu, F. Q. *Chemosphere* 2002, **47**, 981.
47. Oh, C. G.; Ahn, J. H.; Ihm, S. K. *React Funct Polym* 2003, **57**, 103.
48. Pan, B. C.; Xiong, Y.; Su, Q.; Li, A. M.; Chen, J. L.; Zhang, Q. X. *Chemosphere* 2003, **51**, 953.
49. Pan, B. C.; Du, W.; Zhang, W. M.; Zhang, X.; Zhang, Q. R.; Pan, B. J.; Lv, L.; Zhang, Q. X.; Chen, J. L. *Environ Sci Technol* 2007, **41**, 5057.
50. Xu, Z. Y.; Zhang, Q. X.; Chen, J. L.; Wang, L. S.; Anderson, G. K. *Chemosphere* 1999, **38**, 2003.
51. Zhai, Z. C.; Chen, J. L.; Fei, Z. H.; Wang, H. L.; Li, A. M.; Zhang, Q. X. *React Funct Polym* 2003, **57**, 93.
52. Valderrama, C.; Cortina, J. L.; Farran, A.; Gamisans, X.; Lao, C. *J Colloid Interface Sci* 2007, **310**, 35.
53. Valderrama, C.; Gamisans, X.; Heras, F.; Cortina, J. L.; Farran, A. *React Funct Polym* 2007, **67**, 1515.
54. Chang, C. F.; Chang, C. Y.; Hsu, K. E.; Lee, S. C.; Holl, W. J. *Hazard Mater* 2008, **155**, 295.
55. Valderrama, C.; Cortina, J. L.; Farran, A.; Gamisans, X.; Heras, F. *React Funct Polym* 2008, **68**, 718.
56. Xu, Z. W.; Zhang, W. M.; Pan, B. C.; Hong, C. H.; Lv, L.; Zhang, Q. J.; Pan, B. J.; Zhang, Q. X. *J Colloid Interface Sci* 2008, **319**, 392.

57. Pan, B. C.; Xiong, Y.; Li, A. M.; Chen, J. L.; Zhang, Q. X.; Jin, X. Y. *React Funct Polym* 2002, **53**, 63.

58. Pan, B. C.; Zhang, Q. X.; Meng, F. W.; Li, X. T.; Zhang, X.; Zheng, J. Z.; Zhang, W. M.; Pan, B. J.; Chen, J. L. *Environ Sci Technol* 2005, **39**, 3308.

59. Sun, Y.; Chen, J. L.; Li, A. M.; Liu, F. Q.; Zhang, Q. X. *React Funct Polym* 2005, **64**, 63.

60. Sun, Y.; Chen, J. L.; Li, A. M.; Liu, F. Q.; Zhang, Q. X. *Adsorpt Sci Technol* 2005, **23**, 335.

61. Fontanals, N.; Galia, M.; Cormack, P. A. G.; Marce, R. M.; Sherrington, D. C.; Borrull, F. *J Chromatogr A* 2005, **1075**, 51.

62. Penner, N. A.; Nesterenko, P. N.; Rybalko, M. A. *J Anal Chem* 2001, **56**, 934.

63. Laatikainen, M.; Sainio, T.; Davankov, V.; Tsyurupa, M.; Blinnikova, Z.; Paatero, E. *React Funct Polym* 2007, **67**, 1589.

64. Tsyurupa, M. P.; Davankov, V. A. *Doklady Chem* 2004, **398**, 184.

65. Nesterenko, P. N.; Kebets, P. A.; Volgin, Y. V. J. *Anal Chem* 2001, **56**, 715.

66. Sidorov, S. N.; Bronstein, L. M.; Davankov, V. A.; Tsyurupa, M. P.; Solodovnikov, S. P.; Valetsky, P. M.; Wilder, E. A.; Spontak, R. J. *Chem Mater* 1999, **11**, 3210.

67. Bronstein, L. M.; Goerigk, G.; Kostylev, M.; Pink, M.; Khotina, I. A.; Valetsky, P. M.; Matveeva, V. G.; Sulman, E. M.; Sulman, M. G.; Bykov, A. V.; Lakina, N. V.; Spontak, R. J. *J Phys Chem B* 2004, **108**, 18234.

68. Sidorov, S. N.; Volkov, I. V.; Davankov, V. A.; Tsyurupa, M. P.; Valetsky, P. M.; Bronstein, L. M.; Karlinsey, R.; Zwanziger, J. W.; Matveeva, V. G.; Sulman, E. M.; Lakina, N. V.; Wilder, E. A.; Spontak, R. J. *J Am Chem Soc* 2001, **123**, 10502.

69. Sulman, E.; Doluda, V.; Dzwigaj, S.; Marceau, E.; Kustov, L.; Tkachenko, O.; Bykov, A.; Matveeva, V.; Sulman, M.; Lakina, N. *J Mol Catal A Chem* 2007, **278**, 112.

70. van den Berg, A. W. C.; Arean, C. O. *Chem Commun* 2008, 668.

71. Germain, J.; Hradil, J.; Fréchet, J. M. J.; Svec, F. *Chem Mater* 2006, **18**, 4430.

72. Lee, J. Y.; Wood, C. D.; Bradshaw, D.; Rosseinsky, M. J.; Cooper, A. I. *Chem Commun* 2006, 2670.

73. Germain, J.; Svec, F.; Fréchet, J. M. *J Chem Mater* 2008, **20**, 7069.

74. Wood, C. D.; Tan, B.; Trewin, A.; Su, F.; Rosseinsky, M. J.; Bradshaw, D.; Sun, Y.; Zhou, L.; Cooper, A. I. *Adv Mater* 2008, **20**, 1916.

75. Ilinitch, O. M.; Lapkin, A. A. *Polymer* 2002, **43**, 3209.

76. Ilinitch, O. M.; Fenelonov, V. B.; Lapkin, A. A.; Okkel, L. G.; Terskikh, V. V.; Zamaraev, K. I. *Microporous Mesoporous Mater* 1999, **31**, 97.

77. Masuda, T.; Isobe, E.; Higashimura, T.; Takada, K. *J Am Chem Soc* 1983, **105**, 7473.

78. Nagai, K.; Masuda, T.; Nakagawa, T.; Freeman, B. D.; Pinnau, I. *Prog Polym Sci* 2001, **26**, 721.

79. Srinivasan, R.; Auvil, S. R.; Burban, P. M. *J Membr Sci* 1994, **86**, 67.

80. Yu, A.; Shantarovich, V.; Merkel, T. C.; Bondar, V. I.; Freeman, B. D.; Yampolskii, Y. *Macromolecules* 2002, **35**, 9513.

81. Tanaka, K.; Okano, M.; Toshino, H.; Kita, H.; Okamoto, K. I. *J Polym Sci B Polym Phys* 1992, **30**, 907.

82. McKeown, N. B.; Budd, P. M. *Chem Soc Rev* 2006, **35**, 675.

83. McKeown, N. B.; Budd, P. M.; Msayib, K. J.; Ghanem, B. S.; Kingston, H. J.; Tattershall, C. E.; Makhseed, S.; Reynolds, K. J.; Fritsch, D. *Chem Eur J* 2005, **11**, 2610.
84. Budd, P. M.; Ghanem, B. S.; Makhseed, S.; McKeown, N. B.; Msayib, K. J.; Tattershall, C. E. *Chem Commun* 2004, 230.
85. Ghanem, B. S.; McKeown, N. B.; Budd, P. M.; Selbie, J. D.; Fritsch, D. *Adv Mater* 2008, **20**, 2766.
86. McKeown, N. B.; Makhseed, S.; Budd, P. M. *Chem Commun* 2002, 2780.
87. Weber, J.; Antonietti, M.; Thomas, A. *Macromolecules* 2008, **41**, 2880.
88. Weber, J.; Su, O.; Antonietti, M.; Thomas, A. *Macromol Rapid Commun* 2007, **28**, 1871.
89. Eastmond, G. C.; Paprotny, *J Chem Lett* 1999, 479.
90. Eastmond, G. C.; Paprotny, J.; Steiner, A.; Swanson, L. *New J Chem* 2001, **25**, 379.
91. Msayib, K.; Makhseed, S.; McKeown, N. B. *J Mater Chem* 2001, **11**, 2784.
92. Budd, P. M.; Elabas, E. S.; Ghanem, B. S.; Makhseed, S.; McKeown, N. B.; Msayib, K. J.; Tattershall, C. E.; Wang, D. *Adv Mater* 2004, **16**, 456.
93. Du, N. Y.; Song, J. S.; Robertson, G. P.; Pinnau, I.; Guiver, M. D. *Macromol Rapid Commun* 2008, **29**, 783.
94. Kricheldorf, H. R.; Lomadze, N.; Fritsch, D.; Schwarz, G. J. *Polym Sci A Polym Chem* 2006, **44**, 5344.
95. Kricheldorf, H. R.; Fritsch, D.; Vakhtangishvili, L.; Lomadze, N.; Schwarz, G. *Macromolecules* 2006, **39**, 4990.
96. Kricheldorf, H. R.; Fritsch, D.; Vakhtangishvili, L.; Schwarz, G. *Macromol Chem Phys* 2005, **206**, 2239.
97. Ghanem, B.; McKeown, N. B.; Harris, K. D. M.; Pan, Z.; Budd, P. M.; Butler, A.; Selbie, J.; Book, D.; Walton, A. *Chem Commun* 2007, 67.
98. Heuchel, M.; Fritsch, D.; Budd, P. M.; McKeown, N. B.; Hofmann, D. *J Membr Sci* 2008, **318**, 84.
99. Horvath, G.; Kawazoe, K. *J Chem Eng Jpn* 1983, **16**, 470.
100. de Miranda, R. L.; Kruse, J.; Ratzke, K.; Faupel, F.; Fritsch, D.; Abetz, V.; Budd, P. M.; Selbie, J. D.; McKeown, N. B.; Ghanem, B. S. *Phys Stat Sol Rapid Res Lett* 2007, **1**, 190.
101. Staiger, C. L.; Pas, S. J.; Hill, A. J.; Cornelius, C. J. *Chem Mater* 2008, **20**, 2606.
102. McKeown, N. B.; Ghanem, B.; Msayib, K. J.; Budd, P. M.; Tattershall, C. E.; Mahmood, K.; Tan, S.; Book, D.; Langmi, H. W.; Walton, A. *Angew Chem Int. Ed* 2006, **45**, 1804.
103. MacKenzie, H.; Budd, P. M.; McKeown, N. B. *J Mater Chem* 2008, **18**, 573.
104. McKeown, N. B.; Hanif, S.; Msayib, K.; Tattershall, C. E.; Budd, P. M. *Chem Commun* 2002, 2782.
105. Budd, P. M.; Ghanem, B.; Msayib, K.; McKeown, N. B.; Tattershall, C. *J Mater Chem* 2003, **13**, 2721.
106. Ghanem, B. S.; McKeown, N. B.; Budd, P. M.; Fritsch, D. *Macromolecules* 2008, **41**, 1640.
107. Kesting, R. E.; Fritzsche, A. K. *Polymeric Gas Separation Membranes*, Wiley-Interscience, New York, 1993.
108. Madkour, T. M. *Polymer* 2000, **41**, 7489.
109. Robeson, L. M. *J Membr Sci* 1991, **62**, 165.

110. Budd, P. M.; Msayib, K. J.; Tattershall, C. E.; Ghanem, B. S.; Reynolds, K. J.; McKeown, N. B.; Fritsch, D. *J Membr Sci* 2005, **251**, 263.
111. Robeson, L. M. *J Membr Sci* 2008, **320**, 390.
112. Freeman, B. D. *Macromolecules* 1999, **32**, 375.
113. Cecopieri-Gomez, M. L.; Palacios-Alquisira, J.; Dominguez, J. M. *J Membr Sci* 2007, **293**, 53.
114. Budd, P. M.; McKeown, N. B.; Ghanem, B. S.; Msayib, K. J.; Fritsch, D.; Starannikova, L.; Belova, N.; Sanfirovad, O.; Yampolskii, Y.; Shantarovich, V. *J Membr Sci* 2008, **325**, 851.
115. Maffei, A. V.; Budd, P. M.; McKeown, N. B. *Langmuir* 2006, **22**, 4225.
116. Adymkanov, S. V.; Yampolskii, Y.; Polyakov, A. M.; Budd, P. M.; Reynolds, K. J.; McKeown, N. B.; Msayib, K. *J Polym Sci A* 2008, **50**, 444.
117. Meunier, B. *Acc Chem Res* 1992, **92**, 1411.
118. McKeown, N. B.; Budd, P. M.; Book, D. *Macromol Rapid Commun* 2007, **28**, 995.
119. Budd, P. M.; Butler, A.; Selbie, J.; Mahmood, K.; McKeown, N. B.; Ghanem, B.; Msayib, K.; Book, D.; Walton, A. *Phys Chem Chem Phys* 2007, **9**, 1802.
120. Eddaoudi, M.; Moler, D. B.; Li, H.; Chen, B.; Reinke, T. M.; O'Keefe, M.; Yaghi, O. M. *Acc Chem Res* 2001, **34**, 316.
121. Eddaoudi, M.; Li, H. L.; Yaghi, O. M. *J Am Chem Soc* 2000, **122**, 1391.
122. Ferey, G. *Chem Soc Rev* 2008, **37**, 191.
123. El-Kaderi, H. M.; Hunt, J. R.; Mendoza-Cortes, J. L.; Coté, A. P.; Taylor, R. E.; O'Keeffe, M.; Yaghi, O. M. *Science* 2007, **316**, 268.
124. Coté, A. P.; El-Kaderi, H. M.; Furukawa, H.; Hunt, J. R.; Yaghi, O. M. *J Am Chem Soc* 2007, **129**, 12914.
125. Mastalerz, M. *Angew Chem Int Ed* 2008, **47**, 445.
126. Hunt, J. R.; Doonan, C. J.; LeVangie, J. D.; Coté, A. P.; Yaghi, O. M. *J Am Chem Soc* 2008, **130**, 11872.
127. Tilford, R. W.; Mugavero, S. J.; Pellechia, P. J.; Lavigne, J. J. *Adv Mater* 2008, **20**, 2741.
128. Tilford, R. W.; Gemmill, W. R.; zur Loye, H. C.; Lavigne, J. J. *Chem Mater* 2006, **18**, 5296.
129. Han, S. S.; Furukawa, H.; Yaghi, O. M.; Goddard, W. A. *J Am Chem Soc* 2008, **130**, 11580.
130. Garberoglio, G. *Langmuir* 2007, **23**, 12154.

CHAPTER 2

Porous Polymers from Self-Assembled Structures

ERIC M. TODD
Monmouth College, Monmouth, Illinois, USA

MARC A. HILLMYER
University of Minnesota, Minneapolis, Minnesota, USA

2.1 INTRODUCTION AND OVERVIEW

Self-assembly is a general chemical principle according to which molecules interact with one another in specific ways to form ordered structures. Examples of self-assembly include the formation of DNA duplexes, protein tertiary structures, and zeolites. Self-assembly provides an avenue for Feynman's vision of "bottom-up" fabrication of materials, in which molecular building blocks are designed to assemble themselves into a prescribed structure [1]. Typically, this final structure is thermodynamically stable, although there are examples of kinetically controlled self-assembly.

Porous materials are an important and growing technology used predominantly for separation applications, including filtration and chromatographic media. In addition, porous materials are emerging as high-surface-area supports for catalysis and as templates for other nanomaterials. In this chapter we focus on porous organic polymers; however, there are other classes of porous materials of which the reader should be aware, including porous inorganic materials [2], metal–organic frameworks [3], and porous organic materials that are not polymeric [4]. Although all of these materials have their advantages and disadvantages, porous organic polymers are particularly appealing as they combine the processability and mechanical properties of polymers with the molecular control afforded by organic synthesis [5].

There are both "top-down" and "bottom-up" approaches to fabricating porous polymeric materials, and the choice of method is largely determined by the porous material being prepared. Simpler starting materials are typically used in the

Porous Polymers, Edited by Michael S. Silverstein, Neil R. Cameron, and Marc A. Hillmyer

"top-down" methods; however, it can be difficult to control chemical composition at a molecular level. For "bottom-up" methods the challenge is the design and synthesis of building blocks, as well as the elucidation of conditions that lead to the desired assembly. The "top-down" and "bottom-up" techniques are often complementary and have been combined to yield interesting materials [6]. Here, we focus on porous polymeric materials generated by the "bottom-up" approach of self-assembly and highlight the molecular-level control provided by this method.

2.2 SELF-ASSEMBLY AND POLYMERIZATION OF AMPHIPHILIC MONOMERS

2.2.1 Covalent Capture of Liquid Crystalline Phases

Lyotropic liquid crystal (LLC) phases can be formed from lipids (mesogens) or other amphiphiles—discrete molecules that contain a polar head group and one or more long hydrophobic chains (Fig. 2.1a) [7]. When exposed to water, these

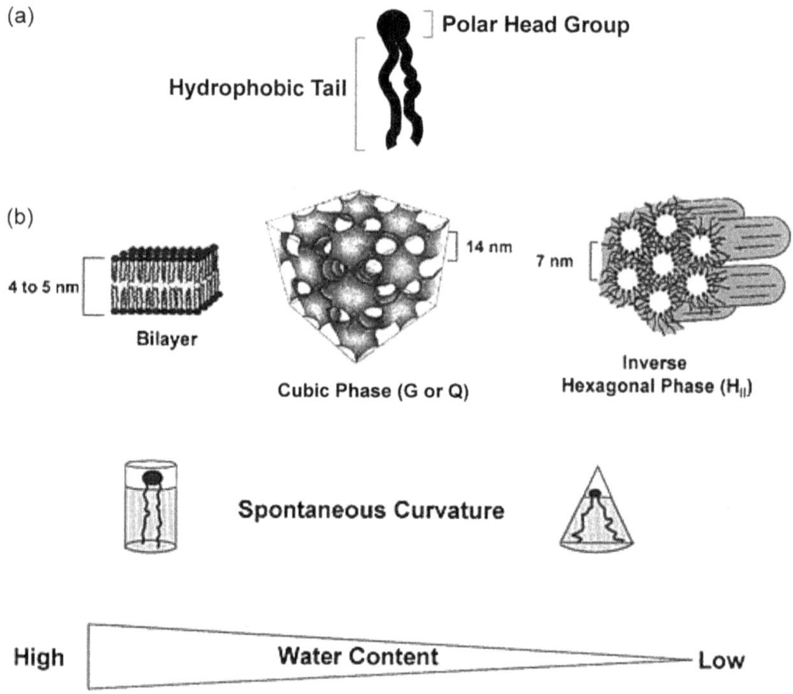

FIGURE 2.1 General characteristics of lyotropic liquid crystals (LLCs). (**a**) Schematic of a typical lipid with polar head group and hydrophobic tails. (**b**) Overview of the LLC phases relevant to forming porous materials and how they relate to the shape of the lipid and the water content. (Reprinted with permission from Reference 10e. Copyright 2002 American Chemical Society.)

amphiphiles self-assemble to minimize contact between the hydrophobic portions and the solvent. The equilibrium structure depends on the size and shape of the polar and nonpolar portions of the lipid and the amount of water present. These equilibrium structures typically have dimensions on the order of 1 to 10 nm (Fig. 2.1b). Lipids were first investigated in regard to their biological roles of forming lipid bilayers; however, as research has progressed, an increasing number of LLC structures have been discovered. The most important structures, as far as porous materials are concerned, include the bicontinuous cubic phase (G or Q) and the inverse hexagonal phase (H_{II}). The cubic phases consist of interconnected hydrophobic and hydrophilic domains that are interwoven, while in the inverse hexagonal phase, water-filled pores are arranged as parallel cylinders. The designation "inverse" refers to the relationship between the hydrophobic and hydrophilic domains in a liquid crystal. In "normal" lipid bilayers water is the major component and the hydrophobic components of the lipids are segregated. For example, a normal hexagonal phase would consist of hydrophobic cylinders surrounded by hydrophilic domains and water. It is possible to predict the liquid crystal structure that will form based on the shape of the lipid [8] and on the spontaneous curvature of the system [9]. Although quantitative methods exist for this prediction, typically only qualitative shape arguments are used in lipid design.

LLCs offer many advantages as self-assembly platforms for porous materials, including the wide range of structures that can form by modifying the lipid structure and modulating the water content. Liquid crystals can be thought of as ordered liquids, and while a fluid structure can be desirable in responding to environmental stimuli, it generally limits their use to mild operating conditions. To provide mechanical support, techniques have been developed to capture the liquid crystal structures by crosslinking one or both of the LLC domains covalently [10]. Figure 2.2 shows some of the common functional groups used for crosslinking/polymerization in liquid crystal phases, of which the dienoyl, acryloyl, and styryl groups have been employed most extensively for the work highlighted in this section. The functional group tolerance of radical polymerization makes it the method most widely used for LLC crosslinking, and both thermally and photochemically activated initiators are used. While the diacetylenyl group has been useful in covalently capturing lamellar structures, the

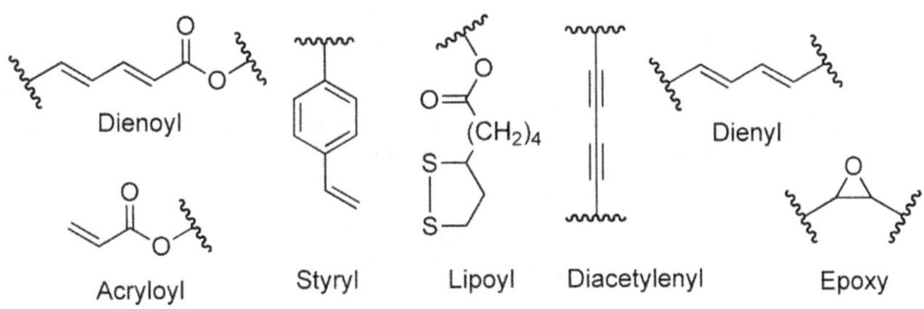

FIGURE 2.2 Common functional groups used in the polymerization/covalent capture of LLC phases.

FIGURE 2.3 Limiting examples of the locations for crosslinking/polymerization in lyotropic liquid crystal assemblies.

topographic constraints introduced by its cycloaddition product have been shown to destroy cubic and inverted hexagonal structures during polymerization [11].

The structural stability of the polymerized LLCs is dependent upon the location of the crosslinking functional groups in the lipids. Figure 2.3 highlights the three limiting cases. Crosslinking could occur at the polar head group of the lipid; however, this technique is not used in the materials described here because the polar head group and water comprise the pores. It has been proposed that the best location for crosslinking is where the polar head group and hydrophobic tail meet. This location has the advantage of providing the most structural stability to the LLC phases while allowing the hydrophobic tails full mobility to form the hydrophobic phase. Placing the crosslinking functionality at the end of the hydrophobic tail is often easier to accomplish synthetically; however, this can lead to easily deformable pores, destroying their long-range order [12].

X-ray scattering data are essential for identifying the ordered-state structures of the LLCs and for ensuring that the morphologies persist after covalent capture. In addition, polarized light microscopy is a simple method for identifying liquid crystalline phases, and microscopy data are often included as supporting evidence of morphology. Occasionally, transmission electron microscopy (TEM) data are used to support the structural assignments; however, due to the small size of the LLC structural features and the difficulty in achieving adequate contrast between the hydrophobic and hydrophilic regions, TEM data are provided less often in these studies.

Finally, it should be pointed out that the inverted hexagonal phase shown in Fig. 2.1 is an idealized structure, but the pores rarely have the long-range order depicted in that diagram. Typically, long-range order is achieved by employing an alignment procedure, such as shear aligning or extrusion. Most of the materials presented here exhibiting H_{II} structures are used as a powder, which maximizes the number of pores that are exposed to the surface. To avoid the need for alignment, materials exhibiting cubic morphologies have received a great deal of interest because their bicontinuous hydrophilic and hydrophobic regions do not require alignment for long-range order.

2.2.1.1 Using External Crosslinking Agents. Generally, it is easier to synthesize lipids with a single polymerizable group; however, the covalently captured LLC phases comprised by these lipids will be less mechanically stable than their counterparts formed from lipids with a large number of polymerizable groups. An approach to circumvent this problem is to include a crosslinking monomer that

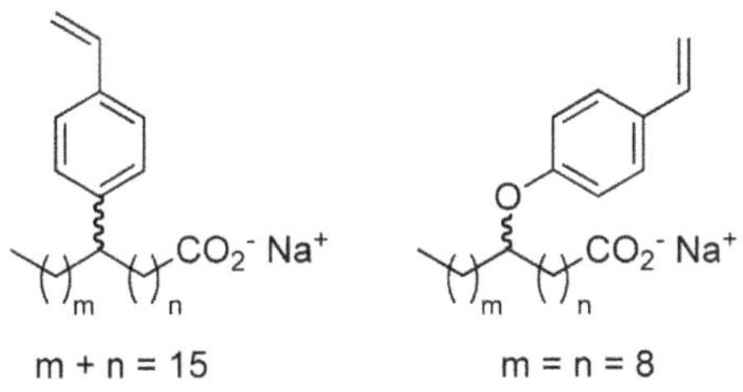

1 2

FIGURE 2.4 Mesogens developed by Gin and coworkers that can form inverted hexagonal liquid crystal phases (H_{II}) when mixed with divinyl benzene (DVB) in water. Crosslinking leads to an improved robustness in the resultant porous material without affecting the H_{II} structure.

segregates into the hydrophobic phase of the LLC and polymerizes with the lipids. In 1963 Herz et al. used this technique to covalently capture an H_{II} LLC phase formed from an isomeric mixture of calcium *p*-styrylundecanoates and 5%–20% divinyl benzene (DVB) [13]. Remarkably, there is one example of a robust LLC structure without a polymerizable moiety. Anderson and Ström demonstrated this with the covalent capture of cubic mesophases of didodecyldimethylammonium bromide and dodecyl hexaethyleneglycol ether by the polymerization of methyl methacrylate (initiator: AIBN) and acrylamide (initiator: hydrogen peroxide), respectively [14]. The authors suggested that polymerization leads to "mechanical crosslinks" between the newly formed polymer and the hydrophobic lipid tails. Anderson and Ström also demonstrated that glucose oxidase could retain its activity after being isolated in the aqueous domains of the dodecyl hexaethyleneglycol ether mesophases.

The styryl mesogens in Fig. 2.4 were developed by Gin and coworkers to form H_{II} phases when mixed with water, DVB, and a radical initiator at ambient temperatures. Mesogen **1** was developed first and has been useful in several applications, which are discussed later [15]. Unfortunately, a key step in the synthesis of **1** involves attaching the styryl group through a carbocation intermediate that can undergo rearrangement. This causes **1** to be prepared as a mixture of regioisomers where the styryl group can be anywhere from the sixth to the seventeenth carbon. These regioisomers do not impede the use of **1** to prepare porous materials; however, their presence inhibits systematic control of the final liquid crystal structure by synthetic modification of the mesogen. The synthesis of **2** addresses this drawback by providing a method of systematically varying the position of the styryl group along the lipid backbone [16]. Using this capability, Reppy et al. were able to vary the interchannel distance of the crosslinked porous material between 29 and 54 Å by adding the styryl moiety closer to the polar head group. In addition, by changing the counterion from sodium to various tetraalkylammonium cations, the authors effectively increased the size of the polar head group, which allowed H_{II} phases to be formed with interchannel distances of up to 65 Å and also allowed the formation of lamellar morphologies.

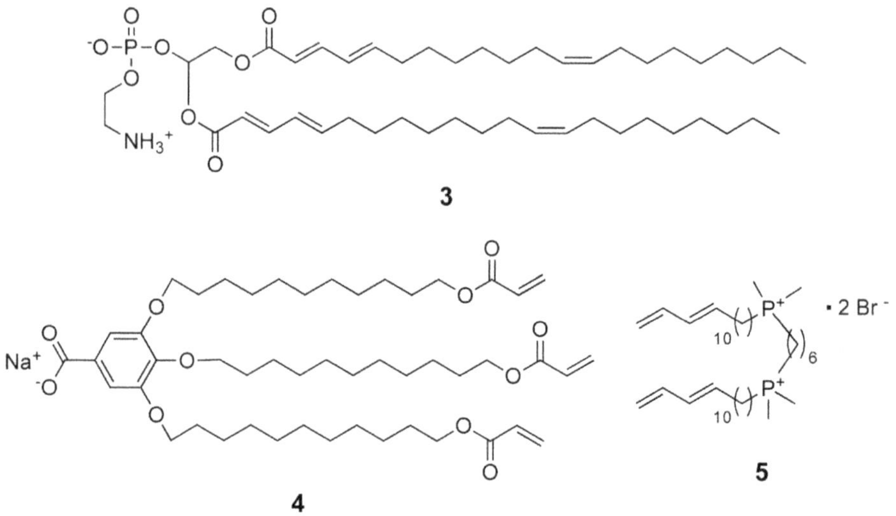

FIGURE 2.5 Examples of multifunctional lipid monomers capable of forming LLCs that can be polymerized with retention of morphology without the addition of crosslinking agents.

2.2.1.2 Using Multifunctional Mesogens. To simplify the covalent capture of porous materials prepared from LLC systems, an effort has been made to prepare lipids that did not require external monomers or crosslinking agents to form a mechanically robust material. Phosphoethanolamine **3** (Fig. 2.5), which was modeled after biologically occurring lipids, contains two dienoyl groups, allowing for multiple chemical connections between neighbors [17]. When **3** was mixed with water at room temperature a LLC formed that adopts a (primarily) H_{II} phase with a small amount of a bicontinuous cubic phase. Heating the LLC above 40°C ensured that only H_{II} was present. The H_{II} phase was covalently captured using either redox or thermal radical initiators at 60°C. The resulting materials were insoluble in common organic solvents and maintained the H_{II} phase even at temperatures less than 40°C, indicating the structure was crosslinked and robust. Varying the temperature can control the diameter of the aqueous channels prior to crosslinking, allowing for control over pore size by selecting a temperature and subsequently polymerizing the material to capture the structure. Srisiri et al. pointed out that the important design parameters for **3** included the small poorly solvated head group and the long hydrophobic chains, which fill space due to their length and *cis* double bonds.

The multifunctional mesogen most extensively investigated is gallate derivative, **4**, which forms H_{II} phases at room temperature and has been polymerized to conversions of greater than 90% using a radical photoinitiator [18]. The tapered shape of **4** makes it ideal for forming H_{II} phases. Although in previous H_{II} systems the hexagonal structure is randomly ordered, it was demonstrated with **4** that long-range alignment of the cylinders could be achieved by heating and pressing the LLC between glass slides before photopolymerizing the material. The size of the aqueous channels formed from **4** can also be controlled by changing the metal counterion. While monovalent (Na^+) and divalent cations (Ni^{2+}, Co^{2+}, Cd^{2+}) provided structures with similar unit cell dimensions, trivalent cations (Eu^{3+}, Ce^{3+}) caused the pores to decrease in size by approximately 5 Å. This contraction of the pores was proposed to depend on the nature of the carboxylate–cation interaction and not the size of the cation. For example, the lanthanide cations Eu and Ce form bonds with carboxylate that have substantial covalent character, resulting in a shorter bond length compared with purely ionic bonds. Shorter bond lengths in turn shrink the size of the aqueous channels by causing the polar head group to be smaller. Analogs of **4** have been prepared that are capable of forming mesophases with the pore size varying with the length of the hydrophobic tail [19] and that use diene functional groups for crosslinking as opposed to acrylates [20].

More recently, Pindzola et al. investigated the LLC behavior of phosphonium gemini amphiphiles [21]. Compound **5** is an important example of this class of compounds because it forms bicontinuous cubic LLCs at room temperature that can be polymerized by irradiation with UV light in the absence of a radical photoinitiator. The resulting materials were isolated as freestanding films that were stable in air up to 300°C. The extent of diene conversion was determined to be approximately 50%, and these materials were insoluble in water and common organic solvents. The primary disadvantage of **5** is its high reactivity; it needs to be stored under inert atmosphere at 0°C.

2.2.1.3 Blending with Crosslinkable Polymers. An alternative approach to capture the structure of LLCs is to mix the lipid monomers with a polymer that is capable of forming chemical crosslinks. This method has been tested with **4** [22] and **5** [23] despite their ability to form robust covalently captured LLC phases without additives, because crosslinking them with a polymer allows for a combination of the polymer and LLC properties.

Blends of butyl rubber (BR) and **4** were formed by first generating a hexagonal phase with **4** and then adding a 15.6 wt% solution of BR in hexanes. The sample was then annealed and the hexanes evaporated. The resulting waxy samples were spread on a glass support and photopolymerized with the aid of a radical photoinitiator to yield a brittle solid. Infrared spectroscopy indicated that approximately 80% of the acrylate groups had reacted, and extraction with organic solvents led to a negligible amount of small-molecule leaching. Remarkably, the LLC was able to accommodate up to 75 wt% BR and still maintain a LLC phase; however, the unit cell of the hexagonal phase stopped expanding after 25 wt% BR, and indications of macrophase separation were observed by polarized light microscopy.

Compound **5** has also been blended with BR, resulting in bicontinuous cubic materials. To form this phase, butyl rubber (up to 35 wt%) was mixed with **3** in water, resulting in a lamellar phase. Upon heating to 70°C the lamellar phase converted to the desired bicontinuous structure, which was then crosslinked using a radical photoinitiator. This bicontinuous/rubber composite was advantageous because there is no need for long-range alignment, in contrast to the H_{II} phases.

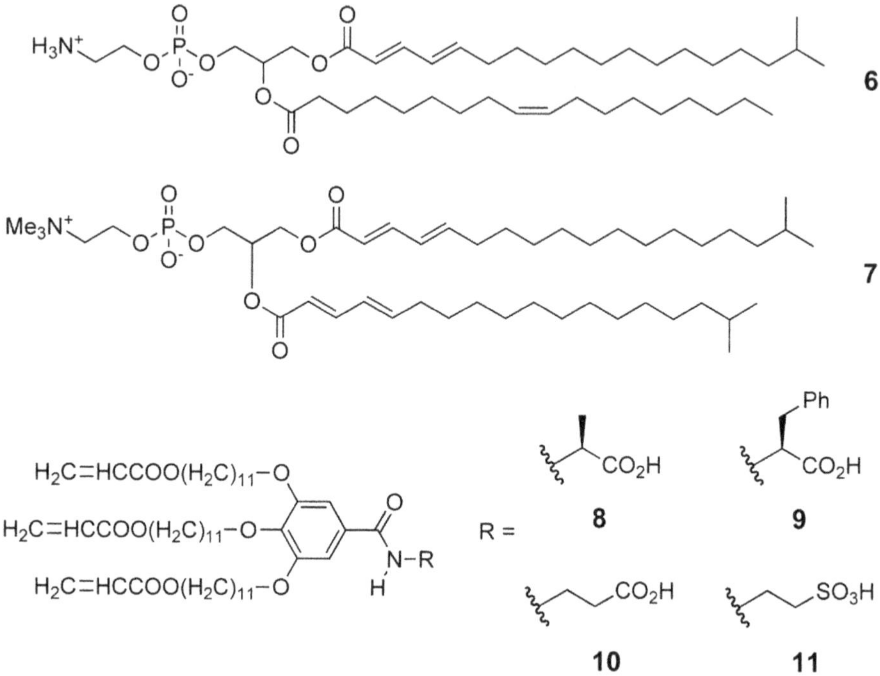

FIGURE 2.6 Reactive lipids that have been mixed to form LLCs with unique structures and functional groups that would not be possible by any of the individual lipids.

2.2.2 Controlling Pore Functionality and Structure by Mesogen Blending

The first example of a covalently captured bicontinuous cubic phase was prepared by blending two phosphoethanolamines, **6** and **7** (Fig. 2.6), in a 3:1 molar mixture in water [24]. Neither mesogen formed a pure bicontinuous phase on its own, but once they were mixed and heated to 60°C, a H_{II} phase formed that slowly converted to a bicontinuous phase. Both the hexagonal and bicontinuous phases were crosslinked using a water-soluble radical initiator; however, the exclusive presence of the crosslinking moieties in the hydrophobic phases indicated that radicals formed in the aqueous phase diffused into the hydrophobic regions. For conversions less than 60%, the materials were completely soluble, while conversions greater than 80% were required for a material that did not dissolve in common organic solvents. The relationship between conversion and properties is strongly dependent on the amount of **7** present in the system, which is responsible for the crosslinking. The higher the percentage of **7**, the lower the conversions can be and still yield a robust, insoluble material.

Brønsted acids, while important for many catalysis applications, do not typically form LLCs when incorporated into amphiphilic structures; therefore, placing this functionality in the aqueous channels of LLCs is an interesting challenge. There

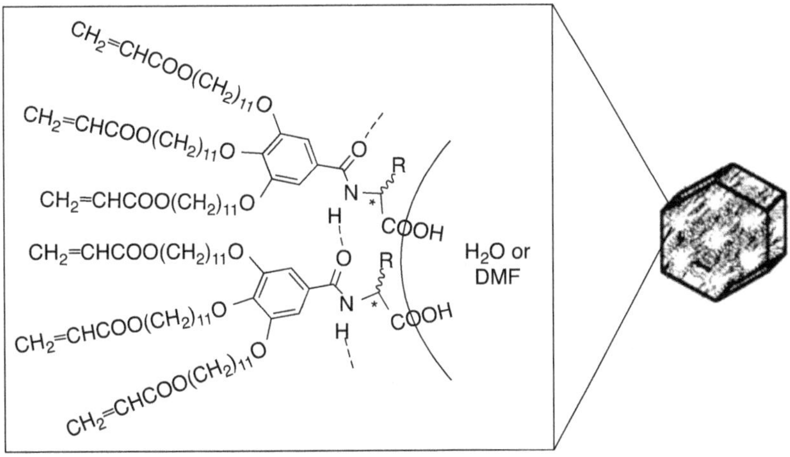

FIGURE 2.7 Schematic of the LLCs formed from **8–11**, highlighting the hydrogen bonding network that directs the H_{II} structure. (Reprinted with permission from Reference 26. Copyright 2003 American Chemical Society.)

have been a couple of examples incorporating Brønsted acids in LLCs; however, the formation of hydrogen bonding networks within the channels can prevent the uptake of water [25]. Zhou et al. were able to prepare H_{II} phases containing Brønsted acid mesogens derived from condensing the acid of **4** with various amino acids (**8–11**) [26]. Compound **8**, which resisted taking up water, adopted the H_{II} phase when mixed with DMF, and this phase was covalently captured using a radical polymerization. Monomer conversions of 80% were achieved, and the covalently captured phase was insoluble in common organic solvents. The propensity of **8** to form H_{II} phases is proposed to derive from the hydrogen bonding ability of the amide groups in this class of mesogens (Fig. 2.7). Direct evidence of the hydrogen bonding came from a shift in N–H stretching in the infrared spectra. Compound **9** formed a lamellar phase with DMF, while compound **10** showed no liquid crystalline behavior. A blending strategy was undertaken to form different H_{II} phases by blending mesogen **8** with **9** or **10** to direct the structure of the LLC by hydrogen bonding interactions. Ratios of **8** to the second mesogen ranging from 3 to 1/3 still formed a H_{II} phase; however, with ratios of **8** less than 1 the hexagonal phase appeared to be less ordered, as evidenced by a broadening of reflections in the SAXS data. Compound **8** was further demonstrated to direct the structure of **11** generating a sulfonic acid H_{II} LLC phase in ratios ranging from 5 to 1/5 in water. Overall, these phases show promise as heterogeneous Brønsted acid catalysts after photochemical radical crosslinking.

2.2.3 Applications

2.2.3.1 Separations. Butyl rubber is an excellent material for protecting humans from chemical agents; however, it is also impervious to water, causing those who

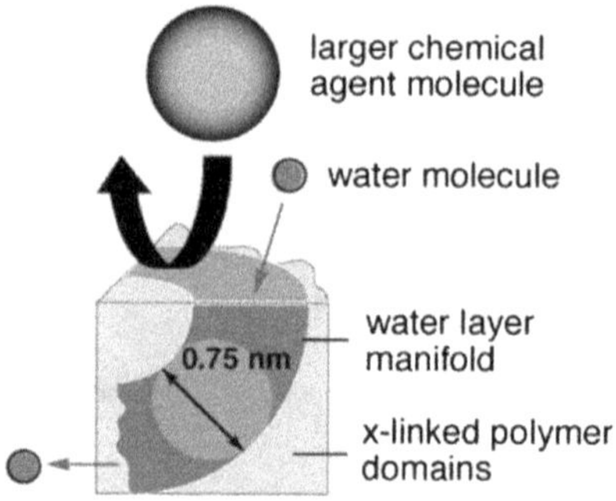

FIGURE 2.8 Schematic of a membrane based on the covalent capture of LLCs. The aqueous channels permit the diffusion of water while blocking larger, undesirable molecules. (Reprinted with permission from Reference 27.)

wear BR garments to overheat. The composites of LLCs and BR previously discussed addressed this problem by introducing small water-filled channels that permit moisture to pass, cooling the wearer by evaporation while protecting him or her from external chemicals (Fig. 2.8). The freestanding, covalently captured LLC films are not strong enough to be used in this manner alone, so the films were either coated or hot pressed into a porous polymer support prior to photochemical crosslinking. The captured H_{II} phase formed from BR/**4** permitted water diffusion while rejecting compounds designed to mimic chemical warfare agents, although not to the levels required for practical applications. The low performance was suggested to be due to a lack of nanochannel orientation [22]. In contrast, the BR/**5** system circumvented this problem by forming a bicontinuous phase, where alignment is not necessary for the aqueous channels to span the entire membrane. Covalently captured LLCs produced from BR/**5** exhibited much-improved characteristics in regard to both water diffusion and the rejection of chemical warfare mimics [23, 27].

This technology has also been adapted for water filtration membranes. Mesogen **4** was applied to polymeric porous supports and then crosslinked to form an H_{II} covalently captured film [28]. The resulting membrane was able to completely reject analytes that were the same size or larger than the 1.2-nm pores. Unfortunately, the flux through this membrane was prohibitively low, which was again suspected to be the result of lack of orientation of the hexagonal channels. Consequentially, mesogen **5** was hot pressed and polymerized onto a porous polymer support [29]. The bicontinuous phase formed by **5** produced membranes that exhibited fluxes comparable with commercial reverse osmosis and nanofiltration membranes. The most promising membranes exhibited pore diameters of approximately 0.75 nm, which is small enough to filter out metal ions and desalinate water.

Membranes prepared from mesogen **4** have also been investigated for use in light gas separations [30]. Both membranes with H_{II} order and isotropic variants were investigated. The H_{II} membranes were prepared on a porous support as described earlier, whereas isotropic membranes where prepared by using the same procedure as the H_{II} LLC except that they were crosslinked at a higher temperature to produce an isotropic phase. The gases investigated were CO_2, N_2, O_2, CH_4, and H_2. The hexagonally organized aqueous channels in the H_{II} membranes lead to a higher solubility for CO_2 compared with the isotropic membranes, while the solubility for the remaining gases remained constant between the two membranes. Diffusion decreased in the ordered membrane relative to the isotropic membrane due to the hexagonal channels acting as barriers to the diffusion of the gases. This decrease was constant among the nonpolar gases, and no separation was observable between them. However, the increased solubility of CO_2 in the aqueous domains afforded an advantage in crossing the membrane, and effective separations were demonstrated between CO_2 and the other gases.

2.2.3.2 Catalysis. Covalently captured LLC porous materials are useful for the heterogeneous catalysis of organic reactions. Heterogeneous catalysts are easier to remove than their soluble counterparts, and porous materials inherently have a large surface area, which is ideal for heterogeneous catalysis. The size of the channels is on the molecular scale, influencing transition states by the confined environment leading to control over the stereochemical outcomes of reactions [31]. Table 2.1 summarizes many of the applications of covalently captured LLCs in catalysis.

The first entry in Table 2.1 illustrates the use of the carboxylic acids inherently present in the covalently captured hexagonal structure of **1** to catalyze the Knoevenagel condensation [32]. Titration experiments reveal that approximately 85% of the carboxylate anions were accessible, and they were estimated to be five orders of magnitude more basic than free aqueous carboxylates due to confinement effects.

Mesogen **12** (Fig. 2.9) was prepared from **2** and used to form a polymerized H_{II} phase, which subsequently underwent ion exchange to replace the sodium with scandium (Table 2.1, entry 2) [33]. The scandium salt of **12** did not form a LLC phase due to scandium's ability to catalyze the cationic polymerization of the styrene moieties; however, ion exchange was very effective in generating a material that was able to act as a Lewis acid catalyst for aldol reactions. In addition, the confined space of the nanochannels led to diastereoselectivities that were not observed in unconfined control experiments.

Mesogen **13** (Fig. 2.9) does not show liquid crystal behavior; however, protonation leads to a compound that will form LLCs in water. Although covalent captured mesophases of protonated **13** were able to catalyze Diels–Alder reactions, there was little difference in the stereochemical outcomes compared to the corresponding homogeneously catalyzed control reactions (Table 2.1, entry 3). These results mirrored those reported for other analogous solid-supported organocatalysts, and the authors suggested using chiral anions to further influence the environment of the nanochannels [34].

The polymerized hexagonal LLC derived from **4** was subjected to ion exchange with a Pd(II) salt and subsequent reduction of the Pd with H_2(g) to yield a LLC with Pd(0) nanoparticles imbedded in the nanopores (Table 2.1, entry 4) [12]. Reduction of the Pd(II) forms nanoparticles several times larger than the original nanochannels, indicating that there was a loss of order in the LLC as the nanoparticles formed. This

TABLE 2.1 Summary of Covalently Captured LLCs That Have Been Applied to Heterogeneous Organic Catalysis

Entry	Catalyst	Reaction in H_{II} Nanochannels	Properties/ Effects	Ref.
1	**1**[a]	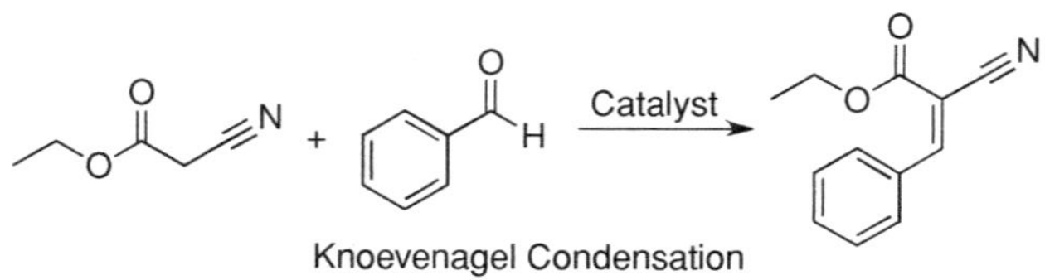	• ~85% of H_{II} carboxylates are accessible	32
2	**12**[a] Na exchanged with Sc(III)	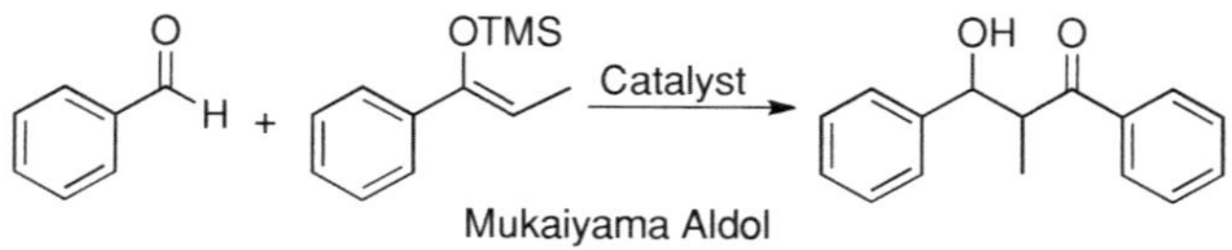	• Carboxylates more basic under H_{II} confinement • Increased diastereo-selectivity	33
3	**13**		• No significant improvement over catalyst in solution	34
4	**4** Na exchanged with Pd(II) reduced to Pd(0)	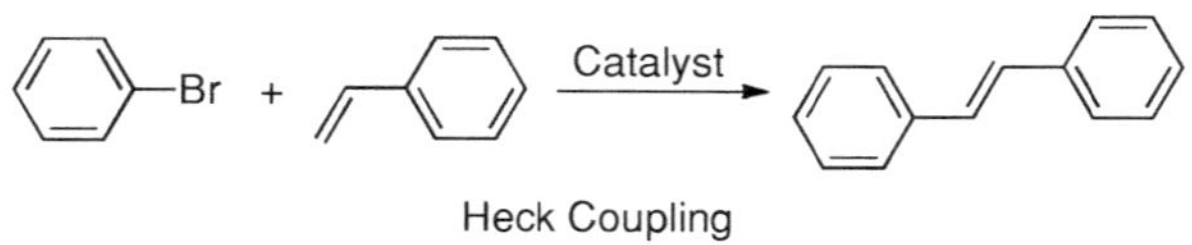	• Catalyst lost order during reduction of Pd(II) to Pd(0) but maintained catalytic activity	12
5	**11/8** (5:1)	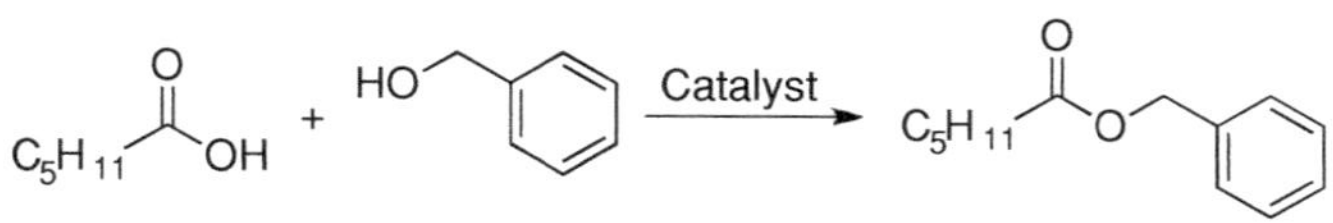	• Slightly slower rate compared to amorphous heterogeneous catalysts • Large decrease in amount of ether byproduct	35

[a]DVB was added to aid in crosslinking.

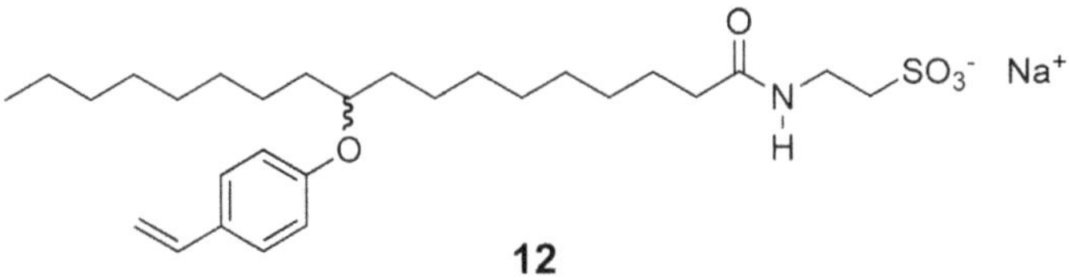

12

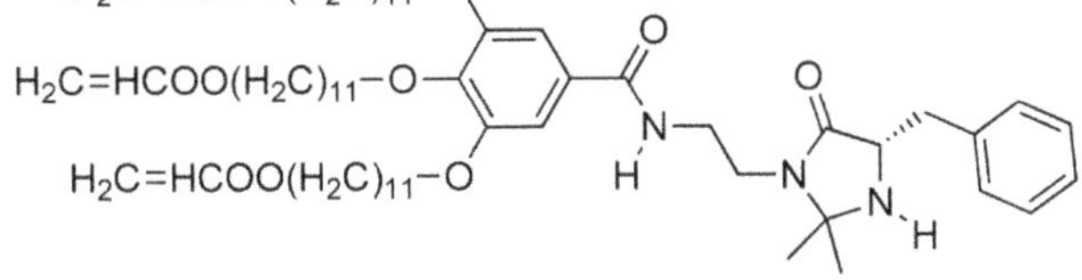

13

FIGURE 2.9 Lipids designed specifically to form covalently captured LLC porous materials for use in confined heterogeneous catalysis.

may be due in part to the crosslinking of **4** at the ends of the hydrophobic tails instead of closer to the aqueous channels. Despite the loss of order, the nanoparticles were still accessible to solvent and catalyzed the Heck reaction effectively.

The structure-directing ability of **8** was used to form a hexagonal LLC with **11**, which is not able to form a hexagonal LLC structure on its own. After covalent capture, this material was an effective heterogeneous catalysis for esterifications (Table 2.1, entry 5) [35]. Compared with commercially available amorphous heterogeneous esterification catalysts, this material exhibited a slightly decreased rate of reaction; however, this was accompanied by a dramatic decrease in the amount of ether byproduct formed.

2.2.3.3 Templating.

The noncovalent interactions required for the self-assembly of LLCs limit this supramolecular approach to particular classes of compounds. Specifically for LLCs, precursors containing a polar head group and a hydrophobic tail are required. This limit can be circumvented by using the porous materials discussed previously to template the formation of other materials that cannot be controlled by self-assembly. Table 2.2 summarizes some of the materials that have been prepared using porous LLCs as templates.

The first entry in Table 2.2 summarizes the formation of a covalently captured LLC self-assembled from **4** with poly(*p*-phenylenevinylene) (PPV) confined to the nanochannels [18]. A polymeric, water-soluble PPV precursor was added during the initial LLC formation where it segregated to the aqueous channels. The hydrophobic domains were then crosslinked using a radical photoinitiator, and subsequent heating generated PPV with unique photophysical properties due to its confined state.

Entry 2 of Table 2.2 describes the formation of a hexagonal LLC from a solution of **1**, DVB, a hydrophobic radical photoinitiator, a hydrophilic photoacid generator, and

TABLE 2.2 Summary of Materials Prepared by Using LLCs as Templates to Control Size and Aggregation[a]

Entry	Monomer	Reaction in H_{II} Nanochannels	Properties/Effects	Ref.
1	**4**	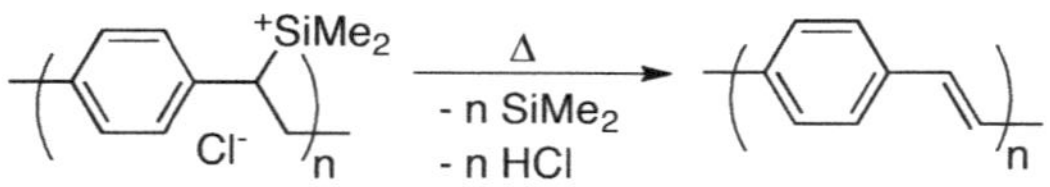	• Higher quantum yield • Blue shift in emission	18
2	**1**[b]	$n\ Si(OEt)_4 + 2n\ H_2O \xrightarrow[\text{photoacid}]{h\upsilon} (SiO_2)_n + 4n\ EtOH$	• Limited degree of SiO_2 conversion	15
3	**1**[b]	$Cd^{2+} + H_2S \longrightarrow CdS + 2\,H^+$	• Control CdS particle size	36

[a]Adapted with permission from Reference 10d. Copyright 2001 American Chemical Society.
[b]DVB was added to aid in crosslinking.

an aqueous silica solution. Irradiation of this covalently captured mesophase caused the simultaneous radical crosslinking of the hydrophobic tails and the generation of acid to catalyze silica condensation. Silicon-29 MAS NMR indicated that the condensation did not proceed to completion, even after subsequent heating under vacuum. These results suggest that the size and orientational constraints of the channels limit silica formation [15].

During an investigation of the effects of exchanging the sodium cation in **1** with different counterions, an H_{II} phase was generated and crosslinked that contained Cd(II) in the aqueous nanopores [36]. Upon exposing the material to H_2S vapor, CdS nanoparticles formed in the channels, and the carboxylic acids in the channels were protonated. The resulting nanoparticles were less than 40 Å in diameter owing to their confinement in the polymerized LLC. The LLC became partially disordered upon nanoparticle formation as a result of the presence of the nanoparticles or carboxylate protonation.

2.3 ETCHABLE BLOCK POLYMERS

A block copolymer is a macromolecule that consists of a continuous run of one monomer linked to a continuous run of a different monomer. In a similar fashion to oil and water, binary polymer mixtures typically macrophase separate, forming heterogeneous disordered structures. Likewise, when the two blocks of a copolymer consist of incompatible segments, the two components prefer to segregate but are unable to as a result of the chemical linkage holding them together. Enthalpically the individual polymer chains prefer to stretch out in an effort to get as far away from each other as possible. However, this stretching is entropically unfavorable, and the system balances these energetic factors to achieve a thermodynamically stable structure. A number of microphase separated structures have been theoretically predicted and experimentally observed in polymer thin films and in bulk polymer samples (Fig. 2.10) [37,38]. The particular structure formed depends on the relative size of the two different polymer blocks and their degree of incompatibility.

Similar to the LLCs phases previously discussed, the cylindrical and bicontinuous cubic/gyroid structures are most pertinent to porous materials; however, with block copolymers, one of the blocks typically needs to be removed (etched) to generate a porous structure (Fig. 2.11). This requires that the polymer block to be etched is accessible to the particular etching technique and that the remaining polymer matrix is able to support the porous structure. Block copolymer self-assembly typically yields materials with features ranging from 5 to 50 nm, which complements the 1- to 10-nm features of the LLC materials. In recent years, polymer chemistry has expanded vastly to include a number of facile and versatile methods to prepare many polymeric structures, and as a result, block copolymers have seen wide application in many fields, including porous materials. This section will highlight some general approaches to using block copolymers as precursors to porous materials; the interested reader is directed to the literature for more comprehensive reviews [39].

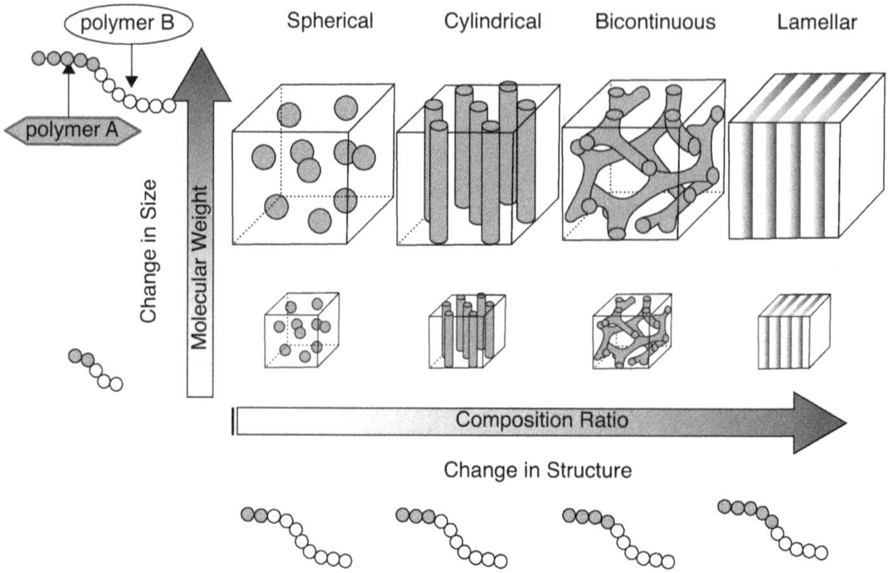

FIGURE 2.10 Morphologies formed by diblock copolymers containing two incompatible blocks. The exact structure formed depends on the volume fractions of the two blocks (composition ratio) and their level of incompatibility. (Reprinted from Reference 45a with permission from APEX/JJAP.)

As precursors to porous materials, block copolymers are primarily used as thin films or macroscopic polymer monoliths. Block copolymer thin films are formed by casting or spin-coating solutions on a substrate and allowing the solvent to evaporate. The process of drying often orders the features of the block copolymer by generating solvent gradients; however, secondary thermal or solvent annealing can lead to higher

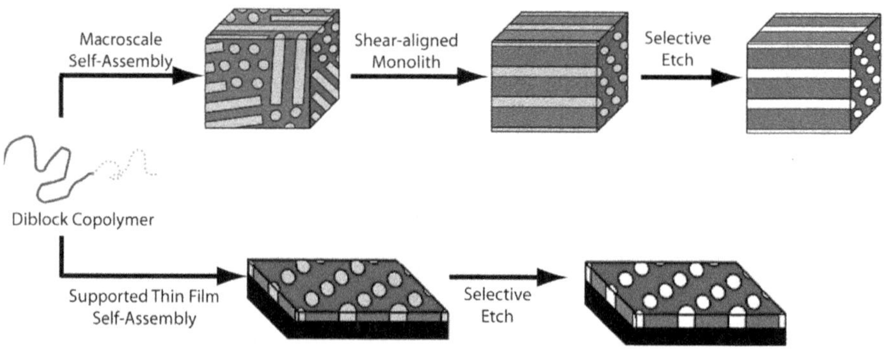

FIGURE 2.11 Schematic example of generating porous macroscopic polymer monoliths and thin films by self-assembly followed by etching of the hexagonally aligned cylindrical domains. (Adapted with permission from Reference 50b. Copyright 2002 American Chemical Society.)

degrees of long-range order [40]. The substrate used to support the thin film can also influence the block copolymer morphology, and methods have been developed to control these effects [41]. These methods include grafting random copolymers containing the same monomers as the block copolymer to the surface of the substrate to prevent preferential interactions between one of the polymer blocks and the substrate surface. Thin films are particularly useful for lithography and surface templating applications. Larger-scale separations and supported catalysis are better suited to macroscopic block copolymer monoliths, which are ordered polymeric materials that do not have a secondary support. As with the LLC materials, the hexagonal cylindrical phases formed by block copolymers often need intervention to achieve long-range order. Methods of alignment include exposure to an electric field [42], exposure of the polymer to various types of flow [43], and the already mentioned thermal and solvent annealing.

The self-assembly of LLCs resulted in water-filled pores; in contrast, generating porosity in materials from block copolymers requires the removal of a sacrificial block. This etching step necessitates additional characterization than described for LLC porous materials. To quantify the removal of the sacrificial component, gravimetric analysis is often used. Various forms of spectroscopy (NMR, IR, UV) are also used to confirm the composition of the etched material, particularly if the final product will dissolve in suitable solvents. The porosity of these materials can be measured by nitrogen adsorption experiments, which is not possible for water-filled pores of the LLC materials. The larger pore sizes and higher contrast between matrix material and empty pores means microscopy (TEM and SEM) often plays a larger role in characterization than for LLC porous materials. Small-angle X-ray scattering is also important for morphology characterization.

2.3.1 Block Polymer Architectures and Etching Techniques

Porous materials from block copolymer precursors have been an active area of research since the work of Nakahama et al. in 1988 [44], and new materials and techniques are published regularly. Table 2.3 summarizes typical methods and materials for generating porous materials from block copolymer precursors. These etching techniques can be classified as wet or dry, depending on particular details of the process. The hydrolysis of PLA, degradation/cleavage of PEO, and ozonolysis of polydienes are all examples of wet methods, where the block copolymer material must interact with a specific reagent to induce degradation of the sacrificial block. The only dry method included in the table is the thermal decomposition of PPG under reduced pressure, where the PPG decomposes as it is heated and the byproducts are removed under vacuum. Reactive ion etching is another common form of dry etching, in which the copolymer (typically as a thin film) is bombarded with reactive ions. The difference in reactivity of the two polymer blocks toward the reactive ions allows the sacrificial block to be completely etched away before the matrix material is consumed [45].

Once etching is complete, the matrix must be able to support the generated porous structure. In the case of the PS matrix materials, the glassy nature of PS allows it to support the porous structure without modification. However, the polydienes (PI and

TABLE 2.3 Typical Materials and Methods Used to Prepare Nanoporous Materials from Block Copolymer Precursors

Polymers	Sacrificial Block	Etching Conditions	Ref.
PS-PMMA	PMMA	Deep UV exposure followed by acetic acid rinsing	46
PS-PI	PI		47
PVP-PI	PI	Ozonolysis	48
PS-PBD	PBD		49
PS-PLA			50
PCHE-PLA	PLA	0.5 M NaOH in 40:60 (v/v) methanol/water mixture (65°C, 5–10 days)	51
PS-PLLA	PLLA		52
PS-PEO	PEO	57 wt% $HI_{(aq)}$ (60°C, 5 days)	53
PI-PDMS			54
PB-PDMS	PDMS	1.0 M TBAF in THF (RT, 36–48 h)	55
PS-PDMS		Anhydrous HF (0°C, 2 h)	56
P*t*BA-PCEMA	[a]	0.05 M TMSI in CH_2Cl_2 (RT, 2–14 days) followed by 5:95 (v/v) water/methanol (RT, 2 days)	57
PE-PS	PS	Fuming nitric acid (RT, 30 min)	58
P(Imide)-PPG	PPG	Heat to 240°C under vacuum	59

[a] P*t*BA is converted to PAA by cleavage of the *t*-butyl group.

PB) require crosslinking before etching to generate a robust polymer network. Related examples in Table 2.3 include PVP, which was crosslinked with an alkyl dihalide, and PCEMA, which was photocrosslinked before etching. While crosslinking increases mechanical stability and resistance to solvent, it typically adds another step in material preparation. The uncrosslinked PS porous materials can be removed by dissolution, and this is a desirable property for lithography and templating applications, where the template will ultimately need to be removed. Uncrosslinked PS porous materials are prepared from the PS-PLA precursors using basic etching conditions; however, the UV irradiation used for etching PS-PMMA and ozonolysis used for etching the PS-polydienes typically cause crosslinking in the PS matrix material, resulting in some degree of solvent resistance.

In addition to selectively removing one block, several methods have been developed for generating porous materials from block copolymer precursors without removing any of the polymeric material. Block copolymers containing a CO_2-philic block (PDMS, PFMA) can be made porous by exposure to supercritical carbon dioxide ($scCO_2$) [60]. The copolymer film or monolith is exposed to $scCO_2$ at a temperature above the glass transition temperature T_g of the matrix material and the $scCO_2$ preferentially segregates to the CO_2-phillic blocks, expanding the structure. Lowering the temperature rigidifies the matrix, and subsequent removal of the $scCO_2$

leaves a porous structure. This approach is particularly interesting because it uses an environmentally friendly compound to generate porosity and the size of the pores can be tuned by modulating the temperature and rate of depressurization. Jeong et al. reported a related method using cylinder-forming PS-PMMA block copolymer thin films [61]. The PS matrix was crosslinked with ozone below the T_g of PMMA. Crosslinking caused a contraction in the PS matrix and increased stress in the glassy PMMA domains. After warming of the thin films above the T_g of PMMA, the PMMA domains relaxed, forming 6-nm pores down the center of the cylindrical domains. Xu et al. also demonstrated that PS-PMMA thin films could be made porous by simply soaking the supported films in acetic acid with no prior UV irradiation [62]. The acetic acid solubilized the PMMA chains and caused them to migrate to the surface of the thin film, increasing the film thickness while forming pores in the process. Remarkably, the pores were similar in size to the pores generated when the PMMA was completely removed by prior UV irradiation and washing. Since the PMMA was still covalently linked to the PS blocks, drying and annealing the film restored the original PS-PMMA cylindrical morphology.

All of the examples in Table 2.3 involve etching techniques that break most, if not all, of the bonds in the polymer block being removed; however, in principle, only the bond connecting the two blocks needs to be broken to remove the sacrificial component. The most straightforward application of this approach is illustrated by copolymer **14**, developed by Lee et al. (Fig. 2.12), for which basic hydrolysis leads to cleavage of the single ester, allowing removal of the PPO [63]. Goldbach et al. developed a PS-PMMA block copolymer linked by an anthracene dimer, **17** [64]. The anthracene dimer is cleavable by both heat and light; however, the thermal stability of the dimer is rather low, limiting the processing methods that can be used. Table 2.3 includes etching conditions for monolithic PS-PEO; however, these conditions use a very strong acid that destroys thin films. The incorporation of a degradable linkage allows PEO, a polymer that resists chemical degradation, to be removed using comparatively mild conditions. Zhang et al. used an acid-labile linker, triphenylmethyl (trityl) ether, to prepare a PS-PEO block copolymer, **15**, that could be etched using

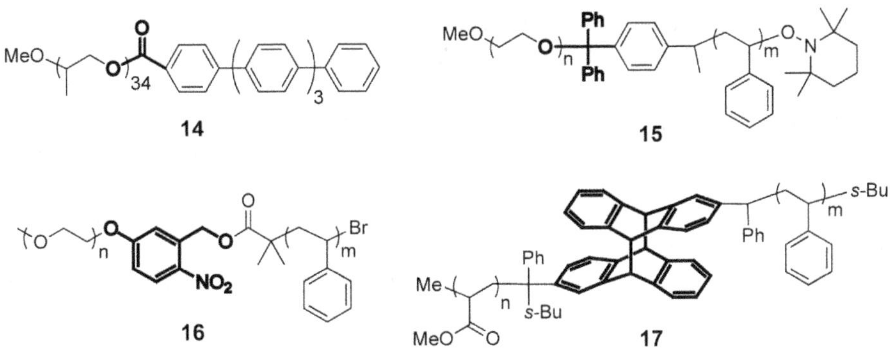

14 **15** **16** **17**

FIGURE 2.12 Block copolymers containing a cleavable unit between the two blocks (shown in bold). Etching can be achieved by breaking these linkages instead of degrading an entire sacrificial block.

TFA vapor followed by rinsing to remove the cleaved PEO. In addition, both cleavage and PEO removal could be achieved by using a 10:1 methanol/HCl (v/v) solution [65]. Alternatively, Kang and Moon and developed a PS-PEO polymer containing a photocleavable linkage [66]. A thin film containing cylindrical PEO domains was made porous by irradiation at 350 nm for 2 h followed by a methanol/water rinse.

Triblock terpolymers—polymers with three different blocks connected in sequence—can be used to expand the functionality of porous materials (see later discussion). Thin films with remarkable long-range order can be generated from PS-PEO copolymers, and we have already described how PEO can be difficult to remove in thin films, which led to the incorporation of the cleavable linkers described in the previous paragraph. Bang et al. approached this problem by preparing a PS-PMMA-PEO triblock terpolymer [67]. The relatively small PMMA block was present as a cleavable linkage between the PS and PEO segments, where the unfavorable interaction between PS and PEO caused long-range order in the thin film. The utility of this method was limited by the morphological details. While thin films containing cylindrical PEO domains were easily obtainable, cleavage was efficiently performed only in thin films with a discrete, phase-separated PMMA domain.

2.3.2 Controlling Pore Size and Functionality

2.3.2.1 Blending Approaches. The pore size and morphology of self-assembled block copolymers can be controlled by varying the identity and molecular weight of the constituent polymer blocks. While this is advantageous, it implies that a new copolymer needs to be synthesized for every modification desired in the material characteristics. In addition, there are limits to how large a block copolymer can be and still achieve a well-structured morphology. High-molecular-weight copolymers have a large kinetic barrier to the formation of well-ordered structures. Jeong et al. developed methods to expand the utility of a specific block copolymer for the formation of porous materials by mixing it with various amounts of homopolymer [68]. Thin films of a PS-PMMA block copolymer mixed with different amounts of PMMA homopolymer formed larger cylindrical pores because the homopolymer PMMA segregated to the PMMA domains of the thin film. The presence of the homopolymer also provided a facile method to generate porosity, by soaking the thin film in acetic acid, thereby extracting the PMMA homopolymer. If larger pores are desired, the UV/acetic acid etching treatment (Table 2.3) could be used to remove the PMMA block of the copolymer along with the homopolymer. Thin films prepared from the same PS-PMMA/PMMA system generated porous materials with either 6- or 22-nm pores, depending on whether the thin films were UV irradiated before the acetic acid wash. Kim et al. went on to demonstrate this method with PS-PEO/PEO and PS-PEO/PMMA systems, although solvent annealing was required to achieve homogeneous, equilibrium morphologies [69]. There are of course limitations to this approach, including macrophase separation from the addition of too much homopolymer.

Hillmyer and coworkers developed a blending strategy for controlling the hydrophilicity of pores in a PS matrix [70]. In this approach, PS-PLA and PS-PEO diblock copolymers were mixed to give cylindrical and gyroid structures, which upon removal of PLA using basic etching conditions yielded porous PS monolithic materials with PEO-lined pore walls. Typical porous PS materials did not readily take up water; however, the PEO-lined pores allowed for easy wettability. In addition, this approach allowed for a straightforward method of tuning the pore properties by changing the ratio of PS-PEO to PS-PLA.

2.3.2.2 Triblock Terpolymers. Triblock terpolymers with a central sacrificial block to facilitate etching were already discussed; however, triblock terpolymers can also be used to introduce functionality into the pores of monolithic materials. The porous materials we have discussed have functional groups lining the pore walls. After etching there is typically some functionality remaining from where the two blocks were covalently connected. For example, after hydrolytic etching of PS-PLA monoliths there are alcohols lining the pore walls that can be functionalized with appropriate reagents [50b]. The surface density of these functional groups is limited by the block copolymer interface and typically is not sufficiently high to impact the surface properties of the pores; the porous PS monoliths derived from PS-PLA still contain hydrophobic pores even though alcohols are present. To increase and control the functional group density inside the nanopores, materials can be made from A-B-C triblock terpolymers, where A as the matrix component, B is a nondegradable mid-block, and C is the sacrificial block that is eventually etched. Figure 2.13 compares the nanoporous materials prepared from diblock and triblock copolymers. The A-B diblock copolymer (e.g., PS-PLA) is aligned and etched to give a hexagonal array of cylinders, where the dots represent the residual functional groups from the linkage between the two blocks. In the A-B-C triblock terpolymer case the polymer is again aligned into hexagonal cylinders, and block C is removed. This leaves a porous structure where the pores are coated with polymer B. Modification of the identity and molecular weight of the B block can control the nature and density of the functional groups present in the pore.

Rzayev and Hillmyer used this approach to prepare porous PS monoliths with hydrophilic PDMA lining the pore walls from PS-PDMA-PLA triblock terpolymers [71, 72]. PS monoliths containing hexagonally arranged cylinders of PDMA and PLA were formed after shear aligning in a channel die. The PLA was removed by hydrolysis (Table 2.3) over a period of 2 days to 2 weeks, depending on the amount of PDMA present. Varying the molecular weight of the PDMA mid-block allowed control over the functional group density inside the pores. Once the PLA was removed, the resulting monolith would readily take up water. In addition, the PDMA could itself be hydrolyzed to give PAA, and they demonstrated that these acids could be functionalized by various amines using standard diimide coupling chemistry.

Bailey et al. also developed a related nanoporous system from PS-PI-PLA triblock terpolymers in which the PLA could be degraded to yield PI-lined pores in a PS matrix [73]. The PI provided alkenes that can be functionalized to modify the pore wall properties. Although this design was realized, the PS-PI-PLA monoliths were not as versatile as those derived from PS-PDMA-PLA. The hydrophobic nature of PI significantly affected the access of hydrophilic etching solutions to the PLA domains.

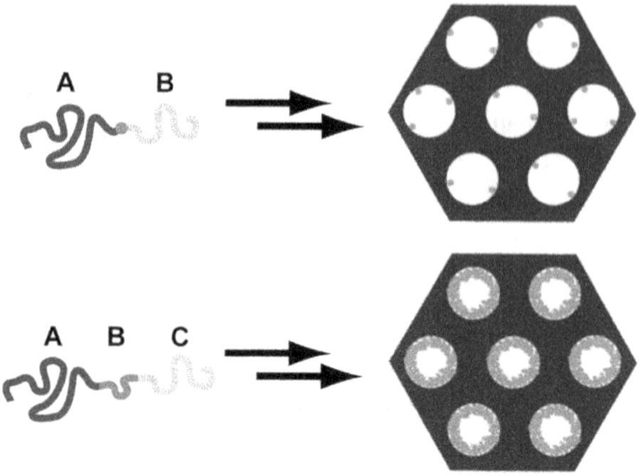

FIGURE 2.13 Schematic representation of cylindrical porous materials derived from A-B and A-B-C block polymers. A-B-C triblocks provide pores lined with polymer B after removal of the C component. (Adapted with permission from Reference 71a. Copyright 2005 American Chemical Society.)

The standard conditions for etching PLA in Table 2.3 were ineffective; however, the addition of sodium dodecyl sulfate (SDS) led to near-complete PLA removal as a result of improved wetting of the pores. There also appeared to be a molecular weight limit for this system, as larger triblock terpolymers resisted PLA etching. For the monolith materials that were successfully prepared, efficient epoxidation of the alkenes was demonstrated using mCPBA in *tert*-butanol. Maximum epoxidation efficiencies of 70%–85% were observed independent of the reaction time. This suggests that some of the alkenes were not accessible to the peroxide reagent—most likely the alkenes located near the PS-PI phase boundary. Guo et al. also investigated the thin-film properties of this and a related polymer in an attempt to generate arrays of nanorings and nanoposts [74].

2.3.3 Applications

2.3.3.1 Separations. Porous materials derived from block copolymer precursors have several features that make them attractive separation media, including the preparation of materials containing pores with well-defined sizes and chemistries. In principle, high fluxes can be attained as a result of the high density of pores that is achievable through self-assembly. This combination of high pore density and relatively monodisperse pore sizes is difficult to realize in other porous membranes for separation purposes.

The transport properties of nanoporous membranes prepared from block polymers containing hexagonally packed cylinders derived from PS-PLA and PS-PDMA-PLA monoliths have been investigated. The diffusion of NaOH (a reactive etchant) in methanol and water was used to probe the properties of porous PS monoliths derived from PS-PLA [75]. Initial wetting of the pores appeared to be a significant barrier to diffusion as indicated by the breakthrough time—the amount of time until the first NaOH makes it across the membrane. The breakthrough time was fastest in methanol, even though NaOH should have diffused faster in water, which was rationalized by more efficient wetting of the PS pores by methanol. This wetting phenomenon was further demonstrated by prewetting the monoliths in water before exposure to NaOH. In this case the breakthrough time was much faster than before, although still not as fast as in the experiments with methanol. Gas diffusion was measured through the porous PS-PDMA monoliths derived from PS-PDMA-PLA triblock terpolymers [76]. Knudsen diffusion was observed for the gases crossing the membrane, meaning that the gas particles are more likely to collide with the pore walls than with each other. The calculated pore sizes from the diffusion measurements were a couple of nanometers less than those measured by SEM and SAXS, and it was proposed that this is a result of the PDMA layer swelling during diffusion, decreasing the effective size of the pores.

Yang et al. used a thin film of PS-PMMA mixed with PMMA homopolymer for filtration of viruses [77]. The thin film was prepared on a sacrificial silicon oxide layer and then transferred onto a polysulfone membrane that acted as a support. Extracting the homopolymer PMMA using acetic acid, with no prior UV irradiation, generated pores measuring 15 nm in diameter. The membrane was then used to filter human rhinovirus type 14 (HRV14), which is approximately 30 nm in diameter, from aqueous solutions (Fig. 2.14). The homogeneity of the pores ensured a sharp molecular weight cutoff, excluding HRV14, and the high areal density of pores and relatively thinness of the film led to fluxes that were much higher than those with commercially available membranes.

2.3.3.2 Templating. Similar to polymerized LLCs, block copolymers have been used to template other nanoscopic materials. Block copolymer thin films have been extensively utilized as masks in lithography; however, we are focusing on applications that are more comparable to the other porous materials described here. Those interested in lithographic applications of block copolymers are referred to the relevant literature and Chapter 10 in this book [78].

Johnson et al. used PHCE and PS monoliths with cylindrical pores to template the growth of polypyrrole and CdS nanoparticles, respectively [79]. In both cases the size of the templated material was dictated by the pore size of the monoliths, and facile isolation was achieved by simply dissolving the monoliths in a good solvent for the polymer matrix. In contrast to the polymerized LLCs, the monolithic porous materials maintained their structure throughout the CdS nanoparticle synthesis. Russell, Emrick, and coworkers deposited preformed CdSe nanoparticles in the cylindrical pores of PS thin films prepared from PS-PMMA [80]. The silicon wafer–supported PS thin film was dipped into a heptane solution of nanoparticles, and capillary forces drew the nanoparticles into the pores of the film as it was withdrawn from the solution. The composite material was modified by controlling the concentration of the solution and the rate of removal of the film and by changing the size of the nanoparticles or pores in the thin film. Indeed, by modifying the

FIGURE 2.14 SEM of porous membrane derived from PS-PMMA block copolymer blended with PMMA homopolymer. The membrane was able to filter human rhinovirus type 14 (HRV14), which is approximately 30 nm in diameter, from aqueous solutions (lighter spheres in image). (Reprinted with permission from Reference 77. Copyright Wiley-VCH Verlag GmbH & Co. KGaA.)

size of the nanoparticles, a composite material was prepared that contained predominantly one nanoparticle in each pore. Bandyopadhyay et al. demonstrated a more controlled nanoparticle deposition method [81]. DNA functionalized gold nanoparticles were deposited into the cylindrical pores of a silicon wafer–supported PS thin film derived from PS-PMMA. The exposed silicon wafer at the bottom of the pores was functionalized with either ammonium cations or a DNA oligonucleotide, and the nanoparticle deposition was driven by either electrostatic interactions or DNA complexation, respectively. By designing the pores and nanoparticles to be similar in size, it was possible to achieve one particle per pore.

Several of the monolithic materials mentioned here have also been used for the confined crystallization of various organic compounds [82]. Crystallization begins with the formation of crystal nuclei, from which mature crystals develop, and crystallization under nanoscale confinement facilitates investigation of crystals with sizes not much larger than these nuclei (Fig. 2.15). Examination of the thermotropic properties of the nanocrystals embedded in the monoliths revealed that pore wall chemistry affects the melting temperature and heat of fusion of the nanoscale crystals, a phenomenon that has not been emphasized in previous examinations of the melting of confined crystals. Furthermore, nanoscale confinement can also lead to the formation and stabilization of crystal forms (i.e., polymorphs) that are not stable under macroscopic conditions, the appearance of previously unobserved polymorphs, and the

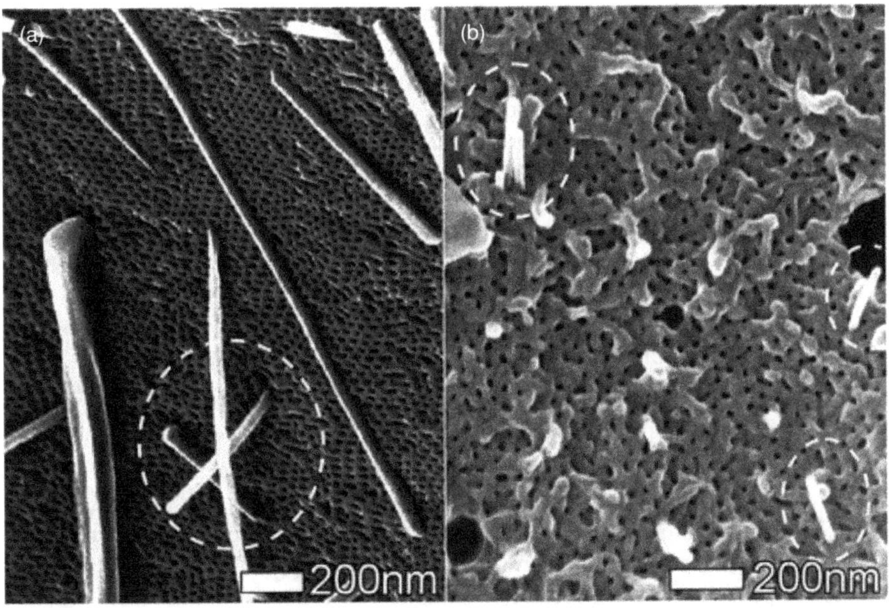

FIGURE 2.15 SEM of glycine nanocrystals on the surface of a porous PS-PDMA monolith. Circles indicate nanocrystals protruding from the monolith pores. (Reprinted with permission from Reference 82d. Copyright 2009 American Chemical Society.)

suppression of crystal growth in favor of an amorphous phase. Having such control over crystallization is useful for understanding crystal nucleation and growth phenomenon, but in addition, these techniques suggest protocols for polymorph screening and control for industries such as the pharmaceutical sector, where possessing the ability to ensure reproducible crystallization outcomes is critical.

2.4 HIERARCHICAL SELF-ASSEMBLY OF BLOCK POLYMERS

In this section we discuss ordered materials that form via hierarchical self-assembly consisting of several sequential self-assembly steps. The defining feature of hierarchical self-assembly is that initial supramolecular structures are capable of more complex self-assembly, which could not be achieved without the initial self-assembly event. In this way, hierarchical systems form complex molecular architectures from relatively simple building blocks [83]. The same mechanisms of self-assembly previously discussed are applicable here, particularly in regard to block copolymer thin films, and all of the following examples contain a block copolymer that drives one of the self-assembly events. The nature of many of the supramolecular interactions required for hierarchical self-assembly means that these approaches are not as general as those previously discussed using block copolymers. This lack of generality comes with the ability to control molecular architectures on multiple length scales and often provides facile methods of generating porosity by breaking up noncovalent interactions, often by simply using a competing solvent. The addition of a second component, either polymeric or small molecule, can also change the morphology of the polymer system. Characterization methods are the same as those already mentioned and include SEM, TEM, AFM, and SAXS. As discussed in connection with the previous porous materials, alignment of cylindrical morphologies can be accomplished using the same techniques used for alignment in block copolymer thin films.

2.4.1 Solution Self-Assembly Followed by Secondary Assembly

Rod-coil polymers (Fig. 2.16) can form micelles in solvents that are selective for the coil block, typically polystyrene (PS). These micelles can further self-assemble into hexagonal porous ("honeycomb") structures when the solution is placed on a surface while the solvent evaporates under a flow of moist gas (air or nitrogen) [84, 85]. The honeycomb structure has spherical voids on the order of micrometers. The initial steps of the proposed mechanism are illustrated in Fig. 2.17. Rapid evaporation cools the polymer solution surface, leading to the condensation of water droplets. Although these droplets typically coalesce, it has been proposed that the micelles present in the solution precipitate around the water droplets, preventing them from combining. Currents in the evaporating solvent then move the polymer-encapsulated water droplets into a close-packed arrangement on the underlying surface as the solvent continues to evaporate, eventually leaving a hexagonally packed layer of water-filled polymer spheres. Subsequent evaporation of the water leads to the final

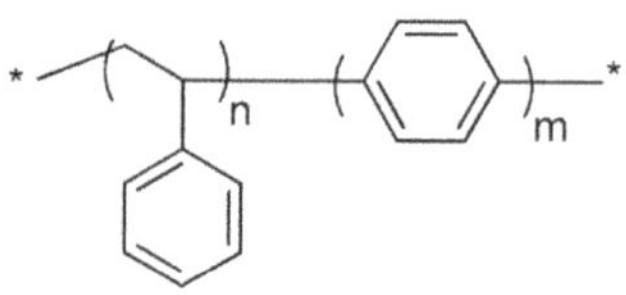

18

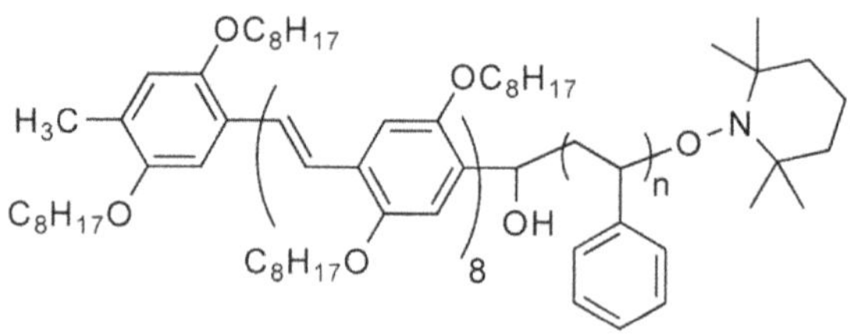

19

FIGURE 2.16 Examples of rod-coil block copolymers used in the water-templated formation of honeycomb porous polymer films.

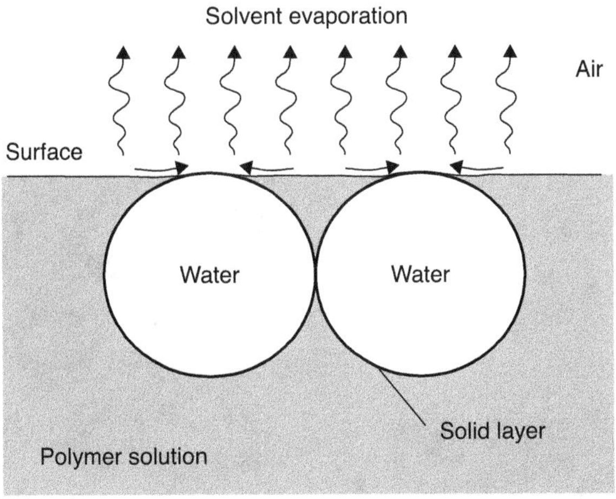

FIGURE 2.17 Schematic representation of the mechanism of the initial stages of solvent evaporation–driven, water-templated porous material formation. The water droplets are covered with precipitated polymer and eventually close pack on the substrate below due to currents in the solvent. (Reprinted with permission from Reference 84c.)

porous structure. Moisture has been shown to be essential for the formation of these structures, as evaporation in dry air or a gas saturated with methanol or ethanol does not lead to porous structures. It is proposed that the quasi-spherical micelles initially formed are necessary for porous film formation, and it is believed that the micelle structure enhances precipitation during the evaporation process. This structure formation argument is further supported by the result that amphiphilic hyperbranched polymers, similar in structure to the micelles, can also form porous honeycomb structures [86]. In addition, the viscosity of the solution increases during the evaporative process, which also serves to lock in the porous structure. Carbon disulfide (CS_2), which is a good solvent for polystyrene and poor solvent for the typical rodlike blocks, is most commonly used for honeycomb film formation [87]. Those interested in these methods are referred to Chapter 4, where water droplet templating is described in more detail.

Jenekhe and Chen developed an alternative approach using the rod-coil diblock copolymer poly(phenylquinoline)-polystyrene (PPQ-PS), **20**, which did not require water condensation (Fig. 2.18) [88]. PPQ-PS formed hollow spheres in CS_2, a good solvent for PS, due to the difficulty of close-packing rigid rods. Evaporation of the solvent led to close packing of these hollow spheres into a hexagonal porous thin film with thickness between 4.5 and 35 μm (one to eight layers of close-packed spheres). Defect free areas larger than 1 cm^2 were observed, and the mechanical properties of the films suggest entanglement of the PS domains. Fullerenes were also incorporated into the films; however this led to destabilization, most likely because the fullerenes segregated to the PS domains.

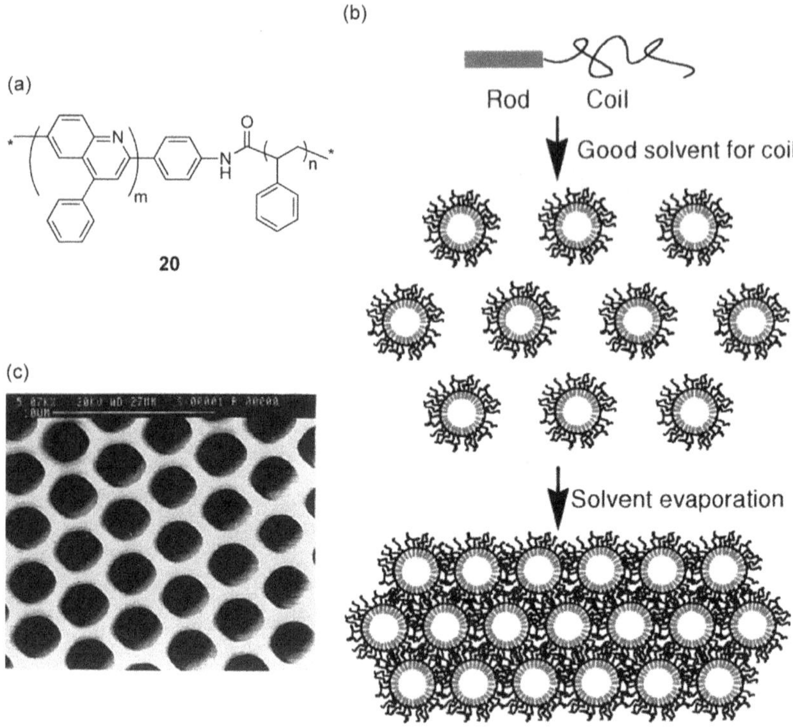

FIGURE 2.18 **(a)** Poly(phenylquinoline)-polystyrene, PPQ-PS, forms porous films by hierarchical self-assembly. **(b)** Schematic of the mechanism of porous film formation. **(c)** SEM of a typical film. (Reprinted with permission from Reference 88.)

2.4.2 Polymer:Polymer and Polymer:Small Molecule Self-Assembled Structures

We have discussed how polymer:polymer interactions can be useful through the blending strategies highlighted in Section 2.2.2. This section focuses on blending that is governed by specific interactions leading to an additional level of organization in the polymeric material, which is in contrast to the earlier blending strategies, which only relied on phase separation. Poly(4-vinylpyridine)-block-polystyrene (P4VP-PS) is most commonly used for materials based on polymer:small molecule interactions [89]. The two blocks are incompatible and self-assemble as discussed in Section 2.3, and the nitrogen present in the P4VP block allows small molecules to form complexes through hydrogen bonding and metal coordination (Fig. 2.19a).

Valkama et al. prepared thin films that organized on two different length scales by complexing zinc dodecylbenzenesufonate ($Zn(DBS)_2$), **21**, with the nitrogens of the P4VP block of a PS-P4VP copolymer in chloroform and then allowed the solvent to evaporate [90]. The P4VP and PS self-assembled into a lamellar structure, and

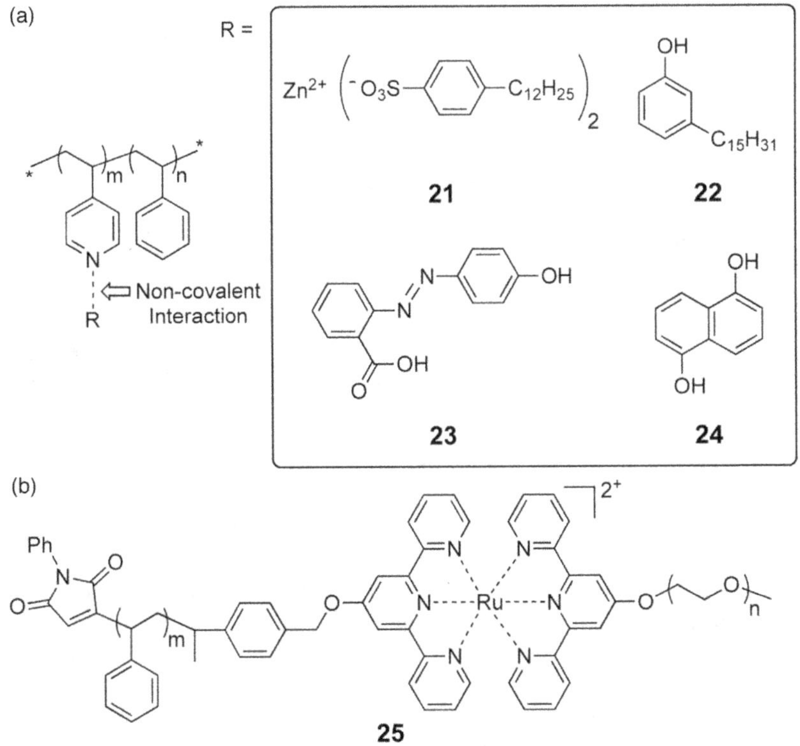

FIGURE 2.19 (**a**) Polymer:small molecule systems that have been used to prepare porous polymer films. (**b**) PS-PEO block copolymer formed by self-assembly.

the P4VP block, complexed with $Zn(DBS)_2$, formed a separate lamellar structure with a smaller domain spacing. This lamellar-within-lamellar structure was rendered porous by extracting the $Zn(DBS)_2$ with methanol to yield 20-nm pores. Remarkably, a stable porous structure was generated from the lamellar structure as a result of the large number of defects (no effort at alignment was made), which stabilized the porous structure in the glassy PS matrix. By varying the volume fractions and molecular weight of the P4VP-PS block copolymer and using a different small-molecule additive, **22**, Mäki-Ontto et al. were able to obtain a morphology containing P4VP cylinders within a PS matrix [91]. The P4VP cylinders were themselves phase separated into P4VP:**22** lamellar structures. Alignment of the cylinders was achieved by thermal annealing and shear orientation, and the structure was made porous by extracting **22** with methanol.

Sidorenko et al. demonstrated that the addition of a small compound complexing with the P4VP block of a PS-P4VP copolymer could result in remarkable control of the thin-film orientation [92]. Thin films were prepared by mixing a 2-(4′-hydroxybenzeneazo)benzoic acid (HABA), **23**, solution with a P4VP-PS copolymer solution using either dioxane or chloroform (both heated near their boiling points).

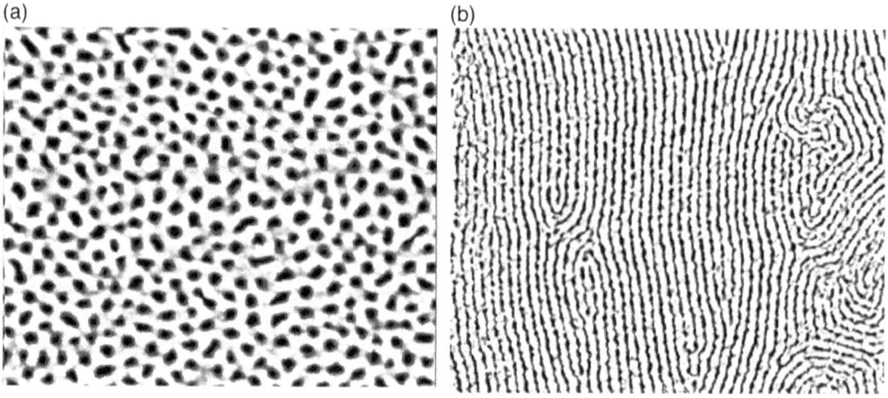

FIGURE 2.20 AFM images of porous thin films formed by dip coating PS-P4VP/**23** solutions and then removing **23** using methanol: (a) dioxane solvent (b) chloroform solvent. (Adapted with permission from Reference 92. Copyright 2003 American Chemical Society.)

After stirring these solutions overnight at room temperature, substrates were dip coated, resulting in thin films of cylindrical P4VP domains in a PS matrix. The cylindrical pores were oriented either parallel (chloroform) or perpendicular (dioxane) to the substrate, depending on the solvent (Fig. 2.20). HABA has both a carboxylic acid and a phenol capable of forming hydrogen bonds to the nitrogens in P4VP. IR measurements indicated that the mode of hydrogen bonding was different in the respective solvents, providing a possible cause for the solvent-dependent behavior. Extraction of the HABA with methanol led to films containing 8-nm pores.

Thin films prepared by dip coating substrates into solutions of P4VP-PS complexed with **24** adopt morphologies containing a PS matrix surrounding nodules of P4VP/**24**, even in cases in which the P4VP/**24** portion was the majority fraction [93]. This reversal in what would normally be predicted was attributed to the THF used in casting the film, as THF is a good solvent for both PS and P4VP but prefers PS. The authors proposed that THF preferentially swells the PS phase, making it effectively larger than the P4VP/**24** phase during film formation. Once the solvent is evaporated, the PS phase shrinks, leaving the observed nodules extending above the surface of the film. The size of the nodules was between 15 and 26 nm and was controlled based on the ratio of **24** to P4VP nitrogens (1:2 to 4:1). The resulting films were made porous by extraction of **24** with methanol.

It is also possible to form a block copolymer using noncovalent interactions. PS and PEO polymers containing terpyridine end-groups have been prepared and subsequently linked together using ruthenium chemistry, resulting in a PS-Ru(II)-PEO block copolymer [94]. Casting of this polymer formed a thin film containing cylindrical PEO domains in a PS matrix. The film was subsequently exposed to UV irradiation, which crosslinked the PS phase. Placing the film in an acidic solution of $Ce(SO_4)_{(aq)}$ for 1 h cleaved the block copolymer by oxidizing Ru(II), which can bind two terpyridine ligands, to Ru(III), which can only bind one. Once freed, the water-soluble PEO diffused out of the film. While this was a unique approach to preparing porous thin films from PS-PEO precursors, the film formed in this case contained a range of different pore sizes centered around 33 nm.

2.4.3 Physical Properties and Applications

In the secondary assembly examples (**18–20**), porosity is derived directly from the self-assembly process after complete removal of the solvents. The nature of the polymer:small molecule interactions simplifies etching by allowing the use of a suitable solvent to break up noncovalent bonds, eliminating the need for complicated or harsh etching techniques. In addition, the removal of the small molecule leaves behind the functionality with which it was associated, typically pyridines, providing a functionalized surface that is available for additional chemistry. Since the identity and amount of the small molecule additive have a large effect on the resulting morphology, a simpler degree of control and tunability in the final structures is available using this approach. Unfortunately, the reversible nature of the polymer:small molecule interactions means that there is uncertainty in the amount of small molecule that is present after thin-film formation.

The porous materials described in this section are similar to those prepared from block copolymer thin films and share many of the same potential applications and physical properties. The thin films prepared from secondary assembly around water droplets possess pores on the micrometer scale, making them less than ideal for separation applications; however, they have been used to template various inorganic particles. In particular, thin films of **19** have been used to template the formation of bowl-shaped aluminum particles [85]. The bowl shape is a result of the original spherical shape of the cavities. Related nanoporous thin films have also been used to template gold particles [95].

The large pores in the thin films formed by secondary assembly coupled with the mechanical properties of the PS matrix makes these materials very brittle. Using higher molecular weight polymers could improve the mechanical properties of the films; however, self-assembly is hindered by overly large molecular weight polymers. Using a blending strategy, higher molecular weight homopolymers can be blended with rod-coil polymers and used to form the porous film [84b]. The added homopolymers improve mechanical properties, while the lower molecular weight rod-coil polymers are still able to efficiently self-assemble.

2.5 SELF-ASSEMBLED STRUCTURES AS POROGENS

2.5.1 Polymerization and Etching of Gels

Gels typically consist of solvent-swollen, crosslinked supramolecular or polymeric structures. The high degree of crosslinking generates a solid-like material even though the majority of the material is solvent. For these systems the component that forms the crosslinked network is called the gelator. In this section we explore molecules that

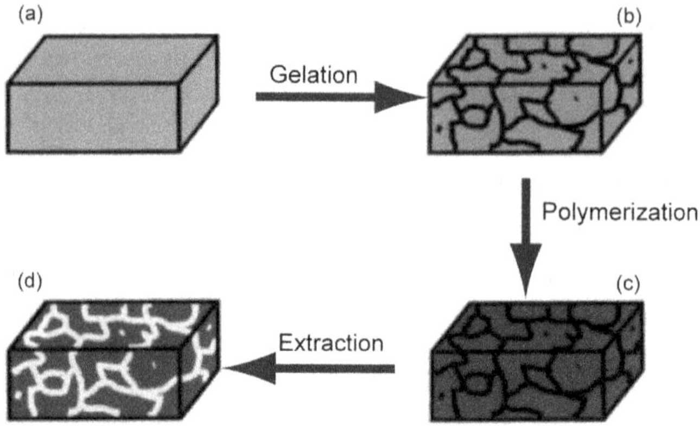

FIGURE 2.21 Schematic of the process to generate porous materials from gels. **(a)** Homogeneous solution of monomers, crosslinkers, and gelator. **(b)** Self-assembly of the gelator forms a crosslinked network that gels the solvent and will serve as a template for the pores. **(c)** Polymerization of the monomers and crosslinkers leads to a mechanically stable material. **(d)** Extraction of the templating molecules generates porosity. (Adapted with permission from Reference 96b. Copyright Wiley-VCH Verlag GmbH & Co. KGaA.)

self-assemble to form networks capable of gelling polymerizable organic liquids (e.g., styrenes and methacrylates). The structure of the gel is locked in by polymerization, and the gelator is subsequently removed to yield a porous structure. In this context the gelator acts as a porogen, a sacrificial substance used to generate porosity in a material (Fig. 2.21) [96]. Previously, we focused on systems in which self-assembly alone generated the structured material. In the case of gel etching, self-assembly is used to construct a template for the pores, which will then typically be removed after hardening of the matrix. Since the porogens form supramolecular structures, only noncovalent bonds need to be broken to generate porosity, and the porogen can often be reused if it can be recovered after the extraction process. In addition, gelators are typically used in very small amounts, meaning that the majority of the material used in this process is commercially available monomers, making this approach a cost-effective route to porous materials.

Several conditions need to be met for this approach to be an effective method of generating porous materials. First, the supramolecular templates need to be stable under the polymerization conditions of the matrix; otherwise the templated pores would be destroyed before the matrix hardened. In addition, the porogen needs to be accessible to extraction techniques after polymerization, and the matrix needs to be able to support the porous structure. In the following sections we discuss self-assembled systems that have been used as porogens through the process of gelation, organized by the type of supramolecular structure formed.

2.5.1.1 Fibril Structures as Porogens. Fibril structures, in the examples chosen here, are formed by a hierarchical self-assembly processes in which components self-assemble into tubes and then further aggregate into larger fibers. While steps can be taken to control the size of these fibers, all of these materials tend to have larger pore diameters and a broader distribution of pore sizes.

Relatively simple surfactants can be used as fiber-forming porogens (Table 2.4). For example, ammonium surfactant **28** was able to gel solutions of styrene/DVB or MMA/DVB [97]. The components were mixed and heated to 90°C to yield a homogeneous solution. A gel formed upon cooling, and the structure was captured by photopolymerization. While **28** was mostly removed by refluxing the resulting material in *tert*-butanol, the ammonium gelator could not be completely removed from the bulk samples. This suggests that not all of the fibers connect to the surface of the gel. Microscopy indicated that pores ranging from 240 nm to 1.4 μm were formed, depending on the cooling rate during gel formation. In an effort to align the pores, the unpolymerized gel was reheated to approximately 90°C and allowed to slowly cool in the presence of a strong magnetic field. No alignment was observed after polymerization; however, the fibers were more homogeneously distributed in the polymeric matrix, suggesting that temperature annealing might be a useful technique in preparing porous materials in this way.

Neutral surfactant compounds like fluorocarbon-hydrocarbon, **33**, are able to undergo crystallization-induced gelation in methacrylate solvents. Möller and coworkers screened a number of methacrylate mixtures to ascertain which combinations gave the best mechanical properties and chemical resistance [98]. Once they had settled on a monomer solution, gels were prepared with **33** containing hydrocarbon sections of $n = 11$ or $n = 19$. Not only did the size of the aggregates formed vary with the length of hydrocarbon section, but also the size of the aggregates was controlled by modifying the thermal treatment before polymerization such as the quench and initiation temperatures and the rate of cooling. The fluorocarbon-hydrocarbon compounds, **33**, were removed by sublimation at 100°C under vacuum or by Soxhlet extraction with *n*-hexane. The pore surface was hydrolyzed with 0.2 M NaOH to yield carboxylic acids or functionalized with an amine as indicated in Fig. 2.22a. The pore modifications affected pore properties such as effective pore size due to solvent swelling and diffusional transport of ions.

A more complicated neutral surfactant, benzamide, **26**, was used to template 5- to 15-nm pores in a photopolymerized matrix originally containing a mixture of several different methacrylates and diphenylethanone as an initiator (Fig. 2.23) [99]. While gels did form at room temperature, the material was further cooled to –50°C for the initial photopolymerization and a second round of irradiation was administered at room temperature to increase crosslinking in the polymeric matrix. Porogen **26** was removed by Soxhlet extraction using ethanol, which was effective in removing greater than 80% of the template, as evidenced by gravimetric analysis. Again, this suggests that not all of the fibers percolate through the material. One of the methacrylates used was 2-(*N*,*N*-diethylamino)-ethyl methacrylate, a portion of which was expected to reside on the pore walls after removal of **26**. The exposed nitrogens were functionalized with methyl iodide, yielding a pore surface containing ammonium cations (Fig. 2.22b). Ion transport measurements indicated selectivity from these pores, most likely a result of interactions with the positively charged pore walls.

TABLE 2.4 Gelators That Form Fibril Structures in the Indicated Polymerizable Solvents

Self-Assembling Porogen	Polymerizable Solvent	Pore Diameter	Ref.
26 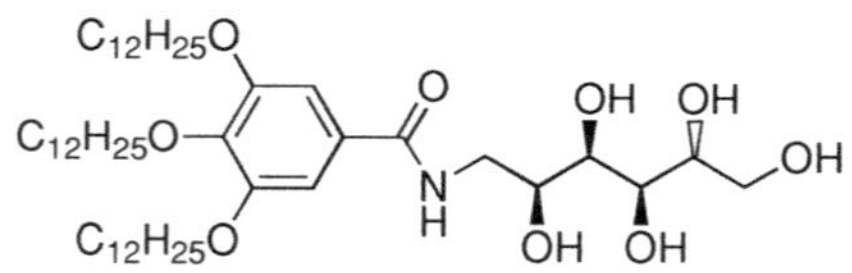	2-Ethyl-hexyl methacrylate; 2-*N*,*N*-diethylaminoethyl methacrylate; 1,4-buanediol dimethacrylate	6–15 nm	99
27 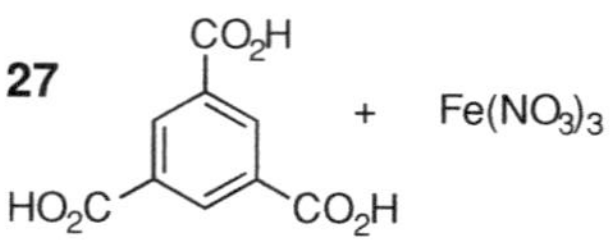	MMA	1–10 μm	103
28 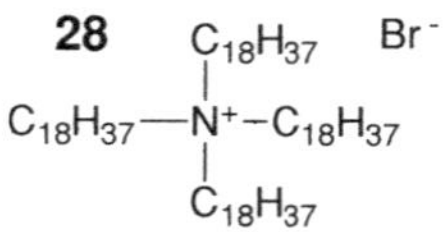	MMA and styrene	240 nm to 1.4 μm	97
29 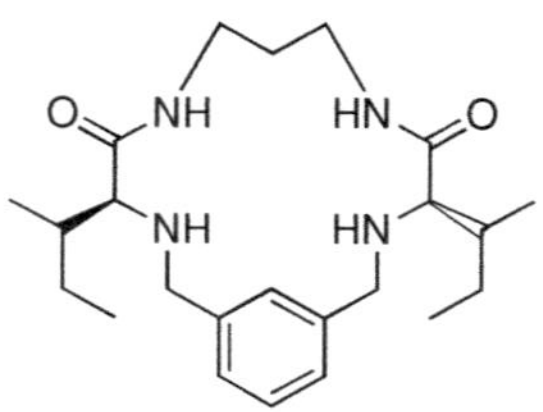	MMA; ethylene glycol dimethacrylate	10- to 100-μm spherulites connected by filaments	100

30 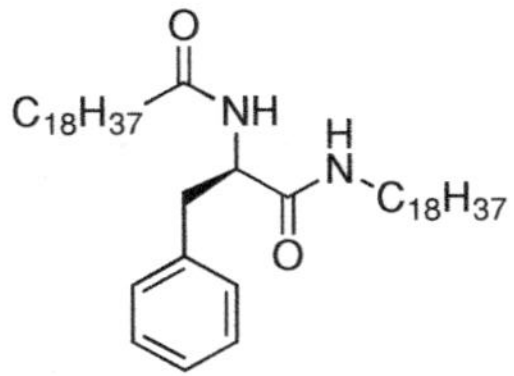	HEMA and PEG diacrylate	100–200 nm	101
31 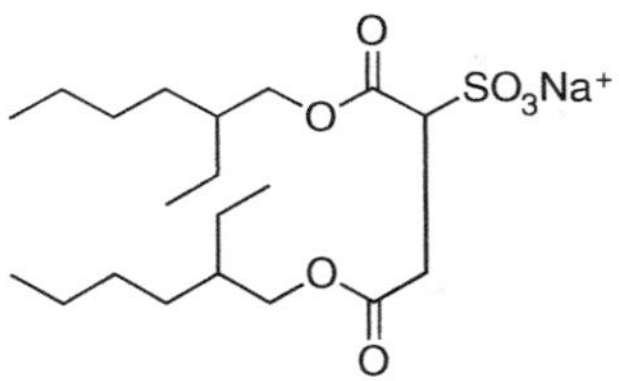 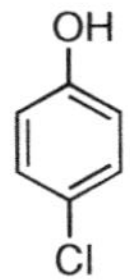 **32**	Divinylbenzene	20–200 nm	102
33 $F_3C-(CF_2)_{11}-(CH_2)_n-CH_3$ n = 11 or 19	Mixture of various methacrylates	70–145 nm ($n = 11$), 6–50 nm ($n = 19$)	98

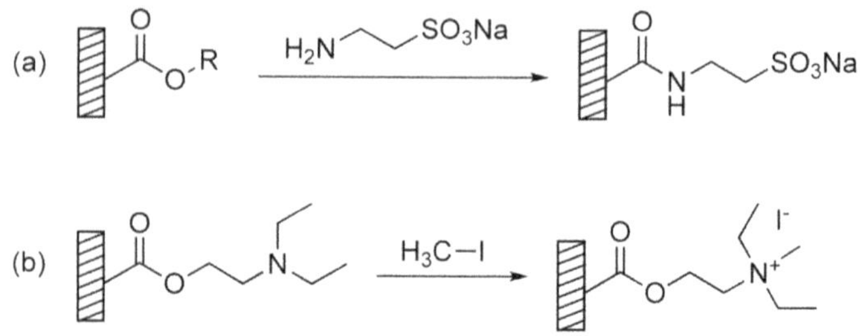

FIGURE 2.22 Methods used to modify the functionality of pore walls generated through a "polymerization and etching of gels" strategy.

A large number of gelators have been derived from peptides in an effort to utilize their hydrogen bonding interactions for gel formation in nonpolar solvents. Cyclic pseudopeptide **29** causes gelation in a 1:1 mixture of MMA and EGDMA at room temperature after first homogenizing the solution by heating [100]. A nucleation and growth mechanism led to a fibrillar network consisting of 10- to 100-μm globular structures connected by filaments of different diameters. Methanol extraction was able to remove greater than 75% of template **29**, generating a porous structure that was investigated by fluorescence spectroscopy. A related approach used **30**, which was derived from L-phenylalanine, to gel a solvent mixture of HEMA and PEG diacrylate [101]. Pores measuring 100–200 nm in diameter were observed after photopolymerization and template extraction with ethanol. Enhanced absorption was observed for L-phenylalanine, as opposed to D-phenylalanine, in these pores and this selectivity was attributed to the chiral nature of gelator **30**.

Combining additional elements of hierarchical self-assembly, **31** and **32** were used in a 1:1 ratio to gel divinylbenzene. The phenolic protons formed hydrogen bonds with the sulfate groups and assisted in gelation through aromatic stacking interactions

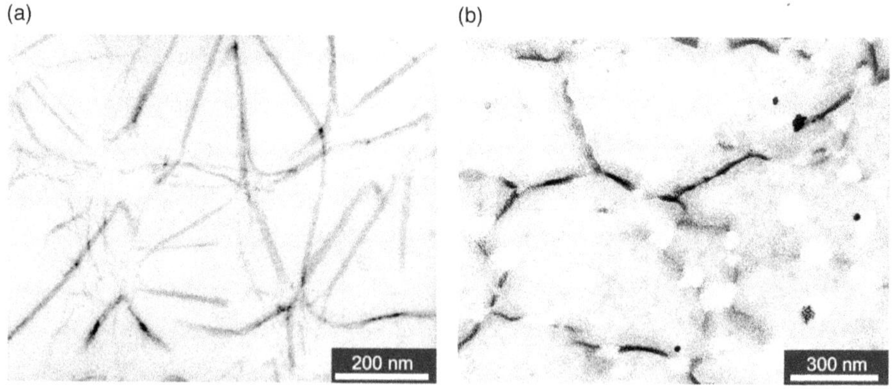

FIGURE 2.23 TEM of gel formed from benzamide, **26**, and a mixture of several different methacrylates and diphenylethanone as an initiator. **(a)** Before removal of porogen, **26**. **(b)** After removal of porogen, followed by filling the pores with $HAuCl_4$. (Adapted with permission from Reference 99.)

[102]. These columnar structures then aggregate into a hexagonal lattice followed by further self-assembly into fibers. After photopolymerization using AIBN as an initiator, the self-assembled template was removed by first using water to break up the hydrogen-bonded assemblies and then isooctane to remove the unassembled components. The uncontrolled nature of fibril formation led to pore sizes from 20 to 200 nm; however, the pores strongly reflected the original topology of the fiber network.

Metallo-supramolecular structures can also be used as porogens. Benzotricarboxylic acid, **27**, formed a polymeric network when complexed with ferric nitrate in a 3:2 molar ratio [103]. A gel resulted from forming this metallo-supramolecular polymer in a solution of ethanol and MMA (1:1 v/v). Subsequent photopolymerization of MMA using AIBN as an initiator, followed by removal of the metallo-supramolecular polymer by treatment with 1 M HCl led to a powder with a spongelike structure. SEM indicated pores ranging from 1 to 10 μm (Fig. 2.24).

2.5.1.2 Micellar Porogens. The nonionic surfactant copolymer PPO-PEO-PPO, **34**, has been used to form micellar structures in polymerizable media, leading to porous materials after polymerization and removal of the micelle components (Table 2.5). Jang and Bae formed micelles in a solution of acrylonitrile and *N*-methyl-2-pyrrolidone using PPO-PEO-PPO copolymers [104]. Thermal polymerization at 70°C followed by extraction of PPO-PEO-PPO by treatment with refluxing water for 24 h yielded materials with 6- to 11-nm pores. The pore sizes were controlled by using different copolymers of varying compositions and by altering the copolymer surfactant to monomer ratio. Meng et al. also used the micelle-forming capabilities of PPO-PEO-PPO to form porous materials; however, they utilized evaporation-induced self-assembly to achieve more ordered structures [105]. An ethanol solution was prepared containing low-molecular-weight phenol/formaldehyde oligomers (M_W = 500–5000 kDa) and PPO-PEO-PPO copolymer surfactants, where the hydrophilic PEO block of the copolymer interacts with the hydroxyls of the phenol/formaldehyde oligomers. Evaporation of the ethanol led to ordering of the copolymers into micellar structures surrounded by polymerizable oligomers. The structures formed depended

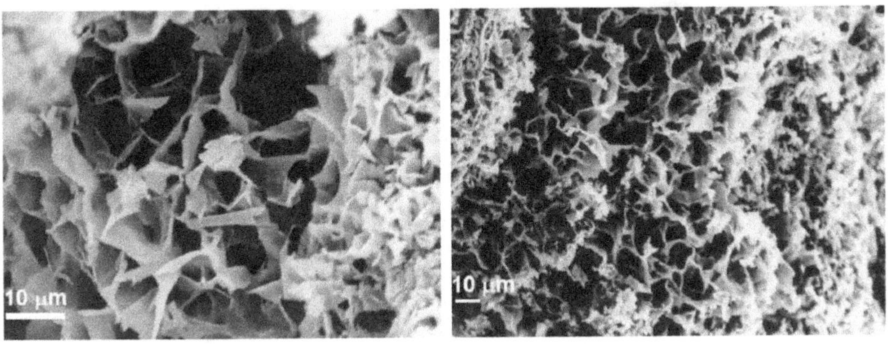

FIGURE 2.24 Spongelike porous PMMA formed after removal of Fe:**27** metallo-supramolecular polymer from polymerized MMA:Fe:**27** gel. (Adapted from Reference 103 by permission of The Royal Society of Chemistry.)

TABLE 2.5 Gelators That Form Micellar Structures in the Indicated Polymerizable Solvents

Self-Assembling Porogen	Polymerizable Solvent	Pore Diameter, nm	Ref.
34 PPO-PEO-PPO	Low-molecular-weight phenol/formaldehyde oligomers	7	105
34 PPO-PEO-PPO	Acrylonitrile	6–11	104
31 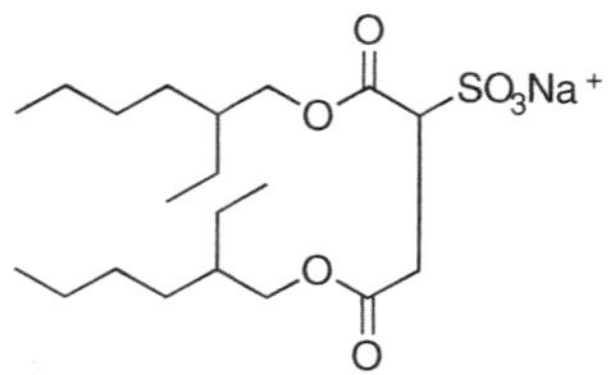 + H_2O	Styrene and divinylbenzene	7.5–36	106

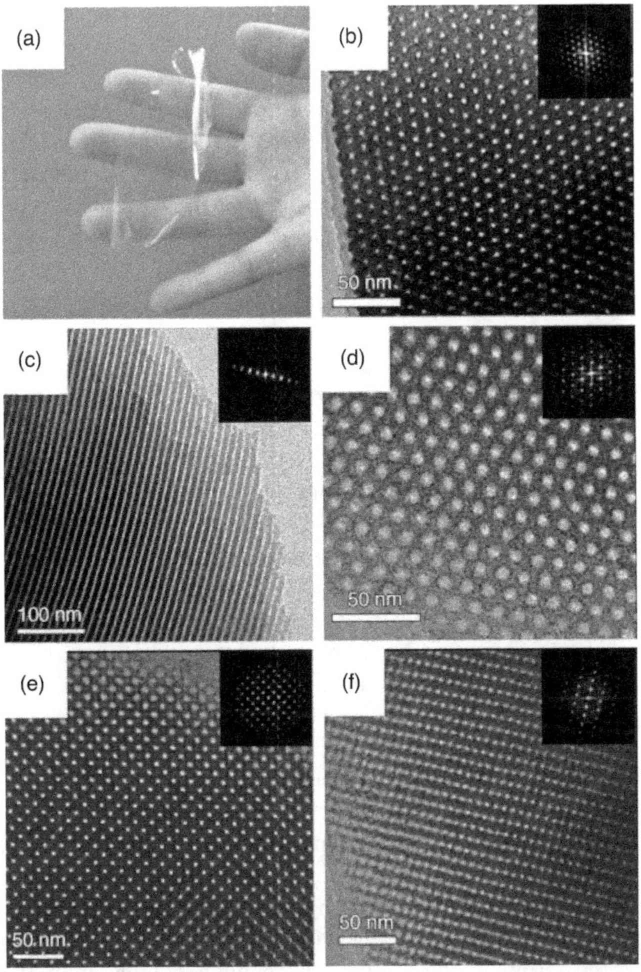

FIGURE 2.25 **(a)** Photograph of polymer containing hexagonally arranged cylinders of PPO-PEO-PPO. **(b, c)** TEM of hexagonal pores in phenol/formaldehyde matrix after removal of PPO-PEO-PPO porogen through calcination at 350°C. **(d–f)** TEM of cubic pores in phenol/formaldehyde matrix after removal of PPO-PEO-PPO porogen through calcination at 350°C. (Adapted with permission from Reference 105. Copyright Wiley-VCH Verlag GmbH & Co. KGaA.)

on the ratio of oligomers to surfactant copolymer and included hexagonally packed cylinders and a bicontinuous cubic phase. Thermal polymerization of the oligomers was conducted at 100°C–140°C, and the copolymer was removed by calcination at 300°C–500°C under a nitrogen atmosphere (Fig. 2.25). The pores generated by this method were approximately 7 nm in diameter, and further heating led to carbonization of the structure, which was stable up to 1400°C.

Surfactant **31**, which was used in conjunction with chlorophenol, **32**, to form fibril gels, can also be used in micellar self-assembly [106]. The addition of water to a solution of **31** in styrene and DVB led to the formation of micelles. Photopolymerization using AIBN as an initiator resulted in a solid material. The micelles formed did not percolate through the entire structure, so the material was ground in a ball mill before removal of the surfactant and water by Soxhlet extraction with toluene, ethanol, and then ether. The pore sizes were controlled from 7.5 to 36 nm by varying the amount of water added, although the larger micelles also displayed a larger polydispersity in sizes, a common feature of micelles.

2.5.1.3 Porogens from Well-Ordered Supramolecular Polymers. Compound **37**, shown in Table 2.6, formed helical structures in ethylene glycol diacrylate at relatively low concentrations (5 wt%) [107]. Photopolymerization captured the resulting assembly, which can be made porous by extracting **37** with chloroform. TEM indicated that the 31-nm pores retained the helical shape of the original template (Fig. 2.26), which is in contrast to the structures templated from **30**, which showed enantioselectivity in the absorption of phenylalanine but contained pores that were not visibly chiral. In an effort to increase the porosity of the final material, photopolymerization was carried out on a gel containing 50 wt% of **37**. While this approach did have the desired effect, the pores formed a cylindrical instead of helical structure.

The aromatic and hydrogen bonding functionally of **36** allowed it to assemble into tubes in solutions of styrene and DVB, although the size of these aggregates suggested discrete supramolecular polymers instead of the fibril structures discussed previously [108]. Although compound **36** has terminal alkenes that could potentially be polymerized, photopolymerization led to a crosslinked network consisting of only styrene and DVB, which was demonstrated by removing **36** from the polymerized matrix by extraction with a THF-methanol solvent mixture. The reactivity of the terminal alkene was low enough compared to the conjugated alkenes of styrene and DVB that they are unreactive under these polymerization conditions.

Compound **38** is a unique gelator in this class of compounds because it is not removed from the final polymerized material. The wedge shape and large aromatic surface lead to **38** arranging in hexameric circular disks that stack to form long tubes [109]. The porosity was built into this gelator by the crown ether and from the central gap that is formed with the hexameric assembly. The alkyl chains of **38** were functionalized with acrylate groups to be polymerized with the matrix, which consisted of a commercially available solution of various methacrylates. Photopolymerization followed by TEM analysis identified pores ranging from 3.5 to 4.5 nm. Cation transport through the crown ether–functionalized pores mirrored the interaction strength of the investigated cations. Lithium interacted the least with the crown ethers and showed the fastest transport, while sodium transport was the slowest as a result of its stronger interactions with the crown ethers.

TABLE 2.6 Gelators That Form Well-Ordered Supramolecular Polymers in the Indicated Polymerizable Solvents

Self-Assembling Porogen	Polymerizable Solvent	Pore Diameter, nm	Ref.
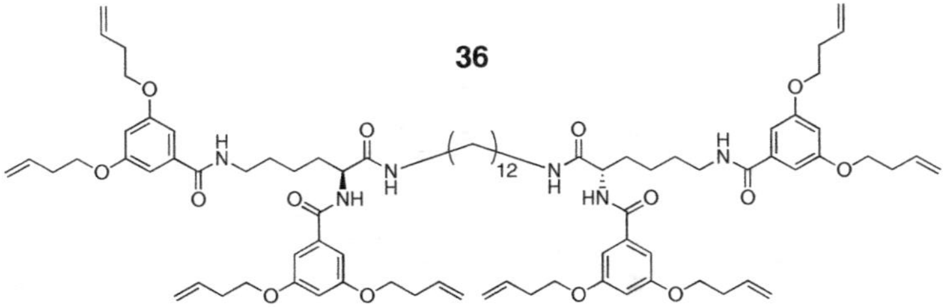 36	Styrene and divinylbenzene	<10	108
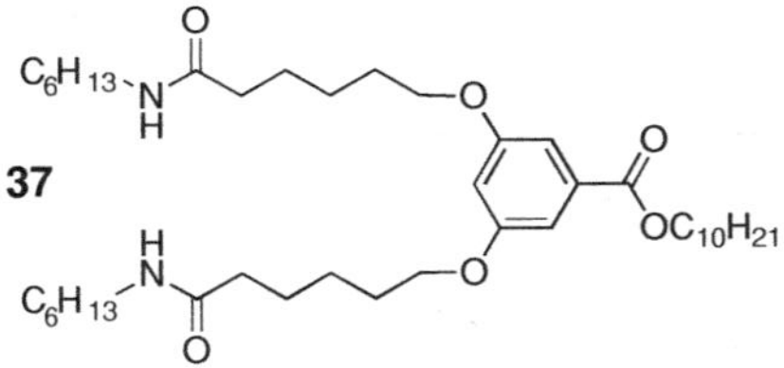 37	Ethylene glycol diacrylate	31	107
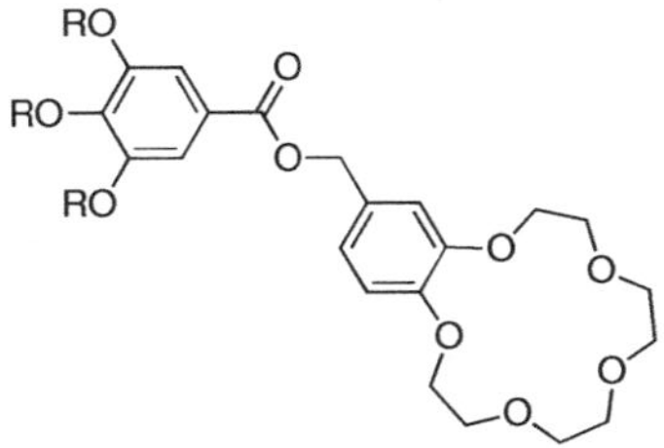 38 R = 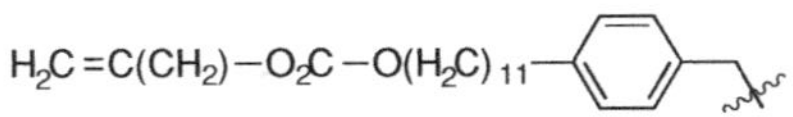	Mixture of various methacrylates	3.5–4.5	109

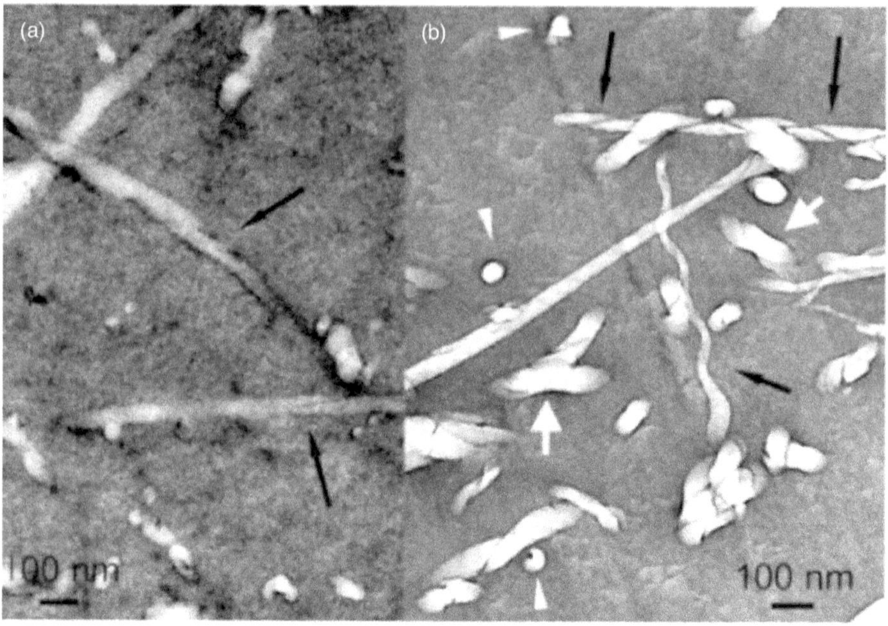

FIGURE 2.26 TEM of microsections of polymerized gel formed from ethylene glycol diacrylate and **37**. (**a**) Helical structures before removal of **37**. (**b**) Helical pores after removal of **37** by extraction with chloroform. Black arrows: helical tape. White arrow: tilted helical tap. White arrowheads: helical pores, viewed along their axis. (Reprinted with permission from Reference 107. Copyright 2007 American Chemical Society.)

2.6 CONCLUSIONS

In this chapter we introduced the concept of self-assembly as it applies to the "bottom-up" formation of porous organic polymers. This class of porous materials is particularly appealing as a result of the combination of the mechanical and processing characteristics of polymers with the molecular-level control of organic synthesis. Molecular design is an important principle that can be used to dictate the ultimate structure of the polymeric materials. In this way, self-assembly strategies take their inspiration from the methods that nature uses to generate complex structures. These highlighted approaches provide starting points for the generation of porous polymeric materials by self-assembly, and new methods will certainly be developed in the future.

ABBREVIATIONS

The following abbreviations are used throughout this chapter: AFM, atomic force microscopy; AIBN, 2,2′-azobis(2-methylpropionitrile); BR, butyl rubber; DMF, dimethyl formamide; DNA, deoxyribonucleic acid; DVB, divinyl benzene; EGDMA, ethylene glycol dimethacrylate; HABA, 2-(4′-hydroxybenzeneazo)benzoic acid; HEMA, hydroxyethyl methacrylate; IR, infrared; LLC, lyotropic liquid crystals; MAS NMR, magic angle spinning nuclear magnetic resonance; mCPBA, *m*-chloroperoxybenzoic acid; MMA, methyl methacrylate; M_n, number-average molecular weight; M_w, weight-average molecular weight; NMR, nuclear magnetic resonance; P4VP, poly(4-vinylpyridine); PAA, poly(acrylic acid); PAN, poly (acrylonitrile); PB, poly(1,2-butadiene); PBD, poly(butadiene); PCEMA, poly(2-cinnamoylethyl methacrylate); PCHE, poly(cyclohexylethylene); PDI, polydispersity index; PDMA, poly(*N*,*N*-dimethylacrylamide); PDMS, poly(dimethylsiloxane); PE, polyethylene; PEG, poly(ethylene glycol); PEO, poly(ethylene oxide); PI, polyisoprene; PLA, poly(D,L-lactide); PLLA, poly(L-lactide); PMMA, poly(methyl methacrylate); PPG, poly(propylene glycol); PPQ, poly(phenylquinoline); PPO, poly(propylene oxide); PPV, poly(*p*-phenylenevinylene); PS, polystyrene; P*t*BA; poly(*tert*-butyl acrylate); RIE, reactive ion etch; RT, room temperature; SAXS, small-angle X-ray scattering; $scCO_2$, supercritical carbon dioxide; SDS, sodium dodecyl sulfate; SEM, scanning electron microscopy; TBAF, tetrabutylammonium fluoride; TEM, transmission electron microscopy; TFA, trifluoroacetic acid; T_g, glass-transition temperature; THF, tetrahydrofuran; TMSI, trimethylsilane iodide; UV, ultraviolet.

ACKNOWLEDGMENTS

The authors gratefully acknowledge support for this work from the National Science Foundation (NIRT-CMMI-0707610). We thank Mark Amendt, Benjamin Hamilton, and Ligeng Yin for reviewing this manuscript prior to submission. We also appreciate comments from Prof. Doug Gin during the preparation of this chapter.

REFERENCES

1. Examples of materials formed from self-assembly: (a) Kazmaier, P.; Chopra, N. *MRS Bull* 2000, **25**, 30–35. (b) Mio, M. J.; Moore, J. S. *MRS Bull* 2000, **25**, 36–41. (c) Stupp, S. I.; Pralle, M. U.; Tew, G. N.; Li, L.; Sayar, M.; Zubarev, E. R. *MRS Bull* 2000, **25**, 42–48.
2. Feng, Q.; Kanoh, H.; Ooi, K. *J Mater Chem* 1999, **9**, 319–333.
3. (a) Eddaoudi, M.; Moler, D. B.; Li, H.; Chen, B.; Reineke, T. M.; O'Keeffe, M.; Yaghi, O. M. *Acc Chem Res* 2001, **34**, 319–330. (b) James, S. L. *Chem Soc Rev* 2003, **32**, 276–288. (c) Rowsell, J. L. C.; Yaghi, O. M. *Micropor Mesopor Mater* 2004, **73**, 3–14.
4. (a) Langley, P. J.; Hulliger, J. *Chem Soc Rev* 1999, **28**, 279–291. (b) Nangia, A. *Curr Opin Solid State Mater Sci* 2001, **5**, 115–122.
5. Hawker, C. J.; Wooley, K. L. *Science* 2005, **309**, 1200–1205.
6. (a) T. Matsuura. *Synthetic Membranes and Membrane Separation Processes*, CRC Press, Boca Raton, FL, 1993. (b) M. Cheryan. *Ultrafiltration Handbook*, Technomic, Lancaster, PA, 1986.

7. LLC general reviews: (a) Tiddy, G. J. T. *Phys Rep* 1980, **57**, 1–46. (b) Seddon, J. M. *Biochim Biophys Acta* 1990, **1031**, 1–69. (c) Tate, M. W.; Eikenberry, E. F.; Turner, D. C.; Shyamsunder, E.; Gruner, S. M. *Chem Phys Lipids* 1991, **57**, 147–164.
8. (a) Israelachvili, J. N. *Intermolecular and Surface Forces: With Applications to Colloidal and Biological Systems*, Academic Press, New York, 1985. (b) Percec, V.; Heck, J.; Johansson, G.; Tomazos, D.; Kawasumi, M.; Chu, P.; Ungar, G. *J Macromol Sci Pure Appl Chem* 1994, **31**, 1719–1758.
9. Gruner, S. M. *Proc Natl Acad Sci* 1985, **82**, 3665–3669.
10. Materials made from polymerizable LLCs: (a) Ringsdorf, H.; Schlarb, B.; Venzmer, J. *Angew Chem Int Ed* 1988, **27**, 113–158. (b) O'Brien, D. F.; Armitage, B.; Benedicto, A.; Bennett, D. E.; Lamparski, H. G.; Lee, Y.-S.; Srisiri, W.; Sisson, T. M. *Acc Chem Res* 1998, **31**, 861–886. (c) Miller, S. A.; Ding, J. H.; Gin, D. L. *Curr Opin Colloid Interfac Sci* 1999, **4**, 338–347. (d) Gin, D. L.; Gu, W.; Pindzola, B. A.; Zhou, W.-J. *Acc Chem Res* 2001, **34**, 973–980. (e) Mueller, A.; O'Brien, D. F. *Chem Rev* 2002, **102**, 727–758.
11. Lopez, E.; O'Brien, D. F.; Whitesides, T. H. *J Am Chem Soc* 1982, **104**, 305–307.
12. Ding, J. H.; Gin, D. L. *Chem Mater* 2000, **12**, 22–24.
13. Herz, J.; Reiss–Husson, F.; Rempp, P.; Luzzati, V. *J Polym Sci C* 1963, **4**, 1275–1290.
14. Anderson. D. M.; Ström. P. In *Polymer Association Structures*, El-Nokaly, M. A., Ed., American Chemical Society, Washington, DC, 1989, p. 204.
15. Gray, D. H.; Hu, S.; Juang, E.; Gin, D. L. *Adv Mater* 1997, **9**, 731–736.
16. Reppy, M. A.; Gray, D. H.; Pindzola, B. A.; Smithers, J. L.; Gin, D. L. *J Am Chem Soc* 2001, **123**, 363–371.
17. Srisiri, W.; Sisson, T. M.; O'Brien, D. F.; McGrath, K. M.; Han, Y.; Gruner, S. M. *J Am Chem Soc* 1997, **119**, 4866–4873.
18. Smith, R. C.; Fischer, W. M.; Gin, D. L. *J Am Chem Soc* 1997, **119**, 4092–4093.
19. (a) Deng, H.; Gin, D. L.; Smith, R. C. *J Am Chem Soc* 1998, **120**, 3522–3523. (b) Resel, R.; Leising, G.; Markart, P.; Kriechbaum, M.; Smith, R.; Gin, D. *Macromol Chem Phys* 2000, **201**, 1128–1133.
20. Hoag, B. P.; Gin, D. L. *Macromolecules* 2000, **33**, 8549–8558.
21. Pindzola, B. A.; Jin, J.; Gin, D. L. *J Am Chem Soc* 2003, **125**, 2940–2949.
22. Jin, J.; Nguyen, V.; Gu, W.; Lu, X.; Elliott, B. J.; Gin, D. L. *Chem Mater* 2005, **17**, 224–226.
23. Lu, X.; Nguyen, V.; Zhou, M.; Zeng, X.; Jin, J.; Elliott, B. J.; Gin, D. L. *Adv Mater* 2006, **18**, 3294–3298.
24. Lee, Y.-S.; Yang, J.-Z.; Sisson, T. M.; Frankel, D. A.; Gleeson, J. T.; Aksay, E.; Keller, S. L.; Gruner, S. M.; O'Brien, D. F. *J Am Chem Soc* 1995, **117**, 5573–5578.
25. Gin, D. L.; Gu, W. *Adv Mater* 2001, **13**, 1407–1410.
26. Zhou, W.; Gu, W.; Xu, Y.; Pecinovsky, C. S.; Gin, D. L. *Langmuir* 2003, **19**, 6346–6348.
27. Lu, X.; Nguyen, V.; Zeng, X.; Elliott, B. J.; Gin, D. L. *J. Membr Sci* 2008, **318**, 397–404.
28. Zhou, M.; Kidd, T. J.; Noble, R. D.; Gin, D. L. *Adv Mater* 2005, **17**, 1850–1853.
29. Zhou, M.; Nemade, P. R.; Lu, X.; Zeng, X.; Hatakeyama, E. S.; Noble, R. D.; Gin, D. L. *J Am Chem Soc* 2007, **129**, 9574–9575.
30. Bara, J. E.; Kaminski, A. K.; Noble, R. D.; Gin, D. L. *J. Membr Sci* 2007, **288**, 13–19.

31. (a) Li, C.; Zhang, H.; Jiang, D.; Yang, Q. *Chem Commun* 2007, 547–558. (b) Thomas, J. M.; Raja, R. *Acc Chem Res* 2008, **41**, 708–720. (c) Koblenz, T. S.; Wassenaar, J.; Reek, J. N. H. *Chem Soc Rev* 2008, **37**, 247–262.

32. Miller, S. A.; Kim, E.; Gray, D. H.; Gin, D. L. *Angew Chem Int Ed* 1999, **38**, 3021–3026.

33. Gu, W.; Zhou, W.–J.; Gin, D. L. *Chem Mater* 2001, **13**, 1949–1951.

34. Pecinovsky, C. S.; Nicodemus, G. D.; Gin, D. L. *Chem Mater* 2005, **17**, 4889–4891.

35. Xu, Y.; Gu, W.; Gin, D. L. *J Am Chem Soc* 2004, **126**, 1616–1617.

36. Gray, D. H.; Gin, D. L. *Chem Mater* 1998, **10**, 1827–1832.

37. Reviews pertaining to block copolymer self-assembly: (a) Fredrickson, G. H.; Bates, F. S. *Annu Rev Mater Sci* 1996, **26**, 501–550. (b) Bates, F. S.; Fredrickson, G. H. *Phys Today* 1999, **52**, 32–38. (c) Lodge, T. P. *Macromol Chem Phys* 2003, **204**, 265–273. (d) Hamley, I. W. *Angew Chem Int Ed* 2003, **42**, 1692–1712. (e) Park, C.; Yoon, J.; Thomas, E. L. *Polymer* 2003, **44**, 6725–6760.

38. Books pertaining to block copolymer self-assembly: (a) Hamley, I. W. *The Physics of Block Copolymers*, Oxford University Press, Oxford, 1998. (b) Hadjichristidis, N.; Pispas, S.; Floudas, G. A. *Block Copolymers: Synthetic Strategies, Physical Properties, and Applications*, Wiley-Interscience, New York, 2003.

39. Reviews of nanoporous materials derived from block copolymer precursors: (a) Hillmyer, M. A. *Adv Polym Sci* 2005, **190**, 137–181. (b) Olson, D. A.; Chen, L.; Hillmyer, M. A. *Chem Mater* 2008, **20**, 869–890.

40. Kim, S. H.; Misner, M. J.; Xu, T.; Kimura, M.; Russell, T. P. *Adv Mater* 2004, **16**, 226–231.

41. (a) Mansky, P.; Liu, Y.; Huang, E.; Russell, T. P.; Hawker, C. *Science* 1997, **275**, 1458–1460. (b) Ryu, D. Y.; Shin, K.; Drockenmuller, E.; Hawker, C. J.; Russell, T. P. *Science* 2005, **308**, 236–239. (c) Niemz, A.; Bandyopadhyay, K.; Tan, E.; Cha, K.; Baker, S. M. *Langmuir* 2006, **22**, 11092–11096.

42. Mansky, P.; DeRouchey, J.; Russell, T. P.; Mays, J.; Pitsikalis, M.; Morkved, T.; Jaeger, H. *Macromolecules* 1998, **31**, 4399–4401.

43. (a) Hadziioannou, G.; Mathis, A.; Skoulios, A. *Colloid Polym Sci* 1979, **257**, 136–139. (b) Bates, F. S.; Koppi, K. A.; Tirrell, M.; Almdal, K.; Mortensen, K. *Macromolecules* 1994, **27**, 5934–5936. (c) Chen, Z.-R.; Kornfield, J. A.; Smith, S. D.; Grothaus, J. T.; Satkowski, M. M. *Science* 1997, **277**, 1248–1253. (d) Albalak, R. J.; Thomas, E. L.; Capel, M. S. *Polymer* 1997, **38**, 3819–3825. (e) Villar, M. A.; Rueda, D. R.; Ania, F.; Thomas, E. L. *Polymer* 2002, **43**, 5139–5145.

44. Lee, J. S.; Hirao, A.; Nakahama, S. *Macromolecules* 1988, **21**, 274–276.

45. (a) Asakawa, K.; Hiraoka, T. *Jpn J Appl Phys*, 2002, **41**, 6112. (b) Naito, K.; Hieda, H.; Sakurai, M.; Kamata, Y.; Asakawa, K. *IEEE Trans Magn* 2002, **38**, 1949–1951. (c) Olayo-Valles, R.; Lund, M. S.; Leighton, C.; Hillmyer, M. A. *J Mater Chem* 2004, **14**, 2729–2731.

46. Thurn-Albrecht, T.; Steiner, R.; DeRouchey, J.; Stafford, C. M.; Huang, E.; Bal, M.; Tuominen, M.; Hawker, C. J.; Russell, T. P. *Adv Mater* 2000, **12**, 787–791.

47. Urbas, A. M.; Maldovan, M.; DeRege, P.; Thomas, E. L. *Adv Mater* 2002, **14**, 1850–1853.

48. Okumura, A.; Nishikawa, Y.; Hashimoto, T. *Polymer* 2006, **47**, 7805–7812.

49. Mansky, P.; Harrison, C. K.; Chaikin, P. M.; Register, R. A.; Yao, N. *Appl. Phys Lett* 1996, **68**, 2586–2588.

50. (a) Zalusky, A. S.; Olayo-Valles, R.; Taylor, C. J.; Hillmyer, M. A. *J Am Chem Soc* 2001, **123**, 1519–1520. (b) Zalusky, A. S.; Olayo-Valles, R.; Wolf, J. H.; Hillmyer, M. A. *J Am Chem Soc* 2002, **124**, 12761–12773. (c) Olayo-Valles, R.; Guo, S.; Lund, M. S.; Leighton, C.; Hillmyer, M. A. *Macromolecules* 2005, **38**, 10101–10108.
51. Wolf, J. H.; Hillmyer, M. A. *Langmuir* 2003, **19**, 6553–6560.
52. Ho, R.-M.; Chiang, Y.-W.; Tsai, C.-C.; Lin, C.-C.; Ko, B.-T.; Huang, B.-H. *J Am Chem Soc* 2004, **126**, 2704–2705.
53. Mao, H.; Hillmyer, M. A. *Macromolecules* 2005, **38**, 4038–4039.
54. (a) Cavicchi, K. A.; Zalusky, A. S.; Hillmyer, M. A.; Lodge, T. P. *Macromol Rapid Commun* 2004, **25**, 704–709. (b) Hansen, M. S; Vigild, M. E.; Berg, R. H.; Ndoni, S. *Polym Bull* 2004, **51**, 403–409.
55. Guo, F.; Andreasen, J. W.; Vigild, M. E.; Ndoni, S. *Macromolecules* 2007, **40**, 3669–3675.
56. Ndoni, S.; Vigild, M. E.; Berg, R. H. *J Am Chem Soc* 2003, **125**, 13366–13367.
57. (a) Liu, G.; Ding, J.; Stewart, S. *Angew Chem Int Ed* 1999, **38**, 835–838. (b) Liu, G.; Ding, J.; Hashimoto, T.; Kimishima, K.; Winnik, F. M.; Nigam, S. *Chem Mater* 1999, **11**, 2233–2240.
58. Uehara, H.; Yoshida, T.; Kakiage, M.; Yamanobe, T.; Komoto, T.; Nomura, K.; Nakajima, K.; Matsuda, M. *Macromolecules* 2006, **39**, 3971–3974.
59. (a) Chung, C.-M.; Lee, J.-H.; Cho, S.-Y.; Kim, J.-G.; Moon, S.-Y. *J Appl Polym Sci* 2006, **101**, 532–538. (b) Hedrick, J. L.; Labadie, J. W.; Volksen, W.; Hilborn, J. G. *Adv Polym Sci* 1999, **147**, 61–111.
60. (a) Yokoyama, H.; Sugiyama, K. *Macromolecules* 2005, **38**, 10516–10522. (b) Yokoyama, B. H.; Li, L.; Nemoto, T.; Sugiyama, K. *Adv Mater* 2004, **16**, 1542–1546. (c) Li, L.; Yokoyama, H.; Nemoto, T.; Sugiyama, K. *Adv Mater* 2004, **16**, 1226–1229.
61. Jeong, U.; Ryu, D. Y.; Kim, J. K.; Kim, D. H.; Russell, T. P.; Hawker, C. J. *Adv Mater* 2003, **15**, 1247–1250.
62. Xu, T.; Stevens, J.; Villa, J. A.; Goldbach, J. T.; Guarini, K. W.; Black, C. T.; Hawker, C. J.; Russell, T. P. *Adv Funct Mater* 2003, **13**, 698–702.
63. Lee, M.; Park, M.-H.; Oh, N.-K.; Zin, W.-C.; Jung, H.-T.; Yoon, D. K. *Angew Chem Int Ed* 2004, **43**, 6465–6468.
64. (a) Goldbach, J. T.; Lavery, K. A.; Penelle, J.; Russell, T. P. *Macromolecules* 2004, **37**, 9639–9645. (b) Goldbach, J. T.; Russell, T. P.; Penelle, J. *Macromolecules* 2002, **35**, 4271–4276.
65. Zhang, M.; Yang, L.; Yurt, S.; Misner, M. J.; Chen, J. T.; Coughlin, E. B.; Venkataraman, D.; Russell, T. P. *Adv Mater* 2007, **19**, 1571–1576.
66. Kang, M.; Moon, B. *Macromolecules* 2009, **42**, 455–458.
67. Bang, J.; Kim, S. H.; Drockenmuller, E.; Misner, M. J.; Russell, T. P.; Hawker, C. J. *J Am Chem Soc* 2006, **128**, 7622–7629.
68. Jeong, U.; Kim, H. C.; Rodriguez, R. L.; Tsai, I. Y.; Stafford, C. M.; Kim, J. K.; Hawker, C. J.; Russell, T. P. *Adv Mater* 2002, **14**, 274–276.
69. Kim, S. H.; Misner, M. J.; Russell, T. P. *Adv Mater* 2004, **16**, 2119–2123.
70. (a) Mao, H.; Hillmyer, M. A. *Macromol Chem Phys* 2008, **209**, 1647–1656. (b) Mao, H.; Hillmyer, M. A. *Soft Matter* 2006, **2**, 57–59. (c) Mao, H.; Arrechea, P. L.; Bailey, T. S.; Johnson, B. J. S.; Hillmyer, M. A. *Faraday Discuss* 2005, **128**, 149–162.

71. (a) Rzayev, J.; Hillmyer, M. A. *Macromolecules* 2005, **38**, 3–5. (b) Rzayev, J.; Hillmyer, M. A. *J Am Chem Soc* 2005, **127**, 13373–13379.
72. A related system is described in: Guo, F.; Jankova, K.; Schulte, L.; Vigild, M. E.; Ndoni, S. *Macromolecules* 2008, **41**, 1486–1493.
73. Bailey, T. S.; Rzayev, J.; Hillmyer, M. A. *Macromolecules* 2006, **39**, 8772–8781.
74. Guo, S.; Rzayev, J.; Bailey, T. S.; Zalusky, A. S.; Olayo-Valles, R.; Hillmyer, M. A. *Chem Mater* 2006, **18**, 1719–1721.
75. Cooney, D. T.; Hillmyer, M. A.; Cussler, E. L.; Moggridge, G. D. *Crystallogr Rev* 2006, **12**, 13–24.
76. Phillip, W. A.; Rzayev, J.; Hillmyer, M. A.; Cussler, E. L. *J. Membr Sci* 2006, **286**, 144–152.
77. Yang, S. Y.; Ryu, I.; Kim, H. Y.; Kim, J. K.; Jang, S. K.; Russell, T. P. *Adv Mater* 2006, **18**, 709–712.
78. Hawker, C. J.; Russell, T. P. *MRS Bull* 2005, **30**, 952–966.
79. Johnson, B. J. S.; Wolf, J. H.; Zalusky, A. S.; Hillmyer, M. A. *Chem Mater* 2004, **16**, 2909–2917.
80. (a) Misner, M. J.; Skaff, H.; Emrick, T.; Russell, T. P. *Adv Mater* 2003, **15**, 221–224. (b) Zhang, Q.; Xu, T.; Butterfield, D.; Misner, M. J.; Ryu, D. Y.; Emrick, T.; Russell, T. P. *Nano Lett* 2005, **5**, 357–361.
81. Bandyopadhyay, K.; Tan, E.; Ho, L.; Bundick, S.; Baker, S. M.; Niemz, A. *Langmuir* 2006, **22**, 4978–4984.
82. (a) Ha, J.-M.; Wolf, J. H.; Hillmyer, M. A.; Ward, M. D. *J Am Chem Soc* 2004, **126**, 3382–3383. (b) Ha, J.-M.; Hillmyer, M. A.; Ward, M. D. *J. Phys Chem B* 2005, **109**, 1392–1399. (c) Hamilton, B. D.; Hillmyer, M. A.; Ward, M. D. *Cryst Growth Des* 2008, **8**, 3368–3375. (d) Hamilton, B. D.; Weissbuch, I.; Lahav, M.; Hillmyer, M. A.; Ward, M. D. *J Am Chem Soc* 2009, **131**, 2588–2596.
83. Review articles: (a) Ikkala, O.; ten Brinke, G. *Science* 2002, **295**, 2407–2409. (b) Elemans, J. A. A. W.; Rowan, A. E.; Nolte, R. J. M. *J Mater Chem* 2003, **13**, 2661–2670. (c) Ikkala, O.; Brinke, G. t. *Chem Commun* 2004, 2131–2137.
84. (a) Widawski, G.; Rawiso, M.; Francois, B. *Nature* 1994, **369**, 387–389. (b) François, B.; Pitois, O.; François, J. *Adv Mater* 1995, **7**, 1041–1044. (c) Pitois, O.; François, B. *Eur Phys J B* 1999, **8**, 225–231.
85. de Boer, B.; Stalmach, U.; Nijland, H.; Hadziioannou, G. *Adv Mater* 2000, **12**, 1581–1583.
86. Liu, C.; Gao, C.; Yan, D. *Angew Chem Int Ed* 2007, **46**, 4128–4131.
87. Examples using other solvents: Karthaus, O.; Maruyama, N.; Cieren, X.; Shimomura, M.; Hasegawa, H.; Hashimoto, T. *Langmuir* 2000, **16**, 6071–6076.
88. Jenekhe, S. A.; Chen, X. L. *Science* 1999, **283**, 372–375.
89. PI-P4VP has also been used to prepare materials with polymer:small molecule interactions; however, this material was never made porous: Bondzic, S.; de Wit, J.; Polushkin, E.; Schouten, A. J.; ten Brinke, G.; Ruokolainen, J.; Ikkala, O.; Dolbnya, I.; Bras, W. *Macromolecules* 2004, **37**, 9517–9524.
90. Valkama, S.; Ruotsalainen, T.; Kosonen, H.; Ruokolainen, J.; Torkkeli, M.; Serimaa, R.; ten Brinke, G.; Ikkala, O. *Macromolecules* 2003, **36**, 3986–3991.

91. Mäki-Ontto, R.; Moel, K. d.; Odorico, W. d.; Ruokolainen, J.; Stamm, M.; Brinke, G. t.; Ikkala, O. *Adv Mater* 2001, **13**, 117–121.
92. Sidorenko, A.; Tokarev, I.; Minko, S.; Stamm, M. *J Am Chem Soc* 2003, **125**, 12211–12216.
93. Laforgue, A.; Bazuin, C. G.; Prud'homme, R. E. *Macromolecules* 2006, **39**, 6473–6482.
94. Fustin, C.-A.; Lohmeijer, B. G. G.; Duwez, A. S.; Jonas, A. M.; Schubert, U. S.; Gohy, J. F. *Adv Mater* 2005, **17**, 1162–1165.
95. Shimomura, M.; Koito, T.; Maruyama, N.; Arai, K.; Nishida, J.; Gråsjö, L.; Karthaus, O.; Ijiro, K. *Mol Cryst Liq Cryst* 1998, **322**, 305–312.
96. (a) Gankema, H.; Hempenius, M. A.; Möller, M.; Johansson, G.; Percec, V. *Macromol Symp* 1996, **102**, 381–390. (b) Beginn, U. *Adv Mater* 1998, **10**, 1391–1394. (c) Wan, Y.; Shi, Y.; Zhao, D. *Chem Mater* 2008, **20**, 932–945.
97. Gu, W.; Lu, L.; Chapman, G. B.; Weiss, R. G. *Chem Commun* 1997, 543–544.
98. Gankema, H.; Hempenius, M. A.; Möller, M. *Rec Trav Chim Pays-Bas* 1994, **113**, 241–249.
99. Beginn, U.; Möller, M.; Keinath, S. *Chem Eng Comm* 2005, **192**, 1116–1128.
100. Burguete, M. I.; Galindo, F.; Gavara, R.; Izquierdo, M. A.; Lima, J. C.; Luis, S. V.; Parola, A. J.; Pina, F. *Langmuir* 2008, **24**, 9795–9803.
101. Fu, X.; Yang, Y.; Wang, N.; Wang, H.; Yang, Y. *J Mol Recognit* 2007, **20**, 238–244.
102. Tan, G.; Singh, M.; He, J.; John, V. T.; McPherson, G. L. *Langmuir* 2005, **21**, 9322–9326.
103. Wei, Q.; James, S. L. *Chem Commun* 2005, 1555–1556.
104. Jang, J.; Bae, J. *Chem Commun* 2005, 1200–1202.
105. Meng, Y.; Gu, D.; Zhang, F.; Shi, Y.; Yang, H.; Li, Z.; Yu, C.; Tu, B.; Zhao, D. *Angew Chem Int Ed* 2005, **44**, 7053–7059.
106. Zhu, X. X.; Banana, K.; Yen, R. *Macromolecules* 1997, **30**, 3031–3035.
107. Simon, F.-X.; Khelfallah, N. S.; Schmutz, M.; Diaz, N.; Mesini, P. J. *J Am Chem Soc* 2007, **129**, 3788–3789.
108. Moffat, J. R.; Seeley, G. J.; Carter, J. T.; Burgess, A.; Smith, D. K. *Chem Commun* 2008, 4601–4603.
109. (a) Beginn, U.; Zipp, G.; Möller, M. *Adv Mater* 2000, **12**, 510–513. (b) Beginn, U.; Zipp, G.; Mourran, A.; Walther, P.; Möller, M. *Adv Mater* 2000, **12**, 513–516.

CHAPTER 3

Porogen Incorporation and Phase Inversion

LEI QIAN and HAIFEI ZHANG

Department of Chemistry, University of Liverpool, Liverpool, United Kingdom

3.1 INTRODUCTION

Porous polymers have been widely used in many applications, such as tissue engineering, drug delivery, separation science, photonics, and electronics. There is a range of methods that can be utilized to prepare porous polymers. Here, we introduce the methods of porogen incorporation and phase inversion for the production of porous polymeric structures.

A "porogen" is a substance that can be used as a template and then removed to generate pores. A porogen can be incorporated into a polymer matrix and then removed to produce porous polymers. For example, micrometer-sized particles can be mixed with a polymer solution. After solidifying the polymer solution and removing the particles, a porous polymeric structure is produced. This is the method known as "particle leaching," which can also be combined with other techniques to prepare multiple porous structures [1, 2]. In this chapter, we focus on systems starting as a homogeneous solution. Different types of solvents are incorporated as porogens. Phase separation is induced during polymerization within the solutions by tuning the solvent properties or by freezing the solutions. The removal of the solvent porogen leaves behind cavities and generates pores in the polymers.

Typically, solvent porogens are used in suspension polymerization to make permanently porous microparticles [3, 4]. The monomer and the crosslinker (e.g., styrene and divinyl benzene) are suspended in an aqueous medium. The solvent porogen is used to dissolve the monomer and the crosslinker. The porogen should be (1) soluble in the monomer and crosslinker (2) phase separated upon polymerization of the monomers, and (3) not soluble in the dispersive medium. There is a wide range of organic solvents, including hydrophobic solvents such as toluene and polar solvents

Porous Polymers, Edited by Michael S. Silverstein, Neil R. Cameron, and Marc A. Hillmyer

such as alcohols with different chain lengths, which have been used as porogens to obtain desired pore surface areas, hydrophilicity, and proportion of micropores or macropores.

There are other ways to induce phase separation in polymer solutions or gels. Supercritical fluids (SCFs) are solvents used at the temperature and pressure above their critical point. SCFs exhibit liquid-like density, gas-like diffusivity, and very low surface tension. Their solvent properties can be simply tuned by changing the temperature and pressure. In recent years, SCFs have been regarded as "green" solvents and widely used for chemical reactions and extractions in which a large volume of organic solvents is typically used. SCFs are also used as porogens or foaming agents to produce porous polymers or used as nonsolvents in a phase-inversion process to prepare porous membranes. These topics are also covered in this chapter.

In addition to inducing phase separation by bringing a polymer solution into contact with a nonsolvent, thermally induced phase separation is utilized to fabricate porous membranes or microcellular foams. This method is based on the principle that a single homogeneous polymer solution made at elevated temperature is converted via the removal of thermal energy into two phase-separated domains composed of a polymer-rich phase and a polymer-poor phase. Subsequent freeze-drying of the liquid–liquid phase-separated polymer solution produces microcellular structures as a result of solvent removal [5]. For polymer solutions prepared at either room or elevated temperature, they can also be frozen into a solid by contacting with a cold liquid such as liquid nitrogen. The frozen solvent is removed by freeze-drying directly to produce porous polymeric materials. The size and orientation of frozen solvent crystals such as ice may be tuned by adjusting the freezing conditions. This could result in a facile control of porous structure [6].

This chapter covers the preparation of porous polymers by solvent porogens, supercritical (SC) CO_2, and freeze-drying. The application of phase inversion for the preparation of porous polymers is also described. For each topic, the basic scientific principle is introduced first, and the latest progress in the area is then described.

3.2 SOLVENT POROGENS AND PHASE INVERSION

3.2.1 Solvent Porogens

Polymerization of styrene (St) and divinyl benzene (DVB) has been extensively investigated. A suspension polymerization of St and DVB is used here as an example to elucidate the basic principle of the use of solvent porogens to prepare porous polymers.

A suspension polymerization of St and DVB in an aqueous medium can produce crosslinked polystyrene (PS) microparticles. When the monomer mixture consists only of St and DVB without any organic solvent, nonporous hard PS spheres are produced. The polymer chains are in molecular contact with each other, and the materials have a very low surface area in the dry state. However, when a solvent (or an inert diluent) is included in the monomer mixture, it can give rise to phase separation with the progress of the polymerization. The removal of the solvent or porogen at the end of the polymerization generates PS microparticles with permanent pores.

Droplets of monomers (e.g., St and DVB) and their solvent (e.g., toluene) are suspended in an aqueous medium with a stabilizer. After the polymerization is initiated, there is a local formation of small volumes of microgel (microgelation), as proposed by Sherrington [3]. The mass of growing polymer molecules dissolved in the solution becomes crosslinked into one finite network. In the end, the whole-polymer chains are all crosslinked into a monolithic soft gel with an infinite network. Figure 3.1 shows the scheme illustrating how the porous polymer particles are formed via the incorporation and removal of the solvent porogen. The polymerization is initiated in the droplet phase of the monomers (Fig. 3.1a), and a polymer network is formed with time (Fig. 3.1b). When the crosslinked polymer is precipitated from the solution, the phase separation starts between the solvent and the network (Fig. 3.1c). The solvent phase (porogen) acts as a pore template. After removal of the porogen, porous microparticles are produced (Fig. 3.1d). Although the illustration is based on polymerizing droplets in a suspension, bulk solutions can be polymerized in a similar manner to produce porous polymeric monoliths.

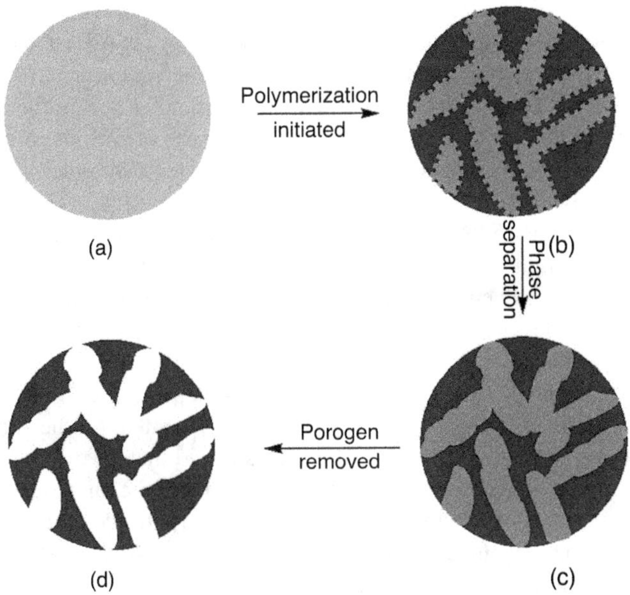

FIGURE 3.1 A schematic showing the formation of porous polymeric microparticles via the incorporation and removal of a porogen. (**a**) The droplet containing a solvent and monomers in a suspension. (**b**) Polymerization initiated in the droplet phase. (**c**) Phase separation between the solvent and the formed polymer network. (**d**) The porogen removed to produce porous microparticles.

To act as a porogen, (1) the solvent should be able to dissolve monomers to form a homogeneous solution, (2) but it should be unable to dissolve the formed polymers, and (3) in the case of a suspension polymerization, it should not be soluble in the aqueous medium. The phase separation is the crucial step to generate the porous structure. Therefore, in order to adjust pore sizes and surface areas of the formed porous polymer, it is important to control the point of phase separation, which depends on the nature and amount of the porogen, its compatibility with the incipient polymer matrix, and the level of crosslinking. The polymerization in the porogen phase after phase separation could fuse microgel particles together and infill the pores between them, which leads to the loss of surface area [3]. It should be noted that nonporous polymers can be made in the presence of a porogen, provided that the porogen is present at a low level, which does not cause phase separation of the growing polymer from the solution.

3.2.2 Porous Polymers via the Use of Solvent Porogens

Different types of solvent have been used as porogens to prepare a wide range of porous polymers [3, 4]. We present the progress in this area based on the types of solvent porogens, which include hydrophobic solvents, hydrophilic solvents, solvent mixtures, and oligomers or polymers.

Among the hydrophobic solvents, toluene was widely used for the synthesis of polystyrene [3, 4]. Recently, toluene was used as a porogen to prepare highly crosslinked monodisperse porous poly(divinylbenzene) (PDVB) microspheres in a process of distillation–precipitation polymerization in acetonitrile [7]. The porous microspheres were formed through a precipitation manner during the distillation of the solvent from the reaction system. In another study, α-cyclodextrin was used as a molecular porogen to increase the micropore volume of the microporous carbons derived from polyimide [8]. The N_2 sorption results showed that the micropore volume increased from 0.15 to 0.24 cm^3/g when the amount of α-cyclodextrin rose from 0% to 15% by weight, whereas the micropore diameter remained almost constant. A range of organic solvents, including butyl acetate, isobutyl acetate, ethyl isobutyrate, propyl acetate, and xylene, were investigated as porogens to synthesize porous polyacrylate [9]. In order to improve the specific surface area, PDVB microspheres were prepared via a suspension polymerization in the presence of different types of porogens [10]. The effect of porogen type and DVB content on the specific surface area and porosity was investigated. The results showed that, as the solubility parameter of the porogen increased, the specific surface area of the microspheres increased initially and then fell, while the average pore size initially declined and then rose. Toluene as a porogen yielded highly microporous microspheres with a large specific surface area (700 m^2/g). A higher specific surface area (720 m^2/g) and porosity (68%) could be achieved by using 1-chlorodecane as a porogen.

Hydrophilic organic solvents were also used as porogens to produce porous polymers. For example, Poinescu and Beldie found that cyclohexanol was one of the most efficient porogens for building up a higher porosity even at a low level of DVB [10]. Porous poly(2-hydroxyethyl methacrylate–methyl methacrylate) (PHEMA-PMMA)

FIGURE 3.2 SEM images of (**a**) porous PHEMA-PMMA microparticles and (**b**) the porous surface. (Reprinted with permission from Reference 11, sample C in Table 1. Copyright 2007 Wiley Periodicals, Inc.)

particles crosslinked with ethylene glycol dimethacrylate were synthesized by a free radical suspension copolymerization in an aqueous phase initiated by an oil-soluble initiator, 2,2-azobisisobutyronitrile (AIBN) [11]. 1-Octanol was used as a pore-forming agent. The crosslinker created the three-dimensional (3D) networks, whereas the porogen is a good solvent for monomers but a nonsolvent for PHEMA. As an illustration, Fig. 3.2 shows the scanning electronic microscopy (SEM) images of the porous microparticles and their porous surface.

Since the pore structure of porous polymers can be affected by the choice of solvent porogen, mixtures of solvent porogens were further investigated to achieve desirable surface areas or porous structures. Okay [12, 13] investigated porosity formation and porous structure stability of styrene–divinyl copolymers prepared with the mixture of toluene–cyclohexanol as a porogen. It was found that the total porosity of the copolymers increases with DVB concentration in the monomer mixture. Rigid porous poly(glycidyl methacrylate–DVB) microspheres were synthesized through the suspension polymerization with a mixture of isooctane and 4-methyl-2-pentanol as the porogen. These microspheres were used as column packing materials for protein separation after the further coupling reaction with polyethylene glycol [14]. Hydrophobic porous copolymer resins based on DVB and acrylonitrile (AN) were prepared in the presence of 1,2-dichloroethane and heptanes as porogens. Compared with the resins based on DVB and methyl acrylate (MA), the AN/DVB resins thus obtained were water swellable over a wider range of copolymer compositions [15]. In situ polymerization of butyl methacrylate and ethylene dimethacrylate monomers was conducted in fused-silica capillaries with 1,4-butanediol and 1-propanol as the porogens [16]. The flowthrough porosity of the capillary columns increased with the increasing concentration of the binary porogen solvent mixture. It was found that the concentration of 1,4-butanediol was the main factor enhancing flowthrough pore formation.

Oligomers and polymers were also used as porogens. For example, macroporous poly(dicyclopentadiene) beads were produced in a suspension polymerization [17].

Phase separation was promoted by the enthalpic and entropic changes induced by the polymerization of dicyclopentadiene. By using poly(1,2-butylene glycol) monobutyl ether (M_n = 500 g/mol) as the porogen, the stabilized droplets of the monomer/porogen mixture were polymerized. The phase separation occurred, resulting in solid, biphasic microspheres. The porogen was extracted with methanol to produce porous microparticles. The major problem in using polymeric porogens is that the viscosity of the comonomer mixtures rises sharply as the weight percentage of these porogens increases. This leads to low efficiency in dispersing and retaining spherical droplets in the suspension medium. Oligomers with low molecular weight could be selected as porogens. Macintyre and Sherrington investigated mixtures of toluene and various oligomers, including poly(propylene glycol) 1000, poly(propylene glycol) 4000, and poly(dimethylsiloxane) (PDMS), as porogens to prepare porous PDVB [18]. The oligomeric porogen could induce an early phase separation, yielding in due course "macrogel" particles and macropores, while the toluene would phase separate much later in the normal manner, yielding microgel and micropores. The use of toluene yielded a highly microporous resin with a large dry-state surface area ($\sim$650 m^2/g), whereas the presence of oligomeric porogen yielded a highly macroporous resin with rather low surface area ($\sim$15–50 m^2/g). By using the mixture of toluene and oligomer PDMS as the porogen, porous microspheres with a bimodal pore size distribution with one population of pores in the micropore region (from toluene templating, characterized by N_2 adsorption) and one in the macropore region (from oligomer templating, characterized by Hg intrusion porosimetry) were formed (Fig. 3.3). In another study, a volatile solvent (toluene or diglyme) in combination with a linear polymer porogen [poly(vinyl acetate)] was used to prepare molecularly imprinted polymers in both thin-film and bulk monolith formats [19]. The mixture of diglyme and poly(vinyl acetate) resulted in a porous morphology and a 60-fold enhancement in the binding capacity for ^{3}H-labeled S-propranolol compared to a nonporous film prepared with a highly volatile solvent but in the absence of the polymer porogen.

3.2.3 Phase Inversion

In a phase inversion process, a homogeneous polymer solution is placed in contact with a nonsolvent. The subsequent exchange of solvent and nonsolvent occurs at the interface. The polymer should be soluble in one solvent and insoluble in the other, while the two solvents should be miscible. The polymer solution can be coated on a substrate as a thin film and then immersed in the nonsolvent to allow the solvent exchange process to occur. It is also possible to cast the polymer solution in the nonsolvent bath. The solvent exchange results in a polymer-rich phase and a polymer-poor phase. In both cases, the polymer in the solution is coagulated by the nonsolvent. The amount of nonsolvent should be sufficient to dissolve the polymer solvent and replace it in the polymer-rich phase so that a solidified membrane can be formed. The removal of the nonsolvent (porogen) generates porous membranes.

Figure 3.4 shows a schematic ternary phase diagram of polymer/solvent/nonsolvent [20]. At high polymer concentrations, phase separation is continued

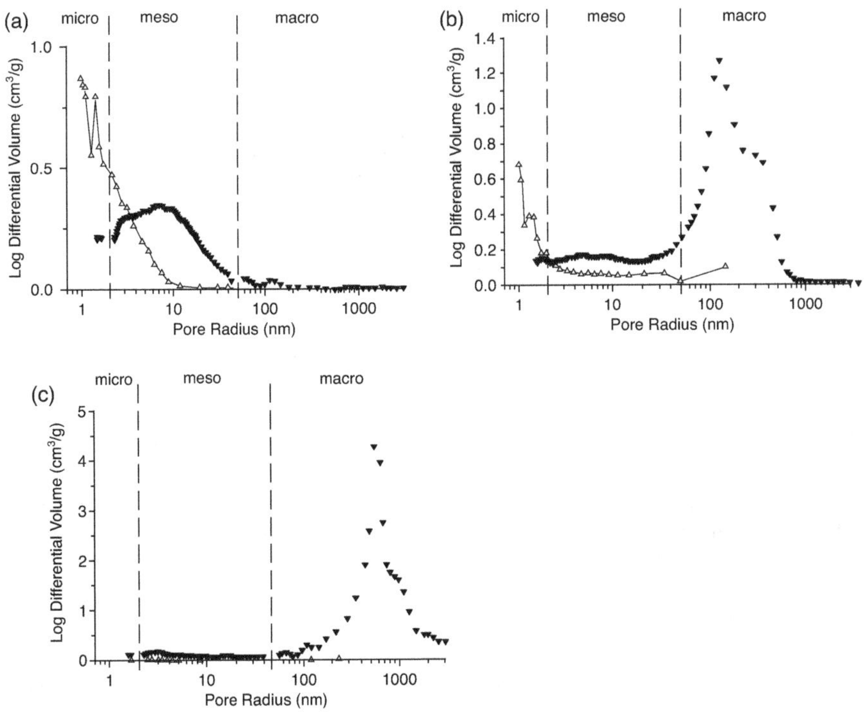

FIGURE 3.3 Combined N_2 adsorption-derived (△) and Hg intrusion-derived (▼) pore size distribution for PDVB resins: **(a)** toluene porogen 100 vol%; **(b)** PDMS coporogen 20 vol%; **(c)** PDMS porogen 100 vol%. (Reprinted with permission from Reference 18. Copyright 2004 American Chemical Society.)

until the polymer-rich phase is solidified. The final morphology is influenced by the solidification process, which can be achieved by gelation or crystallization of the polymer-rich phase. At low polymer concentrations, liquid–liquid phase separation can occur either by spinodal decomposition or by nucleation and growth, which depends on the throughpath of the phase diagram. Spinodal decomposition occurs immediately if the composition path is through the metastable region before the stable nuclei are formed.

3.2.4 Phase Inversion for Porous Membranes

Functional polymer or polymer composite membranes have been prepared for a wide range of applications. Porous membranes were prepared with thickness in the range 120–145 μm from a commercial resin, polyvinylbutyral, using the phase inversion technique. The membranes were produced with 6%–20% (w/v) polymer concentration using *N,N*-dimethylacetamide (DMAA) as solvent and water as the casting solution at 20°C [21]. It was found that pore size, pore density, water permeation rate,

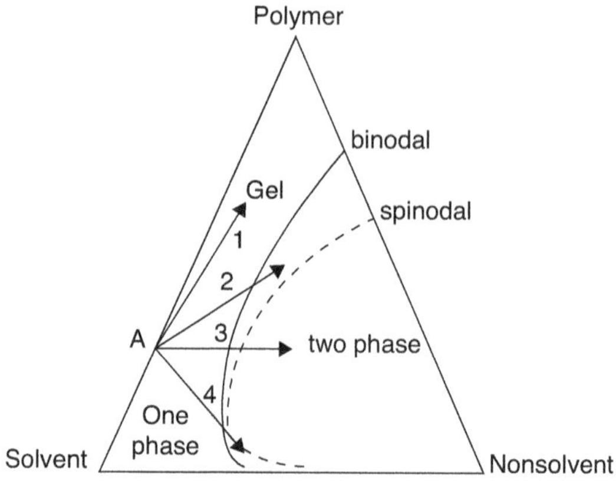

FIGURE 3.4 A schematic representation of a ternary phase diagram of polymer/solvent/nonsolvent showing the liquid–liquid phase separation: 1, solidification; 2, nucleation and growth of polymer-poor phase; 3, spinodal decomposition; 4, nucleation and growth of polymer-rich phase. (Reprinted with permission from Reference 20. Copyright 1999 Wiley Periodicals, Inc.)

water content, and elongation of the membranes decreased, while breakpoint stress increased, with the increase of polymer concentration. The surface of the membrane was activated using glutaraldehyde for bovine serum albumin (BSA) separation. A remarkable adsorption capacity of BSA was achieved.

Pu et al. [22] developed a process to prepare porous poly(acrylonitrile–methyl methacrylate) (PAN-PMMA) membranes by phase inversion techniques using an ultrasonic humidifier. After dissolving PAN-PMMA in *N*,*N*-dimethylformamide (DMF) with mechanical stirring, the formed homogeneous solution was cast onto a clean glass plate. Subsequently, the glass plate was exposed to water vapor produced by an ultrasonic humidifier, inducing the phase inversion. The porous membranes were obtained after washing and drying. The membrane had an ionic conductivity of 2.52×10^{-3} S/cm at room temperature when gelled with 1 M $LiPF_6$/EC-DMC (1:1 v/v) electrolyte solution, which suggested that it could be used as gel electrolyte for lithium batteries. Blanco et al. [23] reported the preparation of polysulfone-based membranes by wet-phase inversion. Nanofiltration membranes were formed by coagulating a concentrated solution (e.g., 40 wt%) of sulfonated polyethersulfone with water.

Fabrication conditions of porous thermoplastic polyurethane (PU) for use in vascular grafts were studied by Khorasani and Shorgashti [24]. The segmented PU was dissolved in DMF and a range of solvents, including water, methanol, ethanol, and isopropanol, were used as coagulants. Figure 3.5a–c shows the porous PU films produced with methanol, ethanol, and isopropanol as nonsolvents at room temperature. Porosity variations due to various factors such as concentration of PU solution, composition of coagulation bath, effect of coagulant temperature, and effect of dissolved air in the

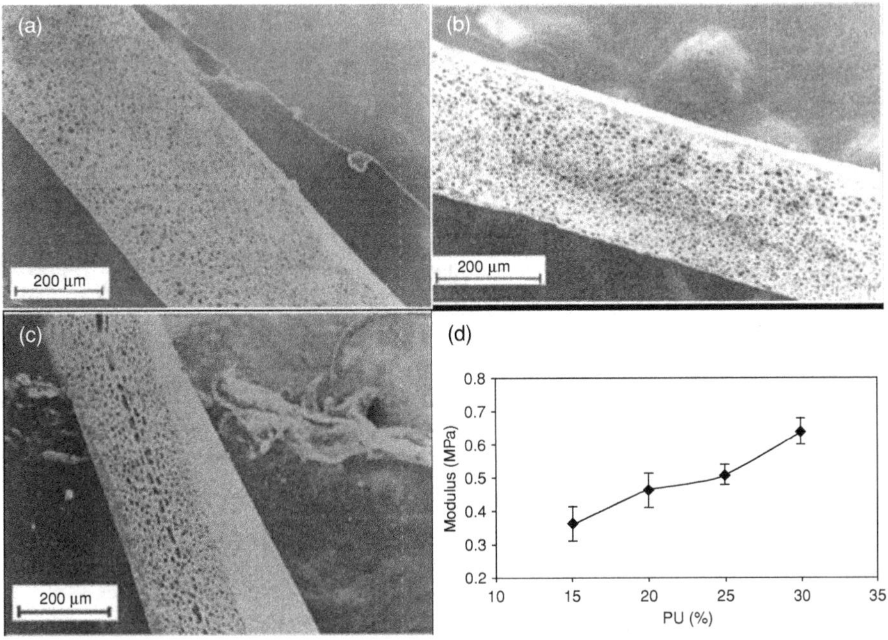

FIGURE 3.5 SEM image of cross-section morphology of PU films prepared in **(a)** methanol, **(b)** ethanol, and **(c)** isopropanol as nonsolvents at room temperature and **(d)** the tensile modulus (100% strain) of PU films as a function of PU concentration. These films were prepared in water at 10°C. (Reprinted with permission from Reference 24. Copyright 2005 Wiley Periodicals, Inc.)

solution were investigated. It was found that the macropore population decreased with increasing the polymer concentration and coagulant temperature (>23°C), which enhanced the tensile modulus of the PU films (Fig. 3.5d). The copolymer membranes, poly(vinylidene fluoride–hexafluoropropylene) (PVDF-HFP), were recently prepared using poly(ethylene glycol) (PEG) as an additive with acetone or DMF as solvent [25]. The morphology of the membranes was varied with the change of the composition of the polymer and solvent. The prepared membranes were employed as separators in $LiCoO_2$/Li cells and were subjected to charge–discharge studies. It was found that the polymer membrane prepared with a ratio of PVDF-HFP (50):PEG (50) using DMF as solvent delivered higher discharge capacity than did the membranes prepared with acetone as solvent.

Modification of porous PVDF was achieved by blending the scaffold with amphiphilic polymers for the adsorption of BSA. The polymers were blended first and then used to prepare porous membranes by a phase inversion process [26]. The antifouling properties of the prepared membranes were evaluated by static and dynamic BSA adsorption. Nanoparticle-filled polymer membranes as electrolytes for lithium rechargeable batteries have been studied extensively to improve mechanical strength and to enhance the lithium ionic transport properties. Al_2O_3 and TiO_2 nanoparticles were used as the filler particles. The characteristics of porous membranes could be important factors in the polymer electrolytes of lithium rechargeable batteries. Effective pores could provide good conduction paths for carrier ions, which are governed by properties of porosity, pore connectivity or tortuosity, and even liquid uptake. Kim et al. investigated the preparation of the composite membranes of PVDF-HFP/SiO_2 by a phase inversion technique with DMF as solvent and water as nonsolvent [27]. Their results showed that the uptake of electrolyte solution by the prepared membranes reached greater than 400%, and the room temperature ionic conductivity was greater than 10^{-3} S/cm. The structure control of asymmetric poly(vinyl butyral)–TiO_2 composite membranes was investigated by Fu et al [28]. The membranes were prepared by phase inversion, which produced a skin layer on both the outer and the inner surface at the initial stage. A further study showed that the outer surface became porous with time as the polymer in the membrane's outer surface was decomposed by the photocatalysis of TiO_2. The water permeability increased with the increase of TiO_2 content and with time (becoming constant after about 15 days).

3.3 SUPERCRITICAL FLUIDS

3.3.1 What Are Supercritical Fluids?

There are two definitions of supercritical fluids. Baiker termed a fluid supercritical when its temperature exceeds the critical temperature (T_c), regardless of the pressure [29]. Darr and Poliakoff described a SCF as "any substance, the temperature and pressure of which are higher than their critical values, and which has a density close to or higher than its critical density" [30]. We use the latter definition for a SCF in this chapter. Figure 3.6 illustrates the SCF region in a pressure–temperature phase diagram, which is above both T_c and critical pressure (P_c).

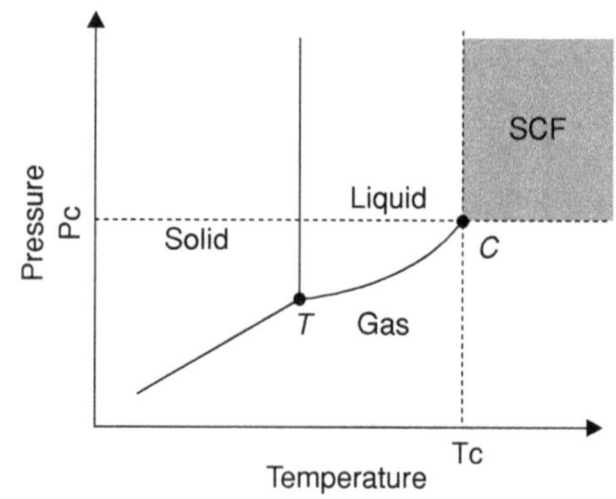

FIGURE 3.6 A scheme illustrating the SCF region in a pressure–temperature phase diagram.

TABLE 3.1 Typical Physical Data of Liquids, Gases, and SCFs near the Critical Region[a]

Physical Quantity	Gas (Ambient)	Supercritical Fluid (T_c, P_c)	Liquid (Ambient)
Density ρ, kg/m^3	0.6–2	200–500	600–1600
Dynamic viscosity η, mPa s	0.01–0.03	0.01–0.03	0.2–3
Kinematic viscosity ν,[b] 10^6 m^2/s	5–500	0.02–0.1	0.1–5
Diffusion coefficient D, 10^6 m^2/s	10–40	0.07	0.0002–0.002

[a]Data taken from Reference 31.
[b]Kinematic viscosity was estimated from dynamic viscosity and density, $\nu = \eta/\rho$.

Many of the physical properties of SCFs are intermediate between those of a liquid and those of a gas. Table 3.1 compares the typical values of physical properties of liquids, gases, and SCFs around the critical region [29, 31]. SCFs have liquid-like density and gas-like diffusivity and viscosity. Therefore, SCFs have sufficient density to give appreciable dissolving power; however, the diffusivity of solutes in SCFs is higher than in liquids, and the viscosity of SCFs is lower. This facilitates mass transport and is a significant advantage for diffusion-controlled chemical reactions or processes. A unique property of SCFs is that their density is pressure dependent and can be continuously adjusted from that of a vapor to that of a liquid. Around the critical point, the isothermal compressibility of a pure SCF is huge. The density and dissolving power of SCFs can thus be tuned sensitively through small changes in pressure. SCFs in the range $1 < T/T_c < 1.1$ and $1 < P/P_c < 2$ have been used mostly for a variety of applications such as chemical reactions and materials processing [32].

Among the SCFs, SC CO_2 has been most investigated. Carbon dioxide has a mild critical point, with $T_c = 31.1°C$ and $P_c = 73$ atm. Therefore, the supercritical state can be reached easily. As a result, the process operation cost is low compared to that for other SCFs. SC CO_2 is widely used as a solvent for organic synthesis, catalysis, polymer synthesis, materials processing, extraction, and cleaning [29,30,32]. When used as a solvent, SC CO_2 can revert to the gaseous state by reducing the pressure and is released or separated from the reaction system. Worldwide, there is a huge quantity of organic solvents being used every day in various processes. With the serious concern about global pollution and climate change, there has been a major drive to replace organic solvents with sustainable or "green" solvents. CO_2 is nontoxic and nonflammable. Although CO_2 is a known greenhouse gas, there is no new CO_2 generated or released in the processes involving SC CO_2. For this reason and its unique properties, SC CO_2 is widely regarded as a "green" solvent.

SC CO_2 has been used in a range of processes to prepare porous materials [33]. This chapter covers SC CO_2 as a porogen or tunable solvent for the fabrication of porous polymers. The advantages of using SC CO_2 are as a replacement for organic solvents, its ease of separation simply by reducing pressure, and the absence of residue of organic solvents in the formed structures.

3.3.2 SC CO_2 as Porogen

A suspension polymerization in water with SC CO_2 as porogen was carried out in a high-pressure reactor with an impeller stirrer [34]. A mixture of monomer trimethylolpropane trimethacrylate (TRIM), initiator AIBN, and an aqueous solution of 0.5% w/v poly(vinyl alcohol) (PVA; as stabilizer to inhibit droplet coalescence) were first charged into the reactor. The reactor was purged and then pressurized with CO_2 until the desired pressure. The reactor was then heated to 60°C to initiate the polymerization. The volumetric ratio of the CO_2 droplet phase to the aqueous phase was controlled at 20:80. Porous polymer microparticles were produced with CO_2 as porogen, while nonporous microparticles were formed when CO_2 was not present in the polymerization. This was confirmed by the mercury intrusion data for the polymer microparticles (Fig. 3.7). It was further shown that the pore volume and surface area of the polymer microparticles could be dramatically tuned simply by conducting the polymerization under different pressures.

In another study, monoliths of permanently porous, highly crosslinked poly (methacrylate) (PMA) resins were prepared using SC CO_2 as the adjustable solvent [35]. The polymerization was carried out in a 10-mL high-pressure view cell at 60°C. After the polymerization, CO_2 was released by reducing the pressure, and dry porous monoliths were obtained directly. The surface area and average pore size

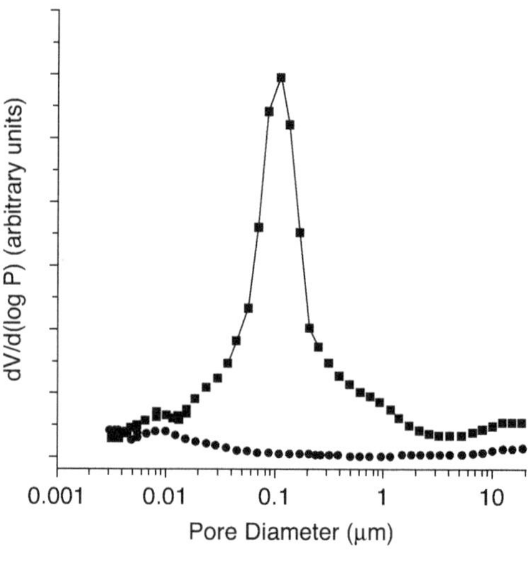

FIGURE 3.7 Mercury intrusion porosimetry data for polymer microparticles. Circles: microparticles synthesized in the absence of CO_2. Squares: equivalent microparticles synthesized using SC CO_2 as porogen. (Reprinted with permission from Reference 34. Copyright 2001 American Chemical Society.)

of the porous monoliths could be tuned just by varying the reaction pressures. The surface areas of the materials, which could be related to the final pressure of the polymerization, were continuously tuned in the range of 90–320 m^2/g. This was rationalized by combining the variation in solvent quality and the resulting influence on the mechanism of nucleation, phase separation, and pore formation [35].

3.3.3 SC CO_2 as Nonsolvent for Phase Inversion Processes

There are two characteristics that make SC CO_2 a suitable candidate for producing porous membranes via a phase inversion process: (1) Most polymers, except for perfluoropolymer [36] and some hydrocarbon polymers [37, 38], are not soluble in SC CO_2, and (2) a number of organic solvents are miscible or partly soluble in SC CO_2. A polymer solution can be coated on a substrate as thin film and then placed in a high-pressure reactor with controlled temperature. The reactor is pressurized with CO_2. The exchange of organic solvent and SC CO_2 takes place across the polymer solution film. SC CO_2 acts as a nonsolvent and leads to the phase inversion. After a defined period, CO_2 is released by reducing the pressure, and the dry porous membrane is obtained directly.

PMMA microcellular membranes were prepared using an SC CO_2-phase inversion process [39]. Solutions of PMMA in dimethylsulfoxide (DMSO), acetone, and tetrahydrofuran (THF) were prepared with concentrations in the range of 1%–25% w/w. It was found that the membrane cell size decreased on increasing the pressure and decreasing the temperature. On increasing the mutual affinity between solvent and nonsolvent, cell sizes decreased and the structure changed from nearly closed to open interconnected cells. An antibiotic (amoxicillin)-loaded PMMA membrane was prepared using a similar process [40]. The antibiotic was either dissolved in the same organic solvent used to solubilize the polymer or suspended in the organic solution formed by the polymer and solvent. The drug release from the structure was then studied.

A shallow PS solution in a dish was placed inside a reactor and CO_2 was introduced to induce the phase separation. Dry, flat microporous membranes were obtained without collapse of the structure after the CO_2 pressure was diminished [41]. Porous poly(L-lactic acid) (PLLA) was prepared using supercritical CO_2 as a nonsolvent. The cellular pores in the samples indicate the occurrence of a liquid–liquid demixing process, followed by crystallization of the polymer-rich phase [42]. In another study, poly(lactide) was dissolved in THF or chloroform and then cast onto a clean glass slide using a blade. The porous membrane was obtained after the phase inversion process by SC CO_2 [43]. Microporous PVDF membranes were prepared from PVDF/DMAA solutions. The membranes exhibited morphological characteristics resulting from both liquid–liquid phase separation and crystallization, that is, cellular pores surrounded by interlinked PVDF particulate crystallites [44]. Reverchon and Cardea investigated the preparations of PVDF-HFP [45]. The effect of pressure (from 80 to 200 bar) and temperature (from 35°C to 55°C) on the membrane formation was studied. In the pressure range of 80–200 bar, the pore size decreased when the

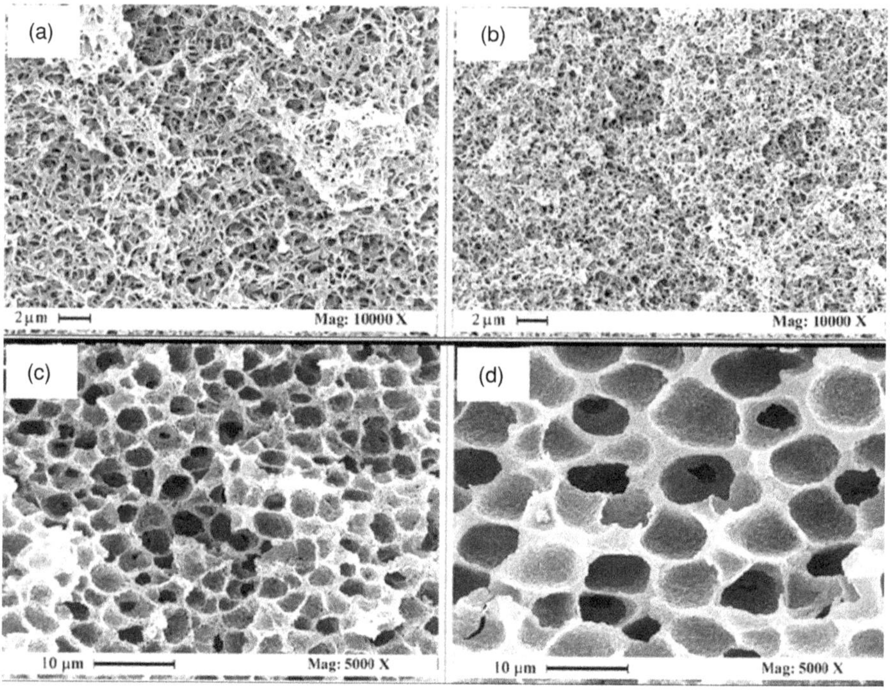

FIGURE 3.8 Effect of pressure on PVDF-HFP membrane morphology and cell size at 35°C and 15% w/w, operating at **(a)** 200, **(b)** 150, **(c)** 100, and **(d)** 80 bar. (Reprinted with permission from Reference 45. Copyright 2006 American Chemical Society.)

pressure increased, and the membrane structure changed correspondingly from cellular to bicontinuous (Fig. 3.8).

Because of the limited affinity between SC CO_2 and water, it was difficult to apply the SC CO_2-phase inversion process to aqueous solutions. To remove this limitation, Reverchon and Cardea proposed to use SC CO_2–expanded ethanol [46]. This process was applied to an aqueous PVA solution, and porous PVA membranes were successfully prepared.

3.3.4 SC CO_2 Foaming Process

3.3.4.1 Principle of the Foaming Process. The phase change of amorphous polymers in continuous processing is the glass-to-rubber transition characterized by a glass transition temperature T_g. The state above T_g is soft and called the rubbery or liquid state. Below T_g, the polymer is rigid and in a glassy state. It is known that the compressed CO_2 can lower the T_g of amorphous polymers, sometimes dramatically, which means the polymers can easily be plasticized. The SC CO_2 foaming process can be divided into two steps [47,48]: (1) The polymer is plasticized and saturated with CO_2. The temperature should be above T_g for amorphous polymers and melting point

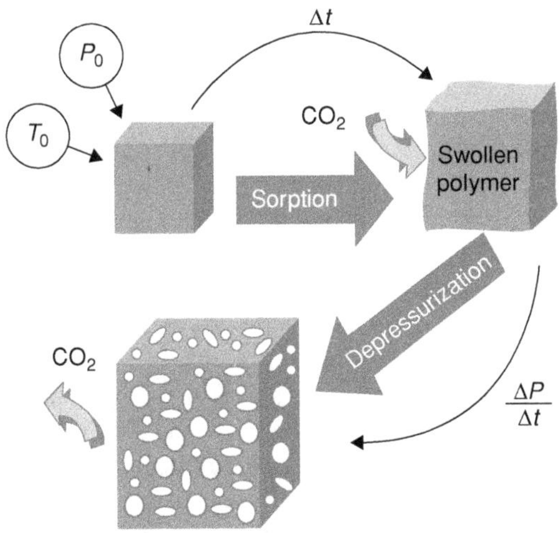

FIGURE 3.9 Schematic illustration of the SC CO_2 foaming process. P_0 and T_0 indicate the ambient pressure and temperature, respectively. (Reprinted with permission from Reference 48. Copyright 2008 Royal Society of Chemistry.)

(T_m) for crystalline polymers. (2) Cell nucleation is induced by a thermodynamic instability (usually a pressure decrease) and cell growth and coalescence. In this step, the polymer is normally expanded. Both steps, together with some important foaming parameters, are schematically shown in Fig. 3.9.

Once the polymer is saturated by SC CO_2, cell nucleation and growth are controlled by the pressure release rate and foaming temperature. If the temperature is below T_g, the polymer matrix is too rigid, and the foaming step will not occur. However, because SC CO_2 can lower the T_g of amorphous polymers, the foaming temperature can sometimes be much lower than their intrinsic T_g. For example, it was demonstrated that PMMA ($T_g = 80°C$) could be produced by SC CO_2 at temperatures as low as 40°C [49]. A foaming process can be carried out in a batch system or in a continuous way. For a batch process, the preshaped samples are placed in a high-pressure vessel and then saturated with SC CO_2. The important factors include CO_2 saturation time, final pressure, pressure release rate, and the temperature at the start of foaming and during the foaming process. A continuous foaming process using an extruder is regarded as more economically favorable because of the high productivity, easier control, and flexible product shaping [50]. In the extrusion process, the polymer is saturated with CO_2 and a single-phase solution is formed by screw rotation and in-line mixer. At the die of the extruder, the pressure is released and the nucleation and cell growth are initiated. The die geometry, pressure release rate, and temperature have a substantial effect on the pore morphology of the final product [48, 51].

For a CO_2 foaming process, it is generally suggested that the phase separation is caused by cell nucleation of a new gas phase from a metastable melt phase, requiring an activation energy barrier to be surmounted to induce the phase separation. Another possible mechanism is spinodal decomposition, which is the spontaneous phase separation from a thermodynamically unstable state. Concentration gradients, pressure gradients, or temperature gradients can drive nucleation. The classical nucleation theory is based on the Gibbs free energy required to create a void in liquid, resulting in a critical bubble. However, this theory does not apply to the SC CO_2 foaming process. Modeling of the nucleation and cell growth has been extensively investigated [48]. Taki carried out the experimental and numerical studies based on the batch foaming of a polypropylene (PP)/CO_2 system [52]. A simple kinetic model for the creation and expansion of bubbles was developed based on the model of Fumerfelt's group [53]. The calculated results of the number density of bubbles and bubble growth rate agreed well with experimental results.

In most foaming applications, either a pressure drop or a temperature increase is used to decrease the gas solubility and make the solution supersaturated. When clusters of gas molecules are greater than the critical size, the activation energy is overcome, and nucleation occurs. Additives such as talc or nano-clay are commonly used to adjust the nucleation rate, presumably enhancing the rate by decreasing the activation barrier. Marrazzo et al. investigated the effect of various nucleating agents (talc, TiO_2 nanoparticles, alumina powders, exfoliated and intercalated clays, and carbon nanotubes) on the foaming of poly(ε-caprolactone) (PCL) in a batch process [54]. Foam densities and morphologies, in terms of number of cells per initial unit volume, were measured and compared.

3.3.4.2 Biodegradable Polymer Foams. The SC CO_2 foam process has been extensively used in fabricating biodegradable polymer foams [55]. There are two reasons for this: (1) SC CO_2 can plasticize the amorphous polymers and lower their T_g. Therefore, the foaming can be performed at lower temperatures, which in principle can reduce operation cost and prevent the polymer from degrading during processing. (2) SC CO_2 is released simply by reducing the pressure. The process is simple and involves no organic solvents. Traces of CO_2 trapped in the polymer foams have minimum impact, as CO_2 is nontoxic and nonflammable.

Among the biodegradable polymer foams prepared using the SC CO_2 foaming process, a large percentage are composed of PLA and poly(lactide-*co*-glycolide) (PLGA). For example, porous 85/15 PLGA foams were produced by the pressure quench method using supercritical CO_2 as the foaming agent [56]. The foaming experiments were carried out at 35°C and 40°C with CO_2 pressure at 10, 14, 15, and 20 MPa for each temperature. Porous PLGA foams were generated with relative densities 0.107–0.232, pore sizes 30–100 μm, and an interconnected porosity of 89%. However, this process was difficult to control with respect to changes in porosity and pore interconnectivity. A depth investigation of polymer CO_2 interaction was carried out by studying CO_2 sorption and swelling isotherms at 35°C and up to 200 bar on a variety of homopolymers and copolymers of lactic and glycolic acids [57]. Figure 3.10 shows the sorption isotherms (quantity of CO_2 absorbed per gram of polymer) of various polymers at 35.0°C as a function of pressure. The sorption increased monotonically with pressure. Moreover, a significant change of slope appeared at a pressure of about 75 bar, which corresponded to the transition

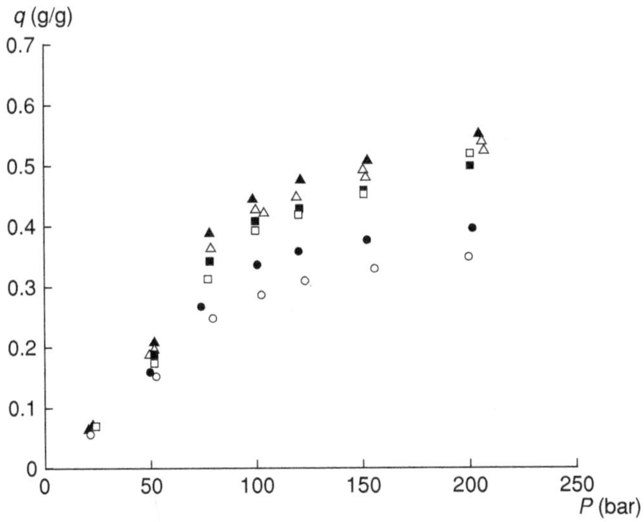

FIGURE 3.10 Sorption isotherms at 35.0°C as a function of pressure for PLA15k (MW 15k, ▲), PLA52k (MW 52k, △), PLGA8515 (85 mol% lactic acid, 15 mol% glycolic acid, MW 77k, ■), PLGA7525 (75 mol% lactic acid, 25 mol% glycolic acid, MW 72k, □), PLGA6535 (65 mol% lactic acid, 35 mol% glycolic acid, MW 52k, •), and PLGA5050 (50 mol% lactic acid, 50 mol% glycolic acid, MW 53k, ○). (Reprinted with permission from Reference 57. Copyright 2008 Wiley Periodicals, Inc.)

of CO_2-rich phase from gas to the denser supercritical state. The Sanchez–Lacombe (SL) equation of state was used to predict the sorption isotherms and then compared with the experimental results [57]. NaCl particle leaching was combined with CO_2 foaming to prepare porous PLGA [58]. PLGA disks and NaCl particles were compression molded at room temperature and subsequently allowed to equilibrate with compressed CO_2. The NaCl particles were leached to yield macropores after the foaming process. PLLA was foamed using SC CO_2 as a porogen and was crystallized under the experimental conditions [59]. The study also addressed the impregnation of biopolymers with an active agent dispersed in SC CO_2. The drug used for impregnation was triflusal, a platelet antiaggregant inhibitor for thrombogenic cardiovascular diseases [59]. Two formulations of PLGA sponges with different copolymer compositions (85:15 and 50:50) were fabricated as novel scaffolds to guide human hepatoma (Hep3B) cell growth in vitro. The PLGA sponges showed desirable biodegradability and exhibited uniform pore size distribution with moderate interconnectivity [60].

Biodegradable PCL foam was prepared from the system of CO_2–molten PCL. CO_2 diffusion and equilibrium solubility in molten PCL in contact with CO_2 were investigated. The nucleation and growth of the bubbles were believed to be related to the sorption equilibrium and mass transport kinetics of the CO_2 in the molten polymer, as well as to the rheological and surface properties of the mixture surrounding the bubble itself. This study focused upon interpreting the effects of the kinetics

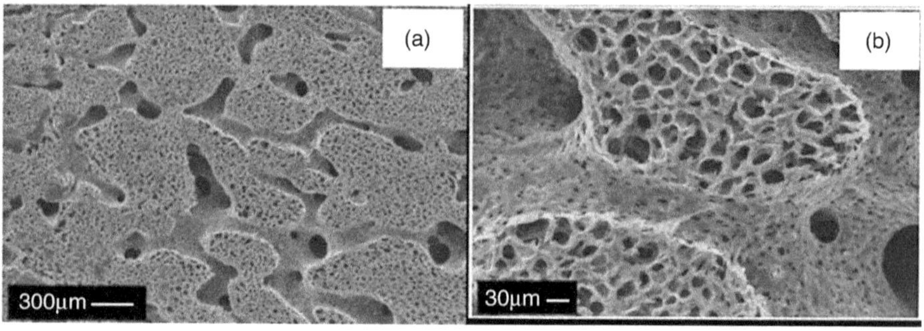

FIGURE 3.11 SEM micrographs of bimodal porous PCL scaffold at different magnification, prepared by a gas foaming/polymer extraction technique from PCL/gelatin (60/40 in mass) blend foamed with 80–20 vol% N_2–CO_2 blowing mixture at the saturation pressure 180 bar, the pressure drop rate 700 bar/s, and the foaming temperature 44°C. (Reprinted with permission from Reference 63. Copyright 2007 Wiley Periodicals, Inc.)

and equilibrium of sorption of CO_2 in PCL. The data were analyzed using the SL lattice model and the Peng–Robinson equation of state (PR-EOS) [61]. The porous structure of PCL foams was characterized systematically through mercury porosimetry, helium pycnometry, scanning electron microscopy (SEM), and X-ray microtomography observations coupled with image analysis [62]. Bimodal porous PCL was prepared with CO_2 and N_2 as foaming agents [63]. This foaming process was combined with a selective polymer extraction technique. PCL was melt mixed with thermoplastic gelatin. The foaming temperature was changed from 38°C to 100°C, with N_2 amount ranging from 0 to 80 vol%. Gelatin was removed by soaking in water after the foaming process. A bimodal porous PCL was produced with the pore sizes in the region of 300 and 30 μm (Fig. 3.11). The foaming behavior of PCL with nitrogen as the only foaming agent was also investigated using a uniquely designed and constructed batch foaming apparatus [64].

Other types of biodegradable polymer foams were also prepared. For example, starch-based foams were prepared using a supercritical fluid extrusion (SCFE) process [65]. Pregelatinized corn starch was extruded with different concentrations of whey protein and different CO_2 injection rates. The formation of polycarbonate foams was investigated [66]. The conditions leading to the formation of cells and to the onset of crystallization of polycarbonate were examined.

3.3.4.3 Other Polymer Foams. In addition to biodegradable polymer foams, the SC CO_2 foaming process has been employed to produce foams from a wide range of polymers. For example, polymer foams of PS [67–69], PMMA [49, 59], PP [51, 68, 70], cellulose acetate [69], cellulose acetate butyrate [71], PVDF [72], and poly(ethylene terephthalate) film [73] were prepared with SC CO_2 as the foaming agent. The foam of the copolymer poly(styrene–*co*-methyl methacrylate) generated with SC CO_2 was investigated [74]. The effect of temperature with a rapid depressurization rate (<1 s) was studied in terms of foam density versus the temperature.

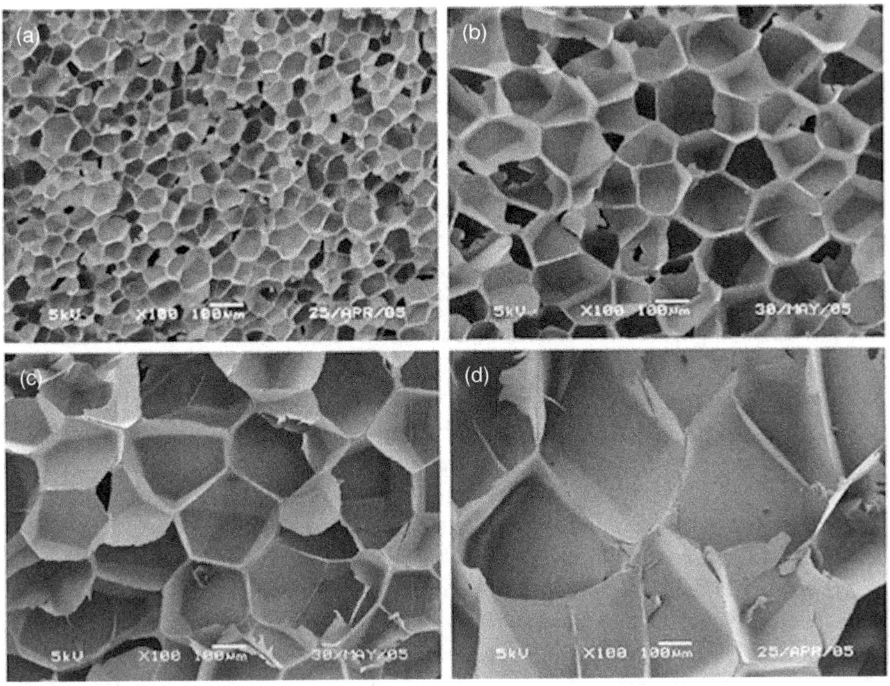

FIGURE 3.12 Porous structure of the polymer foams formed at 200 bar and 118°C with depressurization time of **(a)** <1, **(b)** 35, **(c)** 100, and **(d)** 250 s. (Reprinted with permission from Reference 74. Copyright 2007 Elsevier.)

At 93°C the density of the produced foams was much higher than the density of the foams produced at 118°C, even though the solubility of CO_2 at 90°C is higher. This indicated a premature stop in cell growth. To determine the effect of the depressurization rate on the foam morphology, the time to depressurize the content of the high-pressure vessel was varied from <1 up to 600 s. Figure 3.12 shows the pore structure of the materials produced with depressurization time of <1, 35, 100, and 250 s. It can be seen that the pore size of the polymer foams increased with increasing depressurization rate. A dramatic increase of the cell size up to approximately 400 μm is observed.

Open-cell foams of thermoplastic polymers such as PS, polycarbonate, and high-melt-strength polypropylene (HMS-PP)/metallocene polyethylene (mPE) were produced using a SC CO_2–based extrusion process [68,75]. Cell interconnection could be promoted by inducing (1) a nonhomogeneous melt structure by crosslinking, polymer blending, or filler compounding, (2) cell wall thinning by a high volume expansion ratio while maintaining soft cell walls, (3) cell wall thinning by a high cell-population density, and (4) plasticization of the soft region of the cell walls with a secondary blowing agent. The significant temperature difference between the core and surface of the foam extrudate was induced by a surface cooling method [68].

3.3.4.4 Composite Foams. Composite foams can be produced by blending different polymers or mixing polymers and ceramics and then foaming using SC CO_2. Porous blends of styrene–isoprene–styrene copolymer elastomer (SIS) and tetrahydrofurfuryl methacrylate (THFMA) at three SIS/THFMA compositions were foamed and characterized in terms of their morphological, mechanical, and biological properties [76]. The results showed that the foaming factors were dependent on blend composition and the foaming conditions, demonstrating that the process was tunable. Microcellular foaming of biodegradable and biocompatible PLA/starch composites in supercritical/compressed CO_2 was studied [77]. Sorption and desorption experiments were performed to study the CO_2 uptake rate and equilibrium concentration in PLA/starch composites. The effects of a series of variable factors, such as saturation time and saturation temperature, on the foaming morphology were studied by SEM and density measurements [77]. A series of PP/PS blends were prepared by solvent blending with PS-grafted PP copolymers (PP-*g*-PS) having different PS graft chain length as compatibilizers. The blends were foamed by using a pressure-quenching process and supercritical CO_2 as the blowing agent [78]. Combined with the increased interfacial area, the compatibilized interface led to foams with increased cell density compared to an uncompatibilized control. The increase in interfacial compatibility also decreased the gas escape, resulted in the retention of more gas for cell growth, and facilitated the increase in expansion ratio of PP/PS blend foams [78]. Highly porous (>85%) and well-interconnected scaffolds were obtained in which the size, distribution, and number of pores could be controlled [79]. The polymer used was a blend of poly(ethyl methacrylate) and tetrahydrofurfuryl methacrylate. Under the foaming conditions (saturation for 8 h at 100 bar and 40°C), the effect of the vent time on porous structure was investigated. With the increase of vent time, the pore size was increased [79]. This was confirmed by the SEM imaging and micro X-ray computed tomography (micro-CT) (Fig. 3.13).

The incorporation of ceramic particles into porous polymer can modify the porous structure and enhance the material's mechanical stability. The presence of ceramic

(a) (b)

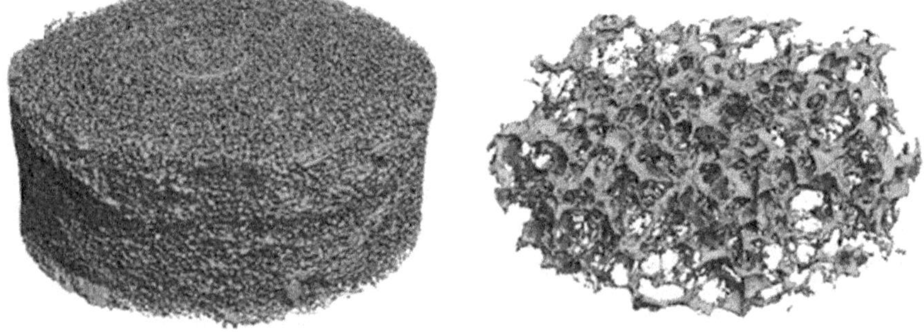

FIGURE 3.13 Micro-CT images of the scaffolds with vent time (**a**) 30 s and (**b**) 60 min. Pore size increases with the increase of vent time. (Reprinted with permission from Reference 79. Copyright 2006 Springer.)

nanoparticles and nano-clay can also adjust the cell nucleation rate [50, 54]. Composites of PCL with different particles were prepared [54]. Intercalated polycarbonate/clay nanocomposites were fabricated using supercritical CO_2 as a foaming agent [80]. The amount of clay, foaming temperature, and pressure were varied in a batch process. A composite PLA–hydroxyapatite (HAP) biomaterial was prepared by in situ polymerization of D,L-lactide monomer and HAP. The porous composites were then produced using the CO_2 foaming process and NaCl particle leaching [81]. A bioresorbable PLLA, two ceramic HAP powders (nanoscale), and β-tricalcium phosphate (micrometer scale) were used in another study [82]. The ceramic particles were dispersed in the PLLA matrix by melt extrusion using a microcompounder with two conical corotating screws. The composite foams were formed in a continuous manner. The effects of foaming parameters (P_{sat}, dP/dt, dT/dt), filler type, and content on foam properties were evaluated.

3.3.4.5 Nanoporous Foams. In this chapter, the pore size of the nanoporous foams is defined as less than 1 μm. Many studies on creating high-performance polymeric foams have focused on increasing the number of cells and decreasing the cell size. The typical cell size of SC CO_2–processed foams ranges from a few micrometers to millimeters, depending on CO_2 pressure, depressurizing rate, temperature, solubility or swellability of the polymer, and other processing parameters. Nuclei of CO_2 in polymeric materials, in general, have a minimum diameter that is thermodynamically determined. Such minimum diameter is at the order of micrometers or just slightly below. However, thermodynamics does not favor small nuclei, which leads to a huge increment of the interfacial area and interfacial free energy. Nanoparticles may serve as nucleation agents and enhance the generation of small bubbles. A small amount of clay nanoparticles was added to PMMA and PS during the SC CO_2 foaming process in order to produce nanocomposite foams with improved properties [83]. A small cell size of 1.7 μm and a high cell density of 1.51×10^{11} cells/cm^3 were reached. A surface-constrained foaming process was developed to prepare nanoporous films [84, 85]. The polymer film was sandwiched in a foaming die. During the depressurization process, CO_2 could only escape from the film edges. This slowed the rate of CO_2 diffusion from the film and reduced the cell coalescence. The pore size of PMMA films could be systematically reduced to less than 100 nm [84]. This surface-mediated foaming process was further explored by the addition of nanoparticles at lower temperatures [85].

Recently, block copolymers were employed to prepare nanoporous foams. Blocks of certain block copolymers (see Chapter 2) can self-assemble into spherical domains that could act as nuclei for nanosize pore formation. These spherical domains could have a good affinity with CO_2 and the ability to absorb CO_2. A polystyrene–block–poly(perfluorooctylethyl methacrylate) (PS-*b*-PFMA) diblock copolymer was synthesized and then used as a template to fabricate polymeric nanocellular monoliths [86]. Films of PS-PFMA were cast from solution and dried under vacuum at room temperature. The film was then placed in a high-pressure vessel and annealed at 60°C for 1 h, and the temperature was then reduced to 0°C. It was essential to reduce the temperature and release CO_2 at a rate of 0.5 MPa/min.

Films were prepared at different CO_2 saturation pressures. Figure 3.14 compares the porous structure of the films prepared by the CO_2 foaming process. The structures were imaged by scanning force microscopy (SFM) and SEM. Reactive ion etching was employed to expose the microstructure of the films. The average diameter of the cells could be tuned by employing different saturation pressures in the range 10–30 nm [86]. This system was further investigated by focusing on crossover from microcellular to nanocellular structures with a controlled depressurization temperature [87]. PS-*b*-PMMA was also processed to prepare porous films with CO_2 localized in the PMMA domains of the film. Ordered nanowells with diameters of approximately 40 nm and depths of 1–2 nm were formed [88].

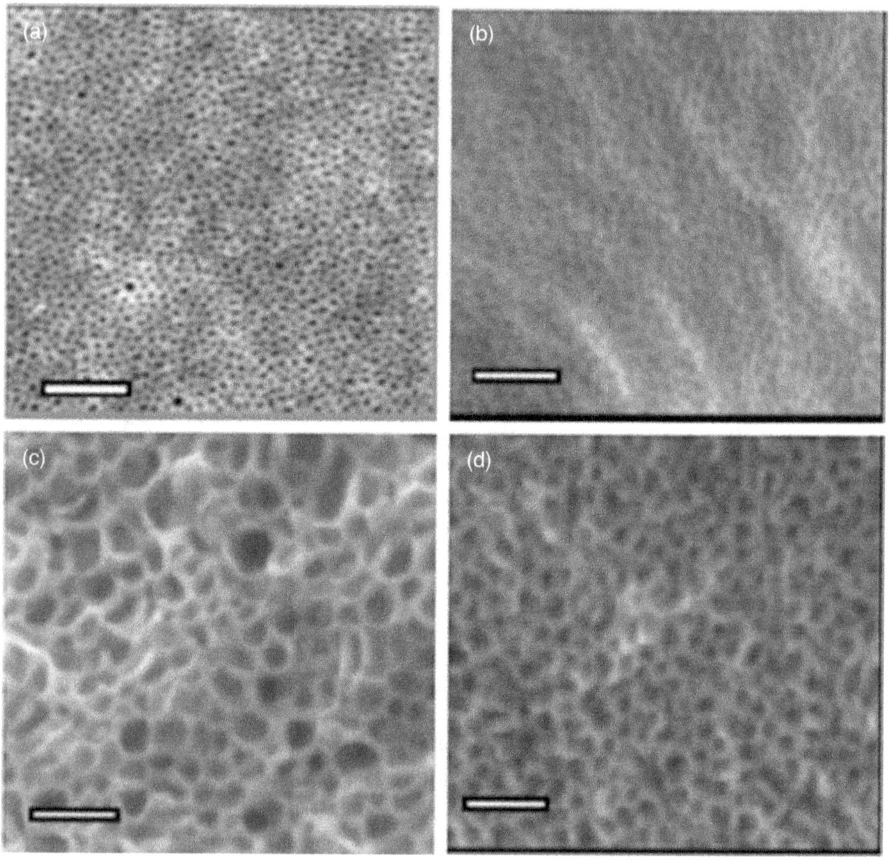

FIGURE 3.14 Nanoporous films of block copolymer PS-*b*-PFMA by SC CO_2 foaming process. **(a)** An SFM topographic image of as-cast film after etching to 30 nm prepared by reactive ion etching with CF_2, scale bar 250 nm. **(b–d)** SEM images of the cryofractured surface of the films processed in a CO_2 saturation pressure of **(b)** 10, **(c)** 20, and **(d)** 30 MPa. Scale bars 100 nm. (Reprinted with permission from Reference 86. Copyright 2004 Wiley Periodicals, Inc.)

Blended polymers were also investigated as a means to form nanoporous foams. A PS/PMMA polymer blend foam was prepared, where bubble nucleation was localized in the PMMA domains [89]. A higher depressurization rate at lower foaming temperature reduced the bubble diameter and increased the bubble density. The nanoporous films with void diameter of 40–50 nm and pore density of 8.5×10^{14} cm^{-3} were obtained [89]. Plastic foams with nanoscale-to-microscale cellular structures were prepared from a poly(propylene) (PP)/propylene–ethylene copolymer (PER) blend by controlling bubble nucleation sites and restricting bubble growth to PER domains [90]. The bubble size and location were highly controlled in disperse PER domains by exploiting the differences in CO_2 solubility and viscoelasticity between the PER domains and the PP matrix. The average cell diameter of PP/PER blend foams can be controlled within 0.5–2 μm [90].

3.4 FREEZE-DRYING

3.4.1 The Principle

Freeze-drying is a process by which a solution is frozen in or in contact with a cold bath and then placed in a freeze-dryer to remove the frozen solvent under vacuum [6]. The frozen solvent acts as a porogen. The porous structure is generated after removing the porogen by sublimation under vacuum, as illustrated in Fig. 3.15. This is somewhat like the method of particle leaching in which the particles are incorporated in a polymer solution (see Chapter 4). For the particle-templating method, the particles are removed by washing or chemical reaction after the composite material is solidified. The key for the success of freeze-drying is that the samples have to remain frozen throughout the procedure to lock in the structure. Alternatively, the solvent sublimation rate must be faster than the melting rate of the frozen solvent.

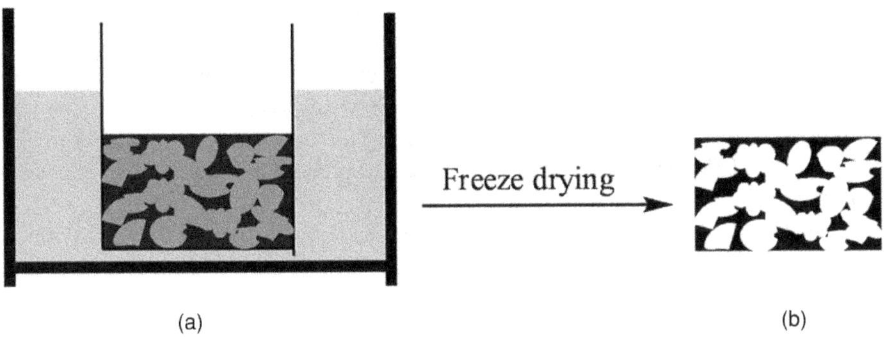

FIGURE 3.15 Freeze-drying for the preparation of porous materials. **(a)** A solution in a vial is first frozen in a cold bath and **(b)** the porous material is obtained after freeze-drying to remove the porogen via sublimation.

In the latter case, once a part of frozen solvent turns to liquid, the liquid is removed immediately under vacuum so that the porous structure is retained.

Freeze-drying has received increasing attention since the first discovery of delicate materials that could be dried from their frozen states under vacuum [91]. It has been widely used to prepare scaffolds for tissue engineering, drug delivery, nanoparticle production, pharmaceuticals, and porous materials [92–100]. Water is an environmental friendly solvent, especially for biological molecules. Thus, ice crystals used as porogens are believed to be safe, and there are no issues of toxicity or pollution related to the process and the final product. In addition, water is very cheap, and there are large amounts of water available for use. Therefore, it is not surprising that many studies have focused on aqueous solutions to make porous structures by freeze-drying [100]. Acidic or basic aqueous solutions and organic polymer solutions can also be processed to make porous scaffolds for tissue engineering because these scaffolds need to be insoluble in an aqueous medium under normal pH [101, 102]. When removing the porogen, freeze-drying does not bring impurities into the samples, and therefore a further purifying process is not necessary [103]. In this respect, freeze-drying can be regarded as a simple and efficient method for making porous materials.

A freeze-drying process consists of three stages: freezing, primary drying, and secondary drying [98, 99]. Freezing is a very important stage, and this directly influences the pore morphology of the prepared materials. Ice crystals can be finely tuned by varying the freezing conditions, including freezing temperature, freezing rate, freezing direction, and solvent and polymer compositions. Many porous materials, such as inorganic, organic, and inorganic–organic hybrid materials with different pore structures, have been produced by freeze-drying [100, 104, 105]. The freezing temperature has a major effect on pore morphology and pore size. When samples are frozen in liquid nitrogen, the extremely low temperature (−196°C) results in rapid formation of ice nuclei and growth of small ice crystals and produces small pores after freeze-drying. If a solution is frozen at a higher temperature—for example, in a freezer at −20°C—ice nucleation is slow, and the nuclei tend to grow into larger ice crystals. The freezing process at relatively high temperature generally produces materials with large, amorphous pores. In addition, the freezing direction is an important factor affecting the morphology of the porous materials. Orientation of ice crystals can be adjusted by controlling the freezing direction. A directional freezing process was developed to orient the growth of ice crystals. A unique pore structure—"aligned porous structure"—was produced after removing the oriented ice crystals by freeze-drying [104, 105]. This method has been used to prepare a wide range of materials with an aligned porous structure.

Primary drying is the step in which the frozen solvent (mainly ice) is sublimed. This sublimation can be achieved by reducing the pressure to a value below the triple point. This step is usually the most time-consuming step in a freeze-drying process. An increase in product temperature during primary drying would greatly reduce the duration of primary drying, provided that the vacuum in the chamber is high enough to remove the solvent efficiently. The purpose of the secondary drying is to desorb the solvent bound to the polymer that is not frozen. The drying temperature can be higher than the freezing point of the solvent but has to be below the glass transition temperature of the matrix. The vacuum level should be even lower than that of the primary drying step [98]. The drying process is usually carried out in a freeze-drier. As a general practice in making porous materials, the samples are kept frozen on a temperature-controlled shelf or in a cold chamber. This means that the freeze-drying temperature should be below the glass transition temperature or the melting point of the frozen sample. This temperature may be much lower than the melting point of the solvent used because the dissolution of a solute in the solvent may decrease the melting point of the mixture dramatically—for example, inorganic salts dissolved in water. Different freeze-drying temperatures and sublimation rates may be selected, which will affect the pore structure in the final products. Although it is possible to use a drying temperature higher than the melting point of the frozen samples, a high solvent removal rate needs to be maintained and the freeze-drying process needs to be closely monitored.

3.4.2 Porous Polymers by Freeze-Drying

This subsection focuses on porous polymers with random porous structures. The materials with ordered pores such as aligned pores are described in Section 3.4.3. Typically, a solution contained in a vessel is immersed into a cold bath or placed in a cold chamber. Ice nuclei are formed, and the ice crystals grow in a random way. However, when a high temperature gradient exists, the growth of ice crystals may be oriented to some extent or in a small area. This is normally the case when a solution is frozen in liquid nitrogen. When a low temperature gradient is employed—for example, freezing a solution in a freezer at –20°C—disordered ice crystals are formed. After freeze-drying, this leads to disordered pore structures.

Aqueous polymer solutions have been frozen and freeze-dried to prepare hydrophilic porous polymeric structures. A range of hydrophilic polymers, including PVA [106, 107], have been used. However, many studies have employed polymers such as poly(vinyl pyrrolidone) (PVP), dextran, alginate, and sugars as excipients or cryoprotectants for active pharmaceutical ingredients (APIs) or proteins [108–110]. Other types of polymers, such as chitosan, chitin, and collagen, have been used to prepare porous scaffolds for tissue engineering. For example, porous chitosan was prepared by controlled freezing and freeze-drying of chitosan solutions and gels [101]. Chitosan solutions were prepared by dissolving in 0.2 M acetic acid. A 2 wt% chitosan solution was frozen in a cylindrical glass tube by immersion into liquid nitrogen. This solution was also frozen in a shallow PS dish by contact with dry ice. Indeed, by freezing the solutions at different temperatures, it was found that the mean pore diameter was increased with the increase of freezing temperature (Fig. 3.16). This porous chitosan could be formed in different shapes, including porous membranes, blocks, tubes, and beads. Mean pore diameters of the materials could be controlled within the range 1–250 μm by varying the freezing conditions [101].

Freeze-drying has been a commonly employed technique for the fabrication of scaffolds for tissue engineering and drug delivery. Glycosaminogylcan (an anionic polysaccharide of the extracellular matrix) was covalently immobilized on chitosan

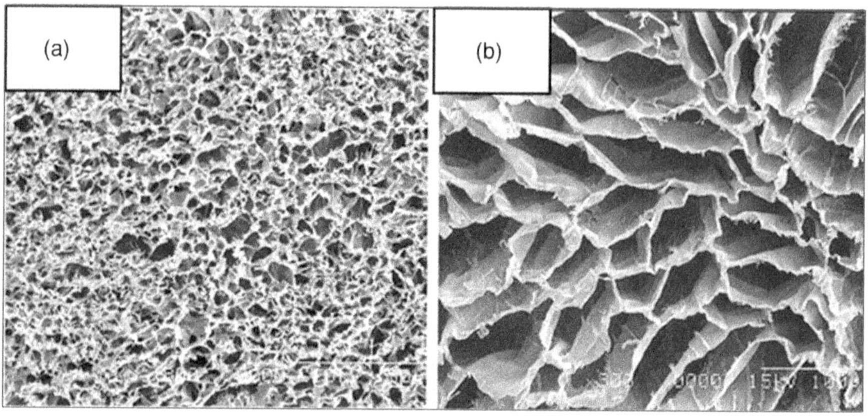

FIGURE 3.16 SEM images of porous chitosan by freeze-drying. **(a)** The cylinder bulk scaffold made from 2 wt% chitosan solution, which was frozen in liquid nitrogen. **(b)** The planar scaffold by contacting 2 wt% chitosan solution in a dish with dry ice and then freeze-drying. (Reprinted with permission from Reference 101. Copyright 1999 Elsevier.)

scaffolds prepared by freeze-drying to form a 3D network. These composite scaffolds were then used for the expansion of human cord blood stem cells in perfusion culture [111]. The spatial spreading and proliferation of endothelial cells and mouse embryonic fibroblasts was investigated on porous chitosan scaffolds [112]. Sodium alginate was dissolved in 1 N NaOH aqueous solution. It was mixed with a chitosan acidic solution at room temperature. The resulting solution was heated and maintained at 70°C, and then the pH was adjusted to 7–7.4. This solution was frozen and freeze-dried to produce chitosan–alginate scaffolds [113]. In another study, silk was boiled for 1 h in an aqueous solution of 0.5 wt% Na_2CO_3 and rinsed thoroughly with water to extract the sericin proteins. The degummed silk was dissolved in $CaCl_2/H_2O/CH_3CH_2OH$ solution at 80°C, and the fibroin solution was obtained after filtering and dialyzing. The method of freeze-drying was then employed to prepare 3D fibroin scaffolds [114].

Aqueous collagen suspensions were processed to prepare porous collagen scaffolds [115, 116]. The collagen extracted from jellyfish could be dissolved in a 0.5 M acetic acid solution at a concentration of 1.5 and 2.0 wt%. This solution was mixed with hyaluronic acid (HA) solution. A novel overrun process was employed to prepare a uniform dual-pore structure of collagen/HA hybrid scaffolds [117]. The same type of collagen solution was processed to form collagen scaffolds. Fibrous PLGA layers were then deposited on the collagen scaffolds by electrospinning. The electrospun fibrous PLGA layer on the surface of a porous tubular collagen scaffold improved the mechanical strength of the collagen scaffolds in both dry and wet states [118].

PCL solutions were prepared by dissolving PCL in THF at 50°C for more than 2 h. The solutions were frozen and freeze-dried at −80°C. Effects of the polymer concentrations on the scaffold properties, such as morphology, porosity, mechanical stability, and degradability, were investigated [119]. Freeze-drying was combined with sugar or salt particle leaching to produce PCL and PLA scaffolds. Polymer solutions were prepared by dissolution in 1,4-dioxane and then added with 92% w/w of particles. Subsequently, the polymer solutions with particles were frozen at −6°C or −25°C. The solvent crystals were removed by freeze-drying for 2—3 days [120]. A 3D nanofibrous PLLA scaffold was fabricated with gelatin microspheres of diameter 250–425 μm. Gelatin microspheres were placed in a Teflon mold and treated with a saturated water vapor atmosphere at 37°C. The PLLA solution in a mixture of water and THF was cast onto the microspheres and was frozen at −76°C after a desired penetration time. The samples were then immersed in cyclohexane for solvent exchange and then frozen at −18°C again before freeze-drying in an ice-salt bath for 5 days [121]. In a previous study, the authors found that nanoscale fibrous PLLA could be fabricated by a procedure involving thermally induced gelation, solvent exchange, and freeze-drying. Figure 3.17 shows one SEM image of the nanofibrous scaffolds. The effects of polymer concentration, thermal annealing, solvent exchange, and freezing temperature on the nanoscale structures were studied. In general, at high gelation temperature, a platelet-like structure was formed. At low gelation temperature, the nanofibrous structure was formed [122].

For the application of bone tissue engineering, the scaffolding materials should be osteoconductive so that osteoblasts and osteoprogenitor cells can adhere, migrate, differentiate, and synthesize new bone matrix. Hydroxyapatite [HAP, $Ca_{10}(PO_4)_6(OH)_2$] makes up the inorganic part of natural hard tissues such as bones. The incorporation of HAP into polymeric scaffolds could enhance the mechanical strength of the

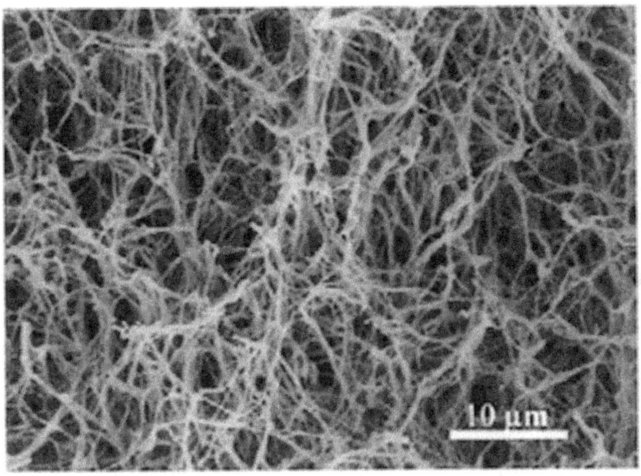

FIGURE 3.17 PLLA nanofibers produced from a 5% w/v PLLA/THF solution at a gelation temperature of 15°C. (Reprinted with permission from Reference 122. Copyright 199 Wiley Periodicals, Inc.)

scaffolds and, more important, provide excellent bioactivity and osteoconductivity, which could lead to potential use in hard-tissue substitutes and regeneratives. HAP powders (platelets ranging from 10 to 100 μm in size) were dispersed in the PLLA–dioxane solution. The mixture was cooled to −18°C to induce solid–liquid phase separation. The phase-separated samples were freeze-dried under vacuum at temperatures between −5°C and −10°C for 7 days to produce the PLLA/HAP hybrid scaffold [123]. HAP particles are highly brittle and stiff; therefore the availability of the particle shape and size is limited. A biomimetic synthetic approach was employed to synthesize HAP in the presence of gelatin [123]. $(NH_3)_2HPO_4$ (P precursor) and $Ca(NO_3)_2{\cdot}4H_2O$ (Ca precursor) were dissolved in a gelatin solution, respectively. The P-precursor gelatin sol was added to the Ca-precursor gelatin sol dropwise at pH 10.0–10.6. The resulting sol was quenched in a freezer at −20°C for 24 h and then freeze-dried for 72 h. The porous scaffolds were then crosslinked and loaded with an antibiotic drug. The release of the drug was evaluated [124]. This approach was also used to synthesize HAP in the presence of the hydrophilic polymer PVA. The hybrid nanocomposites were produced by the process of freezing-thawing and freeze-drying [125].

3.4.3 Aligned Porous Polymers by Directional Freezing

The freezing process can be carried out in a more controlled manner to orient the growth of ice crystals in one direction. A directional freezing process has been developed to orient the ice growth [6, 126]. The removal of oriented ice crystals can generate porous materials with aligned structures. In general, aligned porous materials can be produced by microfabrication, soft lithography, or photolithography. However, the process of directional freezing may produce a wide range of materials with facile control of the pore dimensions and is easy to scale up. For a directional freezing process, a solution in a container or on a substrate can be subjected to a temperature gradient so that the ice crystals can grow from the end of low temperature to the end of high temperature. This can be achieved, for example, by simply lowering a vessel containing the liquid sample into a cold bath (e.g., liquid nitrogen) or placing the substrate between two temperature-controlled plates. A representative scheme is shown in Fig. 3.18. The solvent (in most cases, water) is frozen from the bottom, and the ice crystals grow upward. Around the freezing front, the solution or suspension is concentrated. The growing ice excludes polymeric molecules or particles below a critical freezing rate. As an example, Fig. 3.18b shows an optical image for the directional freezing of gold nanoparticles (size 15 nm). The aligned pattern is clearly seen, which can be transferred into an aligned porous structure after freeze-drying.

Because water is known to form ice crystals upon freezing and the size of the crystals can be adjusted by varying the freezing temperature and rate [127], a water-based sol was first employed to prepare aligned silica fibers by directional freezing [128]. Porous PVA with aligned structure was prepared by the directional freezing of 5 wt% aqueous PVA solution and then freeze-drying (Fig. 3.19a). The spacing between the aligned walls could be tuned by varying the freezing rate. It was found that the width of the aligned channels could be varied from 12 to 50 μm over a range

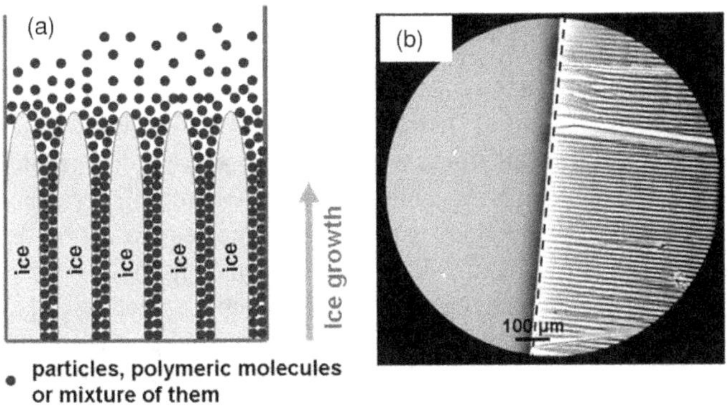

FIGURE 3.18 (**a**) A scheme showing the directional freezing process. (**b**) An optical micrograph showing the directional freezing of a suspension of gold nanoparticles. The dark stripes are aggregates of gold nanoparticles. The white areas are ice, which excludes the gold nanoparticles. The gold nanoparticles are concentrated close to the ice front (*dashed straight line*). (Reprinted with permission from Reference 6. Copyright 2007 Wiley Periodicals, Inc.).

of freezing rates of 10–100 μm/s [126]. This method was then extended to prepare hydrophobic aligned porous structures. For example, a PCL solution was directionally frozen in dichloroethane followed by freeze-drying to produce a porous 3D monolith consisting of closely packed aligned tubular pores of approximately 10–15 μm in diameter (Fig. 3.19b). Honeycomb monolithic PLLA with aligned structure was also successfully prepared. Dehydrated 1,4-dioxane was used as a solvent to dissolve

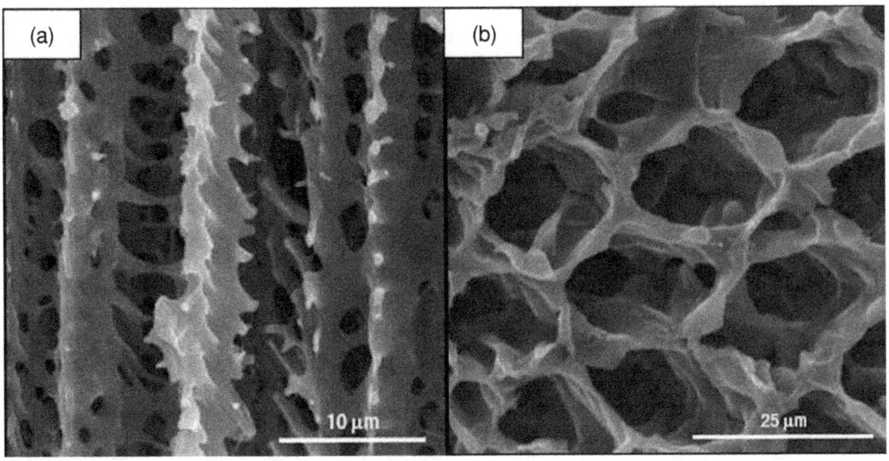

FIGURE 3.19 Aligned porous polymeric structures prepared by directional freezing and freeze-drying: (**a**) aligned porous PVA; (**b**) aligned porous PCL. (Reprinted with permission from Reference 126, copyright 2005 Nature Publishing Group.)

PLLA, and the single-phase mixture was directionally frozen by lowering the sample tube into liquid nitrogen at a constant rate. The porous structure was produced by freeze-drying at −5°C for 4 days [129].

Preparation of aligned porous PVA and its use for drug delivery were investigated by Gutiérrez et al. [130]. The pore sizes were reduced when high-molecular-weight PVA was used. For PVA with the same molecular weight, higher concentrations in the aqueous solutions led to smaller pore sizes. The release of the drug molecule ciprofloxacin was demonstrated with adjustable periods from tens of minutes to several days, corresponding to the porous structure of the PVA scaffolds [130]. Agarose scaffolds with uniaxial channels containing recombinant brain-derived neurotrophic factor (BDNF) protein were prepared and tested in an adult rat model of spinal cord injury. These scaffolds were found to be well integrated with host tissue, individual channels were penetrated by cells, and axons grew through scaffolds in a strikingly linear fashion [131]. Furthermore, the regeneration was significantly augmented by the incorporation of BDNF protein into the walls and lumen of the scaffold [131]. A basic collagen suspension was used to prepare porous collagen scaffolds with aligned structure. Acetic acid or ethanol was added to the collagen suspensions and homogenized at room temperature. The collagen suspension was solidified in a gradient freezing stage. After solidification, the samples were stored for at least 24 h in a freezer at −80°C and subsequently dried under vacuum for 48 h in a freeze-dryer [132].

The directional freezing process has also been widely used to produce composites or inorganic materials with aligned pore structures. In many cases, the hydrophilic polymer PVA was used as a structuring agent to template the formation of aligned structures. For example, PVA solution with silica nanoparticles [126], PVA solution with a silica sol [133], PVA with HAP slurry [134], and chitosan with carbon nanotubes [135] were investigated to prepare polymer/inorganic composites with aligned structures by directional freezing. The aligned porous PVA scaffolds were also used as templates to prepare composite materials. For example, the composites of polypyrrole (PPy) and PVA with 3D aligned microstructures were fabricated via vapor deposition polymerization of pyrrole onto the scaffolds of PVA and $FeCl_3$. The apparent conductivity of the composite containing 20 wt% PPy was measured to be approximately 0.1 S/cm. The ammonia gas sensor based on this 3D composite exhibited high sensitivity [136]. A range of porous ceramics was produced by the method of freeze-casting [137]. It should be noted that aqueous polymer colloidal suspensions could also be employed to produce aligned structures or aligned microwires. For example, PS colloids with negative surface charges and average diameter 450 nm were directionally frozen. Aligned microwires consisting of assemblies of PS colloids were formed after freeze-drying. Of interest, these microwires could be heated in an oven at 150°C. The PS colloids were fused together to generate smooth microwires [138]. Shi et al. synthesized the core-shell nanoparticles silica@poly(*N*-isopropylacrylamide) and silica@polyacrylonitrile, which were then used to produce microgel fibers [139], porous carbons, and other types of microfibers [140] by the method of ice templating.

So far, water and organic solvents have been employed as the solvents to produce porous structures by freeze-drying. In the previous section, supercritical CO_2 was used as a porogen or foaming agent to generate various porous structures. When CO_2

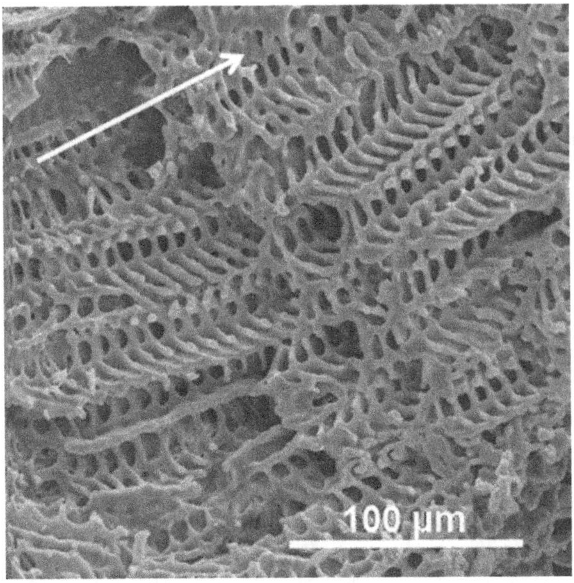

FIGURE 3.20 Aligned porous BGAL produced by directional freezing of a liquid CO_2 solution. The arrow represents the approximate direction of freezing. (Reprinted with permission from Reference 141. Copyright 2005 American Chemical Society.)

is compressed to liquid at room temperature, it may be used to dissolve polymers or organic compounds. The freezing process may be applied to the compressed CO_2 solution to prepare aligned porous structures. To demonstrate this concept, Zhang et al. [141] selected a sugar acetate, 1,2,3,4,6-pentaacetyl β-D-galactose (BGAL), which is highly soluble in CO_2 and is a solid at ambient temperatures [142]. The sugar acetate was dissolved in liquid CO_2 at room temperature at 75 bar in a stainless steel tube. The tube was then slowly lowered into liquid nitrogen to freeze the liquid CO_2. The frozen sample was then left in a fume cupboard with the valve open so that CO_2 could be slowly released. A well-defined aligned porous structure was formed after this process, as shown in Fig. 3.20. This method avoids the use of organic solvents, and there is no need to remove the solvent by freeze-drying [141]. In principle, this method can be applied to any CO_2-soluble or even CO_2-swollen system to produce aligned porous structures of different materials.

3.4.4 Freeze-Drying and Emulsion Templating

Emulsions are heterogeneous mixtures of one immiscible liquid dispersed in another one in the form of droplets. Porous polymers can be produced by polymerizing the monomers in the continuous phase and then removing the droplet phase [143]. Polymerization is thus used as a way to lock in the emulsion structure [144] (see Chapter 4). Alternatively, the emulsion structure can be locked by rapid freezing.

For example, a water-in-oil emulsion was created by homogenizing two layers of ultrapure water and methylene chloride with dissolved PLGA. The emulsion was poured into a cylindrical copper mold, quenched in liquid nitrogen, and freeze-dried. The prepared porous materials were used as scaffolds for protein release [145].

The combination of emulsion templating and freeze-drying to make porous structures may offer several advantages. There is a wide range of recipes that may be used to prepare emulsions involving different types of materials. The materials can be selected to suit target applications—for example, as scaffolds for tissue engineering. Because the freezing process can be very fast, the emulsions do not have to be very stable. The volume percentage of the internal phase in an emulsion can be systematically changed—for example, from 10% to 95% v/v. The concentration of the polymers can also be varied on purpose. This will exert a great control on the pore volume, pore size, and pore interconnectivity of the porous materials. In principle, the technique can be applied to any type of material that is solid at room temperature and can be processed into an emulsion.

In a recent study, an organic dye, oil red, was dissolved in an oil droplet phase of a water-in-oil emulsion with a PVA dissolved in the aqueous continuous phase. After freezing and freeze-drying the emulsion, a highly porous structure with loaded oil red nanoparticles was formed [146]. This material could be dissolved in water within a few seconds. An aqueous organic nanoparticle suspension could be formed instantly by dissolution of the polymeric matrix. A nanoparticle suspension of triclosan was formulated using this technique, which showed enhanced biocidal activity [146]. If a hydrophobic polymer instead of an organic compound was dissolved in the oil droplet phase, a composite of porous microparticles supported in an aligned porous matrix could be produced [147]. In that study, PCL was dissolved in the oil phase. As shown in Fig. 3.21a, porous microparticles are embedded in the aligned porous PVA. The internal porous structure and the porous surface with aligned features are shown in Fig. 3.21b and c. The matrix could be dissolved in water to release the porous microparticles (Fig. 3.21d), which were further tested as scaffolds to support the growth of mouse stem cells.

3.5 CONCLUSIONS

Porogen incorporation and phase inversion have been used to prepare porous polymeric materials. This chapter has described the use of organic solvents, supercritical fluids, and frozen solvent crystals as porogens for the preparation of porous polymers. The porogens could be removed by washing and drying at room temperature, releasing via pressure reduction, or freeze-drying under vacuum. With the concerns about environmental impact and applications in biological areas, SC CO_2 is a “green” alternative solvent as a porogen and as a nonsolvent for phase inversion processes. This is because of the easy separation of supercritical CO_2 from the system and the nontoxic, nonflammable, and economical nature of CO_2.

For many applications, such as tissue engineering, drug delivery, and health care, organic solvent residuals in porous polymers have been a major problem. In many

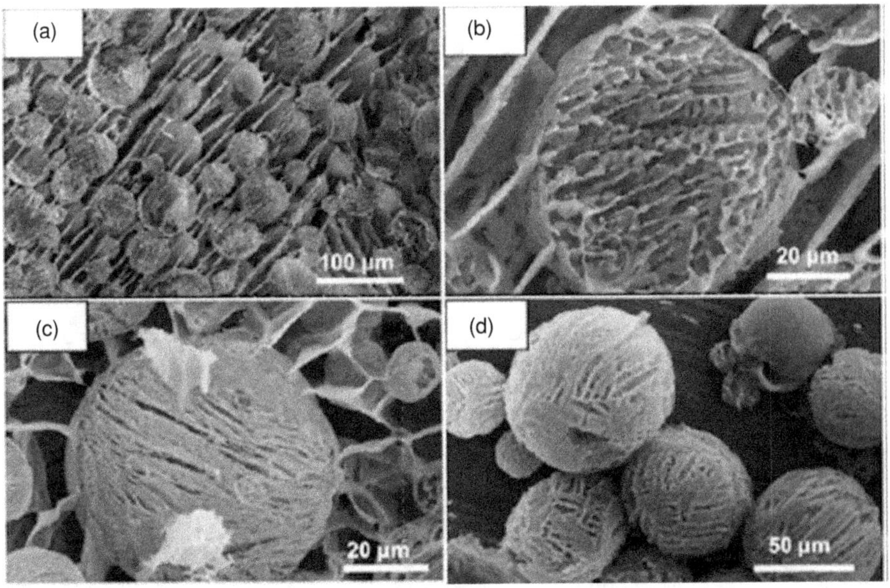

FIGURE 3.21 (**a**) Aligned porous PCL particles entrapped in an aligned polymer matrix. (**b**) Internal aligned porous structure of a single sectioned PCL particle. (**c**) Aligned porous surface of a single PCL particle in the matrix. (**d**) The aligned porous PCL particles. (Reprinted with permission from Reference 147. Copyright 2008 Wiley Periodicals, Inc.).

cases, the limits of the residual solvent have to be very low. For this reason, SC CO_2 foaming processes have been extensively used to produce porous polymers as scaffolds. Freeze-drying has also been employed as a potential clean technique to make porous polymers. A wide range of polymeric materials with various porous structures such as aligned pores has been fabricated. Due to the advantages offered by these techniques, the applications of these porous polymers are expected to be explored continuously and extensively.

ACKNOWLEDGMENTS

HZ is a RCUK academic fellow. Financial support by the EPSRC (EP/F016883/1) and the Royal Society of the United Kingdom is gratefully acknowledged.

NOTATION AND ACRONYMS

AIBN	2,2-Azobisisobutyronitrile
AN	Acrylonitrile
BSA	Bovine serum albumin

DMAA	*N*,*N*-Dimethylacetamide
DMF	*N*,*N*-Dimethylformamide
DMSO	Dimethylsulfoxide
DVB	Divinylbenzene
HA	Hyaluronic acid
HAP	Hydroxyapatite
MA	Methyl acrylate
MMA	Methyl methacrylate
PAN	Poly(acrylonitrile)
P_c	Critical pressure
PCL	Poly(ε-caprolactone)
PDMS	Poly(dimethylsiloxane)
PDVB	Polydivinylbenzene
PEG	Poly(ethylene glycol)
PFMA	Poly(perfluorooctylethyl methacrylate)
PHEMA	Poly(2-hydroxyethyl methacrylate)
PLA	Poly(lactide)
PLGA	Poly(lactide-*co*-glycolide)
PLLA	Poly(L-lactic acid)
PMA	Poly(methyl acrylate)
PMMA	Poly(methyl methacrylate)
PP	Polypropylene
PS	Polystyrene
PU	Polyurethane
PVA	Poly(vinyl alcohol)
PVDF	Poly(vinylidene fluoride)
PVDF-HFP	Poly(vinylidene fluoride-hexafluoropropylene)
PVP	Poly(vinyl pyrrolidone)
SC CO_2	Supercritical carbon dioxide
SCFs	Supercritical fluids
SEM	Scanning electron microscopy
St	Styrene
T_c	Critical temperature
T_g	Glass transition temperature
THF	Tetrahydrofuran
T_m	Melting temperature
TRIM	Trimethylolpropane trimethacrylate

REFERENCES

1. Olah, L.; Filipczak, K.; Jaegermann, Z.; Czigany, T.; Borbas, L.; Sosnowski, S. Ulanski, P.; Rosiak, J. M. *Polym Adv Technol* 2006, **17**, 889.
2. Shastri, V. P.; Martin, I.; Langer, R. *Proc Natl Acad Sci USA* 2000, **97**, 1970.

3. Sherrington, D. C. *Chem Commun* 1998, 2275.
4. Okay, O. *Prog Polym Sci* 2000, **25**, 711.
5. Nam, Y. S.; Park, T. G. *Biomaterials* 1999, **20**, 1783.
6. Zhang, H.; Cooper, A. I. *Adv Mater* 2007, **19**, 1529.
7. Bai, F.; Huang, B.; Yang, X.; Huang, W. *Polymer* 2007, **48**, 3641.
8. Liu, J.; Wang, H. *J Non-Cryst Solids* 2005, **351**, 936.
9. Rohr, T.; Knaus, S.; Gruber, H.; Sherrington, D. C. *Macromolecules* 2002, **35**, 97.
10. Poinescu, I. C.; Beldie, C.; Vlad, C. *J Appl Polym Sci* 1984, **29**, 23.
11. Fang, D.; Pan, Q. Rempel, G. *J Appl Polym Sci* 2007, **103**, 707.
12. Okay, O. *Angew Makromol Chem* 1998, **157**, 1.
13. Okay, O. *Angew Makromol Chem* 1998, **157**, 15.
14. Wang, R. W.; Zhang, Y.; Ma, G. H.; Su, Z. *Colloid Surf B* 2006, **51**, 93.
15. Tang, W.; Wei, J.; Yan, J. *J Appl Polym Sci* 2004, **94**, 2041.
16. Urban, J.; Moravcová, D.; Jandera, P. *J Sep Sci* 2006, **29**, 1064.
17. Martina, A. D.; Graf, R.; Hilborn, J. G. *J Appl Polym Sci* 2005, **96**, 407.
18. Macintyre, F. S.; Sherrington, D. C. *Macromolecules* 2004, **37**, 7628.
19. Schmidt, R. H.; Belmont, A.-S.; Haupt, K. *Anal Chim Acta* 2005, **542**, 118.
20. Kim, J. Y.; Kim, Y. D.; Kanamori, T.; Lee, H. K.; Baik, K.; Kim, S. C. *J Appl Polym Sci* 1999, **71**, 431.
21. Tanyolac, D.; Sonmezisik, H.; Ozdural, A. R. *Biochem Eng J* 2005, **22**, 221.
22. Pu, W. H.; He, X. M.; Wang, L.; Tian, Z.; Jiang, C. Y.; Wan, C. R. *J Membr Sci* 2006, **280**, 6.
23. Blanco, J. F.; Sublet, J.; Nguyen, Q. T.; Schaetzel, P. *J Membr Sci* 2006, **283**, 27.
24. Khorasani, M. T.; Shorgashti, S. *J Biomed Mater Res B Apply Biomater* 2006, **76B**, 41.
25. Hwang, J.; Jeong, S. K.; Nahm, K. S.; Stephan, A. M. *Eur Polym J* 2007, **43**, 65.
26. Zhao, Y. H.; Qian, Y. L.; Zhu, B. K.; Xu, Y. Y. *J Membr Sci* 2008, **310**, 567.
27. Kim, K. M.; Kim, J. C. Ryu, K. S. *Macromol Chem Phys* 2007, **208**, 887.
28. Fu, X. Y.; Matsuyama, H.; Nagai, H. *J Appl Polym Sci* 2008, **108**, 713.
29. Baiker, A. *Chem Rev* 1999, **99**, 453.
30. Darr, J. A.; Poliakoff, M. *Chem Rev* 1999, **99**, 495
31. Taylor, L. *Supercritical Fluid Extraction*, Wiley, New York, 1996.
32. Eckert, C. A.; Knutson, B. L.; Debenedetti, P. G. *Nature* 1996, **383**, 313.
33. Cooper, A. I. *Adv Mater* 2003, **15**, 1049.
34. Wood, C.D.; Cooper, A.I. *Macromolecules* 2001, **34**, 5.
35. Hebb, A. K.; Senoo, K.; Bhat, R.; Cooper, A. I. *Chem Mater* 2003, **15**, 2061.
36. DeSimone, J. M.; Guan, Z.; Elsbernd, C. S. *Science* 1992, **257**, 945.
37. Sarbu, T.; Styranec, T.; Beckman, E. J. *Nature* 2000, **405**, 165.
38. Tan, B.; Cooper, A. I. *J Am Chem Soc* 2005, **127**, 8938.
39. Reverchon, E.; Rappo, E. S.; Cardea, S. *Polym Eng Sci* 2006, **46**, 188.
40. Reverchon, E.; Cardea, S.; Rappo, E. S. *J Membr Sci* 2006, **273**, 97.
41. Matsuyama, H.; Yano, H.; Maki, T.; Teramoto, M.; Mishima, K.; Matsuyama, K. *J Membr Sci* 2001, **194**, 157.

42. Tsivintzelis, I.; Pavlidou, E.; Panayiotou, C. J. *Supercrit Fluids* 2007, **40**, 317.
43. Xu, Q.; Pang, M.; Peng, Q.; Jiang, Y.; Li, J.; Wang, H.; Zhu, M. *J Appl Polym Sci* 2005, **98**, 831.
44. Huang, S. R.; Wu, G. Z.; Chen, S.M. *J Membr Sci* 2007, **293**, 100.
45. Reverchon, E.; Cardea, S. *Ind Eng Chem Res* 2006, **45**, 8939.
46. Reverchon, E.; Cardea, S.; Rappo, E. S. *J Supercrit Fluids* 2008, **45**, 356.
47. Park, C. B.; Baldwin, D. F.; Suh, N. P. *Am Soc Mech Eng* 1994, **53**, 109.
48. Jacobs, L. J. M.; Kemmere, M. F.; Keurentjes, J. T. F. *Green Chem* 2008, **10**, 731.
49. Goel, S. K.; Beckman, E.J. *Polym Eng Sci* 1994, **34**, 1137.
50. Tomasko, D. L.; Li, H.; Liu, D.; Han, X.; Wingert, M. J.; Lee, L. J.; Koelling, K.W. *Ind Eng Chem Res* 2003, **42**, 6431.
51. Lee, P. C.; Kaewmesri, W.; Wang, J.; Park, C. B.; Pumchusak, J.; Folland, R.; Praller, A. *J Appl Polym Sci* 2008, **109**, 3122.
52. Taki, K. *Chem Eng Sci* 2008, **63**, 3643.
53. Shafi, M. A.; Lee, J. G.; Fumerfelt, R. W. *Polym Eng Sci* 1996, **36**, 1950.
54. Marrazzo, C.; Di Maio, E.; Lannace, S. *Polym Eng Sci* 2008, **48**, 336.
55. Davies, W. R.; Lewis, A. L.; Whitaker, M. J.; Tai, H.; Shakesheff, K. M.; Howdle, S. M. *Adv Drug Del Rev* 2008, **60**, 373.
56. Singh, L.; Kumar, V.; Ratner, B. D. *Biomaterials* 2004, **25**, 2611.
57. Pini, R.; Storti, G.; Mazzotti, M.; Tai, H.; Shakesheff, K. M.; Howdle, S. M. *J Polym Sci B Polym Phys* 2008, **46**, 483.
58. Harris, L. D.; Kim, B. S.; Mooney, D. J. *J Biomed Mater Res* 1998, **42**, 396.
59. López-Periago, A. M.; Vega, A.; Subra, P.; Argemí, A.; Saurina, J.; Garcia-González, C. A.; Domingo, C. *J Mater Sci* 2008, **43**, 1939.
60. Zhu, X. H.; Lee, L. Y.; Jackson, J. S. H.; Tong, Y. W.; Wang, C. H. *Biotech Bioeng* 2008, **100**, 998.
61. Cotugno, S.; Di Maio, E.; Mensitieri, G.; Iannace, S.; Roberts, G. W.; Carbonell, R. G.; Hopfenberg, H. B. *Ind Eng Chem Res* 2005, **44**, 1795.
62. Léonard, A.; Calberg, C.; Kerckhofs, G.; Wevers, M.; Jérôme, R.; Pirard, J.-P.; Germain, A.; Blacher, S. *J Porous Mater* 2008, **15**, 397.
63. Salerno, A.; Oliviero, M.; Di Maio, E.; Iannace, S.; Netti, P. A. *J Appl Polym Sci* 2007, **106**, 3335.
64. Marrazzo, C.; Di Maio, E.; Iannace, S.; Nicolais, L. *J Cellular Plastics* 2008, **44**, 37.
65. Mariam, I.; Cho, K.Y.; Rizvi, S. S. H. *Int J Food Properties* 2008, **11**, 415.
66. Mascia, L. Del Re, G.; Ponti, P.P.; Bologna, S.; Di Giacomo, G.; Haworth, B. *Adv Polym Technol* 2006, **25**, 225.
67. Zhai, W. T.; Yu, J.; He, J. S. *Polymer* 2008, **49**, 2430.
68. Lee, P. C.; Li, G.; Lee, J. W. S.; Park, C. B. *J Cellular Plastics* 2007, **43**, 431.
69. Reverchon, E.; Cardea, S. *J Supercrit Fluids* 2007, **40**, 144.
70. Zhai, W.; Wang, H.; Yu, J.; Dong, J. Y.; He, J. *Polymer* 2008, **49**, 3146.
71. Jacobs, L. J. M.; Hurkens, S. A. M.; Kemmere, M. F.; Keurentjes, J. T. F. *Macromol Mater Eng* 2008, **293**, 298.
72. Shieh, Y. T.; Hsiao, T. T.; Chang, S. K. *Polymer* 2006, **47**, 5929.

73. Wirges, W.; Wegener, M.; Voronina, O.; Zirkel, L.; Gerhard-Multhaupt, R. *Adv Funct Mater* 2007, **17**, 324.
74. Jacobs, L. J. M.; Danen, K. C. H.; Kemmere, M. F.; Keurentjes, J. T. F. *Polymer* 2007, **48**, 3771.
75. Park, C. B.; Padareva, V.; Lee, P. C.; Naguib, H. E. *J Polym Eng* 2005, **25**, 239.
76. Barry, J. J. A.; Nazhat, S. N.; Rose, F. R. A. J.; Hainsworth, A. H.; Scotchford, C. A.; Howdle, S. M. *J Mater Chem* 2005, **15**, 4881.
77. Hao, A.; Geng, Y.; Xu, Q.; Lu, Z.; Yu, L. *J Appl Polym Sci* 2008, **109**, 2679.
78. Zhai, W.; Wang, H.; Yu, J.; Dong, J.; He, J. *J Polym Sci B Polym Phys* 2008, **46**, 1641.
79. Barry, J. J. A.; Silva, M. M. C. G.; Cartmell, S. H.; Guldberg, R. E.; Scotchford, C. A.; Howdle, S.M. *J Mater Sci* 2006, **41**, 4197.
80. Ito, Y.; Yamashita, M.; Okamoto, M. *Macromol Mater Eng* 2006, **291**, 773.
81. Teng, X.; Ren, J.; Gu, S. *J Biomed Mater Res B Appl Biomater* 2007, **81B**, 185.
82. . Mathieu, L. M.; Montjovent, M. O.; Bourban, P. E.; Pioletti, D. P.; Månson, J. A. E. *J Biomed Mater Res* 2005, **75A**, 89.
83. Zeng, C.; Han, X.; Lee, L. J.; Koelling, K. W.; Tomasko, D. L. *Adv Mater* 2003, **15**, 1743.
84. Siripurapu, S.; Coughlan, J. A.; Spontak, R. J.; Khan, S. A. *Macromolecules* 2004, **37**, 9872.
85. Siripurapu, S.; DeSimone, J. M.; Khan, S. A.; Spontak, R. J. *Adv Mater* 2004, **16**, 989.
86. Yokoyama, H.; Li, L.; Nemoto, T.; Sugiyama, K. *Adv Mater* 2004, **16**, 1542.
87. Yokoyama, H.; Sugiyama, K. *Macromolecules* 2005, **38**, 10516.
88. Taki, K. Waratani, Y.; Ohshima, M. *Macromol Mater Eng* 2008, **293**, 589.
89. Otsuka, T.; Taki, K.; Ohshima, M. *Macromol Mater Eng* 2008, **293**, 78.
90. Nemoto, T.; Takagi, J.; Ohshima, M. *Macromol Mater Eng* 2008, **293**, 574.
91. Volti, R. *Freeze Drying*, Facts on File, New York, 1999.
92. Lee, J.; Cheng, Y. *J Control Release* 2006, **111**, 185.
93. Mumenthaler, M.; Leuenberger, H. *Int J Pharm* 1991, **72**, 97.
94. Leuenberger, H. *J Nanoparticle Res* 2002, **4**, 111.
95. Shin, H.; Jung, H. S.; Hong, K. S.; Lee, J.-K. *J Solid State Chem* 2005, **178**, 15.
96. Boiadjieva, T.; Cappelletti, G.; Ardizzone, S.; Rondinini, S.; Vertova, A. *Phys Chem Chem Phys* 2003, **5**, 1689.
97. Ma, D.; Schadler, L. S.; Siegel, R. W.; Hong, J.-I. *Appl Phys Lett* 2003, **83**, 1839.
98. Chen, G.; Wang, W. *Dry Technol* 2007, **25**, 29.
99. Liu, J. *Pharm Dev Technol* 2006, **11**, 3.
100. Gutiérrez, M. C.; Ferrer, L.; del Monte, F. *Chem Mater* 2008, **20**, 634.
101. Madihally, S. V.; Matthew, H. W. T. *Biomaterials* 1999, **20**, 1133.
102. Ren, L.; Tsuru, K.; Hayakawa, S.; Osaka, A. *Biomaterials* 2002, **23**, 4765.
103. Lu, L.; Eychmüller, A. *Acc Chem Res* 2008, **41**, 244.
104. Zhang, H.; Hussain, I.; Brust, M.; Butler, M. F.; Rannard, S. P.; Cooper, A. I. *Nat Mater* 2005, **4**, 787.
105. Mukai, S. R.; Nishihara, H.; Shichi, S.; Tamon, H. *Chem Mater* 2004, **16**, 4987.
106. Yokoyama, F.; Masada, I.; Shimamura, K.; Ikawa, T.; Monobe, K. *Colloid Polym Sci* 1986, **264**, 595.

107. Trieu, H. H.; Qutubuddin, S. *Colloid Polym Sci* 1994, **272**, 301.
108. Fini, A.; Cavallari, C.; Ospitali, F. *Eur J Pharm Biopharm* 2008, **70**, 409.
109. Newman, A.; Engers, D.; Bates, S.; Ivanisevic, I.; Kelly, R.C.; Zografi, G. *J Pharm Sci* 2008, **97**, 4840.
110. Santagapita, P. R.; Brizuela, L. G.; Mazzobre, M. F.; Ramirez, H. L.; Corti, H. R.; Santana, R. V.; Buera, M. P. *Biomacromolecules* 2008, **9**, 741.
111. Cho, C. H.; Eliason, J. F.; Matthew, H. W. T. J *Biomed Mater Res* 2008, **86A**, 98.
112. Huang, Y.; Siewe, M.; Madihally, S. V. *J Biotechnol Bioeng* 2006, **93**, 64.
113. Li, Z.; Zhang, M. *J Biomed Mater Res* 2005, **75A**, 485.
114. Lv, Q.; Feng, Q. *J Mater Sci Mater Med* 2006, **17**, 1349.
115. Schoof, H.; Apel, J.; Heschel, I.; Rau, G. *J Biomed Mater Res Appl Biomater* 2001, **58**, 352.
116. Yeong, W. Y.; Chua, C. K.; Leong, K. F.; Chandrasekaran, M.; Lee, M. W. *J Biomed Mater Res Appl Biomater* 2007, **82B**, 260.
117. Lee, S. J.; Kim, S. Y.; Lee, Y. M. *J Biomed Mater Res Appl Biomater* 2007, **82B**, 506.
118. Jeong, S. I.; Kim, S. Y.; Cho, S. K.; Chong, M. S.; Kim, K. S.; Kim, H.; Lee, S. B.; Lee Y. M. *Biomaterials* 2007, **28**, 1115.
119. Gercek, I.; Tiğli, R. S.; Gümüsderelioğlu, M. *J Biomed Mater Res* 2008, **86A**, 1012.
120. Hou, Q.; Grijpma, D. W.; Feijen, J. *J Biomed Mater Res Appl Biomater* 2003, **67B**, 732.
121. Liu, X.; Won, Y.; Ma, P. X. *Biomaterials* 2006, **27**, 3980.
122. Ma, P. X.; Zhang, R. Y. *J Biomed Mater Res* 1999, **46**, 60.
123. Ma, P. X.; Zhang, R.; Xiao, G.; Franceschi, R. *J Biomed Mater Res* 2001, **54**, 284.
124. Kim, H. W.; Knowles, J. C.; Kim, H. E. *J Biomed Mater Res Appl Biomater* 2005, **74B**, 686.
125. Nayar, S.; Pramanick, A. K.; Guha, A.; Mahato, B. K.; Gunjan, M.; Sinha, A. *Bull Mater Sci* 2008, **31**, 429.
126. Zhang, H.; Hussain, I.; Brust, M.; Butler, M. F.; Rannard, S. P.; Cooper, A. I. *Nat Mater* 2005, **4**, 787.
127. Oetjen, G.-W. *Freeze-Drying*, Wiley-VCH, Weinheim, Germany, 1999.
128. Mahler, W.; Bechtold, M. F. *Nature* 1980, **285**, 27.
129. Kim, J. W.; Taki, K.; Nagamine, S.; Ohsima, M. *Chem Eng Sci* 2008, **63**, 2858.
130. Gutiérrez, M. C.; García-Carvajal, Z. Y.; Jobbágy, M.; Rubio, F.; Yuste, L.; Rojo, F.; M. Ferrer, L.; del Monte, F. *Adv Funct Mater* 2007, **17**, 3505.
131. Stokols, S.; Tuszynski, M. H. *Biomaterials* 2006, **27**, 443.
132. Schoof, H.; Apel, J.; Heschel, I.; Rau, G. *J Biomed Mater Res Appl Biomater* 2001, **58**, 352.
133. Gutierrez, M. C.; Jobbagy, M.; Rapun, N.; Ferrer, M.L.; del Monte, F. *Adv Mater* 2006, **18**, 1137.
134. Deville, S.; Saiz, E.; Nalla, R. K.; Tomsia, A. P. *Science*, 2006, **311**, 515.
135. Gutierrez, M. C.; Hortiguela, M. J.; Amarilla, J. M.; Jimenez, R.; Ferrer, M. L.; del Monte, F. *J Phys Chem C* 2007, **111**, 5557.
136. Bai, H.; Li, C.; Chen, F.; Shi, G. *Polymer* 2007, **48**, 5259.
137. Deville, S. *Adv Eng Mater* 2008, **10**, 155.

138. Zhang, H.; Lee, J. Y.; Ahmed, A.; Hussain, I.; Cooper, A. I. *Angew Chem Int Ed* 2008, **47**, 4573.

139. Shi, Q.; An, Z.; Tsung, C.-K.; Liang, H.; Zheng, N.; Hawker, C. J.; Stucky, G. D. *Adv Mater* 2007, **19**, 4539.

140. Shi, Q.; Liang, H.; Feng, D.; Wang, J.; Stucky, G. D. *J Am Chem Soc* 2008, **130**, 5034.

141. Zhang, H.; Long, J.; Cooper, A. I. *J Am Chem Soc* 2005, **127**, 13482.

142. Raveendran, P.; Wallen, S. L. *J Am Chem Soc* 2002, **124**, 12590.

143. Zhang, H.; Cooper, A. I. *Soft Matter* 2005, **1**, 107.

144. Zhang, H.; Cooper, A. I. *Chem Mater* 2002, **14**, 4017.

145. Whang, K.; Goldstick, T. K.; Healy, K. E. *Biomaterials* 2000, **21**, 2545.

146. Zhang, H.; Wang, D.; Butler, R.; Campbell, N. L.; Long, J.; Tan, B.; Duncalf, D. J.; Foster, A. J.; Hopkinson, A.; Taylor, D.; Angus, D.; Cooper, A. I.; Rannard, S. P. *Nature Nanotechnol* 2008, **3**, 506.

147. Zhang, H.; Edgar, D.; Murray, P.; Rak-Raszewska, A.; Glennon-Alty, L.; Cooper, A. I. *Adv Funct Mater* 2008, **18**, 222.

CHAPTER 4

Colloidal Templating

NEIL R. CAMERON
University of Durham, Durham, United Kingdom

PETER KRAJNC
University of Maribor, Maribor, Slovenia

MICHAEL S. SILVERSTEIN
Technion – Israel Institute of Technology, Haifa, Israel

This chapter describes the use of colloidal systems, including emulsion droplets, bicontinuous microemulsion channels, water droplets, and solid particles, as templates to create porous polymer materials (Fig. 4.1). Conceptually, the process is general to all types of colloidal dispersion: A biphasic system is created, and the continuous phase (or, in the case of bicontinuous microemulsions, one of the co-continuous phases) is polymerized. The colloidal entities, which are removed following polymerization, thus serve to create porosity in the final polymeric material. Depending on the nature of the colloidal system employed, the characteristic domain size of the porous network can range from a few nanometers to hundreds of micrometers.

4.1 HIGH INTERNAL PHASE EMULSION TEMPLATING

4.1.1 Introduction

A large variety of porous polymers, generally known as polyHIPEs, have been synthesized using high internal phase emulsions (HIPEs) as templates for the porous structure. These porous polymers, usually synthesized through free radical polymerization, include both hydrophobic polymers synthesized within water-in-oil (W/O) emulsions and hydrophilic polymers synthesized within oil-in-water (O/W) emulsions. The types of materials synthesized include copolymers, interpenetrating polymer networks, biodegradable materials, organic–inorganic hybrids that can be pyrolyzed

Porous Polymers, Edited by Michael S. Silverstein, Neil R. Cameron, and Marc A. Hillmyer

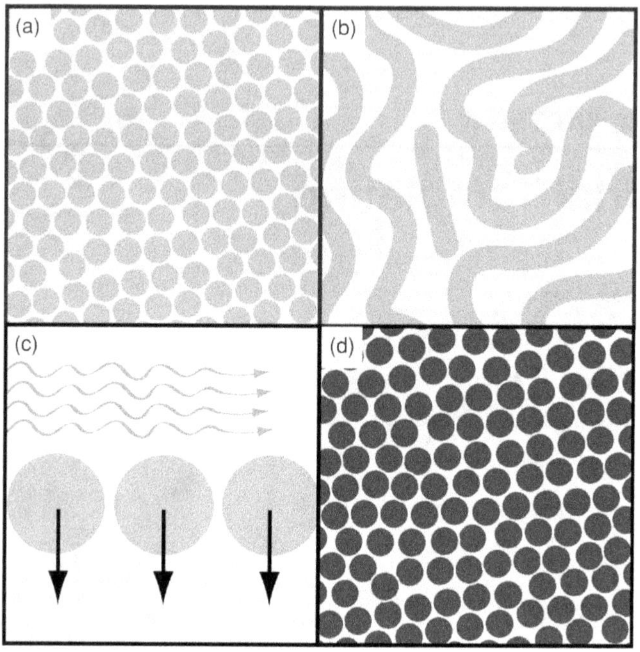

FIGURE 4.1 An overview of colloidal templating: (**a**) high internal phase emulsions; (**b**) microemulsions; (**c**) water droplet templating; (**d**) particle templating.

to porous inorganics, nanocomposites, and hydrophobic–hydrophilic bicontinuous polymers. The large number of methods available to functionalize polyHIPEs has enhanced their utility for such applications as chemical synthesis, chromatography, ion exchange, separation, sensing, tissue engineering, and controlled drug delivery, to name but a few.

HIPEs are highly viscous, paste-like emulsions in which the dispersed, internal phase constitutes more than 74% of the volume. A well-known example of a HIPE is mayonnaise. The major phase in mayonnaise, vegetable oil, is emulsified in the minor phase, vinegar, using the lecithin in egg yolk as the surfactant. HIPEs have a tendency to undergo phase inversion. Only a few of the available emulsifiers are able to keep the major internal phase dispersed within the minor external phase. If the internal phase, external phase, or both phases contain monomers, then a polymer can be synthesized within the HIPE. The two-phase structure of the HIPE can be maintained during polymerization under the right conditions. A concentrated latex results if the discrete, internal phase contains monomers. A polyHIPE—a continuous polymer envelope surrounding the dispersed droplets of the internal phase—results if only the continuous, external phase contains monomers.

There is often a highly significant difference between the structure of a HIPE and the structure of the resulting polyHIPE. Ruptures—termed holes, interconnects, or windows—can develop at the thinnest points of the external phase envelope surrounding the dispersed internal phase under the right conditions (e.g., appropriate surfactant and internal phase contents). The formation of these holes transforms the discrete droplets of the internal phase into a continuous interconnected phase. Removal of the internal phase, which is now continuous, yields an open-pore void structure templated by the droplets that formed the HIPE's internal phase. The holes in the polymer wall yield a highly interconnected porous structure. The porous structure of polyHIPE can be manipulated through variations in the structure of the HIPE. There are several extensive reviews that describe HIPEs, polymerizations within HIPEs, concentrated latices, and polyHIPEs [1–4].

4.1.2 High Internal Phase Emulsions

Lissant was one of the first to describe the structure and properties of the non-Newtonian, thixotropic, Bingham fluids that were termed high internal phase emulsions (or high internal phase ratio emulsions, HIPREs) [5–7]. These emulsions have also been referred to as concentrated emulsions [8–10], gel emulsions [11–13], and hydrocarbon gels [14]. HIPEs were originally investigated as a method for entrapping volatile, toxic solvents in non-Newtonian formulations for such applications as cleaning wax or sulfur from oil and gas wells and as a method for transporting bulk solids [15–18]. Lissant classified HIPEs as emulsions with internal phase contents of 70% or more [19]. More recent definitions of HIPEs refer to internal phase contents of 74% or more, since 74% is the maximum packing fraction for monodisperse hard spheres. For polyHIPE applications, where porosity is to be maximized, internal phase contents of around 90% are typical. The history, structure, and properties of HIPEs are described in great detail elsewhere [1].

4.1.2.1 HIPE Formation and Stability. HIPEs are formed by mixing two immiscible liquids in the presence of an emulsifier, usually a surfactant. One of the liquids is almost always an aqueous solution, and the other is usually hydrophobic. The major phase is usually added slowly, under constant agitation, to a solution consisting of the minor phase and the surfactant. HIPEs can also be formed by applying a centrifugal field to an emulsion [20, 21]. This "creaming" process separates the excess continuous phase from the emulsion, concentrating the dispersed phase into a HIPE. Emulsions, which are not thermodynamically stable systems and tend to coalesce, can also undergo creaming as they phase separate.

Internal phase contents of over 74% can be reached through the deformation of monodispersed droplets into polyhedra, through the formation of polydispersed droplets, or both. Theoretical analysis has shown that a monodisperse system of polyhedra should be favored over a polydisperse system [22]. The original work indicated that droplets should assume rhomboidal dodecahedral packing for dispersed phase contents of 74%–94% and tetrakaidecahedral (truncated octahedral) packing for above 94% [5]. Later work indicated that pentagonal dodecahedra should be formed at dispersed phase contents above 96% [23, 24]. Initial investigations of polystyrene-based polyHIPEs showed that the voids were relatively monodisperse and were polyhedral in shape [7]. Subsequent work with other polyHIPEs has shown that spherical polydisperse systems are also common. The rate and time of stirring have significant effects on the HIPE structure. There is an optimal increase in stability with increasing viscosity. As the viscosity begins to increase, the barrier to coalescence increases. At very high viscosities it becomes difficult to disperse the internal phase. An initially polydisperse system of droplets becomes more monodisperse with prolonged stirring. However, this reduction in droplet polydispersity makes the HIPE more viscous and more difficult to stir [25]. Stirring is more difficult for systems that are more viscous *a priori*, and, therefore, the resulting HIPEs are relatively polydisperse [6].

Typically, HIPE stability is influenced by a number of factors, including the molecular structures of the components comprising the phases and of the surfactant, the surfactant content, the dispersed phase content, the temperature, and the presence of stabilizing salts. The nature of the surfactant is critical to the stability of a major phase dispersed within a minor phase. Thermodynamics would predict that such a system should undergo phase separation or phase inversion, where a W/O emulsion becomes an O/W emulsion or vice versa. The surfactant used to stabilize a HIPE should, therefore, be completely insoluble in the dispersed phase to prevent phase separation and/or phase inversion. The molecular geometry of the surfactant is also important. The assembly of surfactant molecules at the interface should enhance convex interface curvature within the minor phase. A nonionic surfactant with a low hydrophilic lipophilic balance (HLB), such as sorbitan monooleate (Span 80) with its HLB of 4.3, is needed to form a stable W/O HIPE. The effects of the nature of the surfactant, the surfactant concentration, and the presence of cosurfactants in a HIPE whose external phase contained a solution of styrene (S) and divinylbenzene (DVB) were investigated [26–29]. The stability of such HIPEs appeared to vary inversely with the cosurfactant HLB.

To achieve a stable HIPE, the surfactant must lower the interfacial tension between the phases, must form a rigid interfacial film, and must rapidly adsorb at the interface [30]. The stability of a HIPE is kinetic and depends on the repulsion generated between the droplets of the internal phase. Ionic surfactants stabilize HIPEs through electrostatic repulsion, while nonionic surfactants stabilize HIPEs through steric repulsion. The higher the interfacial tension between the phases, the greater is the tendency of the surfactant to concentrate at the interface and form a strong interfacial film. The tendency toward coalescence decreases with the increasing strength of the interfacial film. The strength of the interfacial film is enhanced by the ability of the surfactant to pack closely, and a surfactant blend is often able to pack more effectively. The presence of a salt can also enhance the ability of the surfactant head groups to pack into an ordered structure.

The addition of NaCl was found to enhance the rigidity of the interfacial film in W/O HIPEs, raising the HIPEs' elastic modulus and apparent yield stress and enhancing its stability [31]. Various salts are used to stabilize HIPEs, including calcium chloride hydrate and potassium sulfate. The nature of the salt is important [12]. Salts that reduce the interaction of nonionic surfactants with the aqueous phase, thereby increasing surfactant–surfactant interaction and the formation of a more ordered interface, were more effective at enhancing HIPE stability [32]. The salts can also enhance stability by inhibiting Ostwald ripening, a process by which large droplets grow at the expense of smaller ones. They reduce the solubility of the aqueous phase in the oil phase, thus reducing the attractive forces between the droplets [33].

The stability of HIPEs has been correlated with the water/oil interfacial free energy for both W/O and O/W systems. The larger the interfacial free energy, the greater is the HIPE stability and the greater is the amount of internal phase that can be incorporated [34, 35]. The polarity of the organic component in a W/O HIPE also affects the stability. In general, the greater the hydrophobicity of the oil phase and the greater the hydrophilicity of the aqueous phase, the greater is the interfacial tension and the greater is the stability of the HIPE [34]. Relatively polar organic components require more hydrophobic surfactants, whereas relatively hydrophobic organic components require more hydrophilic surfactants. Thus, it was found to be easier to form HIPEs with aliphatic organics than with aromatic organics [13]. When the oil phase in an O/W emulsion contained aromatic or halogenated organics it was difficult to form a HIPE using a nonionic surfactant due to the interactions between the surfactant and the dispersed phase [35]. The amount of organic that could be added to an O/W HIPE increased with increasing surfactant concentration [36]. Partial polymerization of the monomers in the dispersed phase was also found to enhance the stability of the HIPE [37].

Increasing the viscosity of the continuous phase reduces the amount of dispersed phase that could be incorporated [35]. Increasing the temperature enhances the coalescence of the dispersed droplets and reduces emulsion stability [11, 13]. Poly(butylene oxide)/poly(ethylene oxide) block copolymers were developed that could stabilize HIPEs at concentrations as low as 0.12 wt% of the organic phase without the need for a salt stabilizer [38]. The stability of a HIPE, whether W/O or O/W, is usually related to the presence of relatively large amounts of a surfactant that is highly soluble in the external phase (typically, around 20% of the external phase), is highly insoluble in the internal phase, is easy to pack on the outside of a sphere, but is difficult to pack on the inside of a sphere. There can be exceptions to this rule of thumb. A water-soluble ionic surfactant, cetyltrimethylammonium bromide (CTAB), has been used to form a stable W/O HIPE with S/DVB in the external phase through vigorous stirring [39]. Polymerization produced the typical polyHIPE porous structure, although the voids are larger than those found in typical polyHIPE.

4.1.2.2 Nonaqueous, Carbon Dioxide, and Pickering HIPEs. W/O and O/W HIPEs are by far the most common types and are used for a variety of applications, including the synthesis of polymer spheres through polymerization in the internal phase and porous polymers (polyHIPEs) through polymerization in the external phase. There are only a few examples of nonaqueous oil-in-oil (O/O) HIPEs. An approach to such systems would combine a nonpolar organic with a polar organic using a nonionic surfactant or a block copolymer in which each of the blocks is soluble in a different phase. HIPEs of jet engine fuel in formaldehyde have been investigated, where the surfactant had an HLB of around 12 [40–43]. Similarly, the stability of nonaqueous HIPEs of petroleum ether in a variety of polar aprotic solvents was investigated. Block copolymer surfactants were found to be crucial for HIPE stability [44]. As with HIPEs containing an aqueous phase, the lack of solubility of the surfactant in the internal phase was paramount. The nonaqueous HIPEs were similar in many ways to HIPEs containing an aqueous phase. The internal phase exhibited polyhedral or polydisperse structures. The rheological behavior was that of a non-Newtonian, thixotropic, Bingham fluid.

PolyHIPEs can be synthesized within nonaqueous HIPEs. A polyHIPE was synthesized from a HIPE consisting of petroleum ether that was dispersed within vinyl comonomers that included a maleimide-terminated poly(aryl ether sulfone) using a PEO-PPO-PEO triblock copolymer surfactant containing poly(ethylene oxide) (PEO) and poly(propylene oxide) (PPO) blocks [45]. HIPEs were also formed by mixing polystyrene dissolved in a poor solvent for poly(2-vinylpyridine) with poly(2-vinylpyridine) dissolved in a poor solvent for polystyrene [46, 47]. Polymer blends were produced by casting films from the polymer-containing HIPEs.

Supercritical carbon dioxide ($scCO_2$) has been used as the dispersed phase within an aqueous phase for the formation of an $scCO_2$-in-water (C/W) HIPEs. PolyHIPEs have also been synthesized within C/W HIPEs. The aqueous phase in such C/W HIPEs consisted of a solution of acrylamide (AAm) or 2-hydroxyethyl acrylate (HEA) with *N*,*N*-methylene bisacrylamide (MBAAm) as a crosslinking comonomer [48, 49]. The porous materials resulting from polymerization in the C/W HIPE exhibited typical polyHIPE structures. Block copolymer surfactants were used to enhance the internal phase content of the HIPEs [50]. Emulsion templated porous polymers were also produced using an aqueous polymer solution, instead of a monomer solution, in C/W HIPEs [51].

Pickering emulsions are stabilized by solid particles at the interface rather than by surfactants [52]. Recent work has shown that emulsions with 40% external phase can be stabilized by nanoparticles such as carbon nanotubes through the formation of Pickering emulsions [53]. Very low loadings of titania nanoparticles functionalized with oleic acid yielded stable W/O HIPEs, and these HIPEs were used to synthesize PS-based polyHIPEs [54].

4.1.3 PolyHIPE Preparation and Functionalization

The term "polyHIPE," a Unilever trade name, describes a material produced by polymerization in the external phase of a HIPE in a 1982 patent by Barby and Haq [55]. The wide-ranging research and development of polyHIPEs has been described in several extensive reviews [1, 3, 4]. PolyHIPEs, originally developed as substrates for chemical synthesis, chromatography, ion exchange, or separations [56, 57], are commonly synthesized from hydrophobic monomers in W/O HIPEs. Research done on polymerizing monomers in the external phases of emulsions prior to the 1982 patent [55] includes the synthesis of closed-cell structures from HIPEs containing up to 85% water in the internal phase and with styrene or methyl methacrylate (MMA) in the external phase [58–60]. Other research produced closed-cell polyHIPEs containing up to 90% water in the internal phase and a mixture of unsaturated polyester and styrene in the external phase [61, 62].

4.1.3.1 Chain-Growth Polymerization. Polymerization should only begin following HIPE formation to prevent an increase in viscosity that would impede mixing.

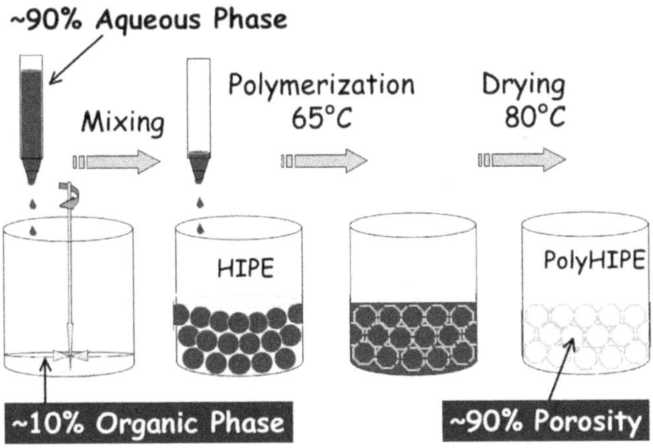

FIGURE 4.2 Schematic illustration of polyHIPE formation.

Chain-grown polymerization using free radical initiation is ideal for such a constraint. Step-growth polymerization, which can involve monomers with hydrophilic end groups, is more problematic for polyHIPE formation. This is especially the case where water is a byproduct of the reaction, can react with one of the components, or can render one of the components inactive. In addition, many step-growth polymerization reactions occur at temperatures above 100°C, temperatures that are not practical for HIPEs containing an aqueous phase.

PolyHIPEs have been successfully synthesized by free radical polymerization of monomers such as styrene, acrylates, and methacrylates. PolyHIPEs are usually crosslinked using a crosslinking comonomer to prevent collapse or disintegration under the capillary stresses generated during drying. DVB is commonly used as a crosslinking comonomer for hydrophobic polyHIPEs. The polyHIPEs most commonly investigated are based on styrene and divinylbenzene, the polymer described in the original patent [55]. The HIPE is formed by adding the aqueous phase to a solution of styrene, divinylbenzene, and a nonionic surfactant, as illustrated schematically in Fig. 4.2. A surfactant HLB range of 2–6 is needed to produce a stable styrene-based W/O HIPE. The surfactant usually chosen for such HIPEs is sorbitan monooleate with an HLB of 4.3. At least 4% surfactant (based on the total mass of the external phase) is usually needed to form an open-cell polyHIPE structure. The optimal surfactant contents are usually 20%–50% of the external phase. Surfactant contents greater than 80% of the external phase can yield a closed-cell structure [26–29]. Block copolymer surfactants have also been used to synthesize PS-based polyHIPEs [63].

The initiator can be added to the organic phase (e.g., 2,2′-azobisisobutyronitrile [AIBN] or benzoyl peroxide [BPO]) or to the aqueous phase (e.g., potassium persulfate [KPS]). Polymerization is usually allowed to take place for around 24 h at elevated temperatures (between 50°C and 80°C). The result is a water-filled, crosslinked

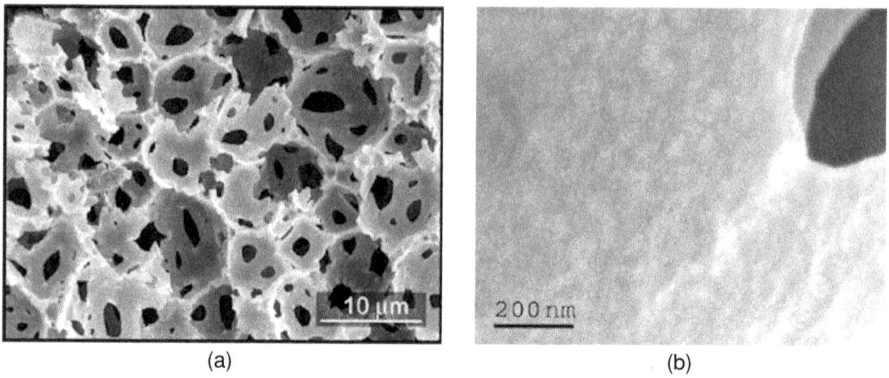

FIGURE 4.3 Scanning electron microscopy (SEM) micrographs of a poly(S-*co*-DVB) polyHIPE. (Reprinted from Reference 64 with permission from John Wiley & Sons.)

monolith in the shape of the reaction vessel. The low-density, brittle, highly porous polymer seen in Fig. 4.3 was produced by removing the internal phase via Soxhlet extraction with a lower alcohol and drying in vacuo or at an elevated temperature [64]. Removing the dispersed phase droplets leaves spherical voids between 5 and 15 μm in Fig. 4.3. The void walls in Fig. 4.3 contain numerous interconnecting holes between 0.1 and 2 μm where the thin polymer envelope has ruptured. The mechanical behavior of polyHIPEs was varied from rubbery to glassy through copolymerization in such systems as styrene/acrylate copolymers [65]. In addition, polyHIPEs with high glass transition temperatures T_gs were synthesized through copolymerization of styrene with maleimides [66, 67].

Initiation takes place within the monomer when the initiator is dissolved in the external phase but takes place at the interface when the initiator is dissolved in the internal phase. The locus of initiation has been shown to be extremely important in determining the molecular structure, the porous structure, and the properties [68]. This is especially true when the monomer and the crosslinking comonomer have different interfacial activities. An initiator located at the interface promotes the reaction of the monomer that is found in excess at the interface, while an initiator located in the bulk promotes a more random reaction.

Free radical polymerization can take place within the external phase of O/W HIPEs containing an aqueous solution of monomer, crosslinking comonomer, and KPS in the continuous phase. Hydrophilic polyHIPEs were synthesized based on such monomers as acrylic acid (AA), AAm, *N*-isopropyl acrylamide (NiPAAm), 2-hydroxyethyl methacrylate (HEMA), and HEA [69–72]. In addition, a hydrosilylation reaction between polymethylhydrosiloxane and vinylsiloxane was also used to synthesize a polyHIPE [73]. A 2-acryloxyethyl-2′-bromoisobutyrate comonomer (a polymerizable initiator for controlled/living atom transfer radical polymerization [ATRP]) was incorporated into a HIPE without compromising emulsion stability and was used to graft polymethacrylates.

4.1.3.2 Step-Growth Polymerization. The synthesis of polyHIPEs using step-growth polymerization has met with limited success. Usually the resulting porous structures are less highly interconnected and have relatively high densities. The reactions studied include base-catalyzed polycondensation of 2-nitroresorcinol with cyanuric chloride [74] and the reaction of an isocyanate with a polyol to produce polyurethane [75]. PolyHIPEs have been synthesized through resorcinol–formaldehyde step-growth polymerization in an O/W HIPE [76]. Other polyHIPE systems synthesized through step-growth polymerization include urea–formaldehyde, phenol–formaldehyde, melamine–formaldehyde, and a system based on a polysiloxane elastomer [77]. Interestingly, pyrolysis of the resorcinol–formaldehyde polyHIPE in an inert atmosphere has led to the formation of high–surface area porous carbon materials of interest for rechargeable battery or supercapacitor applications [77, 78].

In addition, ring-opening metathesis polymerization was used to synthesize a polyHIPE from tetracyclo[6.2.$1^{3,6}$.$0^{2,7}$]dodeca-4,9-diene at −15°C and from dicyclopentadiene and tetracyclododecene at 80°C using ruthenium catalysts [79, 80].

4.1.3.3 Surface Functionalization. A number of reactions have been investigated for the surface functionalization of polyHIPEs (Fig. 4.4) [3]. PolyHIPE surfaces have been functionalized by using 4-vinylbenzyl chloride (VBC) as a comonomer or by using surfactant molecules soluble in the monomer possessing an allyl or an acryloyl group [81–84]. Functionalization has also proceeded via pendent unreacted DVB vinyl groups in poly(S-*co*-DVB) polyHIPEs [85–88]. Glycidyl methacrylate has also been used as a functional monomer [89]. Reactive acrylates such as 4-nitrophenyl acrylate, 2,4,6-trichlorophenyl acrylate, and even *t*-butyl acrylate were also used to enhance functionalization [90–92]. ATRP was used to graft a polymethacrylate to a polyHIPE surface functionalized using a bromoester comonomer [93]. A ruthenium catalyst was also attached to polyHIPEs for alkylidene exchange reactions [94]. Stimuli responsive peptide side-chain polymers prepared by reversible addition–fragmentation chain transfer (RAFT) polymerization were attached to the C=C bonds of residual crosslinker in polyHIPEs based on poly(ethylene glycol) (PEG), leading to materials that were capable of reversible assembly with complementary molecules in solution [95].

4.1.4 PolyHIPE Morphology

4.1.4.1 Open- and Closed-Cell Structures. The droplet diameters in HIPEs usually range from 5 to 100 μm, and these diameters correspond to the typical voids seen in the resulting polyHIPEs. Early work on polyHIPEs indicated that a closed-cell structure was produced at surfactant contents below 5% of the external phase, while a highly interconnected open-cell structure was produced at higher surfactant contents (above 7%) [28]. The surfactant content is often more important in determining the porous structure than the internal phase content. A HIPE with an internal phase content of 97% but with only 5% surfactant in the external phase will yield a closed-cell polyHIPE [28].

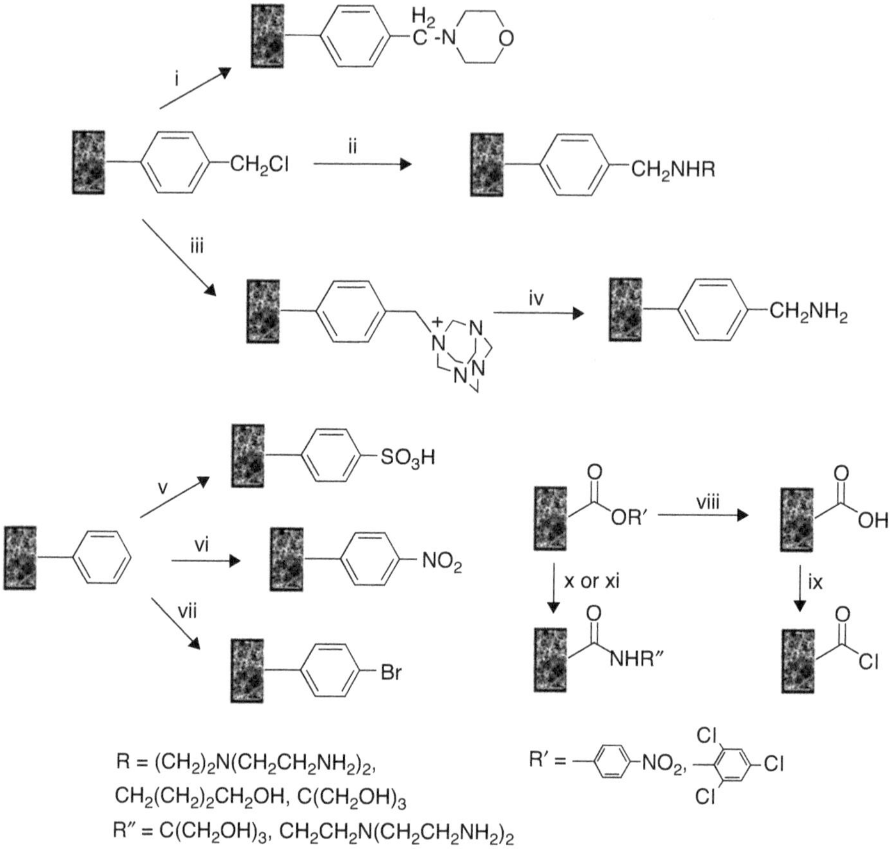

FIGURE 4.4 Schematic illustration of polyHIPE functionalization. Reagents: (i) morpholine, (ii) amine, (iii) hexamethylenetetramine, (iv) HCl/EtOH, (v) $CH_3(CH_2)_{10}CO_2SO_3H$, (vi) $Bu_4NNO_3/(CF_3CO)_2O$, (vii) $Br_2/SnCl_4$, (viii) NaOH, (ix) $SOCl_2$, (x) $H_2NC(CH_2OH)_3$, and (xi) $N(CH_2CH_2NH_2)_3$. (Reprinted from Reference 3 with permission from Elsevier.)

The mechanism proposed to describe the formation of the interconnecting holes emphasizes the importance of the surfactant content. It is clear that these holes are formed as the polymerization produces a "solid" structure; otherwise the connection of the discrete droplets in the internal phase would lead to coalescence and disintegration of the HIPE. The formation of the holes was ascribed to volume contraction, phase separation, and changes in interfacial tension that occur during polymerization near the gel point [96]. Increasing the surfactant concentration reduces the droplet size and thus reduces the thickness of the envelope of external phase surrounding the droplets. As the film thickness decreases, the film's susceptibility to rupture from density changes increases. The changes in density can result in a volume contraction within the thin monomer-swollen polymer film [96]. Some results suggest that the

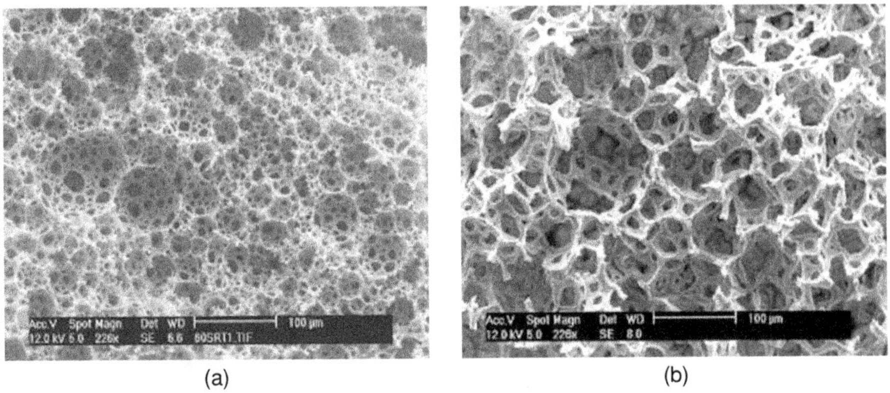

FIGURE 4.5 SEM micrographs of PS-based polyHIPEs produced using different aqueous phase temperatures: (a) room temperature; (b) 80°C. The scale bars are 100 μm. (Reprinted from Reference 99 with permission from the Royal Society of Chemistry.)

holes result from postpolymerization processing such as drying or extraction [97]. Other results indicate that the holes are formed during polymerization since bicontinuous phase structures have been observed in polyHIPE that have not undergone postpolymerization processing [98].

4.1.4.2 Void and Interconnecting Hole Size. Various factors affect the droplet size of the dispersed phase. In the basic P(S-*co*-DVB) polyHIPEs the void size decreased with DVB content, reflecting the greater hydrophobicity of DVB [27]. The addition of a salt can also reduce the void size by an order of magnitude. The porous structure can be tailored by increasing the temperature of the internal aqueous phase (Fig. 4.5) or by adding organic additives that promote HIPE destabilization, yielding larger voids and larger interconnecting holes [99]. The coarsening of the emulsion results from an enhancement in Ostwald ripening, as can be determined from the rates of diffusion of water through the continuous phase.

4.1.4.3 Membranes and Beads. Membranes have been synthesized within HIPEs using several different methodologies. The simplest methodology is the polymerization of monomers within both phases of a HIPE [98]. In the case of a W/O HIPE, this yields a hydrophobic scaffold that is filled with a hydrogel, as seen in Fig. 4.6 and illustrated schematically in Fig. 4.7. Porous polyHIPE membranes for electrochemical sensor applications have been prepared by polymerization of a HIPE sandwiched between two polytetrafluoroethylene (PTFE) plates [100]. Another method for membrane formation is doctor blade casting, which has been used to prepare poly(4-vinylbenzyl chloride) membranes [101].

The polyHIPE beads shown in Fig. 4.8 were synthesized through the suspension polymerization of a HIPE, adding a W/O HIPE to water at elevated temperatures [90, 102, 103]. Sedimentation polymerization in hot oil was introduced for the production

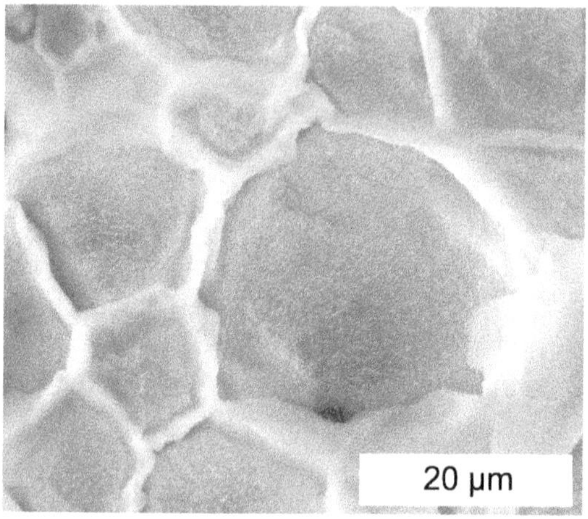

FIGURE 4.6 SEM micrograph of a polyHIPE containing a swollen hydrogel. (Reprinted from Reference 98 with permission from the Royal Society of Chemistry.)

of spherical polymer beads from an O/W HIPE [104, 105]. Hydrogel polyHIPE beads were produced through sedimentation polymerization of an aqueous monomer solution in the HIPE's external phase [69, 106]. The introduction of an inorganic precursor into the external phase produced a hybrid polyHIPE whose calcination yielded silica, alumina, zirconia, or titania beads [106–108].

4.1.5 PolyHIPE Properties

4.1.5.1 Surface Area. PolyHIPEs exhibit only modest surface areas, around 5–20 m^2/g, since the voids in polyHIPEs are tens of micrometers in diameter and the

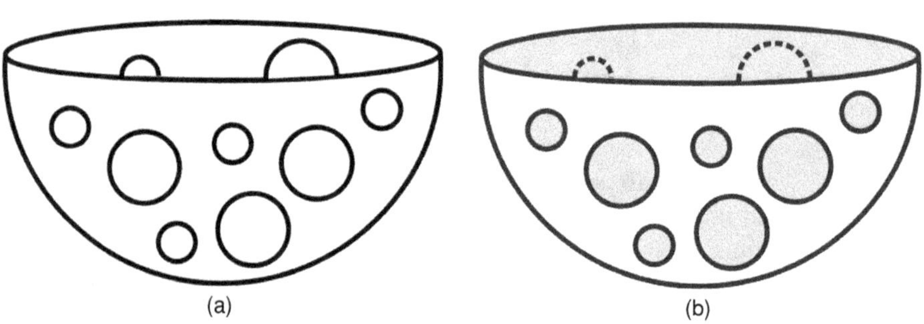

FIGURE 4.7 A schematic illustration of polyHIPE morphologies. **(a)** Cross section of a void that is surrounded by a hydrophobic polymer shell containing interconnecting holes. **(b)** Cross section of a void and its interconnecting holes that are filled with a swollen hydrogel. (Reprinted from Reference 98 with permission from the Royal Society of Chemistry.)

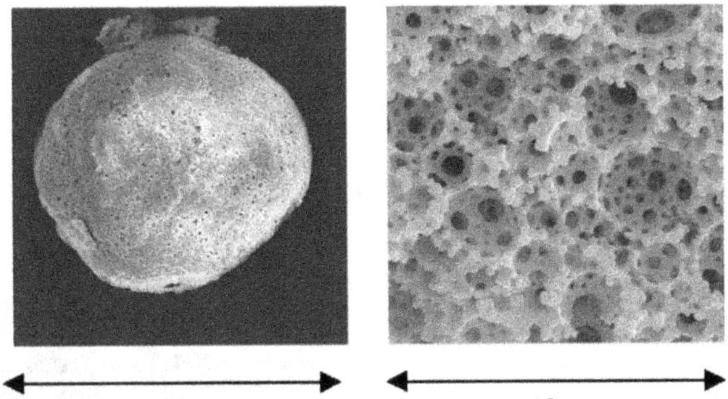

FIGURE 4.8 SEM micrograph of hydrophobic polyHIPE beads. (Reprinted from Reference 102 with permission from Elsevier.)

walls are essentially "solid" [1]. Surface areas of hundreds of square meters per gram have been achieved by adding a porogen to the external phase. Adding a porogen creates a second family of pores due to phase separation within the developing polymer matrix, resulting in surface areas up to 700 m^2/g and reducing the density by over a factor of two [64, 109–112]. While typical polyHIPE densities are around 0.1 g/cm^3, polyHIPEs with densities as low as 0.0126 g/cm^3 have been synthesized through variations in the emulsifier type, the emulsifier content, and the porogen content [113].

The use of a porogen often affects the structure of the wall. The wall structure can resemble the assembly of particles seen in Fig. 4.9 rather than the smooth walls seen in Fig. 4.3. This structure seems to reflect the formation and the phase separation of porogen-swollen microgel particles during polymerization. Eventually, the microgel particles aggregate and react to form the walls of the polyHIPE. This "particle assembly" wall structure is also seen for hydrophilic polyHIPEs synthesized from aqueous solutions of monomers (Fig. 4.10) with water acting as the swelling porogen. Recent work indicates that the porosity of a swollen polyHIPE can be significantly greater than that of a dried polyHIPE [114]. PolyHIPEs can also be "fixed" in the swollen state through hypercrosslinking, generating extremely high surface areas (up to 1200 m^2/g) [115].

4.1.5.2 Liquid Absorption. PolyHIPEs are able to absorb large quantities of liquid through capillary action [55]. Absorption begins when the polyHIPE is immersed in the liquid and can continue until the air has been displaced and the voids are filled with liquid. The tendency of a particular polyHIPE to absorb a particular liquid depends on the interfacial tension between the two, specifically on the contact angle. A PS-based polyHIPE will readily absorb toluene, and the walls will swell. Methanol will be absorbed to a lesser extent, and water will not be absorbed at all. The

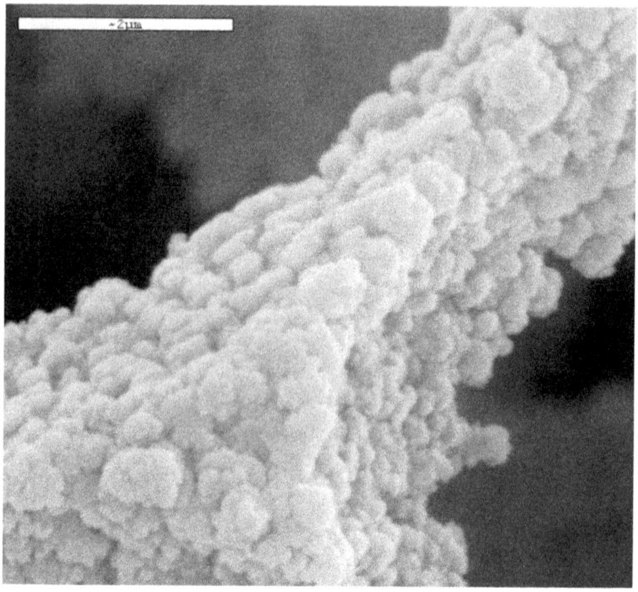

FIGURE 4.9 SEM micrograph of a polyDVB polyHIPE prepared with a porogenic solvent in the organic external phase. The scale bar is 2 μm. (Reprinted from Reference 110 with permission from the Royal Society of Chemistry.)

presence of residual surfactant in a PS-based polyHIPE, however, will lower the contact angle, and significant amounts of water can be absorbed. Hydrogel polyHIPEs will absorb water readily and will undergo extensive swelling. Liquids that are readily absorbed can be easily pumped through polyHIPEs. The back pressures generated by pumping will be relatively low owing to the relatively large pore size. The ability

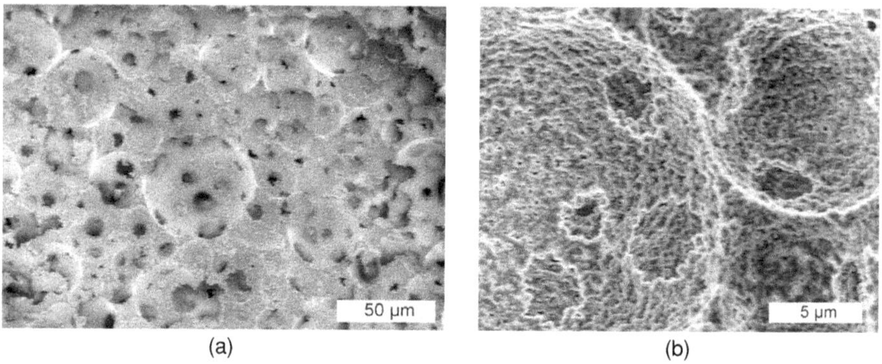

FIGURE 4.10 SEM micrographs of a hydrophilic polyHIPE prepared with an aqueous solution of monomer (HEMA) in the external phase with water acting as a swelling porogen. (Reprinted from Reference 72 with permission from the Royal Society of Chemistry.)

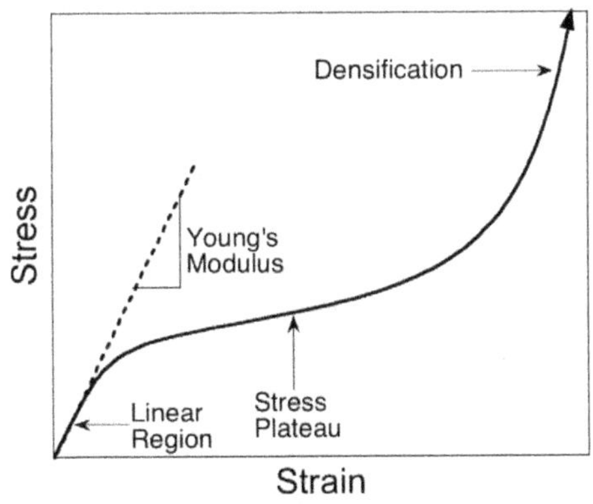

FIGURE 4.11 Schematic illustration of a typical polyHIPE stress–strain curve.

to pump liquids is of importance for many potential applications, including catalysis, liquid chromatography, separation, and tissue engineering.

4.1.5.3 Mechanical Properties. PolyHIPEs generally exhibit compressive stress–strain curves typical of foams. These curves contain three distinct regions: a linear elastic stress–strain region at low strains, from which the Young's modulus can be derived; a stress plateau region; and a densification or crushing region exhibiting a rapid rise in stress (Fig. 4.11). The modulus and the stress plateau decrease with decreasing density, reflecting the volumetric replacement of solid polymer with air. The modulus of a polyHIPE is also dependent upon the modulus of the wall material. The modulus of a stiff polymer far below its T_g can be orders of magnitude larger than the modulus of a flexible elastomer far above its T_g. A linear elastic analysis of foams has resulted in the following equation [116], which describes the relationship between the relative modulus (foam modulus E_f divided by wall modulus E_w) and the relatively density (foam density ρ_f divided by wall density ρ_w) in terms of ϕ, the fraction of the solid material that is found in the edges of the cells surrounding the voids. The fraction of the solid material that is found in the walls of the cells surrounding the voids is $(1 - \phi)$:

$$\frac{E_f}{E_w} = C_1\phi^2\left(\frac{\rho_f}{\rho_w}\right)^2 + C_2(1-\phi)\left(\frac{\rho_f}{\rho_w}\right) \tag{4.1}$$

where C_1 and C_2 are proportionality constants that are approximately equal to 1. For an open-cell foam, ϕ is, by definition, equal to 1 since an open-cell structure is all edges and no walls. Experimental data indicate that ϕ lies between 0.6 and 0.8 for closed-cell foams, which have cell walls in addition to the cell edges [116].

It is often imperative to improve the mechanical properties of polyHIPEs, specifically the modulus and the toughness, to enhance their suitability for particular applications. Equation 4.1 indicates that increasing the density by increasing the volume fraction of monomer will yield a significant increase in modulus. Relatively tough porous polymers were synthesized using relatively low internal phase contents of around 60%. These polyHIPEs had densities of around 0.4 g/cm^3 [117]. PS-based polyHIPEs initiated using KPS in the internal aqueous phase had a higher modulus than identical formulations initiated using AIBN in the external organic phase [27]. Changing the locus of initiation affected both the macromolecular structure and the porous structure, thus affecting the modulus [68, 98, 118]. Increasing the surfactant content in PS-based polyHIPEs yielded a reduction in modulus through plasticization of the polymer [27]. Increasing the crosslinking comonomer content or copolymerization with a relatively stiff monomer produces an increase in modulus, usually at the expense of toughness. On the other hand, copolymerization with a relatively flexible monomer yields an increase in toughness, usually at the expense of modulus. There are a large number of polyHIPE-based systems in which the synthesis of interpenetrating polymer networks, organic–inorganic hybrids, and nanocomposites have been used to enhance polyHIPE mechanical properties.

4.1.6 PolyHIPE-Based Systems

4.1.6.1 Crystallinity, Interpenetrating Polymer Networks, Degradability, and Hydrogels. Many different routes have been used to produce molecular architectures within polyHIPEs that go beyond the typical crosslinked, amorphous, hydrophobic polymers. PolyHIPEs are, for the most part, amorphous owing to the nature of the monomers used and to the relatively high degrees of crosslinking. PolyHIPEs with significant degrees of crystallinity have been synthesized by using monomers (acrylates and methacrylates) with long aliphatic side chains [68, 118, 119]. The movement of the side chains is not completely restricted by the crosslinked network of the polymer backbone, and therefore they are able to crystallize.

Simultaneous interpenetrating polymer networks (IPNs) based on crosslinked polystyrene and crosslinked polyurethane (PU) were synthesized within polyHIPEs [120]. Styrene was polymerized and crosslinked with divinylbenzene while a diol was polymerized and crosslinked with an isocyanate in simultaneous but mutually exclusive reactions. When S/DVB constituted 60 mol% or more of the monomers the result was a typical polyHIPE porous structure with the expected porosity. The polyHIPE tended to collapse at smaller S/DVB contents, reflecting the reduction in molecular stiffness. In addition, hydroxybutyl methacrylate, which has a vinyl group that could react with the styrene as well as a hydroxy group that could react with the isocyanate, was used as a comonomer to create interconnected networks. Interpenetrating polymer networks based on polyHIPEs were synthesized by swelling crosslinked polyHIPEs in solutions of monomer, crosslinking comonomer, and initiator and then polymerizing the monomers imbibed within the polyHIPEs [121, 122].

PolyHIPEs with biodegradable moieties were synthesized by using a vinyl terminated oligomeric polycaprolactone or polylactide as comonomers [75, 123–127]. Ultimately, all of the crosslinking comonomer could be replaced by a degradable crosslinker. Fully biodegradable scaffolds were prepared by polymerization of a poly(propylene fumarate) macromonomer crosslinked with propylene fumarate diacrylate using toluene as a diluent [128].

While the majority of polyHIPEs are hydrophobic polymers synthesized within W/O HIPEs, there are a large number of hydrophilic polyHIPEs that have been synthesized (see previous discussion). O/W HIPEs of cyclohexane in an aqueous solution of HEMA yielded void sizes of 5 to 18 μm, while W/O HIPEs of water in a solution of HEMA in dioxane yielded void sizes of approximately 0.6 μm [71].

4.1.6.2 Composite PolyHIPEs. Nanocomposite polyHIPEs were produced by adding nanoscale inorganics to polyHIPEs or by synthesizing nanoscale inorganics *in situ*. If there are no chemical bonds, then a blend is formed, as illustrated in Fig. 4.12a. If there is one chemical bond, then the inorganic moiety is grafted to the polymer backbone (Fig. 4.12b). If there are multiple chemical bonds, then the inorganic moiety acts as a crosslinker (Fig. 4.12c). Exfoliated montmorillonite was added to the comonomers in HIPEs to produce clay nanocomposite polyHIPEs [129]. Silsesquioxane (SSQ) networks were generated *in situ* through the copolymerization of hydrophobic monomers with vinyl-containing trialkoxysilanes [121, 130, 131]. This copolymerization yielded interconnected organic and inorganic networks. Preformed SSQ capable of forming no bonds, one bond, or multiple bonds were also added to HIPEs to produce nanocomposite polyHIPEs [132–134]. The porous structures of the inorganic monoliths formed by pyrolyzing the inorganic–organic hybrid polyHIPEs are quite similar to those of the original polyHIPEs.

The reaction between tetraethoxyorthosilane (TEOS) and various functional silanes in the aqueous phase of O/W polyHIPEs was used to synthesize inorganic monoliths with highly interconnected porous structures [135, 136]. Silica nanoparticles coated with methacryloxypropyltrimethoxysilane (MPTS) were incorporated into polyHIPEs and, even at 1%, produced a significant increase in surface roughness [137]. A significant enhancement in toughness was produced by adding 1% silica nanoparticles to an emulsion templated polymer with a continuous phase content of 40%. Carbon black was incorporated into polyHIPEs by prepolymerization,

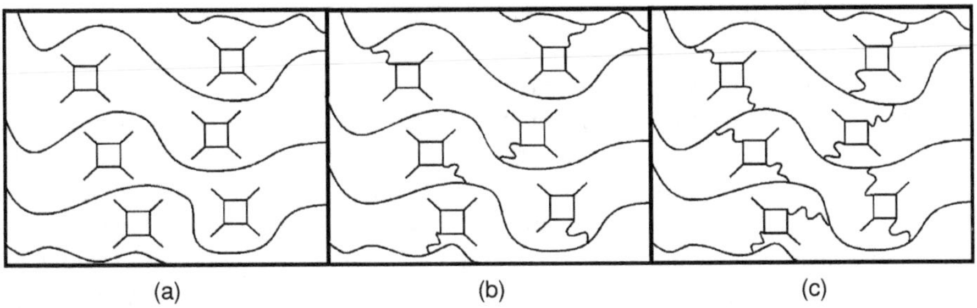

FIGURE 4.12 Schematic illustration of routes for incorporating a nanoscale inorganic within a polyHIPE. The inorganic can be generated *in situ* or preformed. **(a)** Blending; **(b)** grafting; **(c)** crosslinking. (Reprinted from Reference 134] with permission from John Wiley & Sons.)

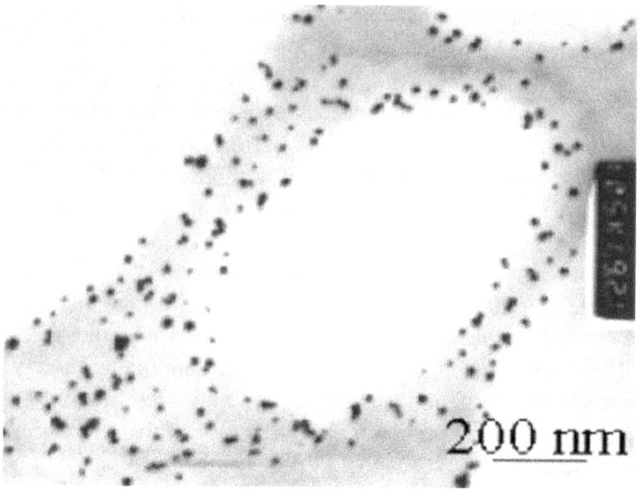

FIGURE 4.13 TEM micrographs of a polyHIPE filled with palladium nanoparticles. (Reprinted from Reference 141 with permission from the American Chemical Society.)

grafting the polymer to the carbon black before HIPE formation [138]. The addition of only 1% carbon black produced a significant increase in surface roughness. Carbon nanotubes and titania nanorods were also incorporated into polyHIPEs [139].

Palladium (Fig. 4.13) and gold particles were synthesized *in situ* within polyHIPEs for applications that include catalysis [140–142]. Scandium oxide, tungsten, and gold particles were incorporated into polyHIPEs to enhance their suitability for inertial confinement fusion experiments [143]. In addition, polyacrylamide polyHIPE beads were shown to be capable of absorbing gold nanoparticles from solution [144].

4.1.7 Applications

The scientific and patent literature is filled with a wide variety of potential applications for polyHIPEs. Much of the early polyHIPE research and development was motivated by high-energy physics applications, specifically inertially confined fusion [29]. Deuterated polystyrene-based polyHIPEs were used for thermal nuclear fusion targets [145]. The majority of applications take advantage of the porous structure and the relative ease with which liquids can be pumped through polyHIPEs. A large proportion of applications involve using polyHIPEs, or more often functionalized polyHIPEs, as a support material. Examples of polyHIPE applications include supports for solid-phase peptide synthesis [57], carriers for flavin in column reactors with high flow rates [146], supports for bicatalyst systems using quaternary ammonium groups [81], carriers for transporting hazardous or flammable liquids [55], and matrices for the immobilization of cells and enzymes [78, 147, 148]. Poly(glycidyl methacrylate)-based polyHIPEs have been functionalized and used as chromatographic columns for protein separation and the flow properties studied [89, 149].

Many applications use the water absorption capabilities of polyHIPEs. Polyvinylpyridine was grafted to poly(S-*co*-VBC) polyHIPEs for the separation of heavy metals from water [150]. Sulfonation of a polyHIPE was used to produce an ion exchange resin [151]. Immobilized catalyst supports were produced by sulfonating poly(S-*co*-DVB), by functionalizing poly(DVB-*co*-VBC) or poly(S-*co*-DVB), or by using a functional acrylate comonomer [152–158]. A PNiPAAm-based hydrogel polyHIPE was used as a particle pump through which organic matter in the internal phase was released on contraction of the hydrogel at the lower critical solution temperature (LCST) in water [159]. Halo-organic contaminants were removed from water using polyHIPEs [64, 160]. A rapid and reversible clathrate hydrate hydrogen storage system was produced by adding a tetrahydrofuran (THF)–water system to a PS-based polyHIPE [161].

The porous structure, ability to absorb water, and well-defined porous morphology make polyHIPEs into materials of interest for biomedical applications. PolyHIPEs were investigated as matrices for the culture of myoblasts [125] and as supports for the degradation of phenol in a bioreactor [162–164]. A commercial polyHIPE support with a polydimethylacrylamide layer, PolyHIPE SU500, was synthesized by Phase Separations (Deeside, UK) and used for the solid-phase synthesis of a 94-mer peptide sequence from the protein tau-2 [165]. The destabilization of HIPEs through the addition of a partitioning agent to enhance Ostwald ripening was used to produce relatively large void sizes for the culture of a wide variety of cell types in three dimensions [166–169] (Fig. 4.14). PolyHIPEs coated with hydroxyapatite were also used for cell growth [170].

The relatively large surface area of polyHIPEs compared to nonporous materials makes conducting polyHIPEs of interest for sensing applications. PolyHIPEs were coated with polythiophene, polypyrrole, and polyaniline to synthesize conductive

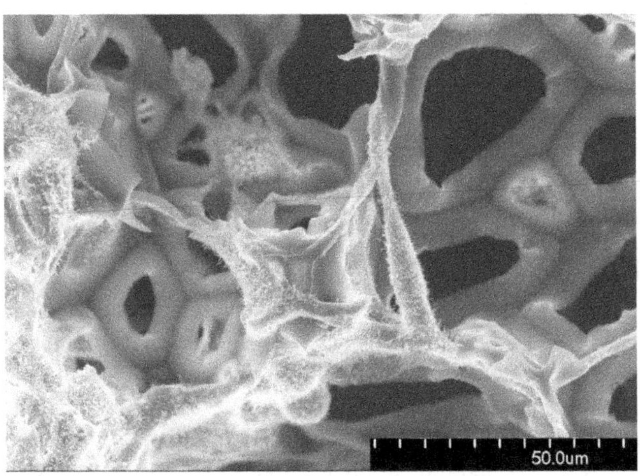

FIGURE 4.14 High-resolution SEM micrograph of hepatocytic cell growth on a polyHIPE. The scale bar is 50 μm. (Reprinted from Reference 168 with permission from Elsevier.)

porous polymers for applications such as electromagnetic interference shielding and sensors [121, 171–175]. In other work, nickel was deposited onto the internal surface of a polyHIPE [176, 177]. Functionalized polyHIPEs, with or without graphite particles, were used for selective electrochemical sensing [100]. PolyHIPEs coated with a conducting polymer were also used as sensors for organic vapors [121]. Alternative electrodes for lead–acid battery applications were produced through the deposition of PbO_2 on polyHIPEs [178]. PolyHIPEs have also been used as templates for the synthesis of porous inorganic monoliths. The pyrolysis of polyHIPEs has been used to produce porous carbon monoliths for a variety of applications [77, 179]. The calcination of hybrid polyHIPEs was used to prepare polyHIPE-templated inorganic monoliths [121, 180].

4.2 BICONTINUOUS MICROEMULSION TEMPLATING

4.2.1 Introduction

Porous polymers can also be obtained by the polymerization of microemulsions. Microemulsions are an intermediate state between micellar solutions and emulsions and can be interpreted as swollen micellar systems. They are one-phase, transparent, thermodynamically stable systems with domain sizes between 10 and 250 nm (some definitions limit the size of the domains to approximately 100 nm). The term "microemulsion" was first used to describe such systems by Hoar and Schulman in 1943 [181], and the first book on the subject appeared in 1977 [182]. Microemulsions consist of at least three components, namely water, oil, and surfactant. The composition of a microemulsion can be figured as a ternary phase diagram, which is usually presented as a triangle with the fractions of water, oil, and surfactant plotted on each side (Fig. 4.15). For a certain system, different areas of the diagram will give different microemulsion structures. The gobular phases are oil in water (O/W) or water in oil (W/O). The *bicontinuous* (sometimes called middle) phase exists as an intermediate state in which oil and water domains are randomly dispersed in two phases. By changing the ratio of components, a microemulsion undergoes a phase inversion, from globular (either O/W or W/O) through bicontinuous to globular (either W/O or O/W) (Fig. 4.16). Apart from globular and bicontinuous microemulsions, phase equilibria, called *Winsor* systems, in which a globular microemulsion coexists with an oil or water phase (especially at lower surfactant concentrations), are also known [183]. An O/W microemulsion in equilibrium with an oil phase is called a *Winsor I* system, a W/O microemulsion in equilibrium with an aqueous phase is called a *Winsor II* system, and a bicontinuous microemulsion in equilibrium with both oil and water phases is called a *Winsor III* system. Considering the small domain size in a globular microemulsion and the number of domains (around 10^{17}/mL), the surface to be stabilized by a surfactant is relatively high (approximately 100 m^2/mL for domain sizes around 18 nm). Therefore, a high amount of a surfactant usually has to be used, as high as 1:1 weight ratio in relation to the total amount of monomer. High surfactant demands can be a drawback when using a microemulsion as a precursor

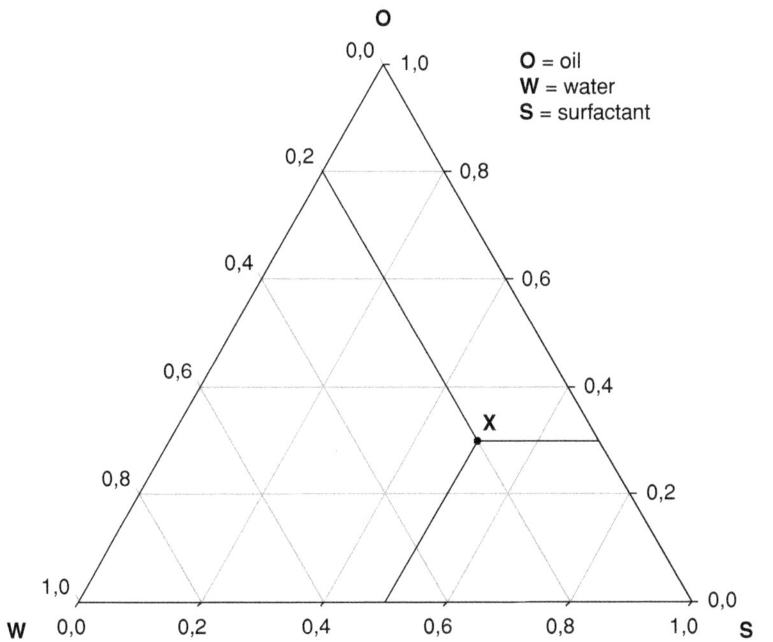

FIGURE 4.15 Ternary phase diagram for a microemulsion system.

for a polymeric material since some surfactants are expensive and can be difficult to remove after the polymerization. This drawback is one of the limiting factors for using microemulsions as material precursors on an industrial scale. Some studies have focused on this problem, such as those conducted by Gan et al. using *Winsor I* type systems with monomer:surfactant ratios up to 14:1 [184, 185]. The efficiency of a surfactant can be expressed as the volume of a microemulsion that is stabilized

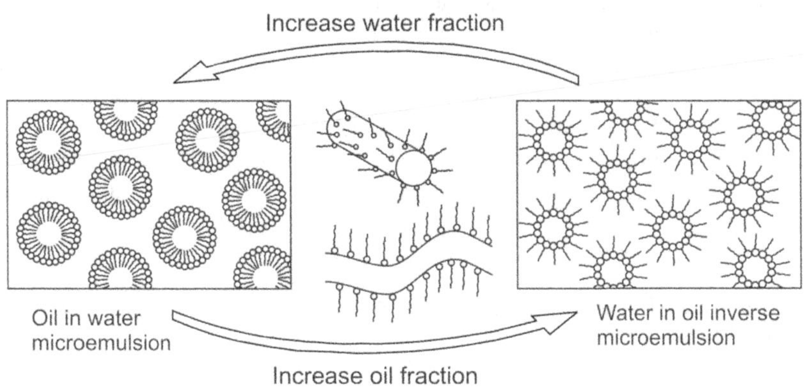

FIGURE 4.16 Phase inversion of a microemulsion.

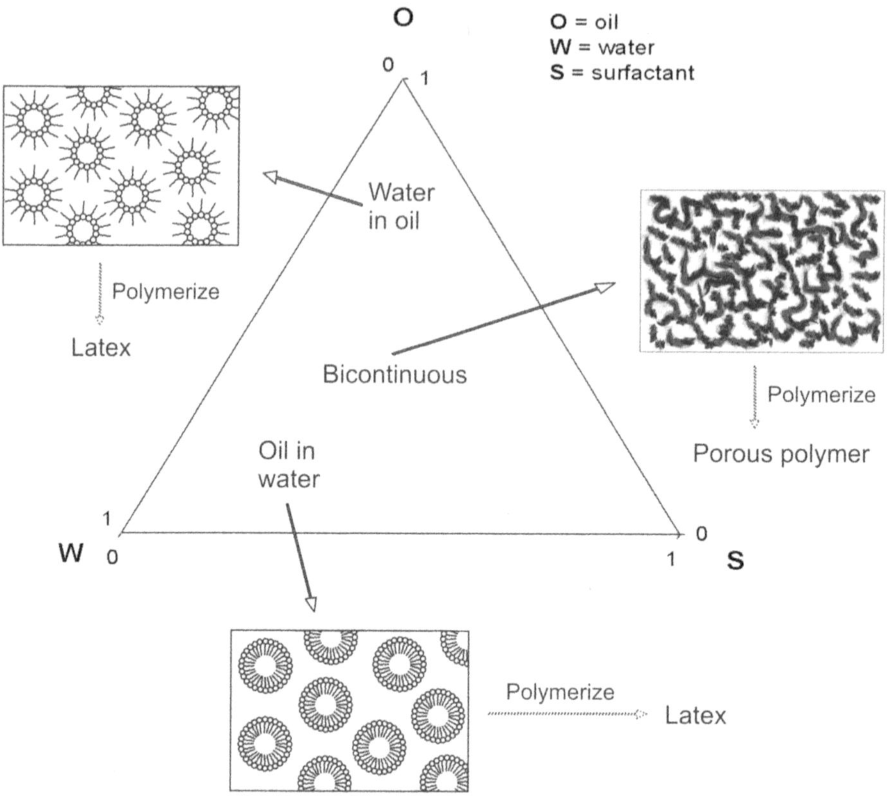

FIGURE 4.17 Materials produced from the polymerization of microemulsions.

per gram surfactant and can vary significantly, depending on the surfactant structure and the system used. A cosurfactant is usually needed when single-chain anionic surfactants (like sodium dodecyl sulfate [SDS]) are used. The use of nonionic, some cationic, or double-chain surfactants normally does not require a cosurfactant.

Materials obtained from microemulsions can be in the form of a latex or a porous, monolithic material. Latex materials can be produced from either globular or bicontinuous microemulsions, while porous polymers are normally derived from bicontinuous compositions, although globular particles can aggregate and yield porous materials as well (Fig. 4.17). While the first production of latex microparticles from a microemulsion was reported in the early 1980's [186, 187], the first reports of microstructured porous materials date to the late 1980s and early 1990s [188, 189].

4.2.2 Porous Polymers Derived from Microemulsions Containing Polymerizable Surfactants

The introduction of polymerizable surfactants or cosurfactants enables the production of transparent microstructured and nanostructured polymers from bicontinuous microemulsions. Actually, monomers that can be polymerized in an inverse or bicontinuous microemulsion, like acrylamide, acrylic acid, HEMA, or sodium acrylate, can be regarded as polymerizable cosurfactants; however, their efficiency is not as high as with "true" surfactants. In addition to superior microemulsion stabilization, polymerizable surfactants also omit the step of surfactant removal after polymerization. Polymerizable surfactants incorporated in the final polymer structure help to retain the bicontinuous structure of the precursor microemulsion. This prevents phase separation, and, therefore, transparent, porous materials can be obtained. Often, up to 4% ethylene glycol dimethacrylate (EGDMA) (based on the monomer mass) is used as a crosslinking comonomer to enhance the mechanical properties of the resulting polymer.

Transparent solid polymers could not be prepared from a MMA/AA/W system using sodium acrylamidoundecanoate and sodium acrylamidostearate as polymerizable surfactants when the water content was lower than 15%, and only opaque solids were obtained [190, 191]. A polymerizable zwitterionic surfactant, AUDMAA (Table 4.1), was used to prepare transparent, porous polymers from MMA/EGDMA. The bicontinuous structure of the precursor microemulsion was retained, and the micrograph in Fig. 4.18 clearly reveals a channel-like microstructure [192].

PHEMA and its copolymers are used in the production of contact lenses, and, therefore, the transparency and porosity of the obtained material is critical. HEMA/MMA/EGDMA microemulsions using a mixture of polymerizable (Na11EAAU, Table 4.1) and nonpolymerizable surfactants (SDS,) were also investigated [193]. Globular and bicontinuous microemulsion compositions were polymerized and the surfactant ratio varied. Both types of microemulsion yielded porous polymers. In the case of globular microemulsions, the particles aggregated; in the case of bicontinuous microemulsions, the polymers were transparent when a larger amount of the polymerizable surfactant was used and were opaque when a larger amount of nonpolymerizable surfactant was used.

Microemulsions containing a polymerizable surfactant were also used to prepare porous polymers in the form of membranes for ultrafiltration [194]. Two surfactants, AUTMAB and AUDMAA (Table 4.1), were used to stabilize the MMA/HEMA/W microemulsion with up to 4% EGDMA. Nanostructured membranes with pore sizes between 1 and 5 nm were obtained using a redox initiator (tetramethylene diamine [TMEDA]). These membranes were transparent and exhibited good water permeability. Increasing the water content in the original microemulsion produced an increase in the pore size. The nature of the surfactant also influenced the pore size, which was smaller when AUDMAA was used.

Microporous polymeric composites were obtained using the aforementioned monomer mixture (HEMA/MMA/EGDMA) by incorporating a nonionic macromonomer C1-PEO-C11-MA-40 (Table 4.1) as a polymerizable surfactant [195, 196]. Again, a transparent, porous material was obtained with pore sizes between 1 and 10 nm. Membranes were also made from this microemulsion, and the separation of PEGs was used to evaluate the pore size. A PEG-based monomer surfactant has also been used to stabilize the microemulsion system consisting of acrylonitrile (AN), 4-vinylbenzenesulfonic acid, EGDMA, and water [197]. PAN

TABLE 4.1 Polymerizable Surfactants Used in the Preparation of Porous Polymers from Microemulsions

Surfactant Name	Type	Formula
(((Acryloyloxy)-undecyl) dimethylammonio) acetate AUDMAA	Zwitterionic	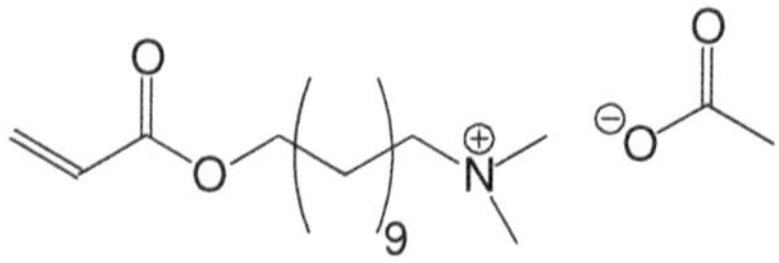
((Acryloyloxy)-undecyl) trimethylammonium bromide AUTMAB	Cationic	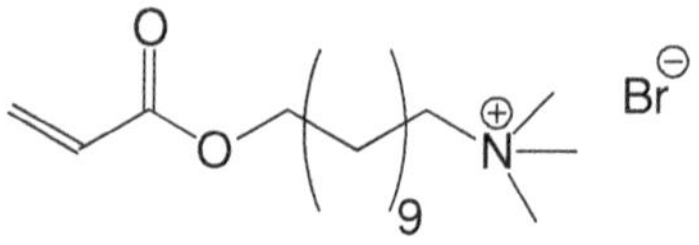
Sodium 11-(*N*-ethylacrylamido) undecanoate Na11-EAAU	Anionic	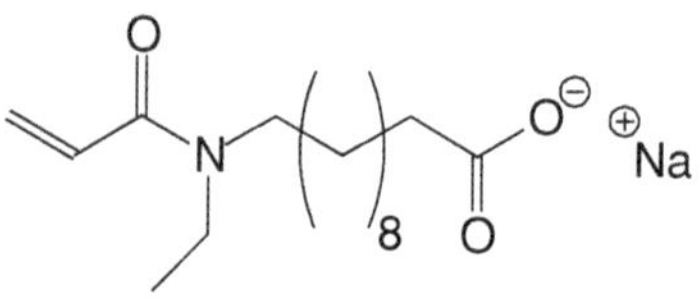
ω-Methoxy poly(ethyleneoxide)$_{40}$-undecyl-α-methacrylate C1-PEO-C11-MA-40	Polymeric	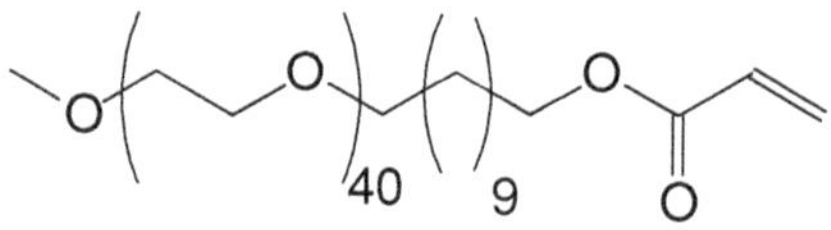
3-((11-Acryloyloxyundecyl) imidazolyl)propyl sulfonate AIPS	Zwitterionic	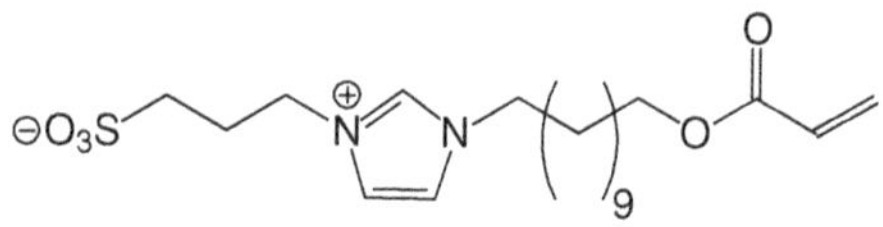
Potassium 10-undecenoate PUD	Anionic	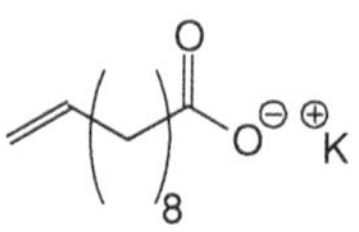

membranes crosslinked with EGDMA were obtained, and their ionic conductivity depended on the water content of the precursor microemulsion.

Recently, the preparation of proton exchange membranes from a system containing a sulfonate monomer [$CH_2{=}C(CH_3)COOCH_2SO_3K$]/MMA/EGDMA/W and using the polymerizable surfactant AIPS (Table 4.1) has been described [198]. Scanning electron microscopy has shown that the membranes consist of winding polymer and aqueous channels, retaining the structure of the bicontinuous microemulsion. The use of an anionic polymerizable surfactant, potassium 10-undecenoate (PUD), in an MMA/EGDMA/W microemulsion system yielded a flake-like structure rather than the "winding" bicontinuous morphology of the precursor microemulsion [199].

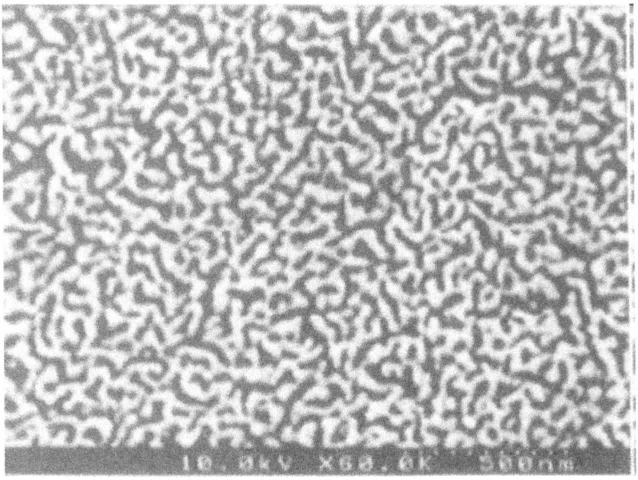

FIGURE 4.18 SEM of a microemulsion-templated, transparent, porous polymer monolith. The scale bar is 500 nm. (Reprinted from Reference 192 with permission. Copyright 1995 American Chemical Society.)

Materials synthesized from microemulsions with higher amounts of water had larger surface areas.

A polymerizable surfactant with an ionic liquid–inducing imidazolium head group has also been synthesized and used to prepare materials by microemulsion templating [200]. Polymerization of the bicontinuous phase gave rise to materials with an interconnected network of pores with sizes in the range 3–8 μm, suggesting that the bicontinuous microemulsion structure was not transferred to the porous polymer.

4.2.3 Porous Polymers Derived from Microemulsions Containing Nonpolymerizable Surfactants

The preparation of transparent microporous solid polymers from a HEMA/MMA/EGDMA/W microemulsion system stabilized by a nonpolymerizable surfactant (*n*-dodecyl trimethylammonium bromide) has been studied [201]. Initially, transparent bicontinuous microemulsions yielded a polymer with a spherical (agglomerated) structure, indicating that the structure of the microemulsion was not retained following polymerization. Transparent materials were obtained using water contents between 30% and 35%; however, after drying, the materials became opaque. The pore size could be controlled by varying the surfactant concentration. The same research group also tested the influence of the alkyl chain length of *n*-alkyl trimethylammonium bromide (C_{12} to C_{16}) on the polymerized product of the same microemulsion system (HEMA/MMA/EGDMA/W) [202]. The use of a shorter-alkyl-chain (C_{12}) surfactant reduced the opacity of the resulting polymer. In the case of C_{12}, an agglomerated globular structure was obtained (globules between 20 and 200 nm). The use of a longer-chain surfactant resulted in elongated, tubular particles that agglomerated to form a monolith, and transparent polymers were not achieved.

The HEMA/MMA/EGDMA/W system was also polymerized using ammonium persulfate/TMEDA as an initiator and SDS (Table 4.2) as a surfactant [203]. The bicontinuous microemulsion yielded a porous polymer made up of agglomerated spherical particles. Polymer membranes were also synthesized using the same microemulsion system using DTAB (Table 4.2) as a surfactant [204]. The membranes prepared from microemulsions with SDS had larger pores (between 100 nm and 3 μm) than the membranes prepared from microemulsions with DTAB (<100 nm). Increasing the concentration of DTAB produced a decrease in the pore size, while the opposite trend was found for SDS. AA/MMA-based microemulsions using AIBN as the initiator and SDS as the surfactant were also studied [205]. The surfactant concentration, the ratio of oil to water, the ratio of AA to MMA, and the amount of crosslinking agent (EGDMA) were varied to probe the influence on polymer morphology. The SDS concentration and the variations in crosslinking agent did not have a profound effect on the morphology, while the AA-to-MMA ratio and the oil-to-water volume ratio did. An open cellular material was obtained at higher water contents (>40%), while a closed-cell morphology was obtained at lower water contents. However, higher water contents produced a wider pore size distribution. Pore sizes were generally larger at higher MMA contents. When the AA content was greater than 60% the resulting polymer was not porous.

DDAB (Table 4.2) was used to stabilize various microemulsions, including MMA/EGDMA/W, S/DVB/W, butyl methacrylate (BMA)/EGDMA/W, AA/EGDMA/hexane, and methacrylic acid (MAA)/EGDMA/hexane [206]. Bicontinuous microemulsions were polymerized to yield porous polymers that were transparent in the case of hydrophilic monomers (AA, MAA) and opaque in the case of hydrophobic monomers (S, BMA). However, even the transparent samples did not retain the microemulsion structure after drying. An attempt was made to improve the properties of HEMA-based hydrogels by using HEMA/W/propanol/methylcyclohexane or HEMA/methylcyclohexane microemulsions stabilized with SDBS (Table 4.2) [207]. These hydrogels exhibited a higher equilibrium water content; however, they became opaque upon swelling. The oxygen permeability was also better than that of standard PHEMA hydrogels.

Embedded graft copolymers of vinyltoluene and AAm/AA were prepared from a microemulsion of vinyltoluene/W stabilized with CTMAB or SDS (Table 4.2) with AA or AAm in the aqueous phase [208]. This led to the polymeric composites having a hydrophilic polymer grafted to the hydrophobic structure. Both an oil-soluble (azobisisobutyronitrile [AIBN]) and a water-soluble initiator ($K_2S_2O_8$) were used. The resulting structure consisted of beads of PAA or PAAm inside the pores of polyvinyltoluene. A polymerized microemulsion based on styrene in the oil phase and AA in the water phase using cetyltrimethylammonium chloride as the surfactant, *m*-diisopropenylbenzene as the crosslinker, and AIBN as the initiator was termed a heterophasic bicontinuous polymer colloid-in-water structure [209]. The resulting polymers exhibited domain sizes ranging from 50 nm to 1 μm, depending on the monomer content. Microemulsions using cocogems (counterion coupled geminis),

TABLE 4.2 Nonpolymerizable Surfactants Used in the Preparation of Porous Polymers from Microemulsions

Surfactant Name	Type	Formula
n-Dodecyltrimethylammonium bromide DTAB	Cationic	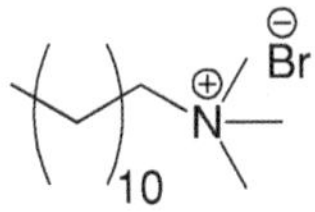
Sodium dodecyl sulfate SDS	Anionic	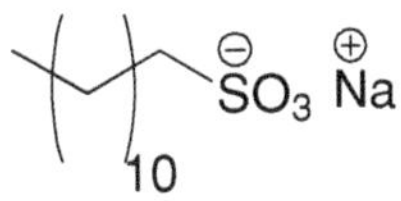
Sodium dodecyl benzene sulfonate SDBS	Anionic	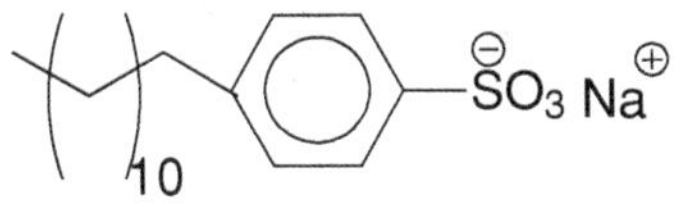
Cetyltrimethylammonium bromide CTMAB	Cationic	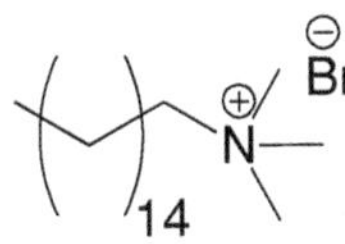
Didodecyldimethylammonium bromide DDAB	Cationic	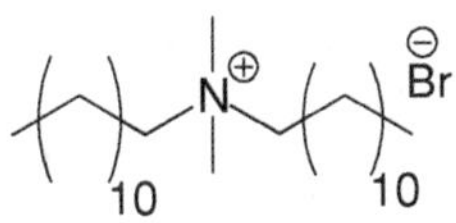
Sodium bis(2-ethylhexyl)sulfosuccinate SBEHSS or AOT	Anionic	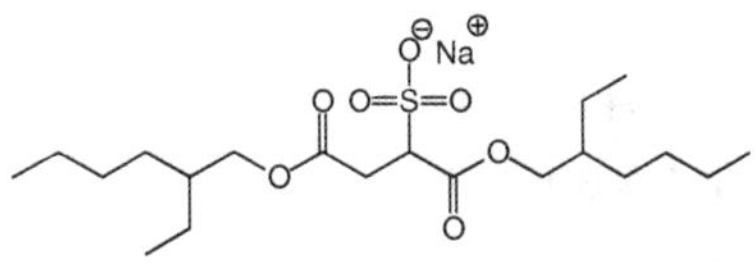
Cetyltrimethylammonium-counterion $CTMA_2$-tartrate, $CTMA_2$-terephthalate, $CTMA_2$-oxalate, CTMA-propanoate	Cocogems (counterion modified surfactants)	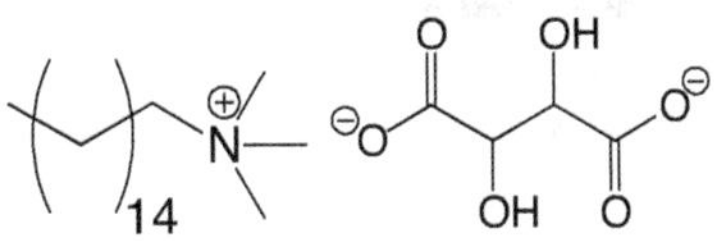 tartrate 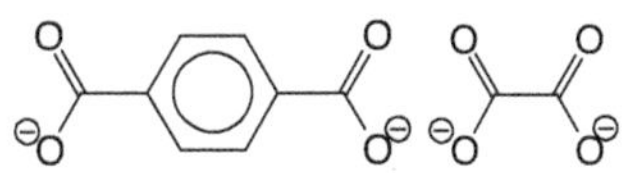 terephthalate oxalate 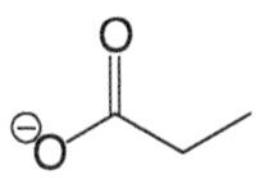 propanoate

CTMA or $CTMA_2$-counterion (Table 4.2), or the block copolymer poly(styrene-*b*-ethyleneoxide) as surfactants were also studied [210]. Cocogems have counterions that are coupled to a micelle or to an aggregate, and this results in higher steric demands and therefore in higher surfactant efficiency (more microemulsion stabilized per mole of surfactant). Variations in the counterion produce variations in the hydrophilic–lipophilic balance (HLB) and variations in the critical micelle concentration (CMC). Sponge-like PAAm gels and porous PS gels were prepared from microemulsions using cocogems.

There are few examples of using step-growth polymerization to prepare porous polymers by microemulsion templating. Porous melamine–formaldehyde resins were synthesized using a nonionic surfactant either alone or mixed with an anionic surfactant [211]. Similarly, porous resorcinol–formaldehyde resins were prepared using CTAB as surfactant [212]. Although porous materials were produced in both cases, the microemulsion structure was not transferred to the polymer.

4.2.4 Applications of Microemulsion-Derived Porous Polymers

Apart from producing latices, microemulsions have been used as precursor templates for preparing porous membranes (Table 4.3). PAN membranes crosslinked with EGDMA were prepared from an AN/EGDMA/W microemulsion stabilized

TABLE 4.3 Applications of Microemulsion-Derived Porous Polymers

Precursor Microemulsion	Material Obtained	Application
MMA/HEMA/EGDMA/W AUTMAB and AUDMAA	Thin film	Ultrafiltration membranes
ST/DVB/VPPBPA/W AOT	Porous monolith, crushed to particles	Catalyst support
AN/EGDMA/W C_1-PEO-C_{11}-MA-40	Thin film	Membranes for nanofiltration
AN/EGDMA/W/4-vinylbenzenesulfonic acid lithium salt C_1-PEO-C_{11}-MA-40	Thin film	Ion conductivity membranes
MMA/EGDMA/W AIPS	Thin film	Proton exchange membranes
HEMA/MMA/EGDMA/NIPAAm/W C_1-PEO-C_{11}-MA-40	Thin film	Membranes for wound dressing and cell grafting
BMA/EDMA/W SDS	Monolithic columns, polymerized inside capillaries	Columns for capillary electrochromatography

with PEO-C11-MA40 (Table 4.1), a macromonomer polymerizable surfactant [213]. The morphological structure of the membranes consisted of nanosized pores ranging from 0.38 to 2.4 nm, depending on the water content. Increasing the water content yielded an increase in the pore size. Membranes prepared from W/O microemulsions were water selective in the pervaporation of high-ethanol-content aqueous solutions. PAN membranes containing the lithium salt of 4-vinylbenzensulfonic acid were used for their ionic conductivity [197].

Porous resorcinol–formaldehyde resins prepared by microemulsion templating were pyrolized to produce monolithic carbon aerogel electrode materials for electrochemical capacitors [200]. The microemulsion templating methods create porous carbons with hierarchies of pores, from micropores (< 2 nm) to mesopores (2–50 nm), to macroporous channels with diameters >50 nm.

Nanostructured membranes prepared from a HEMA/MMA/NIPAAm/EGDMA/W bicontinuous microemulsion stabilized with a polymerizable surfactant (C_1-PEO-C_{11}-MA-40, Table 4.1) were thermosensitive and transparent and had a morphological structure of randomly distributed channels in the range 50–100 nm [214]. The high thermal stability (up to 300°C), appropriate mechanical properties, and appropriate water vapor transmission rate [around 600 g/(m^2 day)] of these membranes make them suitable materials for sterilizable wound dressings that could include controlled drug release.

Research in the field of proton exchange membranes has recently been growing rapidly due to fuel cell applications. Proton exchange membranes can also be prepared from a bicontinuous microemulsion containing MMA/EDGMA/W, a zwitterionic surfactant (AIPS, Table 4.1) and a sulfonic monomer for raising the proton conductivity [198]. The introduction of the zwitterionic surfactant proved to be of critical importance not only for the structure of the resulting polymer, but also for the proton conductivity. Replacing AIPS with C_1-PEO-C_{11}-MA-40 (Table 4.1) resulted in a drop of the proton conductivity by an order of magnitude.

Microemulsion-derived porous polymers have also found use in the field of separation methods. Stationary phases for capillary electrochromatography have been prepared from bicontinuous microemulsions such as BMA/EGDMA/W, using SDS (Table 4.3) as the surfactant and AIBN as the initiator and using 2-acrylamido-2-methyl-1-propane sulfonic acid to provide charges along the polymer backbone for the electroosmotic flow [215, 216]. Microemulsions were polymerized inside capillaries, thus simplifying column preparation (no packing needed). The resulting monoliths were permeable enough to allow flow. A test mixture of phthalates was successfully separated, although with lower efficiency compared to conventional packed columns.

Microemulsions were used to prepare porous polymer supports for catalysts [217]. The researchers synthesized a monomer with phosphonic acid moieties, which they copolymerized with styrene in a microemulsion formulation with AOT as the surfactant. The reactive moieties were located on the pore surface of the copolymer, and the accessibility of the reactive sites was 75%. Palladium was immobilized by ion exchange and the support used a catalyst for the hydrogenation of 1-octene. The polymeric catalyst could be used 10 times without any loss of activity.

4.3 WATER DROPLET TEMPLATING

In an article published in *Nature* in 1994, Widawski et al. demonstrated an elegant method that employs water droplets to create a honeycomb-like regular array of pores in polymer films [218]. A drop of a solution of either star-polystyrene or poly(styrene-*b*-paraphenylene) in carbon disulfide (CS_2) was deposited onto a flat surface and dried in a stream of moist air. The resulting polymer film was observed to possess a regular, hexagonal array of micrometer-sized holes (Fig. 4.19).

The authors made several observations on this process: Regular morphologies were not observed with linear polystyrenes in the absence of moisture or in methanol-saturated conditions; the diameter of the pores increased with the molecular weight of the arms of star polystyrene, although regularity was lost at $M_n > 50k$; regularity increased with the number of arms. The process by which porosity is created is a result of so-called "breath figures," that is, the formation of water droplets that condense on the surface of an evaporating polymer solution, which undergoes significant cooling as solvent is removed [219]. Such breath figure patterns have two important features that result in the formation of the observed porous morphology: They form a hexagonal, close-packed pattern of droplets on the surface, and the droplets do not coalesce on contact since they are separated by a film of solution.

In this initial work, it was claimed that a polymer architecture that resulted in the formation of a nanoscale spherical species, such as a star polymer or a block copolymer micelle, was a requirement for the formation of a regular array of pores. However, subsequent studies have shown that honeycomb-like porous films can in fact be formed from a wide variety of polymer architectures, including, in addition to those mentioned previously, random-coil linear polymers with [220, 221] or without [222] polar end groups, rigid-rod polymers [223], random [224] and block copolymers

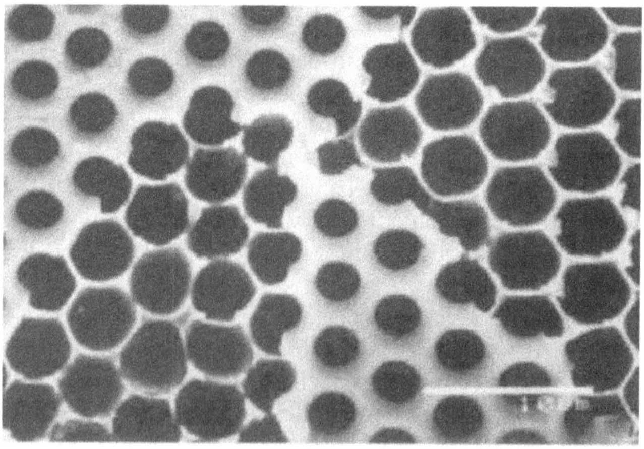

FIGURE 4.19 Honeycomb film prepared from star polystyrene. (Reprinted from Reference 218 with permission from Macmillan Publishers Ltd. Copyright 1994.)

[225], star microgels [226], dendronized polymers [227], and polymer blends [228]. Similar honeycomb films can be prepared by applying a slight vacuum to a polymer solution in a humid atmosphere to accelerate the evaporation process [229]. Porous films, generally with a lower degree of order, can also be produced by dip coating a planar substrate in a polymer solution and drying under relatively dry air [230]. However, the mechanism of bubble formation in this case is suggested to be due to solvent evaporation rather than water droplet templating; consequently, this work is not discussed further here.

4.3.1 Experimental Parameters

To try to elucidate the mechanism(s) that might produce the breath figure effect, several studies on the influence of experimental parameters on the porous morphology of the polymer films have been conducted. Highly ordered honeycomb films were produced from toluene solutions of polystyrene of M_w around 220k, whereas analogs of lower (29k) or higher (1970k) molecular weight gave polydisperse arrays of pores or very few pores, respectively [231]. Solution viscosity may be the cause of these differences; low-molar-mass polymers produce a solution of too low a viscosity to prevent water droplet coalescence, while a very high molar mass gives a polymer solution with such a large viscosity that water droplets cannot penetrate the surface. However, subsequent work did not indicate such a strong influence of solution viscosity and instead pointed toward the presence of polar functional groups as a key factor in promoting honeycomb film formation [221]. Linear polystyrene with either one or two –COOH end groups produced well-ordered porous films when evaporated from CS_2 in moist air, whereas similar polymers with nonpolar end groups produced films with irregular patterns of holes. The hypothesis is that the polar end groups interact with the surface of condensing water droplets, promoting their formation and enabling honeycomb film formation.

Further experiments have examined the role of solvent [221,231]. Key parameters were found to be solvent evaporation rate, water miscibility, and surface tension. The solvent should evaporate sufficiently slowly to allow the water droplets to form a regular array [231] but not too slowly; otherwise droplet coalescence may occur [221]. Furthermore, solvent–water miscibility should be limited; otherwise water droplets may dissolve, as is the case with THF. However, this observation regarding solvent affinity with water is not general; rather it is dependent on the particular system under investigation. Honeycomb films of cellulose acetate butyrate (CAB), carboxy-terminated polystyrene, and poly(methyl methacrylate) have been prepared from THF in humid air or by spin coating from mixtures of THF and water under "dry" conditions (relative humidity less than 30%) [232].

The solvent also has a strong influence on the number of layers of pores formed in the porous films. An examination of the literature reveals some conflicting and counterintuitive results. For example, Widawski et al. observed multilayer pore formation from CS_2 [218], whereas later workers observed only monolayer formation with the same solvent [220]. Several mechanisms have been proposed for multilayer formation (one example is shown in Fig. 4.20) [233], but regardless of which

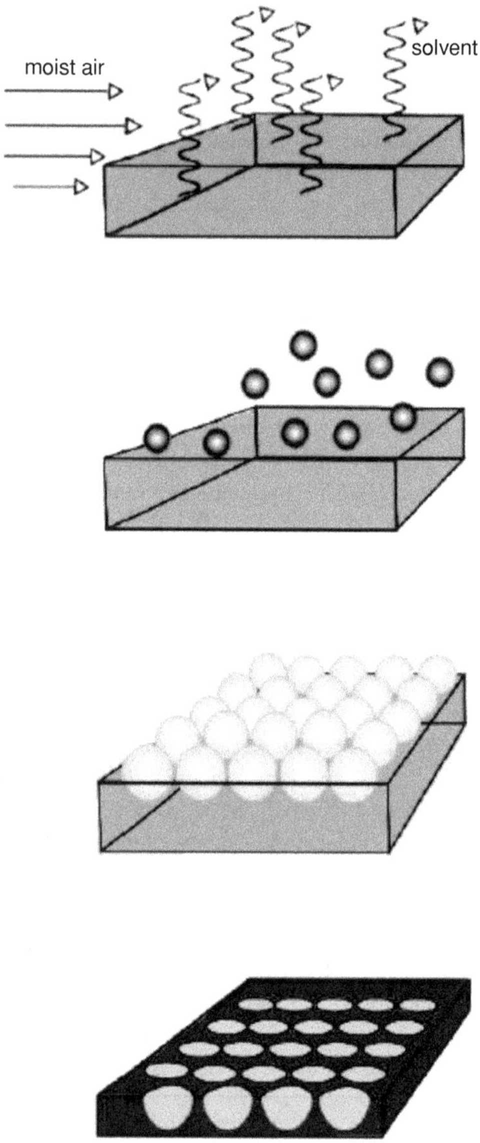

FIGURE 4.20 A suggested mechanism for the formation of three-dimensional porosity in honeycomb films. (Reproduced from Reference 233 with permission. Copyright Wiley-VCH Verlag GmbH & Co. KGaA.)

mechanism operates in any given system, it is clear that formation of a film with three-dimensional porosity requires water droplets to sink into the polymer solution. Factors that are likely to influence this process are solution viscosity and solvent density. It may be the case, therefore, that the different morphologies observed with CS_2

TABLE 4.4 Bolognesi's Interfacial Energy Balance Parameter z_0 and the Cross-Sectional Morphology of Honeycomb Films

Solvent	z_0	Film cross-sectional morphology
CS_2	0.84	Monolayer [220, 221]
$CHCl_3$	1.63	Multilayer [232, 234]
Benzene	1.33	Multilayer [220, 221]
Toluene	1.30	Multilayer [220, 221]
THF	2.76	Monolayer [232, 234]

are due to variations in solution viscosity caused by the use of dissimilar polymers and/or different concentrations rather than solvent density [220].

Later work also seems to contradict the suggestion that density is the dominant factor that dictates film morphology. Chloroform, which is denser than water, was shown to produce a multilayer of pores (albeit lacking three-dimensional order) in films of cellulose acetate butyrate, whereas THF, which is less dense than water, only gives a surface monolayer of well-ordered pores [232]. These results reinforce the view that surface tension plays an important role in determining cross-sectional morphology [221]. Consideration of the interfacial energy balance between water and the solvent allows one to predict whether water droplets will sink into the solution or not, leading to multilayer or monolayer formation, respectively. The key parameter is the interfacial energy balance parameter z_0, given as

$$z_0 = \frac{\gamma_W - \gamma_{S/W}}{\gamma_S} \tag{4.2}$$

where γ_W, γ_S, and $\gamma_{S/W}$ are the surface tensions of water and solvent and the solvent/water interfacial tension, respectively. If z_0 is between –1 and 1, the droplets will float on the surface and a monolayer of pores is formed. If z_0 is greater than 1, the droplets sink into the solvent and a multilayer is formed. Values of z_0 for the solvents mentioned in this section, together with the corresponding film morphologies, are given in Table 4.4. These predictions thus explain the results obtained with CS_2 (for which $z_0 < 1$) [220] and those obtained with chloroform (for which $z_0 > 1$) without the need to consider solution density. In the case of THF, $z_0 > 1$, which suggests that a multilayer pore structure should be produced. The monolayer of ordered pores observed is explained by the solubility of water droplets in THF, which causes the droplets that descend into the bulk solution to coalesce and mix with the solvent [232]. Work in which both the solvent–water interfacial tension was varied (by mixing THF and $CHCl_3$) and the polymer concentration was varied elegantly demonstrated that these factors influenced the morphology [234]. A high polymer concentration counteracts a low interfacial tension since a high solution viscosity inhibits the water droplet from sinking. Clearly, several factors are involved, and the morphology observed depends on the interplay between them.

4.3.2 Polymer Film Substrates

Initial work on honeycomb films prepared by the breath pattern method employed common polymers such as polystyrene and poly(methyl methacrylate), but the palette of chemistries available has since been extended significantly. This is due to the simplicity of the manufacturing process; the only requirement is that the polymer dissolves in a suitable organic solvent. Examples of the types of polymer that have been used to prepare honeycomb films include engineering plastics such as polyimide [235, 236], poly(phenyleneoxide) [237], poly(ether ether ketone) (PEEK) [238], and polycarbonate [239]; biopolymers such as cellulose and cellulose derivatives [232,234,240]; biodegradable synthetic polymers [241,242]; amphiphilic block copolymers [243, 244] and polyionic complexes [245]; ionomers [246]; and conjugated polymers such as polypyrrole, polythiophene, and poly(*p*-phenylene ethynylene) [223, 244]. Furthermore, a variety of polymer architectures can yield well-defined porous films (see earlier discussion). Highly stable crosslinked films have been prepared from solutions containing photocrosslinkable oligomers upon irradiation with ultraviolet (UV) light [247].

4.3.3 Morphology

Porous films produced by the breath pattern method can be processed postformation to give structures with interesting, novel morphologies. For some applications, such as guiding cells to grow in a particular direction, it may be advantageous to produce honeycomb films with nonspherical pores. This can be achieved by stretching a preformed honeycomb film of a viscoelastic polymer [248]. Stretching is performed after water droplet templating but before the film is fully dried. A range of pore shapes can be achieved, depending on the extent of stretching and the direction (uniaxial or biaxial; Fig. 4.21a). Elliptical pores can be produced by having moist air at a velocity of between 20 and 60 m/min across the solution surface [236]. Another process involves peeling the top surface of the porous films by the application of an adhesive tape [249]. The remaining "pin-cushion" structure consists of a regular series of polymer pins, which are the remaining halves of the pillars present between neighboring pores. Thermal treatment before peeling leaves pin cushions, rings, or microdots on the planar substrate, depending on treatment temperature (Fig. 4.21b–d) [250]. The pin-cushion structures can be used as negative molds for the creation of well-defined arrays of dots of another material (see later discussion). The honeycomb films and the resulting pin cushions have also been used as substrates for the creation of metallic microstructures by electroless plating followed by removal of the polymer by calcination [251].

4.3.4 Applications

One application of highly porous polymers in general is as substrates for the culture of cells in three dimensions, either for *in vitro* purposes or for tissue engineering. Honeycomb polymer films have been studied in this regard. The ability of porous films

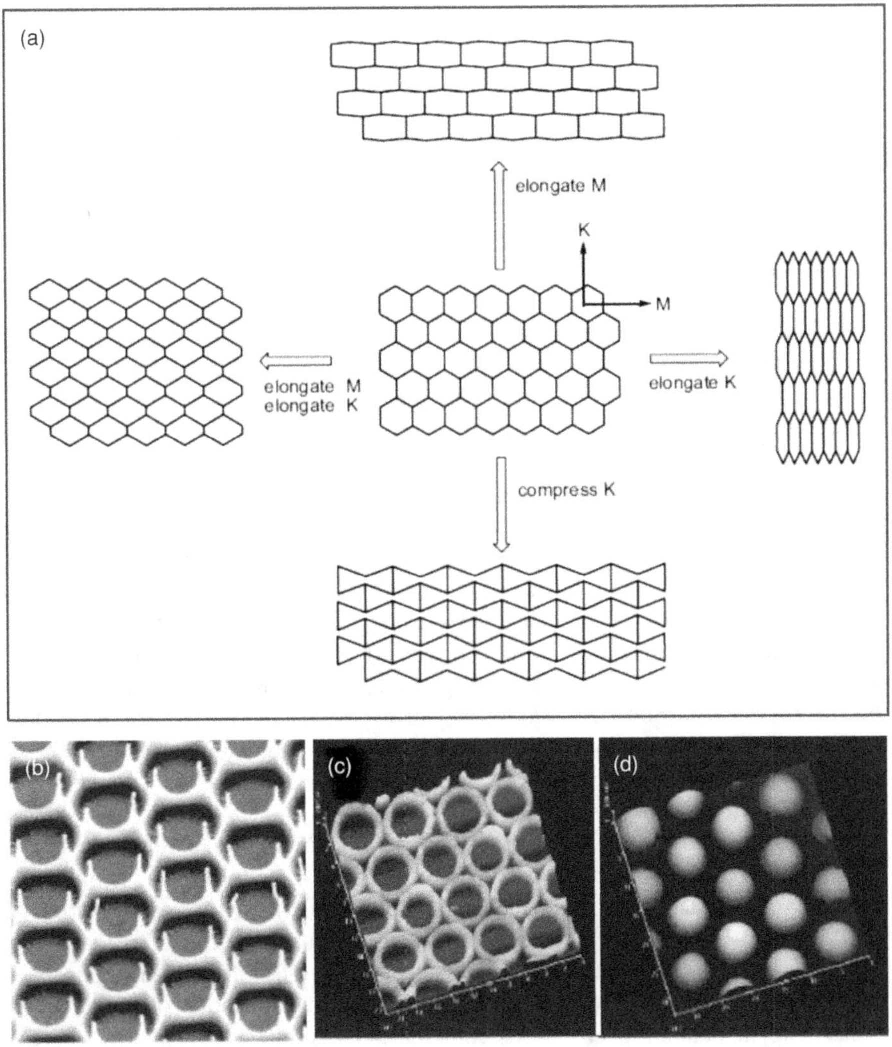

FIGURE 4.21 **(a)** Pore morphologies available through uniaxial or biaxial stretching or compression of honeycomb films. (Reproduced from Reference 233 with permission. Copyright Wiley-VCH Verlag GmbH & Co. KGaA.) Morphologies produced from thermal treatment followed by peeling of the top surfaces: **(b)** pincushions; **(c)** rings; **(d)** microdots. (Reprinted from Reference 250 with permission. Copyright 2006 American Chemical Society.)

prepared from a polypyrrole–amphiphilic copolymer blend to support the growth of mouse fibroblasts was investigated [244]. The surfaces were found to produce a relatively nontoxic cellular response, and cell attachment and growth were optimal (i.e., the same as with tissue culture polystyrene) when the pore diameter was less than 1 μm. Osteoblast-like (MG63) cells were also shown to grow satisfactorily on polylactide and poly(lactic-*co*-glycolic acid) porous films [242]. A hydrophobic copolyacrylamide porous film was used as a substrate for hepatocyte (liver cell) culture and was compared to a flat surface of the same material [252]. The cells were observed to develop different morphologies over time: flattened and stretched on the two-dimensional (2D) substrate, and spheroidal on the porous film. Hepatocytes typically display a flattened morphology when cultured in two dimensions, and this leads to a reduction in cell function. In agreement with this, the production of urea [252] and albumin [253], two liver-specific markers, was upregulated in cells grown on honeycomb films compared to flat substrates. Endothelial cells cultured on the same films were shown to adhere around the rims of the pores, as indicated by focal adhesion kinase (FAK) expression patterns [254]. In contrast, FAK expression in 2D cultures was random. Further work has examined neural stem cell differentiation on such films [255]. Intriguingly, differentiation was found to be dependent on pore size; a film with 3-μm pores was able to maintain cells in an undifferentiated state throughout the culture experiment.

Honeycomb polymer films have been used to prepare well-defined microarrays of proteins [256]. Polystyrene doped with a small amount of amino-terminated polystyrene was used to cast honeycomb films. The amino groups were shown, by staining with a fluorescent dye, to line the inner surfaces of the pores. The top surface of the film was coated with poly(ethylene glycol) (PEG) to prevent nonspecific protein adsorption; then a model protein (bovine serum albumin [BSA]) was attached to the amino groups via a linker. Confocal microscopy indicated the presence of the labeled protein in the interior of the honeycomb film pores. A similar approach was used to prepare microarrays of streptavidin, which was immobilized on the internal surface of porous films by noncovalent interaction with surface-bound biotin [257].

The regular array of micrometer-sized pores found in films prepared by the breath pattern method lends itself to the preparation of structures for diverse applications in materials science. For example, honeycomb films with two layers of pores in the film cross section have been shown to display broadband antireflection behavior in the near-infrared (NIR) region of the spectrum [258]. Molecularly imprinted honeycomb films have also been prepared. Enhanced permeation of the template, (*S*)-5-benzylhydantoin, over both its enantiomer and a phenyl hydantoin derivative were demonstrated [259]. Pin-cushion structures (see earlier discussion) prepared from fluorinated polymer honeycomb films exhibit both superhydrophobic (water contact angle up to 170°) and lipophobic properties [249, 260]. Negative molds of these pin-cushion morphologies can be created from polydimethylsiloxane (PDMS) [221], and the resulting regular array of PDMS hemispheres acts as a rather efficient microlens array (Fig. 4.22) [261]. By stretching the honeycomb film prior to casting the elastomer mold, anamorphic microlens arrays can be fabricated [262]. The PDMS negative molds have been used to create microstructured organic light-emitting diodes (OLEDs) by solvent-assisted microcontact printing [263]. The microstructured OLEDs had improved efficiencies compared to previously prepared small-area polymer LEDs.

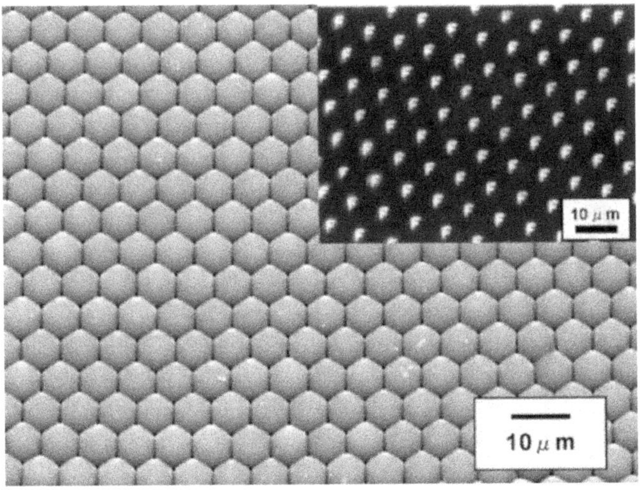

FIGURE 4.22 PDMS microlens array produced from a pin-cushion negative mold. Inset: Multiple copies produced when an image (the letter F) is projected through such an array. (Reprinted from Reference 261 with permission. Copyright 2005 American Chemical Society.)

4.4 PARTICLE TEMPLATING

4.4.1 Preparation of Porous Polymers by Solid-Particle Templating

In addition to soft colloidal templates (emulsion and water droplets, microemulsion domains), "hard" particles, such as silica nanoparticles, have been employed extensively for the creation of porous polymers. Once a population of hard particles of suitable diameter has been obtained, the concept is essentially the same as that involving other template types: Pack the particles into a given volume, fill the interstices with a polymerizable or curable liquid, solidify the nonparticulate phase, and then remove the template. Key challenges are establishing methodologies to create an interconnected network of pores following particle removal and carefully choosing the template material so that it can be removed without adversely affecting the nonparticulate phase. Depending on the size dispersity of the particulate material and the intended application, the creation of a highly ordered array of particles may also be important. It should be noted that there is a considerable body of literature on the creation of porous inorganic materials, such as silica, by this procedure; however, this is outside the scope of this review.

A wide range of hard templates, including salt crystals [264, 265], paraffin wax spheres [266–269], silica colloids [270, 271], and polymer latex particles [272] is available for the creation of porous polymers. When pore size uniformity is not required, particulate templates can be obtained by low-cost methods such as emulsion polymerization and used as prepared or fractionated into reasonably narrow

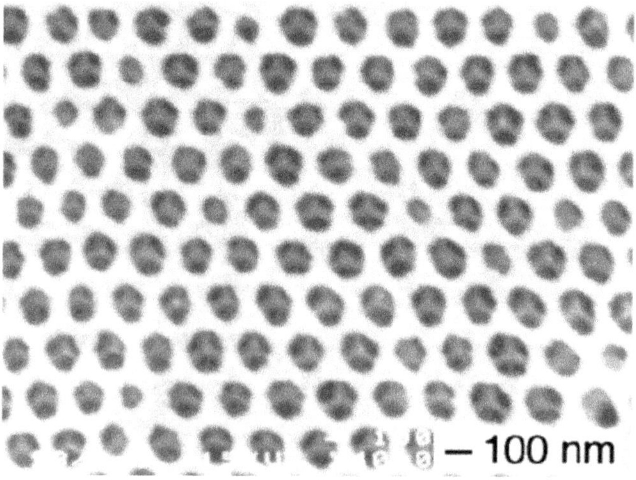

FIGURE 4.23 Highly ordered porous polyacrylate material produced by templating with 200 nm polystyrene beads. (Reprinted from Reference 273 with permission. Copyright 1999 American Chemical Society.)

diameter ranges by sieving. However, the creation of porous polymers with highly ordered 3D networks of monodisperse pores requires access to highly uniform template particles of monodisperse diameter. Fortunately, methods exist for the creation of monodisperse populations of many such hard-particulate templates. Porous materials were produced using monodisperse polymer beads and silica colloids with diameters ranging from 200 nm to 10 μm [273,274]. These were packed into a highly crystalline 3D array under the influence of both fluid flow and sonication. Once the colloidal crystal was assembled, it was dried and the interstices filled by capillary action with a UV-curable mixture (e.g., acrylates or polyurethane precursors). After curing, the template particles were removed by selective dissolution to leave a macroporous material that was an inverse replica of the colloidal crystal (Fig. 4.23).

Using this method, pore diameter and surface density can be tightly controlled, and highly ordered free-standing membranes of up to 1 cm^2 can be prepared. The diameter of the interconnecting "windows" between spherical pores is approximately one-fourth of the particle diameter, although no reason is given for the formation of such a highly interconnected network of pores. Related work on templating using silica particles, however, has shed some light on the formation of the interconnecting pores and the factors that govern their size [270]. Two possible candidate parameters are viscosity and wettability of the particle by the monomer(s). A viscous monomer (a commercial polyurethane prepolymer) was infiltrated into the colloidal crystal at increasing temperatures, giving materials with a concomitant decrease in interconnecting pore diameter. This suggests that viscosity influences pore diameter; a highly viscous monomer solution is less able to permeate the available void space by capillary action. However, a lower limit of 50 nm for interconnecting pore diameters was observed for low-viscosity monomers, suggesting that wetting is also important. Other polymer types, such as phenol–formaldehyde resins, can be prepared by this templating approach [275]. Similarly, macroporous conducting polymers impregnated with metallic and semiconductor nanoparticles have been prepared by electropolymerization of pyrrole within the interstices of a polystyrene bead colloidal crystal deposited on an indium tin oxide (ITO) substrate [276].

A similar templating approach has been used to prepare mesoporous (between 2 and 50 nm) materials [271]. Silica nanoparticles of diameter 35 nm were compressed to produce close packing and then sintered to give slight particle fusion. This allowed the creation of a porous interconnected network. DVB, EGDMA, or a mixture of the two was infiltrated into the fused colloidal crystal and then cured, after which the template was removed with a hydrofluoric acid (HF) solution. The resulting porous materials exhibited highly crystalline arrays of interconnected pores. Interestingly, the pore diameter depended on the chemical nature of the matrix: DVB produced a pore size almost identical to the particle diameter, whereas EGDMA produced a much lower diameter (15 nm) due to shrinkage on template removal. Mesoporous films were produced by layer-by-layer (LbL) assembly of a cationic polyelectrolyte with a blend of anionic silica nanoparticles and poly(acrylic acid) [277]. Highly porous polybenzimidazole (PBI), a high-performance polymer, has also been prepared from silica nanoparticle templates [278]. A crosslinking density of at least 10% was required to prevent pore collapse on template removal. These porous materials had a disordered array of 11-nm mesopores.

4.4.2 Applications

Polymers with hierarchies of pores can be produced by a dual templating method [279]. Poly(methyl methacrylate) (PMMA) beads were used to create macropores in a phenol–formaldehyde resin, with simultaneous creation of mesopores by the inclusion of a surfactant that self-assembled into micelles. Careful control of the synthetic protocol was required to ensure that both templates remained intact until sufficient network formation occurred. The resulting materials were used to prepare mesoporous carbon monoliths.

An elegant method for producing thin membranes with ordered monodisperse pores used the principle of particle-assisted wetting, whereby a water-immiscible liquid spreads on an air–water interface due to the presence of hydrophobic colloids [280]. A solution of hydrophobized silica nanoparticles (average diameter 320 nm) and polymerizable organic liquid (trimethacrylate) was added to a Langmuir trough containing water. The solvent was allowed to evaporate while the particles were compressed into a 2D ordered monolayer, after which the monomer was photopolymerized. The resulting film could be transferred to a variety of solid substrates and the colloids removed by etching to yield free-standing thin porous membranes (Fig. 4.24).

This method has subsequently been extended to the production of porous polymers with 3D ordered arrays of pores [281]. To achieve this, the authors simply increased the quantity of particle/monomer solution, such that a multilayer of particles was formed at the air–water interface. The evaporation rate of the solvent had to be tuned

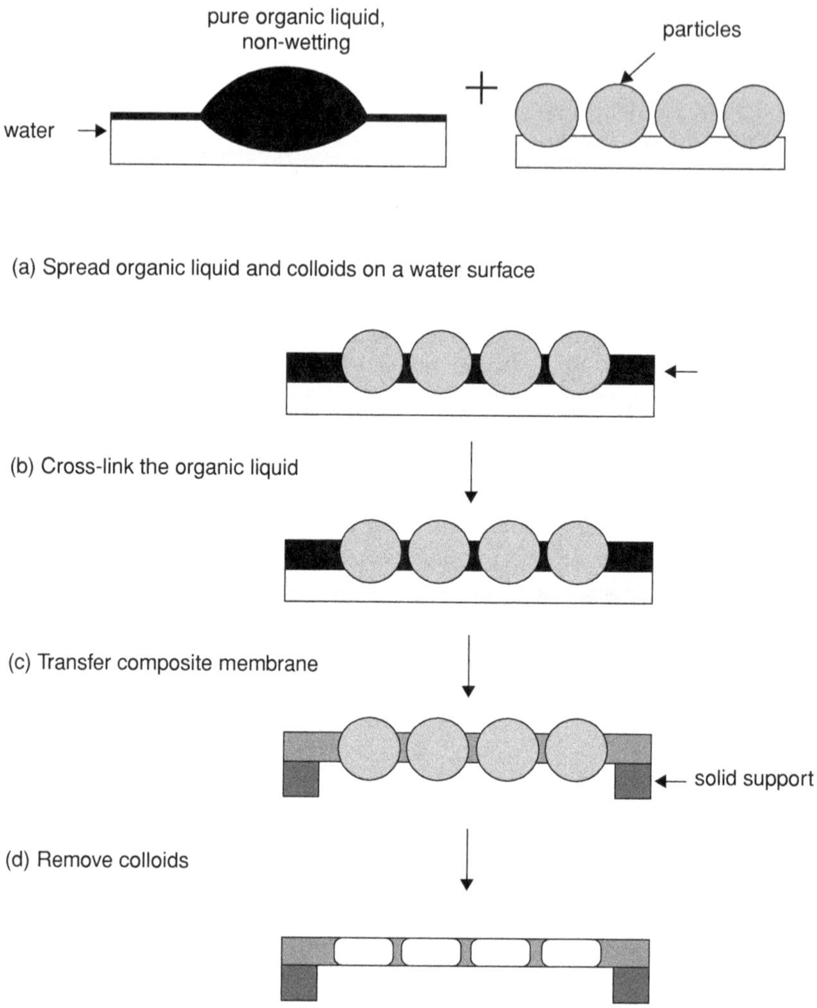

FIGURE 4.24 Creation of free-standing porous films by particle-assisted wetting. (Reproduced from Reference 280 with permission. Copyright Wiley-VCH Verlag GmbH & Co. KGaA.)

carefully to achieve highly ordered particle arrays, while the jump from two to three dimensions allowed the use of a Petri dish rather than a Langmuir trough to contain the system. To address the challenge of large-scale fabrication of such materials, spin coating has been employed to produce materials of up to 80 cm^2 in minutes [282]. Ordering of the particulate template occurs under shear, without the addition of a volatile solvent.

Porous materials prepared by hard-particulate templating have mainly found use as scaffolds for tissue engineering and as photonic band-gap materials. In the former case, pore diameters of tens to hundreds of micrometers together with good pore interconnectivity are required to enable cells to migrate throughout the porous structure. This, together with a tolerance of polydisperse pore sizes, has resulted in the widespread use of salt leaching as a method for the creation of such materials. Porous poly(L-lactic acid) (PLLA) scaffolds were prepared using a variety of salt types as porogens [283]. A slurry of polymer solution and salt particles was dried to produce a composite membrane, from which the salt was leached by immersion in water. Porosities up to 0.93 and a median pore diameter of up to 150 μm were achieved. Alternatively, mixtures of ground polymer particles and salt crystals were compression molded before salt leaching [284]. This work was subsequently extended to other polymer types, such as poly(lactic-*co*-glycolic acid) (PLGA) [285], poly(2-hydroxyethyl methacrylate) (PHEMA) [286], a photocured ethoxylated bis-phenol A derivative [287], and polymer blends such as PLGA with poly(ε-caprolactone) (PCL) [264], poly(ethylene oxide) (PEO) with poly(butylene terephthalate) (PBT) [288], and starch with PLLA [289].

There are, however, some problems associated with this simple salt leaching method. Salt crystals are difficult to remove at low porosities, and the porous scaffolds are fragile at high porosities. A modification to the salt leaching method was used to overcome these problems [284]. The polymer solution–salt mixture was first precipitated into a nonsolvent for the polymer, after which it was filtered, dried, and compressed into disks. The precipitation of the polymer–salt mixture ensured a uniform distribution of salt and polymer particles, leading to a homogeneous porosity. Scaffolds with pore volumes up to 96% were prepared in this manner. An alternative approach involved the use of a salt (ammonium bicarbonate) that degrades to gaseous products on contact with water. This "gas foaming/salt leaching" method led to porous materials with a highly interconnected network of relatively large (300–400 μm) pores [290]. An alternative approach combined supercritical CO_2 foaming with salt leaching to create similar materials [291]. Greater porosity and pore interconnectivity were achieved by utilizing a polymeric porogen in addition to salt crystals [292]. PEO and PCL form a co-continuous immiscible blend, which can be extruded with NaCl particles to form a composite with two water-leachable components. This produces scaffolds with dual porosity: larger pores (hundreds of micrometers) created by the salt crystals and smaller pores (<5 μm) from the PEO phase.

Other types of hard-particulate template have been used to prepare scaffolds for tissue engineering. One of the first examples is gelatin microspheres, which were used to create porous PLGA for bone tissue engineering [293]. Powdered PLGA and gelatin beads were heated under compression, after which the gelatin was removed by immersion in water at 37°C. The scaffold pore diameter was shown to increase with increasing gelatin microsphere diameter. Wax particulates, notably paraffin, have also been used as templates [266, 267]. Wax beads were added to a polymer–solvent mixture and served both to create pores and to precipitate the polymer from solution. The mixture was molded and the wax particles leached with a hydrocarbon solvent. The resulting materials had an interconnected network of pores, the shape of which depended on the shape of the particles, and they could be molded into complex shapes (Fig. 4.25).

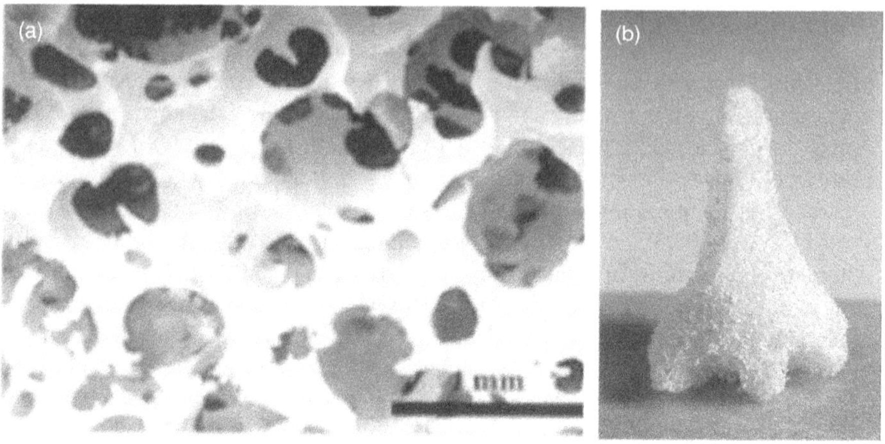

FIGURE 4.25 Paraffin bead–templated scaffolds: (**a**) morphology and (**b**) example of a complex geometry. (Reprinted from Reference 266. Copyright 2000 National Academy of Sciences, U.S.A.)

The extent of interconnectivity in such scaffolds can be improved by thermally bonding the paraffin particles in a separate step and then subsequently infiltrating the sintered particle network with the polymer solution drop by drop [269]. Polymeric particulate templates have also been used in this context [272]. In a further modification of this approach, the polymer solution/paraffin wax mixture is cooled to –70°C prior to particulate leaching to induce phase separation of the polymer [268]. This results in the formation of nanofibrous scaffolds with interconnected pores.

In contrast to tissue engineering scaffolds, photonic band-gap materials require a highly ordered network of monodisperse pores. As described earlier, such porosity can be generated by templating crystalline arrays of monodisperse particulates to produce "inverse opals." Photonic band-gap materials have been prepared using silica colloids with 200 to 400 nm diameters as templates [294]. This template size gives materials with stop-band gaps in the visible region of the spectrum. The preparation method allows control of the number of pore layers in the material, which was shown to have inverse effect on the stop-band gap width. A subsequent modification of this method involved high-temperature sintering of the silica colloidal crystal to improve the efficiency of template removal [295]. In order to tune the band gap, inverse opals were fabricated from an elastomeric polymer. Compression of the resulting rubbery porous network reversibly shifted the stop-band peak position. The material was subjected to thousands of cycles without any noticeable reduction in performance. Semiconducting photonic band-gap materials were prepared by electropolymerization of thiophene [296], pyrrole [296], or aniline [297] in the interstices of a polystyrene bead colloidal crystal deposited on a conducting surface. The thickness of the inverse opal was easily controlled by the electropolymerization time and the applied potential.

Macroscopically, the porous materials produced by this templating approach are iridescent due to the diffraction of selected wavelengths of incident visible light. This property has been used to form the basis of sensors, especially when the physical properties of the material, and thus its characteristic color, change in response to some external stimulus. Porous pH-sensing hydrogels containing acrylic acid were fabricated by a colloidal templating method [298]. The wavelength of diffracted light from the hydrogels changed from around 540 nm at pH < 4 to 850 nm at a pH of around 7 due to swelling of the hydrogel as the acrylic acid –COOH groups become deprotonated. Incorporation of a boronic acid monomer into similar porous hydrogels resulted in materials suitable for sensing glucose [299].

4.5 CONCLUSIONS

Colloidal templating provides a versatile method for the creation of a diverse range of well-defined, functional porous materials. The high degree to which the morphologies of colloid-templated porous polymers can be fine tuned reflects the well-known methodologies used for controlling the phase size and structure in the colloids used as templates. In many cases, the characteristic colloidal domain size is reproduced in the porous material, with the notable exception of bicontinuous microemulsion templating using nonpolymerizable surfactants. Colloid-templated porous polymers are being applied in a very wide range of areas of current scientific, engineering, and societal interest, including 3D cell culture, tissue engineering, supports for catalysts, electronic materials, and separation science.

ABBREVIATIONS

AA	Acrylic acid
AAm	Acrylamide
AIBN	2,2′-Azobisisobutyronitrile
AN	Acrylonitrile
ATRP	Atom transfer radical polymerization
BMA	Butyl methacrylate
BPO	Benzoyl peroxide
BSA	Bovine serum albumin
C/W	$scCO_2$-in-water
C_1	Proportionality constant
C_2	Proportionality constant
CAB	Cellulose acetate butyrate
CMC	Critical micelle concentration
CTAB	Cetyltrimethylammonium bromide
DVB	Divinylbenzene
E_f	Foam modulus
EGDMA	Ethylene glycol dimethacrylate

EtOH	Ethanol
E_w	Wall modulus
FAK	Focal adhesion kinase
HEA	2-Hydroxyethyl acrylate
HEMA	2-Hydroxyethyl methacrylate
HF	Hydrofluoric acid
HIPE	High internal phase emulsion
HIPRE	High internal phase ratio emulsion
HLB	Hydrophilic lipophilic balance
IPN	Interpenetrating polymer networks
ITO	Indium tin oxide
KPS	Potassium persulfate
LbL	Layer-by-layer
LCST	Lower critical solution temperature
LED	Light-emitting diodes
MAA	Methacrylic acid
MBAAm	*N*,*N*-Methylene bisacrylamide
MMA	Methyl methacrylate
MPTS	Methacryloxypropyltrimethoxysilane
NiPAAm	*N*-Isopropyl acrylamide
NIR	Near-infrared
O/W	Oil-in-water
OLED	Organic light-emitting diodes
PBI	Polybenzimidazole
PBT	Poly(butylene terephthalate)
PCL	Poly(ε-caprolactone)
PDMS	Polydimethylsiloxane
PEEK	Poly(ether ether ketone)
PEG	Poly(ethylene glycol)
PEO	Poly(ethylene oxide)
PHEMA	Poly(2-hydroxyethyl methacrylate)
PLGA	Poly(lactic-*co*-glycolic acid)
PLLA	Poly(L-lactic acid)
PMMA	Poly(methyl methacrylate)
polyHIPE	Polymerized high internal phase emulsion
PPO	Poly(propylene oxide)
PS	Polystyrene
PTFE	Polytetrafluoroethylene
PU	Polyurethane
RAFT	Reversible addition–fragmentation chain transfer
S	Styrene
$scCO_2$	Supercritical carbon dioxide
SEM	Scanning electron microscope
SSQ	Silsesquioxane
TEOS	Tetraethoxyorthosilane

T_g	Glass transition temperature
THF	Tetrahydrofuran
TMEDA	Tetramethylene diamine
UV	Ultraviolet
VBC	4-Vinylbenzyl chloride
W	Water
W/O	Water-in-oil
z_0	Interfacial energy balance parameter
ϕ	Fraction of solid material in cell edges
γ_S	Surface tension of solvent
$\gamma_{S/W}$	Solvent/water interfacial tension
γ_W	Surface tension of water
ρ_f	Foam density
ρ_w	Wall density

REFERENCES

1. Cameron, N. R.; Sherrington, D. C. *Adv Polym Sci* 1996, **126**, 163–214.
2. Ruckenstein, E. *Adv Polym Sci* 1997, **127**, 1–58.
3. Cameron, N. R. *Polymer* 2005, **46**, 1439–1449.
4. Zhang, H. F.; Cooper, A. I. *Soft Matter* 2005, **1**, 107–113.
5. Lissant, K. J. *J Colloid Interf Sci* 1966, **22**, 462–468.
6. Lissant, K. J.; Pearce, B. W.; Wu, S. H.; Mayhan, K. G. *J Colloid Interf Sci* 1974, **47**, 416–423.
7. Lissant, K. J.; Mayhan, K. G. *J Colloid Interf Sci* 1973, **42**, 201–208.
8. Kim, K. J.; Ruckenstein, E. *Makromol Chem Rapid Commun* 1988, **9**, 285–290.
9. Ruckenstein, E.; Kim, K. J. *J Appl Polymer Sci* 1988, **36**, 907–923.
10. Ruckenstein, E.; Park, J. S. *J Polym Sci C Polym Lett* 1988, **26**, 529–536.
11. Kunieda, H.; Evans, D. F.; Solans, C.; Yoshida, M. *Colloids Surf* 1990, **47**, 35–43.
12. Kunieda, H.; Yano, N.; Solans, C. *Colloids Surf* 1989, **36**, 313–322.
13. Kunieda, H.; Solans, C.; Shida, N.; Parra, J. L. *Colloids Surf* 1987, **24**, 225–237.
14. Ebert, G.; Platz, G.; Rehage, H. *Ber Bunsenges Phys Chem Chem Phys* 1988, **92**, 1158–1164.
15. Lissant, K. J. US 3732166, 1973.
16. Lissant, K. J. US 3700594, 1972.
17. Lissant, K. J. US 3617095, 1971.
18. Lissant, K. J. US 3523826, 1970.
19. Lissant, K. J. *Emulsions and Emulsion Technology*, Part 1, Chapter 1, Marcel Dekker, New York, 1974.
20. Princen, H. M.; Aronson, M. P.; Moser, J. C. *J Colloid Interf Sci* 1980, **75**, 246–270.
21. Princen, H. M. *J Colloid Interf Sci* 1979, **71**, 55–66.
22. Lissant, K. J. *J Soc Cosmet Chem* 1970, **21**, 141–154.

23. Princen, H. M.; Kiss, A. D. *Langmuir* 1987, **3**, 36–41.
24. Princen, H. M. *Langmuir* 1986, **2**, 519–524.
25. Mannheimer, R. J. *J Colloid Interf Sci* 1972, **40**, 370–382.
26. Williams, J. M. *Langmuir* 1991, **7**, 1370–1377.
27. Williams, J. M.; Gray, A. J.; Wilkerson, M. H. *Langmuir* 1990, **6**, 437–444.
28. Williams, J. M.; Wrobleski, D. A. *Langmuir* 1988, **4**, 656–662.
29. Williams, J. M. *Langmuir* 1988, **4**, 44–49.
30. Ford, R. E.; Furmidge, C. G. L. *J Colloid Interf Sci* 1966, **22**, 331–341.
31. Pons, R.; Solans, C.; Stebe, M. J.; Erra, P.; Ravey, J. C. *Prog Colloid Polym Sci* 1992, **89**, 110–113.
32. Rajagopalan, V.; Solans, C.; Kunieda, H. *Colloid Polym Sci* 1994, **272**, 1166–1173.
33. Kizling, J.; Kronberg, B. *Colloids Surf* 1990, **50**, 131–140.
34. Chen, H. H.; Ruckenstein, E. *J Colloid Interf Sci* 1990, **138**, 473–479.
35. Chen, H. H.; Ruckenstein, E. *J Colloid Interf Sci* 1991, **145**, 260–269.
36. Ruckenstein, E.; Ebert, G.; Platz, G. *J Colloid Interf Sci* 1989, **133**, 432–441.
37. Ruckenstein, E.; Sun, F. *J Appl Polymer Sci* 1992, **46**, 1271–1277.
38. Mork, S. W.; Rose, G. D.; Green, D. P. *J Surfact Deterg* 2001, **4**, 127–134.
39. Zhang, S.; Chen, J. *Polymer* 2007, **48**, 3021–3025.
40. Nixon, J.; Beerbower, A.; Wallace, T. J. *Mech Eng* 1968, **90**, 26.
41. Nixon, J.; Beerbower, A.; Wallace, T. J. *Mech Eng* 1968, **90**, 78.
42. Beerbower, A.; Nixon, J. *Abst Papers Am Chem Soc* 1969, PE32.
43. Nixon, J.; Beerbower, A. *Abst Papers Am Chem Soc* 1969, PE31.
44. Cameron, N. R.; Sherrington, D. C. *J Chem Soc Faraday Trans* 1996, **92**, 1543–1547
45. Cameron, N. R.; Sherrington, D. C. *Macromolecules* 1997, **30**, 5860–5869.
46. Mezzenga, R.; Fredrickson, G. H.; Kramer, E. J. *Macromolecules* 2003, **36**, 4457–4465.
47. Mezzenga, R.; Ruokolainen, J.; Fredrickson, G. H.; Kramer, E. J. *Macromolecules* 2003, **36**, 4466–4471.
48. Butler, R.; Davies, C. M.; Cooper, A. I. *Adv Mater* 2001, **13**, 1459–1463.
49. Butler, R.; Hopkinson, I.; Cooper, A. I. *J Am Chem Soc* 2003, **125**, 14473–14481.
50. Tan, B.; Lee, J. Y.; Cooper, A. I. *Macromolecules* 2007, **40**, 1945–1954.
51. Lee, J. Y.; Tan, B.; Cooper, A. I. *Macromolecules* 2007, **40**, 1955–1961.
52. Pickering, S. U. *J Chem Soc* 1907, **91**, 2001–2021.
53. Menner, A.; Verdejo, R.; Shaffer, M.; Bismarck, A. *Langmuir* 2007, **23**, 2398–2403.
54. Menner, A.; Ikem, V.; Salgueiro, M.; Shaffer, M. S. P.; Bismarck, A. *Chem Commun* 2007, 4274–4276.
55. Barby, D.; Haq, Z. EP 60138, 1982.
56. Sherrington, D. C.; Small, P. W. US 4965289, 1990.
57. Small, P. W.; Sherrington, D. C. *J Chem Soc Chem Commun* 1989, 1589–1591.
58. Bartl, H.; von Bonin, W. *Makromol Chem* 1962, **57**, 74–95.
59. Bartl, H.; von Bonin, W. *Makromol Chem* 1963, **66**, 151–156.
60. Rogez, D.; Marti, S.; Nervo, J.; Riess, G. *Makromol Chem* 1975, **176**, 1393–1409.
61. Horie, K.; Mita, I.; Kambe, H. *J Appl Polym Sci* 1967, **11**, 57–71.

62. Horie, K.; Mita, I.; Kambe, H. *J Appl Polym Sci* 1968, **12**, 13–21.
63. Litt, M. H.; Hsieh, B. R.; Krieger, I. M.; Chen, T. T.; Lu, H. L. *J Colloid Interf Sci* 1987, **115**, 312–329.
64. Sergienko, A. Y.; Tai, H. W.; Narkis, M.; Silverstein, M. S. *J Appl Polym Sci* 2004, **94**, 2233–2239.
65. Cameron, N. R.; Sherrington, D. C. *J Mater Chem* 1997, **7**, 2209–2212.
66. Duke, J. R.; Hoisington, M. A.; Langlois, D. A.; Benicewicz, B. C. *Polymer* 1998, **39**, 4369–4378.
67. Hoisington, M. A.; Duke, J. R.; Apen, P. G. *Polymer* 1997, **38**, 3347–3357.
68. Livshin, S.; Silverstein, M. S. *Soft Matter* 2008, **4**, 1630–1638.
69. Zhang, H.; Cooper, A. I. *Chem Mater* 2002, **14**, 4017–4020.
70. Krajnc, P.; Stefanec, D.; Pulko, I. *Macromol Rapid Commun* 2005, **26**, 1289–1293.
71. Kovacic, S.; Stefanec, D.; Krajnc, P. *Macromolecules* 2007, **40**, 8056–8060.
72. Kulygin, O.; Silverstein, M. S. *Soft Matter* 2007, **3**, 1525–1529.
73. Grosse, M. T.; Lamotte, M.; Birot, M.; Deleuze, H. *J Polym Sci A Polym Chem* 2008, **46**, 21–32.
74. Audouin, F.; Birot, M.; Pasquinet, E.; Deleuze, H.; Besnard, O.; Poullain, D. *J Appl Polym Sci* 2008, **108**, 2808–2813.
75. David, D.; Silverstein, M. S. *J Polym Sci A Polym Chem* 2009, **47**, 5806–5814.
76. Elmes, A. R.; Hammond, K.; Sherrington, D. C. EP 289238, 1994.
77. Even, W. R. J.; Gregory, D. P. *MRS Bull* 1994, **19**, 29–33.
78. Hitchen, D. A. EP 365327, 1993.
79. Deleuze, H.; Faivre, R.; Herroguez, V. *Chem Commun* 2002, 2822–2823.
80. Benmachou, K.; Deleuze, H.; Heroguez, V. *React Funct Polym* 2003, **55**, 211–217.
81. Ruckenstein, E.; Hong, L. *Chem Mater* 1992, **4**, 122–127.
82. Barbetta, A.; Cameron, N. R.; Cooper, S. J. *Chem Commun* 2000, 221–222.
83. Krajnc, P.; Brown, J. F.; Cameron, N. R. *Organic Lett* 2002, **4**, 2497–2500.
84. Leber, N.; Fay, J. D. B.; Cameron, N. R.; Krajnc, P. *J Polym Sci A Polym Chem* 2007, **45**, 4043–4053.
85. Mercier, A.; Deleuze, H.; Mondain-Monval, O. *React Funct Polym* 2000, **46**, 67–79.
86. Mercier, A.; Kuroki, S.; Ando, I.; Deleuze, H.; Mondain–Monval, O *J Polym Sci B Polym Phys* 2001, **39**, 956–963.
87. Mercier, A.; Deleuze, H.; Mondain-Monval, O. *Macromol Chem Phys* 2001, **202**, 2672–2680.
88. Deleuze, H.; Maillard, B.; Mondain-Monval, O. *Bioorg Med Chem Lett* 2002, **12**, 1877–1880.
89. Krajnc, P.; Leber, N.; Stefanec, D.; Kontrec, S.; Podgornik, A. *J Chromatogr A* 2005, **1065**, 69–73.
90. Stefanec, D.; Krajnc, P. *Polym Int* 2007, **56**, 1313–1319.
91. Pulko, I.; Kolar, M.; Krajnc, P. *Sci Total Environ* 2007, **386**, 114–123.
92. Livshin, S.; Silverstein, M. S. *J Polym Sci A Polym Chem* 2009, **47**, 4840–4845.
93. Moine, L.; Deleuze, H.; Maillard, B. *Tetrahedron Lett* 2003, **44**, 7813–7816.
94. Cetinkaya, S.; Khosravi, E.; Thompson, R. *J Mol Catal A Chem* 2006, **254**, 138–144.

95. Fernández-Trillo, F.; van Hest, J. C. M.; Thies, J. C.; Michon, T.; Weberskirch, R.; Cameron, N. R., *Adv Mater* 2009, **21**, 55–59
96. Cameron, N. R.; Sherrington, D. C.; Albiston, L.; Gregory, D. P. *Colloid Polym Sci* 1996, **274**, 592–595.
97. Menner, A.; Bismarck, A. *Macromol Symp* 2006, **242**, 19–24.
98. Gitli, T.; Silverstein, M. S. *Soft Matter* 2008, **4**, 2475–2485.
99. Carnachan, R. J.; Bokhari, M.; Przyborski, S. A.; Cameron, N. R. *Soft Matter* 2006, **2**, 608–616.
100. Zhao, C.; Danish, E.; Cameron, N. R.; Kataky, R. *J Mater Chem* 2007, **17**, 2446–2453.
101. Pulko, I.; Krajnc, P. *Chem Commun* 2008, 4481–4483.
102. Desforges, A.; Arpontet, M.; Deleuze, H.; Mondain-Monval, O. *React Funct Polym* 2002, **53**, 183–192.
103. Stefanec, D.; Krajnc, P. *React Funct Polym* 2005, **65**, 37–45.
104. Ruckenstein, E.; Hong, L. *Polymer* 1995, **36**, 2857–2860.
105. Kitagawa, N. EP 993337, 2000.
106. Zhang, H.; Hardy, G. C.; Rosseinsky, M. J.; Cooper, A. I. *Adv Mater* 2003, **15**, 78–81.
107. Zhang, H.; Hardy, G. C.; Khimyak, Y. Z.; Rosseinsky, M. J.; Cooper, A. I. *Chem Mater* 2004, **16**, 4245–4256.
108. Zhang, H.; Cooper, A. I. *Ind Eng Chem Res* 2005, **44**, 8707–8714.
109. Hainey, P.; Huxham, I. M.; Rowatt, B.; Sherrington, D. C.; Tetley, L. *Macromolecules* 1991, **24**, 117–121.
110. Cameron, N. R.; Barbetta, A. *J Mater Chem* 2000, **10**, 2466–2472.
111. Barbetta, A.; Cameron, N. R. *Macromolecules* 2004, **37**, 3188–3201.
112. Barbetta, A.; Cameron, N. R. *Macromolecules* 2004, **37**, 3202–3213.
113. Richez, A.; Deleuze, H.; Vedrenne, P.; Collier, R. *J Appl Polymer Sci* 2005, **96**, 2053–2063.
114. Jerabek, K.; Pulko, I.; Soukupova, K.; Stefanec, D.; Krajnc, P. *Macromolecules* 2008, **41**, 3543–3546.
115. Schwab, M. G.; Senkovska, I.; Rose, M.; Klein, N.; Koch, M.; Pahnke, J.; Jonschker, G.; Schmitz, B.; Hirscherd, M.; Kaskel, S., *Soft Matter* 2009, **5**, 1055–1059.
116. Gibson, L. J.; Ashby, M. F. *Cellular Solids: Structure and Properties*, Cambridge University Press, Cambridge, 1997.
117. Menner, A.; Haibach, K.; Powell, R.; Bismarck, A. *Polymer* 2006, **47**, 7628–7635.
118. Livshin, S.; Silverstein, M. S. *Macromolecules* 2008, **41**, 3930–3938.
119. Livshin, S.; Silverstein, M. S. *Macromolecules* 2007, **40**, 6349–6354.
120. Lepine, O.; Birot, M.; Deleuze, H. *Polymer* 2005, **46**, 9653–9663.
121. Silverstein, M. S.; Tai, H. W.; Sergienko, A.; Lumelsky, Y. L.; Pavlovsky, S. *Polymer* 2005, **46**, 6682–6694.
122. Tai, H.; Sergienko, A.; Silverstein, M. S. *Polym Eng Sci* 2001, **41**, 1540–1552.
123. Busby, W.; Cameron, N. R.; Jahoda, C. A. B. *Biomacromolecules* 2001, **2**, 154–164.
124. Busby, W.; Cameron, N. R.; Jahoda, A. B. C. *Polym Int* 2002, **51**, 871–881.
125. Lumelsky, Y.; Silverstein M. S. *Macromolecules* 2009, **42**, 1627.

126. Lumelsky, Y.; Zoldan, J.; Levenberg, S.; Silverstein, M. S. *Macromolecules* 2008, **41**, 1469–1474.

127. Lumelsky, Y.; Lalush-Michael, I.; Levenberg, S.; Silverstein, M. S. *J Polym Sci A Polym Chem* 2009, **47**, 7043–7053.

128. Christenson, E. M.; Soofi, W.; Holm, J. L.; Cameron, N. R.; Mikos, A. G. *Biomacromolecules* 2007, **8**, 3806–3814.

129. Lepine, O.; Birot, M.; Deleuze, H. *J Polym Sci A Polym Chem* 2007, **45**, 4193–4203.

130. Tai, H.; Sergienko, A.; Silverstein, M. S. *Polymer* 2001, **42**, 4473–4482.

131. Normatov, J.; Silverstein, M. S. *Polymer* 2007, **48**, 6648–6655.

132. Normatov, J.; Silverstein, M. S. *Macromolecules* 2007, **40**, 8329–8335.

133. Normatov, J.; Silverstein, M. S. *Chem Mater* 2008, **20**, 1571–1577.

134. Normatov, J.; Silverstein, M. S. *J Polym Sci A Polym Chem* 2008, **46**, 2357–2366.

135. Carn, F.; Colin, A.; Achard, M.-F.; Deleuze, H.; Sellier, E.; Birot, M.; Backov, R. *J Mater Chem* 2004, **14**, 1370–1376.

136. Ungureanu, S.; Birot, M.; Laurent, G.; Deleuze, H.; Babot, O.; Julian-Lopez, B.; Achard, M. F.; Popa, M. I.; Sanchez, C.; Backov, R. *Chem Mater* 2007, **19**, 5786–5796.

137. Haibach, K.; Menner, A.; Powell, R.; Bismarck, A. *Polymer* 2006, **47**, 4513–4519.

138. Menner, A.; Powell, R.; Bismarck, A. *Soft Matter* 2006, **2**, 337–342.

139. Menner, A.; Salgueiro, M.; Shaffer, M. S. P.; Bismarck, A *J Polym Sci A Polym Chem* 2008, **46**, 5708–5714.

140. Desforges, A.; Backov, R.; Deleuze, H.; Mondain-Monval, O. *Adv Funct Mater* 2005, **15**, 1689–1695.

141. Desforges, A.; Deleuze, H.; Mondain-Monval, O.; Backov, R. *Ind Eng Chem Res* 2005, **44**, 8521–8529.

142. Feral–Martin, C.; Birot, M.; Deleuze, H.; Desforges, A.; Backov, R. *React Funct Polym* 2007, **67**, 1072–1082.

143. Steckle, W.; Schoonover, J.; Lanier, N.; Nobile, A. *J Mater Sci* 2006, **41**, 4055–4060.

144. Zhang, H. F.; Hussain, I.; Brust, M.; Cooper, A. I., *Adv Mater* 2004, **16**, 27–30.

145. Zhang, L.; Tang, Y. J.; Zhong, C. F.; Luo, X. A.; Zhang, H. Q. *Nucl Instr Meth Phys Res A Accel Spectrom Detect Assoc Equip* 2002, **480**, 242–245.

146. Schoo, H. F. M.; Challa, G.; Rowatt, B.; Sherrington, D. C. *React Polym* 1992, **16**, 125–136.

147. Ruckenstein, E.; Wang, X.-B. *Biotechnol Bioeng* 1994, **44**, 79–86.

148. Ruckenstein, E.; Wang, X. *Biotechnol Bioeng* 1993, **42**, 821–828.

149. Junkar, I.; Koloini, T.; Krajnc, P.; Nemec, D.; Podgornik, A.; Strancar, A. *J Chromatogr A* 2007, **1144**, 48–54.

150. Benicewicz, B. C.; Jarvinen, G. D.; Kathios, D. J.; Jorgensen, B. S. *J Radioanal Nucl Chem* 1998, **235**, 31–35.

151. Wakeman, R. J.; Bhumgara, Z. G.; Akay, G. *Chem Eng J* 1998, **70**, 133–141.

152. Ottens, M.; Leene, G.; Beenackers, A.; Cameron, N.; Sherrington, D. C. *Ind Eng Chem Res* 2000, **39**, 259–266.

153. Krajnc, P.; Leber, N.; Brown, J. F.; Cameron, N. R. *React Funct Polym* 2006, **66**, 81–91.

154. Krajnc, P.; Stefanec, D.; Brown, J. F.; Cameron, N. R. *J Polym Sci A Polym Chem* 2005, **43**, 296–303.

155. Brown, J. F.; Krajnc, P.; Cameron, N. R. *Ind Eng Chem Res* 2005, **44**, 8565–8572.

156. Pierre, S. J.; Thies, J. C.; Dureault, A.; Cameron, N. R.; van Hest, J. C. M.; Carette, N.; Michon, T.; Weberskirch, R. *Adv Mater* 2006, **18**, 1822–1826.

157. Chemin, A.; Mercier, A.; Deleuze, H.; Maillard, B.; Mondain-Monval, O. *J Chem Soc Perkin Trans* 2001, **1**, 366–370.

158. Mercier, A.; Deleuze, H.; Maillard, B.; Mondain-Monval, O. *Adv Synth Catal* 2002, **344**, 33–36.

159. Zhang, H.; Cooper, A. I. *Adv Mater* 2007, **19**, 2439–2444.

160. Sergienko, A. Y.; Tai, H. W.; Narkis, M.; Silverstein, M. S. *J Appl Polym Sci* 2002, **84**, 2018–2027.

161. Su, F.; Bray, C. L.; Tan, B.; Cooper, A. I. *Adv Mater* 2008, **20**, 2663–2666.

162. Erhan, E.; Yer, E.; Akay, G.; Keskinler, B.; Keskinler, D. *J Chem Technol Biotechnol* 2004, **79**, 195–206.

163. Akay, G.; Erhan, E.; Keskinler, B. *Biotechnol Bioeng* 2005, **90**, 180–190.

164. Bokhari, M. A.; Akay, G.; Zhang, S. G.; Birch, M. A. *Biomaterials* 2005, **26**, 5198–5208.

165. Quibell, M.; Packman, L. C.; Johnson, T. *J Am Chem Soc* 1995, **117**, 11656–11668.

166. Hayman, M. W.; Smith, K. H.; Cameron, N. R.; Przyborski, S. A. *Biochem Biophys Res Commun* 2004, **314**, 483–488.

167. Hayman, M. W.; Smith, K. H.; Cameron, N. R.; Przyborski, S. A. *J Biochem Biophys Meth* 2005, **62**, 231–240.

168. Bokhari, M.; Carnachan, R. J.; Cameron, N. R.; Przyborski, S. A. *Biochem Biophys Res Commun* 2007, **354**, 1095–1100

169. Bokhari, M.; Carnachan, R. J.; Cameron, N. R.; Przyborski, S. A. *J Anat* 2007, **211**, 567–576.

170. Akay, G.; Birch, M. A.; Bokhari, M. A. *Biomaterials* 2004, **25**, 3991–4000.

171. Ruckenstein, E.; Park, J. S. *Polym Composites* 1991, **12**, 289–292.

172. Ruckenstein, E.; Park, J. S. *Synthet Metals* 1991, **44**, 293–306.

173. Ruckenstein, E.; Chen, J. H. *J Appl Polym Sci* 1991, **43**, 1209–1218.

174. Ruckenstein, E.; Chen, J. H. *Polymer* 1991, **32**, 1230–1235.

175. Ruckenstein, E.; Park, J. S. *J Appl Polym Sci* 1991, **42**, 925–934.

176. Sotiropoulos, S.; Brown, I. J.; Akay, G.; Lester, E. *Mater Lett* 1998, **35**, 383–391.

177. Brown, I. J.; Sotiropoulos, S. *Electrochim Acta* 2001, **46**, 2711–2720.

178. Blood, P. J.; Brown, I. J.; Sotiropoulos, S. *J Appl Electrochem* 2004, **34**, 1–7.

179. Wang, D.; Smith, N. L.; Budd, P. M. *Polymer Int* 2005, **54**, 297–303.

180. Maekawa, H.; Esquena, J.; Bishop, S.; Solans, C.; Chmelka, B. F. *Adv Mater* 2003, **15**, 591–596.

181. Hoar, T. P.; Schulman, J. H. *Nature* 1943, **152**, 102.

182. Prince, L. H., Ed., *Microemulsions: Theory and Practice*, Academic Press, New York, 1977.

183. Winsor, P. A. *Trans Faraday Soc* 1948, **44**, 376.

184. Gan, L. M.; Lian, N.; Chew, C. H.; Li, G. Z. *Langmuir* 1994, **10**, 2197.

185. Xu, X. J.; Chew, C. H.; Siow, K. S.; Wong, M. K.; Gan, L. M. *Langmuir* 1999, **15**, 8067.

186. Stoffer, J. O.; Bone, T. *J Polym Sci A Polym Chem* 1980, **18**, 2641.

187. Atik, S. S.; Thomas, J. K. *J Am Chem Soc* 1981, **103**, 4279.
188. Haque, E.; Qutubuddin, S. *J Polym Sci C Polym Lett* 1988, **26**, 429.
189. Menger, F. M.; Tsuno, T.; Hammond, G. S. *J Am Chem Soc* 1990, **112**, 1263.
190. Gan, L. M., Chew, C. H. *J Dispers Sci Technol* 1984, **5**, 179.
191. Chew, C. H.; Gan, L. M. *J Polym Sci Polym Chem Ed* 1985, **23**, 2225.
192. Gan, L. M.; Li, T. D.; Chew, C. H.; Teo W. K.; Gan L. H. *Langmuir*, 1995, **11**, 3316.
193. Chieng, T. H.; Gan, L. M.; Chew, C. H.; Ng, S. C.; Pey K. L. *Langmuir* 1996, **12**, 319.
194. Li, T. D.; Gan, L. M.; Chew, C. H.; Teo, W. K.; Gan, L. H. *Langmuir* 1996, **12**, 5863.
195. Gan, L. M.; Liu, J.; Poo, L. P.; Chew, C. H.; Gan, L. H. *Polymer* 1997, **38**, 5339.
196. Liu, J.; Gan, L. M.; Chew, C. H.; Teo, W. K.; Gan, L. H. *Langmuir*, 1997, **13**, 6421.
197. Xu, W.; Siow, K. S.; Gao, Z.; Lee, S. Y.; Chow, P. Y.; Gan, L. M. *Langmuir* 1999, **15**, 4812.
198. Gan, L. M.; Chow, P. Y.; Liu, Z.; Han, M.; Quek, C. H. *Chem Commun* 2005, 4459.
199. Raj, W. R. P.; Sashtav, M.; Cheung, H. M. *Langmuir* 1992, **8**, 1931.
200. Yan, F.; Texter, J. *Soft Matter* 2006, **2**, 109–118.
201. Chieng, T. H.; Gan, L. M.; Chew, C. H.; Lee, L.; Ng, S. C.; Pey, K. L.; Grant, D. *Langmuir* 1995, **11**, 3321.
202. Chieng, T. H.; Gan, L. M.; Chew, C. H.; Ng, S. C.; Pey, K. L. *Polymer*, 1996, **37**, 2801.
203. Chieng, T. H.; Gan, L. M.; Chew, C. H.; Ng, S. C.; Pey, K. L. *J Appl Polym Sci* 1996, **60**, 1561.
204. Chieng, T. H.; Gan, L. M.; Teo, W. K.; Pey, K. L. *Polymer* 1996, **37**, 5917.
205. Davis, E. W.; Mukkamala, R.; Cheung, H. M. *Langmuir* 1998, **14**, 762.
206. Burban, J. H.; He, M.; Cussler, E. L. *AIChE J* 1995, **41**, 907.
207. Bennett, D. J.; Burford, R. P.; Davis, T. P.; Tilley, H. J. *Polym Int* 1995, **36**, 219.
208. Santhanalakshmi, J.; Anandhi, K. *Langmuir* 1996, **12**, 3320.
209. Antonietti, M.; Hentze, H. P. *Colloid Polym Sci* 1996, **274**, 696.
210. Hentze, H. P.; Göltner, C. G.; Antonietti, M. *Ber Bunsenges Phys Chem* 1997, **101**, 1699.
211. Egger, C. C.; du Fresne, C.; Schmidt, D.; Yang, J.; Schadler, V. *J Sol Gel Sci Technol* 2008, **48**, 86–94.
212. Wang, J. B.; Yang, X. Q.; Wu, D. C.; Fu, R. W.; Dresselhaus, M. S.; Dresselhaus, G. *J Power Sources* 2008, **185**, 589–594.
213. Liu, J.; Teo, W. K.; Chew, C. H.; Gan, L. M. *J Appl Polym Sci* 2000, **77**, 2785.
214. Wang, L. S.; Chow, P. Y.; Tan, D. C. W.; Zhang, W. D.; Yang, Y. Y. *Adv Mater* 2004, **16**, 1790.
215. Cameron, N. R.; Flook, K. J.; Wren, S. A. C. *Chromatographia* 2003, **57**, 203.
216. Flook, K. J.; Cameron, N. R.; Wren, S. A. C. *J Chromatogr A* 2004, **1044**, 245.
217. Sundell, M. J.; Pajunen, E. O.; Hormi, O. E. O.; Näsman, J. H. *Chem Mater* 1993, **5**, 372.
218. Widawski, G.; Rawiso, M.; Francois, B. *Nature* 1994, **369**, 387–389.
219. Pitois, O.; Francois, B. *Eur Phys J B* 1999, **8**, 225–231.
220. Srinivasarao, M.; Colling, D.; Philips, A.; Patel, S. *Science* 2001, **292**, 79–83.
221. Bolognesi, A.; Mercogliano, C.; Yunus, S.; Civardi, M.; Comoretto, D.; Turturro, A., *Langmuir* 2005, **21**, 3480–3485.

222. Peng, J.; Han, Y. C.; Fu, J.; Yang, Y. M.; Li, B. Y. *Macromol Chem Phys* 2003, **204**, 125–130.

223. Song, L.; Bly, R. K.; Wilson, J. N.; Bakbak, S.; Park, J. O.; Srinivasarao, M.; Bunz, U. H. F. *Adv Mater* 2004, **16**, 115–118.

224. Zhao, B.; Li, C. X.; Lu, Y.; Wang, X. D.; Liu, Z. L.; Zhang, B. H. *Polymer* 2005, **46**, 9508–9513.

225. Karthaus, O.; Maruyama, N.; Cieren, X.; Shimomura, M.; Hasegawa, H.; Hashimoto, T. *Langmuir* 2000, **16**, 6071–6076.

226. Connal, L. A.; Gurr, P. A.; Qiao, G. G.; Solomon, D. H. *J Mater Chem* 2005, **15**, 1286–1292.

227. Cheng, C. X.; Tian, Y.; Shi, Y. Q.; Tang, R. P.; Xi, F., *Langmuir* 2005, **21**, 6576–6581.

228. Cui, L.; Xuan, Y.; Li, X.; Ding, Y.; Li, B. Y.; Han, Y. C., *Langmuir* 2005, **21**, 11696–11703; Madej, W.; Budkowski, A.; Raczkowska, J.; Rysz, J., *Langmuir* 2008, **24**, 3517–3524.

229. Li, J.; Cheng, J. T.; Zhang, Y.; Gopalakrishnakone, P. *Colloid Polym Sci* 2009, **287**, 29–36.

230. Bormashenko, E.; Pogreb, R.; Musin, A.; Stanevsky, O.; Bormashenko, Y.; Whyman, G.; Barkay, Z. *J Colloid Interf Sci* 2006, **300**, 293–297; Bormashenko, E.; Pogreb, R.; Stanevsky, O.; Bormashenko, Y.; Gendelman, O. *Mater Lett* 2005, **59**, 3553–3557.

231. Peng, J.; Han, Y. C.; Yang, Y. M.; Li, B. Y. *Polymer* 2004, **45**, 447–452.

232. Park, M. S.; Kim, J. K *Langmuir* 2004, **20**, 5347–5352.

233. Bunz, U. H. F. *Adv Mater* 2006, **18**, 973–989.

234. Park, M. S.; Joo, W.; Kim, J. K. *Langmuir* 2006, **22**, 4594–4598.

235. Yabu, H.; Tanaka, M.; Ijiro, K.; Shimomura, M. *Langmuir* 2003, **19**, 6297–6300.

236. Tian, Y.; Liu, S.; Ding, H. Y.; Wang, L. H.; Liu, B. Q.; Shi, Y. Q. *Polymer* 2007, **48**, 2338–2344.

237. Tian, Y.; Jiao, Q. Z.; Ding, H. Y.; Shi, Y. Q.; Liu, B. Q. *Polymer* 2006, **47**, 3866–3873.

238. Gugliuzza, A.; Aceto, M. C.; Macedonio, F.; Drioli, E. *J Phys Chem B* 2008, **112**, 10483–10496.

239. Zhao, B. H.; Zhang, J.; Wu, H. Y.; Wang, X. D.; Li, C. X. *Thin Solid Films* 2007, **515**, 3629–3634.

240. Kasai, W.; Kondo, T. *Macromol Biosci* 2004, **4**, 17–21.

241. Zhao, B. H.; Zhang, J.; Wang, X. D.; Li, C. X. *J Mater Chem* 2006, **16**, 509–513; Tian, Y.; Dai, C.; Ding, H. Y.; Jiao, Q. Z.; Wang, L. H.; Shi, Y. Q.; Liu, B. Q. *Polym Int* 2007, **56**, 834–839; Tian, Y.; Ding, H. Y.; Shi, Y. Q.; Jiao, Q. Z.; Wang, X. L. *J Appl Polymer Sci* 2006, **100**, 1013–1018.

242. Chaudhuri, J. B.; Davidson, M. G.; Ellis, M. J.; Jones, M. D.; Wu, X. J. *Macromol Symp* 2008, **272**, 52–57.

243. Wong, K. H.; Davis, T. P.; Bamer-Kowollik, C.; Stenzel, M. H. *Polymer* 2007, **48**, 4950–4965; Nishikawa, T.; Nishida, J.; Ookura, R.; Nishimura, S. I.; Scheumann, V.; Zizlsperger, M.; Lawall, R.; Knoll, W.; Shimomura, M. *Langmuir* 2000, **16**, 1337–1342; Wang, C. Y.; Mao, Y. D.; Wang, D. Y.; Qu, Q. S.; Yang, G. J.; Hu, X. Y. *J Mater Chem* 2008, **18**, 683–690.

244. Beattie, D.; Wong, K. H.; Williams, C.; Poole-Warren, L. A.; Davis, T. P.; Barner-Kowollik, C.; Stenzel, M. H. *Biomacromolecules* 2006, **7**, 1072–1082.

245. Maruyama, N.; Koito, T.; Nishida, J.; Sawadaishi, T.; Cieren, X.; Ijiro, K.; Karthaus, O.; Shimomura, M. *Thin Solid Films* 1998, **327**, 854–856.

246. Ghannam, L.; Manguian, M.; Francois, J.; Billon, L. *Soft Matter* 2007, **3**, 1492–1499.

247. Yabu, H.; Kojima, M.; Tsubouchi, M.; Onoue, S.; Sugitani, M.; Shimomura, M. *Colloids Surf A Physicochem Eng Aspects* 2006, **284**, 254–256.

248. Nishikawa, T.; Nonomura, M.; Arai, K.; Hayashi, J.; Sawadaishi, T.; Nishiura, Y.; Hara, M.; Shimomura, M. *Langmuir* 2003, **19**, 6193–6201.

249. Yabu, H.; Takebayashi, M.; Tanaka, M.; Shimomura, M. *Langmuir* 2005, **21**, 3235–3237.

250. Yabu, H.; Shimomura, M. *Langmuir* 2006, **22**, 4992–4997.

251. Yabu, H.; Hirai, Y.; Shimomura, M. *Langmuir* 2006, **22**, 9760–9764.

252. Tanaka, M.; Nishikawa, K.; Okubo, H.; Kamachi, H.; Kawai, T.; Matsushita, M.; Todo, S.; Shimomura, M. *Colloids Surf A Physicochem Eng Aspects* 2006, **284**, 464–469.

253. Tsukiyama, S.; Matsushita, M.; Tanaka, M.; Tamura, H.; Todo, S.; Yamamoto, S.; Shimomura, M. *Jpn J Appl Phys* 2008, **47**, 1429–1434.

254. Yamamoto, S.; Tanaka, M.; Sunami, H.; Ito, E.; Yamashita, S.; Morita, Y.; Shimomura, M. *Langmuir* 2007, **23**, 8114–8120.

255. Tsuruma, A.; Tanaka, M.; Yamamoto, S.; Shimomura, M. *Colloids Surf A Physicochem Eng Aspects* 2008, **313**, 536–540.

256. Zhang, Y.; Wang, C. *Adv Mater* 2007, **19**, 913–916.

257. Min, E.; Wong, K. H.; Stenzel, M. H. *Adv Mater* 2008, **20**, 3550–3556.

258. Park, M. S.; Kim, J. K. *Langmuir* 2005, **21**, 11404–11408.

259. Lu, Y.; Zhao, B. H.; Ren, Y.; Xiao, G. S.; Wang, X. D.; Li, C. X. *Polymer* 2007, **48**, 6205–6209.

260. Yabu, H.; Shimomura, M. *Chem Mater* 2005, **17**, 5231–5234.

261. Yabu, H.; Shimomura, M. *Langmuir* 2005, **21**, 1709–1711.

262. Chari, K.; Lander, C. W.; Sudol, R. J. *Appl Phys Lett* 2008, **92**, 111916.

263. Pintani, M.; Huang, J.; Ramon, M. C.; Bradley, D. D. C. *J Phys Condensed Matter* 2007, **19**, 016203.

264. Sung, H. J.; Meredith, C.; Johnson, C.; Galis, Z. S. *Biomaterials* 2004, **25**, 5735–5742.

265. Guarino, V.; Causa, F.; Netti, P. A.; Ciapetti, G.; Pagani, S.; Martini, D.; Baldini, N.; Ambrosio, L. *J Biomed Mater Res B Appl Biomater* 2008, **86B**, 548–557; Gorna, K.; Gogolewski, S. *J Biomed Mater Res A* 2006, **79A**, 128–138; Gao, J.; Crapo, P. M.; Wang, Y. D. *Tissue Eng* 2006, **12**, 917–925; Murphy, W. L.; Dennis, R. G.; Kileny, J. L.; Mooney, D. J. *Tissue Eng* 2002, **8**, 43–52.

266. Shastri, V. P.; Martin, I.; Langer, R. *Proc Natl Acad Sci USA* 2000, **97**, 1970–1975.

267. Zhang, J. C.; Zhang, H.; Wu, L. B.; Ding, J. D. *J Mater Sci* 2006, **41**, 1725–1731.

268. Chen, V. J.; Ma, P. X. *Biomaterials* 2004, **25**, 2065–2073.

269. Shum, A. W. T.; Li, J. S.; Mak, A. F. T. *Polymer Degrad Stabil* 2005, **87**, 487–493.

270. Jiang, P.; Hwang, K. S.; Mittleman, D. M.; Bertone, J. F.; Colvin, V. L. *J Am Chem Soc* 1999, **121**, 11630–11637.

271. Johnson, S. A.; Ollivier, P. J.; Mallouk, T. E. *Science* 1999, **283**, 963–965.

272. Lebourg, M.; Serra, R. S.; Estelles, J. M.; Sanchez, F. H.; Ribelles, J. L. G.; Anton, J. S. *J Mater Sci Mater Med* 2008, **19**, 2047–2053; Diego, R. B.; Olmedilla, M. P.; Aroca,

A. S.; Ribelles, J. L. G.; Pradas, M. M.; Ferrer, G. G.; Sanchez, M. S. *J Mater Sci* 2005, **40**, 6337–6337.

273. Gates, B.; Yin, Y. D.; Xia, Y. N. *Chem Mater* 1999, **11**, 2827–2836.
274. Park, S. H.; Xia, Y. N. *Chem Mater* 1998, **10**, 1745.
275. Yoon, S. B.; Kang, S.; Yu, J. S. *Curr Appl Phys* 2006, **6**, 1054–1058.
276. Yu, A. M.; Meiser, F.; Cassagneau, T.; Caruso, F. *Nano Lett* 2004, **4**, 177–181.
277. Li, Q.; Quinn, J. F.; Wang, Y. J.; Caruso, F. *Chem Mater* 2006, **18**, 5480–5485.
278. Weber, J.; Antonietti, M.; Thomas, A. *Macromolecules* 2007, **40**, 1299–1304.
279. Wang, Z. Y.; Kiesel, E. R.; Stein, A. *J Mater Chem* 2008, **18**, 2194–2200.
280. Xu, H.; Goedel, W. A. *Angew Chem Int Ed* 2003, **42**, 4694–4696.
281. Yan, F.; Goedel, W. A. *Adv Mater* 2004, **16**, 911–915.
282. Jiang, P.; McFarland, M. J. *J Am Chem Soc* 2004, **126**, 13778–13786.
283. Mikos, A. G.; Thorsen, A. J.; Czerwonka, L. A.; Bao, Y.; Langer, R.; Winslow, D. N.; Vacanti, J. P. *Polymer* 1994, **35**, 1068–1077.
284. Hou, Q. P.; Grijpma, D. W.; Feijen, J. *Biomaterials* 2003, **24**, 1937–1947.
285. Lu, L.; Peter, S. J.; Lyman, M. D.; Lai, H. L.; Leite, S. M.; Tamada, J. A.; Uyama, S.; Vacanti, J. P.; Langer, R.; Mikos, A. G. *Biomaterials* 2000, **21**, 1837–1845.
286. Horak, D.; Hlidkova, H.; Hradil, J.; Lapcikova, M.; Slouf, M. *Polymer* 2008, **49**, 2046–2054.
287. Lin-Gibson, S.; Cooper, J. A.; Landis, F. A.; Cicerone, M. T. *Biomacromolecules* 2007, **8**, 1511–1518.
288. Claase, M. B.; de Bruijn, J. D.; Grijpma, D. W.; Feijen, J. *J Mater Sci Mater Med* 2007, **18**, 1299–1307.
289. Ghosh, S.; Gutierrez, V.; Fernandez, C.; Rodriguez–Perez, M. A.; Viana, J. C.; Reis, R. L.; Mano, J. F. *Acta Biomater* 2008, **4**, 950–959.
290. Nam, Y. S.; Yoon, J. J.; Park, T. G. *J Biomed Mater Res* 2000, **53**, 1–7; Yoon, J. J.; Song, S. H.; Lee, D. S.; Park, T. G. *Biomaterials* 2004, **25**, 5613–5620; Yoon, J. J.; Kim, J. H.; Park, T. G. *Biomaterials* 2003, **24**, 2323–2329.
291. Sheridan, M. H.; Shea, L. D.; Peters, M. C.; Mooney, D. J. *J Controlled Release* 2000, **64**, 91–102; Leung, L.; Chan, C.; Baek, S.; Naguib, H. *Biomed Mater* 2008, **3**, 025006.
292. Reignier, J.; Huneault, M. A. *Polymer* 2006, **47**, 4703–4717.
293. Thomson, R. C.; Yaszemski, M. J.; Powers, J. M.; Mikos, A. G. *J Biomater Sci Polym Ed* 1995, **7**, 23–38.
294. Bertone, J. F.; Jiang, P.; Hwang, K. S.; Mittleman, D. M.; Colvin, V. L. *Phys Rev Lett* 1999, **83**, 300–303.
295. Miguez, H.; Meseguer, F.; Lopez, C.; Lopez–Tejeira, F.; Sanchez–Dehesa, J. *Adv Mater* 2001, **13**, 393–396.
296. Cassagneau, T.; Caruso, F. *Adv Mater* 2002, **14**, 34–38.
297. Tian, S. J.; Wang, J. J.; Jonas, U.; Knoll, W. *Chem Mater* 2005, **17**, 5726–5730; Luo, X. L.; Killard, A. J.; Smyth, M. R. *Chem Eur J* 2007, **13**, 2138–2143.
298. Lee, Y. J.; Braun, P. V. *Adv Mater* 2003, **15**, 563–566.
299. Lee, Y. J.; Pruzinsky, S. A.; Braun, P. V. *Langmuir* 2004, **20**, 3096–3106.

SECTION II

CHARACTERIZATION

CHAPTER 5

Surface Area and Porosity Characterization of Porous Polymers

ROLANDO M. A. ROQUE-MALHERBE
University of Turabo, Gurabo, Puerto Rico, USA

5.1 INTRODUCTION

Porous polymers are widely applied as adsorbents, ion exchangers, catalysts, and permeable materials [1–7]. The first resins with some of these properties were obtained by D'Aleleio in 1944 based on the copolymerization of styrene and divinylbenzene. Unfunctionalized polystyrene resins crosslinked with divinylbenzene (Amberlites) and other types of polymeric resins are porous materials that are widely applied as adsorbents [1, 8–19]. In addition, functionalized polystyrene–divinylbenzene resins and other porous materials are extensively used as cation and anion exchangers in industry and science [2–4, 20–29]. Cation and anion-exchange resins are also significantly used as catalysts [1, 30–39]. In addition, polymers are the materials that are most extensively applied for membrane preparation [6, 7, 40–48].

Often, porous polymers are obtained during the crosslinking used to form thermoset networks that do not flow during heating. These are materials with a permanent porous structure produced during their synthesis and preserved in the solid state [13]. These polymers are often synthesized by suspension polymerization, where the polymerization mixture includes a crosslinking monomer, a comonomer, an initiator, and a porogenic agent [16, 19]. Another method applied to produce porous polymers is based on the addition of an inorganic matrix of a known porosity, for example, silica gel or aluminum oxide, to the reacting mixture [49,50]. Subsequent to polymerization the inorganic template is eliminated by dissolution without destruction of the polymer. This chapter discusses the characterization of the surface area and porosity of porous polymers. Examples are used to illustrate some of the characterization methods discussed here using two fulfural polymers [51, 52] and a highly porous, open-cell polyHIPE polymer that was templated within a high internal phase emulsion (HIPE) [19].

Porous Polymers, Edited by Michael S. Silverstein, Neil R. Cameron, and Marc A. Hillmyer

When furfural is heated in the presence of acids it solidifies into a solid thermosetting polymer [51–59]. The polymerization mechanism involves the acid-catalyzed condensation of furfural [60]. Two porous furan resins produced by the polymerization of furfural in the presence of a Brønsted acid catalyst, which are good adsorbents [51, 52], are used to illustrate the estimation of the microporous volume by the Dubinin and the osmotic isotherm equation methods and the measurement of the specific surface area by the Brunauer-Emmet-Teller (BET) method.

A polyHIPE is used to illustrate the determination of the pore size distribution. The organic phase of the HIPE contained styrene, divinylbenzene as a crosslinking comonomer, and toluene as a porogen [19].

5.2 SOME DEFINITIONS AND TERMINOLOGY

5.2.1 Adsorption in Porous Materials

The concept of gas adsorption in solids, in a very general sense, describes the increase in concentration of gas molecules in an adjacent solid surface. The adsorptive, sorptive, or adsorbate is the gas adsorbed by the solid adsorbent [61–65]. Adsorption is an important unit operation in industry and a very useful method for the characterization of porous materials [61–91]. In a general sense, the adsorbents applied for practical purposes are normally porous. The International Union of Pure and Applied Chemistry (IUPAC) classifies the different pore widths of porous adsorbents [61]. IUPAC classifies *micropores* as pores with diameters between 0.3 and 2 nm, *mesopores* as pores with diameters between 2 and 50 nm, and *macropores* as pores with diameters greater than 50 nm [61]. The pore width D_p is defined as the diameter in the case of cylindrical and spherical pores and as the distance between opposite walls in the case of slit-shaped pores.

5.2.2 Physical, Mobile, and Immobile Adsorption

In the application of adsorption for the characterization of porous materials we are interested in physical adsorption [92–102]. For this type of adsorption process there is no need for a reaction involving the exchange of electrons between the solid surface and the gas molecules and the formation of chemical bonds [61, 62, 64, 68]. Physical adsorption is mobile when adsorbed molecules act as a gas in the adsorption space and is immobile when the adsorbed molecule is constrained to vibrate around an adsorption site [62].

5.2.3 Parameters Characterizing Adsorbents

Amount adsorbed: The amount adsorbed normalized by the mass of degassed adsorbent, or the magnitude of adsorption, is measured in mol/g and is given by

$$n_a \approx \frac{n^a}{m_s} \tag{5.1}$$

where n^a is the amount adsorbed, which is typically measured in practical adsorption studies, as will be explained in the next section, and m_s is the mass of degassed adsorbent [61–65].

Adsorption isotherm: The magnitude of adsorption is dependent on the equilibrium adsorption pressure P at constant temperature T. Accordingly, gas adsorption data are in practice expressed by the adsorption isotherm using the function F [61]:

$$n_a = F(P)_T \tag{5.2}$$

It is necessary to emphasize that the adsorption isotherm is the principal tool for the study of adsorption in solid adsorbents.

Monolayer and multilayer adsorption: For open surfaces, the adsorption process is a layer-by-layer filling process, where the first layer is filled when $\theta = \frac{n_a}{N_m} = 1$, where θ is the surface recovery and N_m is the monolayer capacity. As a result we have monolayer adsorption when $\theta = \frac{n_a}{N_m} < 1$ and multilayer adsorption when $\theta = \frac{n_a}{N_m} > 1$.

Specific surface area: This parameter is given by

$$S = \frac{A}{m_s} \tag{5.3}$$

and is measured in m^2/g, where A is the adsorbent surface area and m_s is the mass of degassed adsorbent.

Micropore and mesopore volume of porous materials: The micropore volume, denoted by W^{MP} and measured in cm^3/g, is the measure of the volume of the adsorption space in the micropore range. The pore volume designated by W is the sum of the micropore and mesopore volumes of the adsorbent and is measured in cm^3/g [62, 63].

Pore size distribution (PSD): The PSD is a plot of $\Delta V_p/\Delta D_p$ versus D_p, where V_p is the pore volume per mass accumulated up to the pore of width D_p and is measured in cc-STP/g·Å [61, 62, 64, 65]. The unit cc-STP denotes the quantity adsorbed and is measured in cubic centimeters at standard temperature (273.15 K) and pressure (101.3 kPa) (STP).

5.3 MEASUREMENT OF ADSORPTION ISOTHERMS BY THE VOLUMETRIC METHOD

Possibly the most widely used technique to measure the amount adsorbed and construct the adsorption isotherm is the volumetric adsorption method [61, 62, 64, 66] (see Fig. 5.1). The apparatus to carry out this measurement consists, in general terms, of a temperature-controlled sample cell of volume V_g at the temperature of the adsorption experiment T, a vessel with a very accurately measured volume termed the calibrated volume V_c, a connection to a gas reservoir, and a transducer for pressure measurement. In addition, the volume between stopcock 3 at

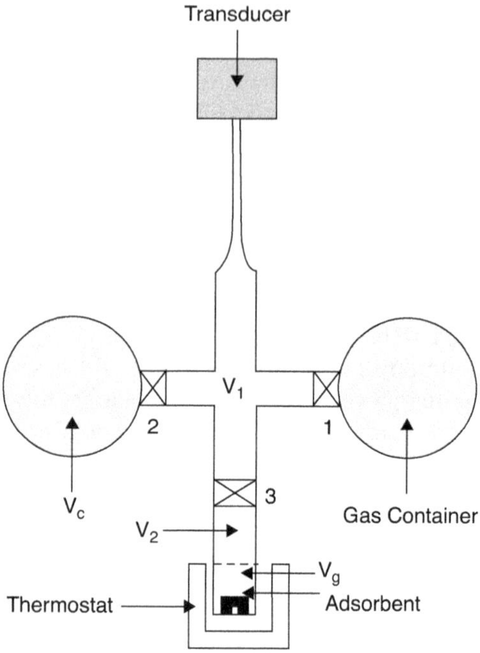

FIGURE 5.1 Schematic representation of a volumetric adsorption experiment.

ambient temperature T_r and the thermostat is V_2 and the volume between stopcocks 1, 2, and 3, also at ambient temperature, is V_I. The procedure to produce an isotherm begins with the determination of the different volumes involved in the experiment using He, a gas not ordinarily adsorbed at the experimental temperature [62,66].

The amount adsorbed as a function of pressure at a constant temperature is measured as follows: Using the values determined for the different volumes, the amount dosed in step i, that is, n^i_{dose}, and the amount left after adsorption in step i, that is, n^i_{final}, are calculated. With these quantities the magnitude adsorbed for the ith isotherm point, $\Delta^i n_a$, is calculated using the following equation [62]:

$$\Delta^i n_a = \frac{n^i_{dose} - n^i_{final}}{m_s} \tag{5.4}$$

where m_s is the mass of degassed adsorbent. Finally, the isotherm is calculated from the sum of the different adsorption steps:

$$n^i_a = \sum_{j=1}^{i} \Delta^j n_a \tag{5.5}$$

where n_a^i is the magnitude of adsorption up to the ith adsorption step. The experimental isotherm is obtained by plotting n_a^i versus P_2^i.

5.4 ADSORPTION INTERACTION FIELDS

When a molecule is within the adsorption space of a solid adsorbent it is subjected to different interactions, such as the dispersion interaction ϕ_D, the repulsion interaction ϕ_R, the electrostatic interactions (polarization ϕ_P, field dipole $\phi_{E\mu}$, and field gradient quadrupole ϕ_{EQ}), the sorbate–sorbate interaction ϕ_{AA}, and other specific interactions such as the acid–base interaction ϕ_{AB}, which is present when the surface contains hydroxyl bridge groups [62–70, 83].

The dispersion or London forces [103] between adsorbed nonpolar molecules and any adsorbent are produced because instantaneous dipoles of the adsorbed nonpolar molecules induce dipoles on the adsorbent atoms. Consequently, both interact to lower the energy of the adsorbate–adsorbent system. The repulsion interaction is generated when the adsorbed molecules are pressed against the solid surface [103, 104]. The mathematical expression normally used to represent the sum of the dispersion and repulsion potentials $\phi_D + \phi_R$ characterizing the interaction of nonpolar molecules with a surface is the Lennard-Jones *6–12* potential [70, 105]:

$$\phi_D + \phi_R = 4\varepsilon\left(\left(\frac{\sigma}{z}\right)^{12} - \left(\frac{\sigma}{z}\right)^{6}\right) \tag{5.6}$$

where ε is the potential energy minimum and σ is the gas–solid separation at maximum interaction. This interaction is experienced by all adsorbate–adsorbent couples. Conversely, the electrostatic interaction ϕ_E [83],

$$\phi_E = \phi_P + \phi_{E\mu} + \phi_{QE} \tag{5.7}$$

is specific, that is, it depends on the adsorption system, on the specific adsorbate–adsorbent couple. The electrostatic potential of the adsorbate–adsorbent system can be calculated with the relation [106, 107]

$$\phi_E = \int V(\overline{r})\rho(\overline{r})d^3r \tag{5.8}$$

where $\rho(\overline{r})$ is the localized charge distribution of the adsorbate and $V(\overline{r})$ is the external electric field generated by the solid adsorbent.

5.5 DETERMINATION OF THE MICROPORE VOLUME

5.5.1 Dubinin Adsorption Isotherm Equation

The Dubinin adsorption isotherm equation is an excellent method for calculating the micropore volume. This isotherm equation can be obtained with the help of Dubinin's theory of volume filling and Polanyi's adsorption potential [62,71,86]. The Dubinin adsorption isotherm equation is given by [71]

$$n_a = N_a \exp\left(-\frac{RT}{E}\ln\left[\frac{P_0}{P}\right]\right)^n \tag{5.9}$$

where n_a is the magnitude of adsorption, P_o is the vapor pressure of the adsorptive at the temperature T of the adsorption experiment, P is the equilibrium adsorption pressure, E is a parameter termed the characteristic energy of adsorption, N_a is the maximum amount adsorbed in the micropore, and n is an empirical parameter ($1 < n < 5$). The Dubinin adsorption isotherm equation in linear form is written as follows [62,83]:

$$\ln(n_a) = \ln(N_a) - \left(\frac{RT}{E}\right)^n \ln\left(\frac{P_0}{P}\right)^n \tag{5.10}$$

The Dubinin plot is the following linear plot:

$$y = \ln(n_a) = \ln(N_a) - \left(\frac{RT}{E}\right)^n \ln\left(\frac{P_0}{P}\right)^n = b - mx \tag{5.11}$$

where

$$y = \ln(n_a),\ b = \ln(N_a),\ m = \left(\frac{RT}{E}\right)^n,\ x = \ln\left(\frac{P_0}{P}\right)^n \tag{5.12}$$

Figures 5.2 and 5.3 are the Dubinin plots of the adsorption isotherms in the range of relative pressures $0.001 < P/P_0 < 0.03$ for the adsorption of CO_2 at 251 K in the furanic resins FH (Fig. 5.2) and FAH (Fig. 5.3) [51,52], where FH was synthesized by the acid-catalyzed condensation of furfural from furfuraldehyde and FAH was synthesized by the acid-catalyzed condensation of furfural from a mixture of acetone and furfuraldehyde [51,52]. The method can also be applied using other adsorbates, such as N_2 at 77 K, in the same range of relative pressures.

The adsorption data used to make both plots were determined volumetrically in a homemade Pyrex vacuum system consisting of a sample holder, a dead volume, a dose volume, a U-tube manometer, and a thermostat [76,85,91].

It is evident that the experimental data can be fit accurately using Equation 5.10, and it was possible to calculate the maximum adsorption capacity in the microporous volume of these furan resins N_a and the characteristic energy of adsorption E (see Table 5.1).

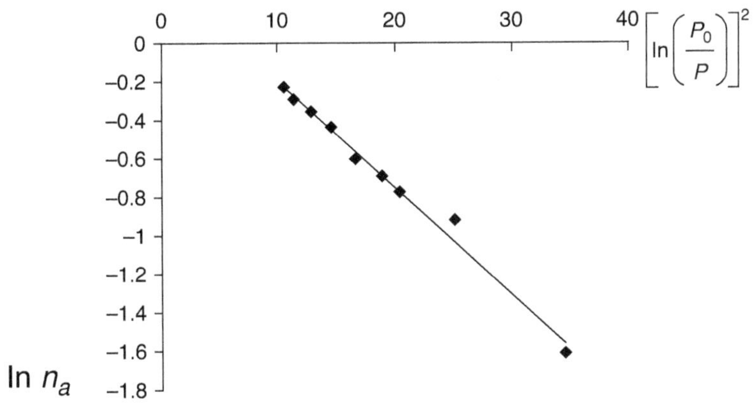

FIGURE 5.2 Dubinin plot for the adsorption of CO_2 at 251 K in the FH furan resin.

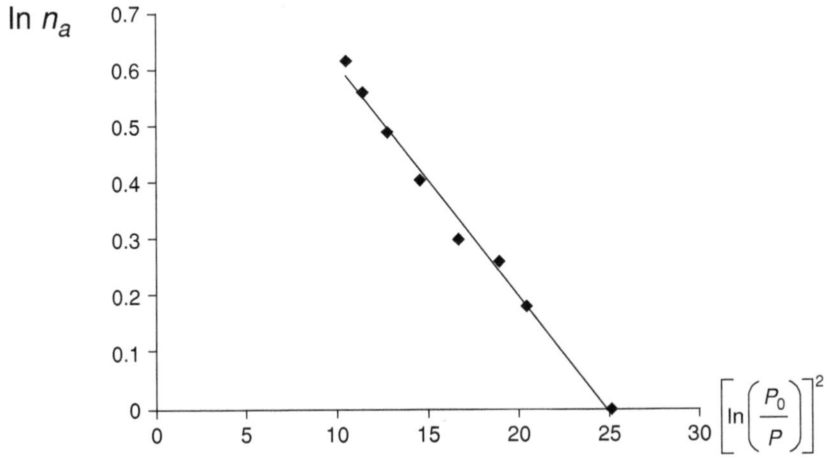

FIGURE 5.3 Dubinin plot for the adsorption of CO_2 at 251 K in the FAH furan resin.

TABLE 5.1 Parameters of the Dubinin Equation for the Adsorption of CO_2 at 251 K in the FH and FAH Furan Resins

Sample	N_a, mmol/g	E, kJ/mol	n
FH	1.40	8.9	2
FAH	2.78	10.3	2

The application of the previously discussed Dubinin methodology in the study of porous polymers is not widespread. However, the methodology is normally applied in the case of coordination polymers and metal–organic porous framework polymers [108–117].

5.5.2 Osmotic Adsorption Isotherm Equation

Another isotherm equation that can be useful for the calculation of the micropore volume is the osmotic isotherm of adsorption. In the frame of the osmotic theory of adsorption, the adsorption process in the microporosity of a porous adsorbent is considered as the "osmotic" equilibrium between two solutions (vacancy plus molecules) of different concentrations [86, 118]. One of these solutions is produced in the micropores and the other in the gas phase. In this case, the role of solvent is performed by the vacancies, that is, by vacuum [86, 118]. Consequently, if we assume that the adsorption process in a micropore system can be described as an osmotic process in which the vacancies are the solvent and the adsorbed molecules are the solute, then it is possible, applying the methods of thermodynamics to the foregoing model, to obtain the so-called osmotic isotherm adsorption equation [86, 118]:

$$n_a = \frac{N_a K_0 P^B}{1 + K_0 P^B} \tag{5.13}$$

where n_a is the magnitude of adsorption in the micropore volume, N_a is the maximum amount or magnitude adsorbed in the volume of the micropore, P is the equilibrium adsorption pressure, K is a constant dependent on temperature, and B is a constant.

Equation 5.13 reduces, for $B = 1$, to a Langmuir-type isotherm equation describing volume filling:

$$n_a = \frac{N_a K_0 P}{1 + K_0 P} \tag{5.14}$$

The linear form of the osmotic equation can be expressed as follows [62, 83]:

$$y = P^B = N_a \left(\frac{P^B}{n_a}\right) + \frac{1}{K_0} = mx + b \tag{5.15}$$

where $y = P^B$, $x = \frac{P^B}{n_a}$, $m = N_a$ is the slope, and $b = \frac{1}{K_0}$ is the intercept.

Figures 5.4 and 5.5 are the linear forms of the osmotic isotherm equation with $B = 1$ using adsorption data in the range $0.001 < P/P_0 < 0.03$ describing the adsorption of CO_2 at 251 K in the furanic resins FH (Fig. 5.4) and FAH (Fig. 5.5) [51, 52]. The method can also be applied using other adsorbates, such as N_2 at 77 K, in the same range of relative pressures.

The adsorption data reported in Figs. 5.4 and 5.5 were also determined volumetrically in the Pyrex vacuum system previously described [76, 85, 91]. With these plots

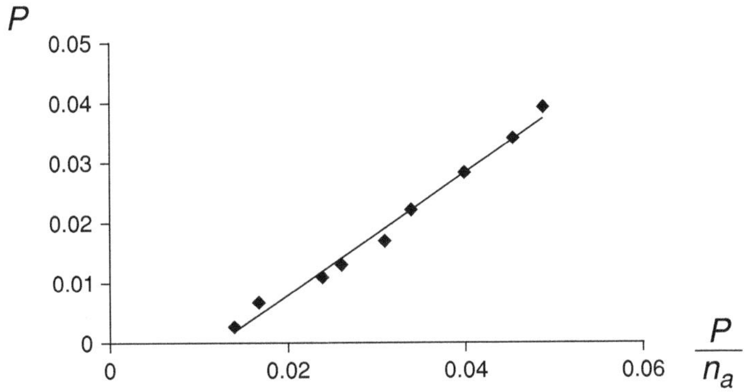

FIGURE 5.4 Linear osmotic plot with $B = 1$ using the adsorption data of CO_2 at 251 K in the FH furan resin.

the maximum adsorption capacity of these furan resins N_a can be calculated (see Table 5.2). The osmotic isotherm equation is not normally applied in the study of porous polymers.

5.5.3 Measurement of the Micropore Volume

Porous materials and, in particular, porous polymers are generally formed by a network of micropores, mesopores, and macropores. The exact value of the micropore volume W^{MP} is particularly difficult to establish through an adsorption study of these materials. In the majority of cases, it is difficult to determine the point where the adsorption process in the micropores finishes and the adsorption process in

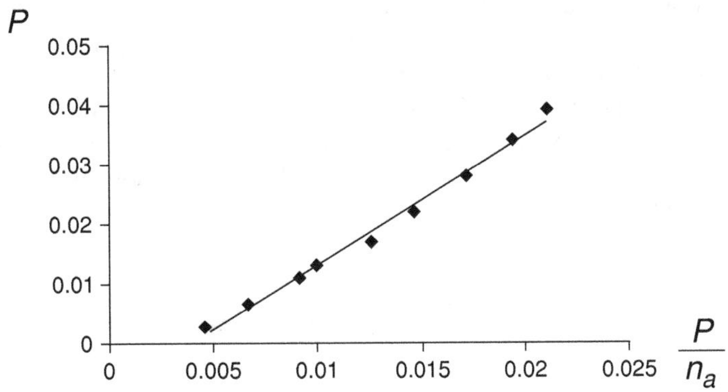

FIGURE 5.5 Linear osmotic plot with $B = 1$ using the adsorption data of CO_2 at 251 K in the FAH furan resin.

TABLE 5.2 Maximum Adsorption Capacity Calculated with the Osmotic Isotherm Equation for the Adsorption of CO_2 at 251 K in the FH and FAH Furan Resins

Sample	N_a, mmol/g	B
FH	1.02	1
FAH	2.20	1

the mesopores begins [62, 80]. Thus, different plots of the adsorption data at low pressures are normally applied to calculate the micropore volume of the porous adsorbent. In general, the Dubinin and osmotic isotherm equations and the t-plot are used. The t-plot method is not discussed here, but this methodology is described in the literature [62, 64, 65].

The mathematical treatment of the Dubinin and osmotic isotherm equations consists [62, 63, 83, 91] of nonlinear regression of the experimental adsorption data to the Dubinin isotherm equation (Eq 5.9) and the osmotic isotherm equation (Eq 5.13) using a least square fit [62, 63, 83, 91, 119] or the fit of linear plots previously discussed. During the fitting process, different parameters are adjusted, specifically, *n*, *E*, K, and *B*, if the Dubinin and osmotic isotherm equations were applied. The fitting process is stopped when all the applied equations yield a similar value for N_a, the maximum adsorption. Then the micropore volume is calculated with the help of the Gurvich rule [120]:

$$W^{\mathrm{MP}} = N_a V_L \tag{5.16}$$

where V_L is the molar volume of the adsorptive used in the adsorption experiment.

This methodology for the determination of the micropore volume can be applied for any material in general and, in particular, for polymers using CO_2 at 251 K. This gas is appropriate because it has a noticeable quadrupole moment Q^{CO_2} of 1.31×10^{13} C m^2 [121] and will consequently interact relatively strongly with the polymer. In addition, its molecular kinetic diameter σ^{CO_2} of 0.394 nm [122] ensures that it easily penetrates through microporosity. In this regard, it is possible to calculate the micropore volumes of both furanic resins (see Table 5.3) using the values obtained for the maximum adsorption magnitude, N_a, applying the osmotic isotherm equation and the Gurvich rule, where the molar volume of CO_2, $V_L^{CO_2}$, is 94.7 cm^3/mol [122].

TABLE 5.3 Micropore Volumes of the Studied Furan Resins

Sample	W^{MP}, cm^3/mol	N_a, mmol/g
FH	0.097	1.02
FAH	0.208	2.20

If the micropore volumes calculated by the Dubinin method were used, the procedure would be the same [62, 83].

To conclude this section, it is necessary to state that fitting the experimental data with the osmotic isotherm equation produced a negative value for the parameter $b = 1/K_0$. Since $K_0 > 0$, this yields an inconsistency in the methodology. However, we are not applying the methodology to calculate thermodynamic parameters; we are applying the technique to calculate the maximum adsorption capacity, and this value coincides fairly well with the value obtained from the Dubinin plot.

5.5.4 Horvath–Kawazoe Approach for the Determination of the Micropore Size Distribution

To create a model of an adsorption system, it is required to give a description of the interaction field and the geometry of the pore system of the adsorbent [83, 84]. In this sense, Horvath and Kawazoe [77] developed a method for the calculation of the micropore size distribution (the HK method) making use of the slit pore geometry potential model of Everett and Powl (Fig. 5.6) [105]. This methodology was later applied by Saito and Foley [78] (the SF methodology) to the case of cylindrical pore geometry and by Cheng and Yang [79] to the case of spherical pore geometry. The slit potential model of Everett and Powl is [105]

$$E(z) = \frac{N_{AS}A_{AS}}{2\sigma^4}\left[\left(-\left(\frac{\sigma}{z}\right)^4 + \left(\frac{\sigma}{z}\right)^{10}\right) + \left(\left(\frac{\sigma}{L-z}\right)^4 - \left(\frac{\sigma}{L-z}\right)^{10}\right)\right] \quad (5.17)$$

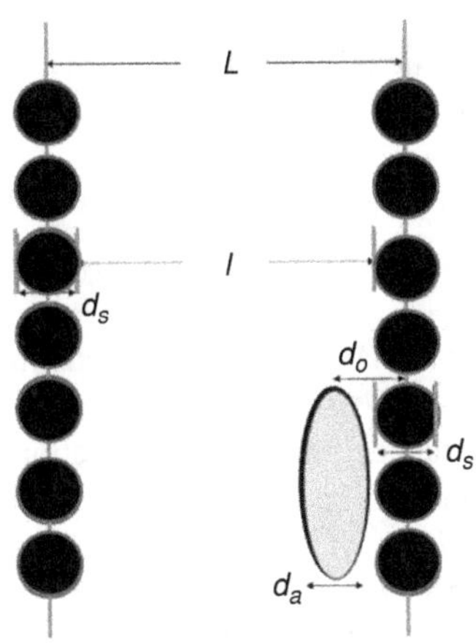

FIGURE 5.6 Schematic representation of a slit pore of diameter L, where a molecule is adsorbed at a distance d_0 from the surface, d_s is the diameter of the adsorbent molecule, and d_a is the diameter of the adsorbate molecule.

a potential that only includes the dispersion and repulsion interactions [62,73,77,83], where N_{AS} is the number of solid molecules per surface unit, L is the distance between the layers (Fig. 5.6), and $\sigma = 0.858d$, where $d = (d_s + d_a)/2$, d_s is the diameter of the adsorbent molecule, and d_a is the diameter of the adsorbate molecule. In addition, z is the internuclear distance between the adsorbate and adsorbent molecules, $(L - d_s)$ is the effective pore width (Fig. 5.6), and A_{AS} is the dispersion constant, which takes into account the adsorbate–adsorbent interaction. A_{AS} is calculated using the Kirkwood–Muller formula [68,73,77–79]. Horvath and Kawazoe proposed that the potential is increased by adsorbate–adsorbate interaction and suggested the following potential [77]:

$$\Phi(z) = \frac{N_{AS}A_{AS} + N_{AA}A_{AA}}{2\sigma^4}\left[\left(-\left(\frac{\sigma}{z}\right)^4 + \left(\frac{\sigma}{z}\right)^{10}\right) + \left(\left(\frac{\sigma}{L-z}\right)^4 - \left(\frac{\sigma}{L-z}\right)^{10}\right)\right] \tag{5.18}$$

where N_{AA} is the number of adsorbed molecules per surface unit, L is the distance between the layers (Fig. 5.6), and $\sigma = 0.858d$. The A_{AA} calculated with the help of the Kirkwood–Muller formula is a constant characterizing the adsorbate–adsorbate interaction [62,73,77]. The next step is to obtain the average interaction energy. This is made by volumetrically averaging the potential expressed by Equation 5.18 in the following fashion [77]:

$$\xi(L) = N_A \int_{d}^{L-d} \frac{\Phi(z)dz}{(L - 2d)} \tag{5.19}$$

where $\Phi(z)$ is the adsorption field inside the slit pore and N_A is Avogadro's number. Integrating Equation 5.18 yields [77]

$$\xi(L) = \left(\frac{N_{AS}A_{AS} + N_{AA}A_{AA}}{\sigma^4(L - 2d)}\right)\left(\frac{\sigma^4}{3(L-d)^3} - \frac{\sigma^{10}}{9(L-d)^9} - \frac{\sigma^4}{3d^3} + \frac{\sigma^4}{9d^9}\right) \tag{5.20}$$

where $\xi(L)$ is the average potential in a given pore obtained by the integration across the effective pore width.

In the case of the cylindrical geometry, the interaction potential averaged over the cylinder enables the calculation of an approximate value for the adsorption field in a cylindrical pore or channel [78]. The adsorption process in this model is described using the potential between a cylindrical pore of infinite length but finite radius r_p [78]. The calculation is carried out with a methodology analogous to that developed by Horvath and Kawazoe for determining the micropore size distribution in the case of the slit pore and includes only the van der Waals interactions calculated using the Lennard-Jones potential. To evaluate the contributions of the dispersion and repulsion energy in the cylindrical case, Everett and Powl [105] applied the Lennard-Jones potential to the case of the interaction of one adsorbate molecule with an infinite cylindrical pore consisting of adsorbent molecules and obtained the potential $E(r)$ [78]:

$$\mathrm{E}(r) = \frac{5}{2}\pi\varepsilon^* \left[\frac{21}{32}\left(\frac{d}{r_p}\right)^{10} \sum_{k=0}^{\infty} \alpha_k \left(\frac{r}{r_p}\right)^{2k} - \left(\frac{d}{r_p}\right)^{4} \sum_{k=0}^{\infty} \beta_k \left(\frac{r}{r_p}\right)^{2k}\right] \quad (5.21)$$

to calculate the interaction of a molecule at a distance r to the pore wall, where

$$\varepsilon^* = \frac{3}{10}\left(\frac{N_{AS}A_{AS} + N_{AA}A_{AA}}{2d^4}\right) \quad (5.22)$$

and N_{AS}, d, d_s, d_a, A_{AS}, N_{AA}, and A_{AA} have the same meaning as in the HK method. In addition,

$$\alpha_k = \left(\frac{\Gamma(-4.5)}{\Gamma(-4.5-k)\Gamma(k+1)}\right)^2 \quad (5.23)$$

$$\beta_k = \left(\frac{\Gamma(-1.5)}{\Gamma(-1.5-k)\Gamma(k+1)}\right)^2 \quad (5.24)$$

where $\alpha_0 = \beta_0 = 1$ and Γ is the gamma function. The next step is to obtain the linear average of the interaction energy by averaging the potential expressed by Equation 5.21 and in this fashion to get molar magnitudes [78]:

$$\xi(r_p) = N_A \frac{\int_0^{r_p-d} \mathrm{E}(r)dr}{r_p - d} \quad (5.25)$$

Therefore

$$\xi(r_p) = \frac{3}{4}\pi N_A \left(\frac{N_{AS}A_{AS} + N_{AA}A_{AA}}{d^4}\right)\left(\sum_{k=0}^{\infty}\left[\frac{1}{2k+1}\left(1-\frac{d}{r_p}\right)^{2k} \times \left\{\frac{21}{32}\alpha_k\left(\frac{d}{r_p}\right)^{10} - \beta_k\left(\frac{d}{r_p}\right)^{4}\right\}\right]\right) \quad (5.26)$$

where $\xi(r_p)$ is the average potential in a given pore obtained by integration across the effective pore width.

It is possible to calculate for both pore geometries [62, 73, 83, 84]

$$RT \ln\left(\frac{P}{P_0}\right) = N_A \xi(\rho) \tag{5.27}$$

where $\rho = L$ or r_p, depending of the methodology applied, that is, L for the HK method and r_p for the SF methodology.

Substituting Equations 5.20 and 5.26 in Equation 5.27 yields the micropore size distribution for the slit pore geometry [77] and cylindrical pore geometry [78], respectively. The HK method states that the relative pressure $x = P/P_0$ needed for filling the micropores of a specific size and shape is directly related to the adsorbate–adsorbent interaction energy [62,73]. This indicates that the micropores are gradually filled with increasing adsorbate pressure. The HK methodology assumes that only pores with dimensions lower than a specific value will be filled for a given relative pressure of the adsorbate [62,73]. Following similar ideas, the SF method enables the calculation of the pore size distribution in the micropore range at low pressures for cylindrical and spherical geometries [73, 78, 79].

The previously described methodology has been infrequently applied for the study of porous polymers. However, the micropore distributions in microporous phthalocyanine network polymers were calculated from the nitrogen adsorption and desorption isotherms at 77 K (Micromeritics ASAP 2020 system) using the Horvath–Kawazoe method [113].

5.6 DETERMINATION OF THE SPECIFIC SURFACE AREA OF POROUS MATERIALS

The approach normally applied for the determination of the specific surface area of porous materials is the BET method [123]. The BET method is based in the theory of multilayer adsorption [123]. The BET theory assumes that the adsorption process is a layer-by-layer process on an energetically homogeneous surface. The molecules are supposed to occupy sites in an adsorption field that has the same intensity at all locations on the surface. Furthermore, the adsorption process is supposed to be immobile, and each molecule is adsorbed at a specific adsorption site on the surface. The first layer of adsorbed molecules has an energy of interaction with the adsorption field E_0^a. The vertical interaction between molecules after the first layer is E_0^L, which is analogous to the liquefaction heat of the adsorbate. The adsorbed molecules do not interact laterally. The BET isotherm equation can be derived using the grand canonical ensemble approach, applying an approach developed by Hill [72]. The specific isotherm equation obtained is [62, 64, 65]

$$\frac{n_a}{N_m} = \frac{Cx}{(1 - x + Cx)(1 - x)} \tag{5.28}$$

where n_a is the magnitude of adsorption, N_m is the monolayer capacity, $x = \frac{P}{P_0}$, and $C = K\exp\left(\frac{E_0^a - E_0^L}{RT}\right)$, where K is a constant. The BET isotherm equation, that is, Equation 5.28, is usually applied to adsorption data in a linear form:

$$y = \frac{x}{n_a(1-x)} = \left(\frac{1}{N_m C}\right) + \left(\frac{C-1}{CN_m}\right)x = b + mx \tag{5.29}$$

where $b = \left(\frac{1}{N_m C}\right)$, $m = \left(\frac{C-1}{CN_m}\right)$, $y = \frac{x}{n_a(1-x)}$, and $x = \frac{P}{P_0}$ in the region $0.05 < x < 0.4$ [62,64,65,83]. If the term $\frac{C-1}{C} \approx 1$ then the slope m of the linear regression can be approximated as $m \approx \frac{1}{N_m}$. As a result, the monolayer capacity N_m can be determined and the specific surface area S can be calculated:

$$S = N_m N_A \sigma \tag{5.30}$$

where σ is the cross-sectional area—the average area occupied by each molecule in a completed monolayer. We have $\sigma(N_2) = 0.162$ nm^2 for N_2 at 77 K and $\sigma(\text{Ar}) = 0.138$ nm^2 for argon at 87 K [62].

In the general case where the condition $(C-1)/C \approx 1$ is not fulfilled, the parameters $b = 1/N_m C$ and $m = (C-1)/CN_m$ must be calculated. This yields two equations with two unknowns that can be solved for N_m and C. Then S is calculated as previously explained.

The BET plots ($0.04 < P/P_0 < 0.3$) for the adsorption of N_2 at 77 K in the furan resins FH and FAH are presented in Figs. 5.7 and 5.8, respectively [51, 52]. The

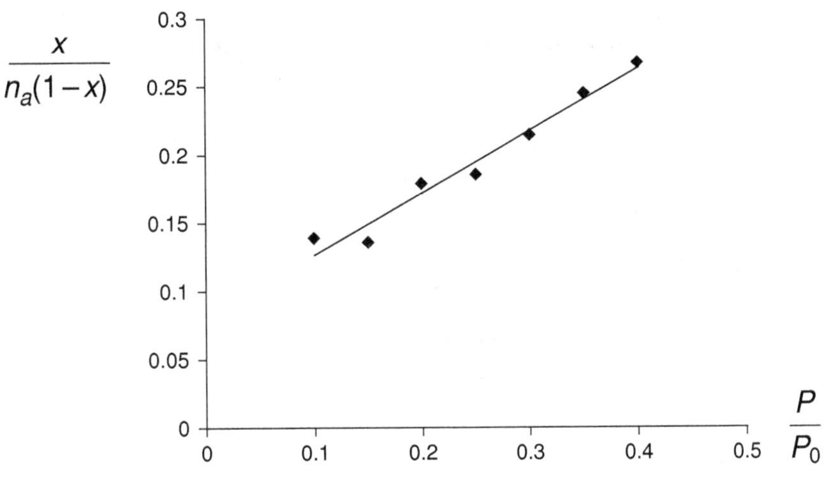

FIGURE 5.7 BET plot of N_2 at 77 K adsorption on the FH furan resin, isothermal data.

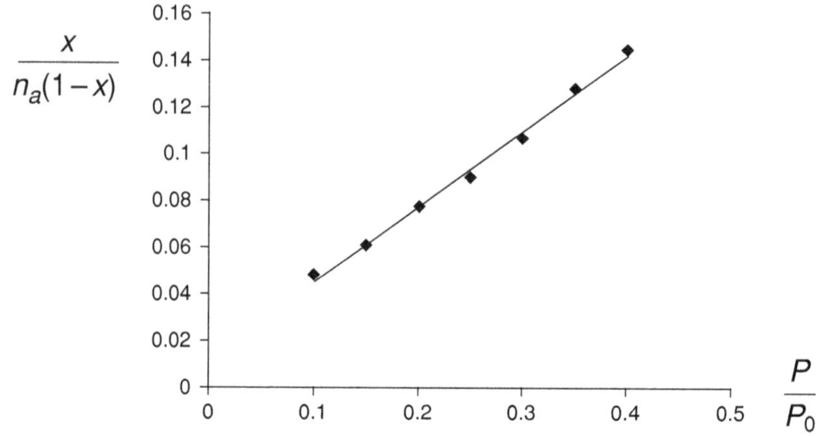

FIGURE 5.8 BET plot of N_2 at 77 K adsorption on the FAH furan resin, isothermal data.

BET isotherm equation can also be fit using a nonlinear regression of Equation 5.28 [62, 119].

The specific surface areas of porous polymers have been commonly measured using the BET method [124–137].

5.7 PORE SIZE DISTRIBUTION INCLUDING THE MESOPORES

We have discussed the application of gas adsorption methods for the calculation of the micropore volume, the specific surface area, and the micropore size distribution of porous materials in general [62, 64, 65] and of porous polymers in particular [108–117, 124–137]. Now we briefly discuss the use of these methods for the calculation of the pore size distribution including the mesopores.

The process of vapor adsorption in complex porous systems generally occurs in the following manner: Initially the micropores are filled, as was described in the study of the Dubinin and osmotic models [62, 63]; then, at higher pressures, external surface coverage consisting of monolayer and multilayer adsorption on the walls of mesopores and open macropores occurs, as discussed in the study of the BET model [62, 64, 65, 80]; finally, capillary condensation in the mesopores occurs [62, 80, 101].

Analysis of the capillary condensation of vapors observed in the adsorption and desorption isotherms is the most important procedure for evaluating the pore size

TABLE 5.4 Specific Surface Area of the Studied Furan Resins

Sample	S, m^2/g
FH	260
FAH	300

distribution (PSD) in the mesopore range [61, 62, 64, 65, 80, 101]. Capillary condensation is linked with a shift of the vapor–liquid coexistence in pores as compared to the bulk fluid. This phenomenon is usually accompanied by hysteresis. During the last five decades, the standard method for determining the PSD in the mesoporous range from the adsorption and desorption isotherms has been the Barret–Joyner–Hallenda (BJH) method [138]. However, this method does not estimate the PSD correctly [101]. Therefore, a novel method of adsorption isotherm assessment based on the nonlocal density functional theory (DFT) has revolutionized the method of PSD calculation for porous materials [62, 101]. The BJH method is based on a simplified macroscopic explanation of the capillary condensation effect with limitations at the microscopic level, and these limitations made the method imprecise. Specifically, the BJH method undervalues the pore size by about 1 nm (10 Å) compared to the DFT method [101].

In order to calculate the PSD of slit and cylindrical pores and of spherical cavities using the DFT method, the experimental isotherm is expressed as a combination of theoretical isotherms in individual pores [62,101, 139–141]. The experimental isotherm is expressed as the integral of the single pore isotherm multiplied by the PSD [62, 73] using the following Fredholm-type integral equation:

$$N_{\text{exp}}\left(\frac{P}{P_0}\right) = \int_{D_{\min}}^{D_{\max}} N_V^{ex}\left(D_p, \frac{P}{P_0}\right)\varphi_V(D_p)dD_p \tag{5.31}$$

where $N_V^{ex}(D_p, P/P_0)$ is the theoretical isotherm in pores of different diameters, $\varphi_V(D_p)$ is the PSD function, and D_p is the pore diameter [62,101]. We calculate the PSD from the experimental isotherm function $N_{\text{exp}}(P/P_0)$ as follows: Initially $N_V^{ex}(D_p, P/P_0)$ is calculated using the DFT theory [62, 101, 139–141]; then the integral Equation 5.31 is represented as a matrix equation and solved to get $\varphi_V(D_p)$, the function describing the pore size distribution [62]. The complete explanation of the process is beyond the scope of this chapter. However, the methodology is implemented in the commercially available software for the automatic measurement of surface area and porosity by the volumetric method.

To illustrate the accuracy of the DFT method and to provide examples of pore size distributions, Fig. 5.9 presents the BJH PSD (Fig. 5.9a) and the DFT PSD (Fig. 5.9b) for an MCM-41 mesoporous molecular sieve (MMS) [62]. This material is unique for the purpose of assessing the validities of the BJH and DFT methods since it is characterized by an ordered hexagonal stacking of highly uniform silica tubes whose diameter is about 3.7 nm in the sample in Fig. 5.9 [62]. The BJH method, as is evident from the comparison of Fig. 5.9a and Fig. 5.9b, underestimates the pore size by about 1 nm. The DFT method yields an average diameter of 3.54 nm [62].

Different authors have applied the BJH methodology for the determination of the pore size distribution of porous polymers [50, 127–130, 137, 142, 143]. However, notwithstanding the fact that the DFT methodology is an exact approach to PSD calculation, it has been applied by only a few researchers [126] in the field of porous polymers.

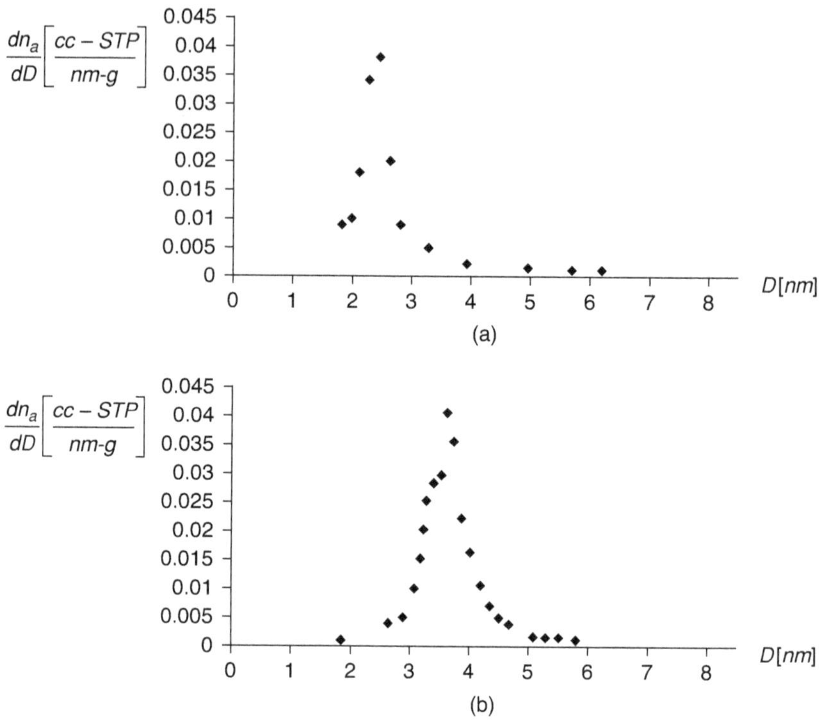

FIGURE 5.9 (**a**) BJH and (**b**) DFT pore size distributions of an MCM-41 mesoporous molecular sieve.

5.8 MACROPOROSITY CHARACTERIZATION

Mercury porosimetry is possibly the most appropriate methodology for the characterization of the pore size distribution in the macropore range that could also be applied in the mesopore range [144–152]. Since porous polymers generally contain macropores, this methodology is widely applied for the characterization of porous polymers [15, 18, 52, 126, 137, 152].

5.8.1 Mercury Porosimetry

To develop the theoretical basis of mercury porosimetry, Washburn [144] applied the Young–Laplace equation:

$$\Delta P = \gamma \left(\frac{1}{r_{\text{I}}} + \frac{1}{r_{\text{II}}} \right) \tag{5.32}$$

where ΔP is the pressure difference across the interface between two contiguous phases at a specific place, γ is the surface tension, and r_{I} and r_{II} are the two radii of curvature describing the surface [145]. Washburn [144, 147], applying the Young–Dupre

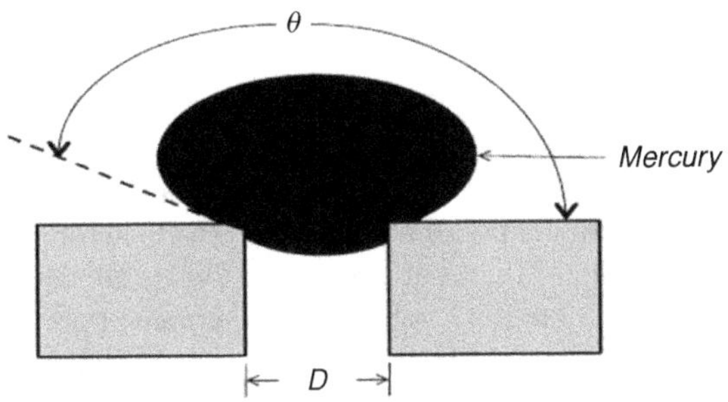

FIGURE 5.10 Mercury making contact with a cylindrical pore.

equation, assumed that, in the case of cylindrical pores of equivalent radius r wetted with a liquid of contact angle θ [146] (Fig. 5.10), a liquid/gas interface enclosed inside a pore will, at equilibrium, adopt the form of uniform average curvature. In a uniform cylindrical pore of radius r or in a parallel-sided slit of width r the mean curvature $1/\bar{r}$ will be equal to [145]

$$\frac{1}{\bar{r}} = -\frac{2\cos\theta}{r}$$

The pressure difference across the interface between two contiguous phases is given by the Washburn equation between the applied pressure and the pore size:

$$\Delta P = -\frac{2\gamma\cos\theta}{r} \tag{5.33}$$

Its mathematical simplicity is related to the selection of the cylindrical pore geometry to evade the complications of dealing with pores of irregular cross sections [147].

Contrary to gas adsorption and pore condensation, in which the pore fluid wets the pore walls (the contact angle is in the range $0 < \theta < \pi/2$), mercury is a nonwetting fluid with a contact angle in the range $\pi/2 < \theta < \pi$ [148]. Consequently, for a nonwetting liquid, it is necessary to apply a positive excess of hydrostatic pressure ΔP to force the fluid to penetrate the pores of radius r.

Mercury is the only fluid truly appropriate for porosimetry-type measurements, given that it does not wet the majority of materials and cannot penetrate pores by capillary action. The surface tension and the contact angle for pure mercury reported in the literature vary; specific values of $\gamma = 484$ mN/m and $\theta = 141°$, 480 mN/m and 140° [150], and 485 mN/m and 139° [153] have been used. For a pressure range of 0.01–200 MPa, the corresponding cylindrical pore radii will be 75 μm to 3.5 nm [145], a broad range that includes macropores and mesopores. Particle size distributions, tortuosity factors, permeabilities, fractal dimensions, compressibilities, and the effects of pore shapes and pore networks on porous solids can be studied with the help of mercury porosimetry [148].

However, it is necessary to take into consideration some effects that influence the measurement when performing mercury porosimetry. A blank correction at high pressures (subtracting an empty-cell run in an actual test) is frequently used to rectify the apparent volume intruded due to compression, to counterweigh the compressibility of mercury and the elastic distortion of the cell and other component parts [145, 147]. The high hydraulic pressure applied on the material during the intrusion can, in some cases, produce structural damage in the specimen. In this event, mercury intrusion can undervalue pore volume and overvalue pore size if the standard Washburn equation is applied [149, 152]. Temperature regulation is also necessary during the experiment because the pressure inside a liquid-filled cell is highly temperature dependent [145, 147]. Another important characteristic of mercury porosimetry curves is the appearance of a hysteresis between the intrusion and extrusion branches [148, 149, 151]. Finally, an additional significant feature is the possible entrapment of mercury within the porous network following extrusion [148, 149, 151].

The specific methodology for the determination of the pore size distribution using mercury porosimetery is by measuring the quantity of nonwetting mercury intruded into pores of the sample as a function of increasing applied pressure and calculating the pore size distribution curves from this information by applying the Washburn equation [146–148]. To carry out the pore size analysis in a mercury porosimeter, the gas contained in the sample cell is evacuated; the sample under test is degassed, since mercury cannot intrude into the sample when pores are filled with another liquid; mercury is transferred into the sample cell under vacuum; and pressure is applied to force mercury into the sample. Throughout the process, the applied pressure P and the intruded volume of mercury V are monitored and an intrusion–extrusion curve is obtained. A pore size distribution curve is obtained by plotting $\Delta V_p/\Delta \log D_p$ versus $\log D_p$ and/or $\Delta V_p/\Delta D_p$versus $\log D_p$ (see Fig. 5.11), where V_p is the accumulated volume of mercury intruded in the pore volume up to the pore of width D_p.

It is also possible to get information from a plot of the cumulative pore areas versus $\log D_p$ (see Fig. 5.12).

The polyHIPE intrusion mercury porosimetry study in Figs. 5.11 and 5.12 was carried out in a Micromeritics AutoPore IV-9500 automatic mercury porosimeter. The sample holder chamber was evacuated up to 5×10^{-5} Torr, and the contact angle and surface tension of mercury applied by the AutoPore software in the Washburn equation to get the pore size distribution were 130° and 485 mN/m, respectively. The equilibration time was 10 s and the mercury intrusion pressure range was from 0.0037 to 414 MPa, that is, the pore size range evaluated was from 335.7 to 0.003 μm.

In the study of porous polymers, mercury intrusion is widely applied [15, 18, 52, 126, 137, 152] since the methodology is fast, the measurement process is automatic, and the measurement provides an assessment of pore sizes over a wide range, from 3 nm to about 200 μm. In addition, other porosity parameters can be calculated. This method is not suitable for measuring small pores due to the compressibility of the polymeric matrix [126]. Therefore, other methods, such as the adsorption methods previously described, are used for the characterization of pores smaller than around 50 nm.

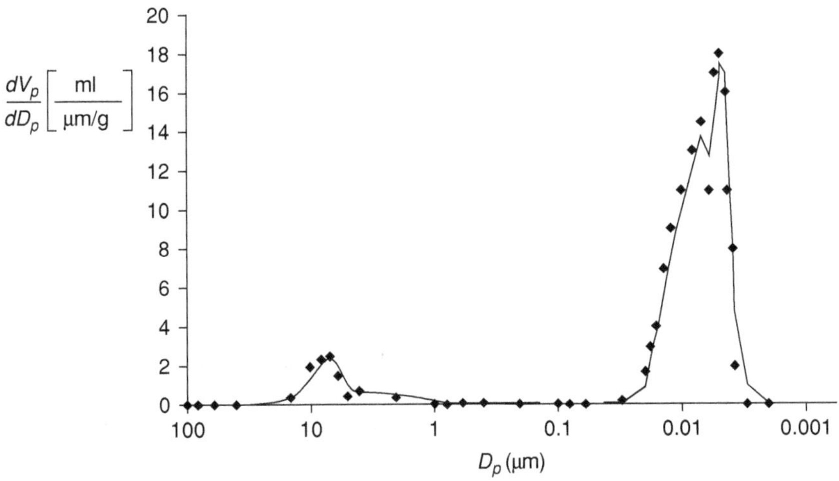

FIGURE 5.11 Differential intrusion of mercury dV_p/dD_p versus log D_p for a highly porous open-cell polyHIPE [19].

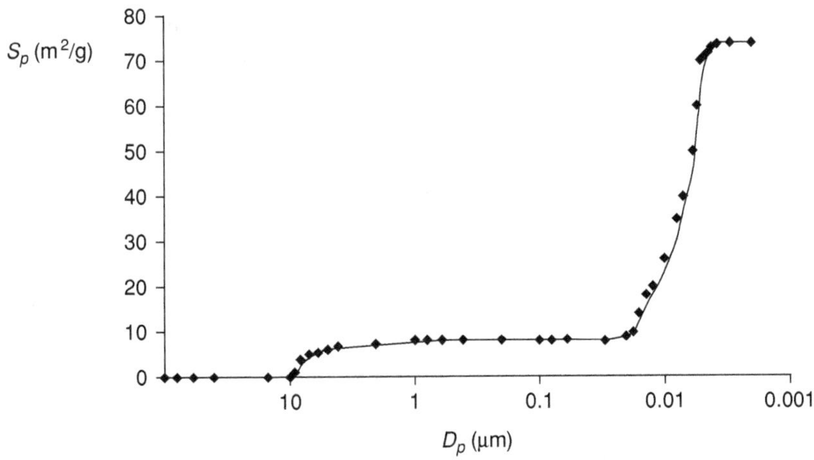

FIGURE 5.12 Cumulative pore area S_p versus log D_p for a highly porous open-cell polyHIPE [19].

5.9 GAS PERMEATION IN POROUS POLYMER MEMBRANES

Membranes have been applied for the treatment of a variety of fluids, including gases, wastewater, seawater, milk, yeast suspensions, and many others [42,48,62,83, 154–159]. In general, a membrane is a barrier that permits selective mass transport between two phases, where the phases on either side of the membrane can be liquid

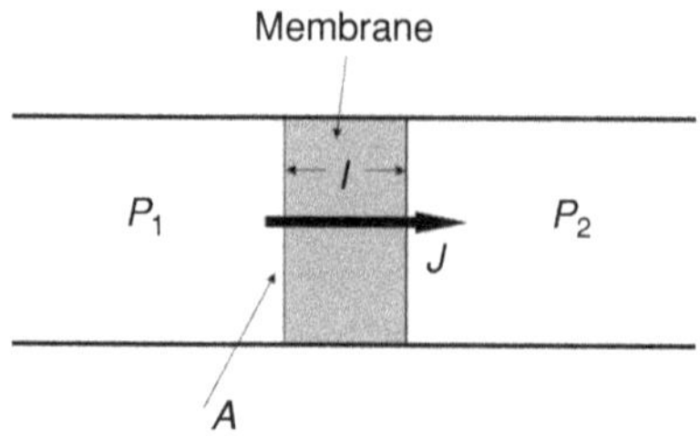

FIGURE 5.13 Membrane permeation.

or gaseous. Polymers are the materials used most often for membranes in science and industry [44, 48, 83, 155, 156]. Polymer membranes are used for gas separation [44], reverse osmosis seawater desalination [159], microfiltration, ultrafiltation, and nanofiltration [48].

We are interested here in gas permeation in porous polymeric membranes, that is, those with pores in the microporous, mesoporous, and macroporous ranges. In the case in which pressure is the driving force of the process, the gaseous molecules will be transported from the high-pressure side to the low-pressure side of the membrane (see Fig. 5.13) [62, 83].

Therefore, for individual gases that do not dissociate during the process, transport follows the Darcy law when a linear pressure drop is applied across the membrane [62, 83, 151]:

$$J = B\left(\frac{\Delta P}{l}\right) = \Pi\,\Delta P \tag{5.34}$$

$$J = \frac{Q}{V_m A} \tag{5.35}$$

$$\Pi = \frac{B}{l} \tag{5.36}$$

where J is the molar gas flow in mol/(m^2·s), A is the effective membrane area in m^3, B is the permeability in mol/(m·s·Pa), l is the membrane thickness, $\Delta P = P_1 - P_2$ is the transmembrane pressure in Pa, Π is the gas permeance in mol/(m^2·s·Pa), Q is the gas filtrate flux in m^3/s, and V_m is the molar volume of the flowing gas in m^3/mol. For an ideal gas,

$$V_m = \frac{V}{n} = \frac{RT}{P} \tag{5.37}$$

In porous materials in general, as well as in porous polymers, three fundamental types of transport process are known: gaseous diffusion, Knudsen diffusion, and configurational diffusion or molecular sieving. For macropores, where, in general, collisions between the molecules take place much more often than collisions with the wall, molecular or gaseous diffusion is the main diffusion mechanism [62]. As the dimensions of the pores become smaller, the number of collisions with the wall increases. At some point Knudsen diffusion becomes predominant [62, 83]. At still smaller pore sizes, in the range of 2 nm or less, when the pore diameter becomes comparable to the size of the molecules, the molecules will constantly undergo interactions with the pore surface. Diffusion in micropores typically takes place in the configurational diffusion or molecular sieving regime [44, 62, 83]. For each mechanism there exists a different relationship between the permeability, the parameters characterizing the membrane, the properties of the flowing gas, the temperature, and the pressure [44, 62, 83].

Specifically, in the case of gaseous or molecular diffusion [62],

$$B = \frac{k}{\eta V_m} \tag{5.38}$$

where k is the permeation factor in m^2 and η is the dynamic viscosity of the gas in Pa·s. The Carman–Kozeny expression should be applied to describe this flow since the Hagen–Poiseuille equation is not valid [160]:

$$k = \frac{\varepsilon\, d_v^2}{16\, C} \tag{5.39}$$

where C is the Carman-Kozeny constant, d_v is the membrane pore diameter, and ε is membrane porosity [62, 83].

For a real mesoporous membrane, which has a complex pore network, the expression for the permeation flux across the membrane is given by [161]

$$J_K = \left(\frac{G}{(2MRT)^{1/2}}\right)\left(\frac{\Delta P}{l}\right) \tag{5.40}$$

where G is a geometrical factor, M is the molar mass of the flowing gas, R is the ideal gas constant, and T is the absolute temperature [62, 83].

In the case of micropores [162, 163]

$$J = \left[\frac{guL}{RT}\exp\left(-\frac{E}{RT}\right)\right]\left(\frac{\Delta P}{l}\right) \tag{5.41}$$

where $g = 1/z$, z is the coordination number, u is the velocity at which the molecule travels ($u = \upsilon l$, where υ is the vibration frequency of the molecule in the adsorption site, for localized adsorption and $u = \left(\frac{8RT}{\pi M}\right)^{1/2}$ for mobile adsorption), and L is the jump distance or diffusional length.

ACKNOWLEDGMENTS

I gratefully acknowledge Dr. Michael Silverstein (Technion-Israel Institute of Technology) for kindly providing the polyHIPE sample. I thank Luis Roberto Cordero (TTC Analytical Service Corp.), Michael L Strickland (Micromeritics Instrument Corp.), and Micromeritics Analytical Services for their support in carrying out the mercury porosimetry intrusion study of the polyHIPE sample.

REFERENCES

1. Schuth, F.; Sing, K. S. W.; Weitkamp, J., Eds. *Handbook of Porous Solids*, Wiley-VCH, Weinheim, Germany, 2002.
2. Zagorodni, A. A. *Ion Exchange Materials: Properties and Applications*, Elsevier, Amsterdam, 2006.
3. Harland, C. H. *Ion Exchange. Theory and Practice*, 2nd ed., Royal Society of Chemistry, London, 1994.
4. Fritz, J. S.; Djerde, D. T. *Ion Chromatography*, 3rd ed., Wiley-VHC, Weinheim, Germany, 2000.
5. Widdecke, H. *Br Polym J* 1984, **16**, 188.
6. Rao, V.; Friedrich, K. A.; Stimming, U. In *Handbook of Membrane Separations: Chemical, Pharmaceutical, and Biotechnological Applications*, Pabby, A. K.; Rizvi, S. S. Sastre, A. N., Eds., CRC Press, Boca Raton, FL, 2008, p. 759.
7. Baker, R. W. In *Membrane Separation Systems. Recent Developments and Future Directions*, Vol. **II**, Baker, R. W.; Cussler, E. L.; Eykamp, W.; Koros, W. J.; Riley, R. L. Strathmann, H., Eds., SciTech Publishing, Raleigh, NC, 1991, p. 100.
8. Wagner, K.; Schulz, S. *J Chem Eng Data* 2001, **46**, 322
9. Rutherford, S. W. *Ind Eng Chem Res* 2001, **40**, 1370.
10. Benson, J. R. *Am Lab*, 2003, **35**, 49.
11. Abrams, M. I.; Millar, J. R. *React Funct Polym* 1997, **35**, 7.
12. Howdle, S. M.; Jerábek, K.; Leocorbo, V.; Marr, P. C.; Sherrington, D. C. *Polymer* 2000, **41**, 7273.
13. Sherrington, D. C. *Chem Commun* 1998, 2275.
14. Xie, S.; Svec, F.; Frechet, J. M. J. *Chem Mater* 1998, **10**, 4072.
15. Hebb, A. K.; Senoo, K.; Bhat, R. Cooper, A. I. *Chem Mater* 2003, **15**, 2061
16. Vivaldo-Lima, E.; Wood, P. E.; Hamielec, A. E.; Penlidis, A. *Ind Eng Chem Res* 1997, **36**, 939.
17. Schwachula, G.; Popov, G. *Pure Appl Chem* 1982, **54**, 2103.
18. Pujari, N. S.; Vishwakarma, A. R.; Pathak, T. S.; Kotha, A. M.; Ponrathnam, S. *Bull Mater Sci* 2004, **27**, 529.
19. Sergienko, A. Y.; Tai, H.; Narkis, M. Silverstein, M. S. *J Appl Polym Sci* 2004, **94**, 2233.
20. Sarzanini, C. *J Chromat* 2002, **956**, 3.
21. Vilensky, M.; Berkowitz, B. B.; Warshawsky, B. *Environ Sci Technol* 2002, **36**, 1851.
22. Wachinski, A. M.; Etzel, J. E. *Enviromental Ion Exchange*, CRC Press, Boca Raton, FL, 1997.

23. Lin, S.H.; Wu, C. L. *Ind Eng Chem Res* 1996, **35**, 553.
24. Economy, J.; Dominguez, L. *Ind Eng Chem Res* 2002, **41**, 6436.
25. Brijmohan, S. B.; Swier, S; Weiss, R. A.; Shaw, M. T. *Ind Eng Chem Res*, 2005, **44**, 8039.
26. Dorfner, K. *Ion Exchangers*, Walter de Gruyter, Berlin, 1991.
27. Alexandratos, S. D.; Crick, D. W. *Ind Eng Chem Res* 1996, **35**, 635.
28. Abrams; I. M.; Millar, J. R. *React Funct Polym* 1997, **35**, 7.
29. Carmona, M.; Warchoł, J.; de Lucas, A; Rodriguez, J. F. *J Chem Eng Data*, 2008, **53**, 1325.3
30. Chakrabarti, A. Sharma, M. M. *React Polym* 1993, **20**, 1.
31. Sharma, M. M. *React Funct Polym* 1995, **26**, 3
32. Chopade, P. C. Sharma, M. M. *React Funct Polym* 1996, **28**, 253
33. Ihm, S.-K.; Ahn, J.-H.; Jo, Y.-D. *Ind Eng Chem Res* 1996, **35**, 2946.
34. Sun, Q.; Harmer, M. A. Farneth, W. E. *Ind Eng Chem Res* 1997, **36**, 5541.
35. Harder, M. A.; Sun, Q. *Appl Catal A General* 2001, **221**, 45.
36. du Toit, E. L. M.Eng.(Chemical) Thesis, University of Pretoria, South Africa, 2003.
37. Gelbard, G. *Ind Eng Chem Res* 2005, **44**, 8468.
38. Gates, B. C. *Catalytic Chemistry*, Wiley Sons, New York, 1992.
39. Gates, B. C. Johanson, L. N. *J Catal* 1969, **14**, 69.
40. Mulder, M. *Basic Principles of Membrane Technology*, Kluver, Dordrecht, Netherlands, 1996.
41. Baker, R.W. *Membrane Technology and Applications*, Wiley, New York, 2004.
42. Goosen, M. F.; Sablani, S. S.; Roque-Malherbe, R. In *Handbook of Membrane Separations: Chemical, Pharmaceutical, and Biotechnological Applications*, Pabby, A. K.; Rizvi, S. S.; Sastre, A. N., Eds.,CRC Press, Boca Raton, FL, 2008, p. 325.
43. Strathmann, H. In *Handbook of Industrial Membrane Technology*, Porter, M., Ed., Noyes Publications, Park Ridge, NJ, 1990, p. 1.
44. Haag, M.-B. In *Handbook of Membrane Separations: Chemical, Pharmaceutical, and Biotechnological Applications*, Pabby, A. K.; Rizvi, S.S.; Sastre, A. N., Eds., CRC Press, Boca Raton, FL, 2008, p. 65.
45. Rautenbach, R. Albrecth, R. *Membrane Processes*, Wiley, New York, 1991.
46. Droste, R. *Theory and Practice of Water and Wastewater Treatment*, Wiley, New York, 1997.
47. Zeman, L. J.; Zydney, A. L. *Microfiltration and Ultrafiltration. Principles and Applications*, CRC Press, Boca Raton, FL, 1996.
48. Sridar, S.; Smitha, B. In *Handbook of Membrane Separations: Chemical, Pharmaceutical, and Biotechnological Applications*, Pabby, A. K.; Rizvi, S. S.; Sastre, A. N., Eds., CRC Press, Boca Raton, FL, 2008, p. 1101.
49. Feibush, B.; Li, N-H. U.S. Patent 4,933,372, 1990.
50. Deryło-Marczewska, A.; Goworek, J. *Langmuir* 2001, **17**, 6518.
51. Sanchez, R.; Hernandez, C.; Roque-Malherbe, R.; Campaña, H. Cuban Patent Certificate CU21644 A1, Cuban Office for the Industrial Property, Havana, 1987.
52. Sanchez, R.; Hernandez, C. *Eur Polym J* 1994, **30**, 51.
53. Tricckey; J. P.; Miner, C. S.; Browniee, H. J. *Ind Eng Chem Res* 1923, **15**, 65.

54. Killefer, D. H. *Ind Eng Chem Res* 1926, **18**, 1217.
55. Dunlop, A.; Peters, F. N. *The Furans*, American Chemical Society, Washington, DC, 1958.
56. Sanchez, R. Ph.D. Dissertation, National Center for Scientific Research, Havana, Cuba, 1988.
57. Roque-Malherbe, R.; Onate-Martinez, J.; Navarro, E. J. *Mat. Sci Lett* 1993, **12**, 1037.
58. Sanchez, R.; Hernandez, C.; Jalsorvskii, G.; Czira, G. *Eur Polym J* 1994, **30**, 37.
59. Budinova, T.; Savova, D.; Petrov, N.; Razvigorova, M.; Minkova, V.; Ciliz, N.; Apak, E.; Ekinci, E. *Ind Eng Chem Res*, 2003, **42**, 2223.
60. Kaminskii, I. V.; Ungurean, N. V.; Ilinskii, V. I. *Plat Massy* 1960, **12**, 9.
61. Sing, K. S. W.; Everett, D. H.; Haul, R. A. W.; Moscou, L.; Pirotti, R. A.; Rouquerol, J.; Siemieniewska, T. *Pure Appl Chem* 1985, **51**, 603.
62. Roque-Malherbe, R. *Adsorption and Diffusion in Nanoporous Materials*, CRC Press, Boca Raton, FL, 2007.
63. Roque-Malherbe, R. *Micropor Mesopor Mater* 2000, **41**, 227.
64. Rouquerol, F.; Rouquerol, J.; Sing, K. S. W. *Adsorption by Powder Porous Solids*, Academic Press, New York, 1999.
65. Gregg, S. J.; Sing, K. S. W. *Adsorption Surface Area and Porosity*, Academic Press, London, 1982.
66. Ross, S.; Olivier, J. P. *On Physical Adsorption*, Wiley, New York, 1964.
67. Adamson, A. W.; Gast, A. P. *Physical Chemistry of Surfaces*, 6th ed., Wiley, New York, 1997.
68. Ruthven, D. W. *Principles of Adsorption and Adsorption Processes*, Wiley, New York, 1984.
69. Yang, R. T. *Adsorbents: Fundamentals and Applications*, Wiley, New York, 2003.
70. Rudzinski, W.; Everett, D. H. *Adsorption of Gases in Heterogeneous Surfaces*, Academic Press, London, 1992.
71. Dubinin, M. M. *Prog Surf Membr Sci* 1975, **9**, 1.
72. Hill, T. L. *Introduction to Statistical Thermodynamics*, Dover, New York, 1986.
73. Rege, S. U.; Yang, R. T. In *Adsorption. Theory, Modeling and Analysis*, Toth, J. Ed., Marcel Dekker, New York, 2002, p. 175.
74. Bering, B. P.; Dubinin, M. M.; Serpinskii, V. V. *J Colloid Interf Sci* 1972, **38**, 185.
75. Roque-Malherbe, R. J. *Thermal Anal* 1987, **32**, 1361.
76. Roque-Malherbe, R. *Physical Adsorption of Gases*, ENPES-MES, Havana, 1987.
77. Horvath, G.; Kawazoe, K. *J Chem Eng Jpn* 1983, **16**, 470.
78. Saito, A.; Foley, H. C. *AIChE J* 1991, **37**, 429.
79. Cheng, L. S.; Yang, R. T. *Chem Eng Sci* 1994, **49**, 2599.
80. Thommes, M. In *Nanoporous Materials: Science and Engineering*, Lu, G. Q.; Zhao, X. S., Eds., Imperial College Press, London, 2004, Chapter 11, p. 317.
81. Roque-Malherbe, R.; Polanco, R.; Marguez-Linores, F. *J Phys Chem* 2010, **114**, 17773.
82. Marquez-Linares, F.; Lopez, V.; Morant, C.; Roque-Malherbe, R.; Domingo, C.; Elizalde, E.; Zamora, F. *J Nanomat* 2010, Article ID 189214, 7 pages, doi 10.1155/2010/189214.

83. Roque-Malherbe, R. *Physical Chemistry of Materials. Energy and Environmental Applications*, CRC Press, Boca Raton, FL, 2009.
84. Roque-Malherbe, R.; Diaz-Castro, F. *J Mol Catal A* 2008, **280**, 194.
85. Roque-Malherbe, R. *Physical Chemistry of Zeolites*, ENPES-MES, Havana, 1988.
86. Dubinin, M. M. *Am Chem Soc Symp Ser* 1977, **40**, 1.
87. Breck, D. W. *Zeolite Molecular Sieves*, Wiley, New York, 1974.
88. Barrer, R. M. *Zeolites and Clay Minerals as Sorbents and Molecular Sieves*, Academic Press, London, 1978.
89. Vansant, E. F. *Pore Size Engineering in Zeolites*, Wiley, New York, 1990.
90. Karge, H. G.; Hunger, M.; Beyer, K. K. In *Catalysis and Zeolites*, Weitkamp, J.; Puppe, L., Eds., Springer, Berlin, 1999, p. 198
91. Roque-Malherbe, R.; Lemes-Fernandez, L.; Lopez-Colado, L.; de las Pozas, C.; Montes-Caraballal, A. In *Natural Zeolites '93 Conference Volume*, Ming, D. W.; Mumpton, F.A., Eds., International Committee on Natural Zeolites, Brockport, NY, 1995, p. 299.
92. Corma, A. *Chem Rev* 1995, **95**, 559.
93. Tsitsisvili, G. V.; Andronikashvili, T. G.; Kirov, G. N.; Filizova, L .D. *Natural Zeolites*, Ellis Horwood, New York, 1992.
94. Roque-Malherbe, R. In *Handbook of Surfaces and Interfaces of Materials*, Vol. 5, Nalwa, H. S., Ed., Academic Press, New York, 2001, Chapter 12, p. 495.
95. Guisnet, M.; Gilson, J.-P. Eds. *Zeolites for Cleaner Technologies*, Imperial College Press, London, 2002.
96. Payra, P.; Dutta, P. K. In *Handbook of Zeolite Science and Technology*, Auerbach, S. M.; Corrado, K. A.; Dutta, P. K. Eds., Marcel Dekker, New York, 2003, p. 1.
97. Roque-Malherbe, R.; Marquez-Linares, F. *Facets-IUMRS J* 2004, **3**, 8.
98. Li, Y.; Yang, R. T. *J Phys Chem B* 2006, **110**, 17175.
99. Nguyen, C.; Sonwane, C. G.; Bhatia, S. K.; Do, D. D. *Langmuir*, 1998, **14**, 4950.
100. Zhao, X. S.; Ma, Q.; Lu, G. Q. *Energy Fuels* 1998, **12**, 1051.
101. Neimark, A. V.; Ravikovitch, P. I. *Micropor Mesopor Mater* 2001, **44**, 697; *J. Phys. Chem B* 2001, **105**, 6817; *Langmuir* 2002, **18**, 1550.
102. Marquez-Linares, F.; Roque-Malherbe, R. J. *Nanosci Nanotechnol* 2006, **6**, 1114.
103. Kaplan, I. G. *Intermolecular Interactions: Physical Picture, Computational Methods and Model Potentials*, Wiley, New York, 2006.
104. Atkins, P. *Physical Chemistry*, 6th ed., W. H. Freeman, New York, 1998.
105. Everett, D. H.; Powl, J. C. *J Chem Soc Faraday Trans I* 1976, **72**, 619.
106. Jackson, J. D. *Classical Electrodynamics*, 2nd ed., Wiley, New York, 1975.
107. Matta, C F.; Boyd, R. J. In *The Quantum Theory of Atoms in Molecules*, Matta, C. F.; Boyd, R. J., Eds., Wiley-VCH, Weinheim, Germany, 2007, p. 1.
108. Seki, K.; Mori, W. *J Phys Chem* 2002, **106**, 1380.
109. Balmaseda, J.; Reguera, E.; Gómez, A.; Díaz, B.; Autie, M. *Micropor Mesopor Mater* 2033, **54**, 285.
110. Roque, J.; Reguera, E.; Balmaseda, J.; Rodríguez-Hernández, J.; Reguera, L.; del Castillo, L. F. *Micropor Mesopor Mater* 2007, **103**, 51.
111. Reguera, L.; Balmaseda, J.; Krap, C. P.; Reguera, E. *J Phys Chem C* 2008, **112**, 10490.

112. Li, H.; Eddaoudi, M.; Groy, T. L.; Yaghi, O. M. *J Am Chem Soc* 1998, **120**, 8511.
113. Eddaoudi, M.; Li, H.; Yaghi, O. M. *J Am Chem Soc* 2000, **122**, 1391.
114. Rosi, N. L.; Kim, J.; Eddaoudi, M.; Chen, M. B.; O'Keeffe, M; Yaghi, O. M. *J Am Chem Soc* 2005, **127**, 1504.
115. Rowsell, J. L. C.; Yaghi, O. M. *J Am Chem Soc* 2006, **128**, 1304.
116. Chandler, B. D.; Yu, J. O.; Cramb, D. T.; Shimizu, G. K. H. *Chem Mater* 2007, **19**, 4467.
117. Poirier, E.; Chahine, R.; Benard, P.; Lafi, L.; Dorval-Douville, G.; Chandonia, P. A. *Langmuir* 2006, **22**, 8784.
118. Bering, P. A.; Serpinskii, V. V. *Izv Akad Nauk SSSR Ser Chim* 1974, 2427.
119. Draper, N. R.; Smith, H. *Applied Regression Analysis*, 3rd ed., Wiley, New York, 1998.
120. Gurvich, L. J. *Phys Chem Russ* 1915, **47**, 805.
121. Calleja, C.; Pau, J.; Calels, J. A. *J Chem Eng Data* 1998, **43**, 994.
122. Poling, B. E.; Praunitz, J. M.; O'Connell, J. P. *The Properties of Gases and Liquids*, 5th ed., McGraw-Hill, New York, 2001.
123. Brunauer, S.; Emmett, P. H.; Teller, E. *J Am Chem Soc* 1938, **60**, 309.
124. Maffei, A. V.; Budd, P. M.; McKeown, N. B. *Langmuir* 2006, **22**, 4225.
125. Pujari, N. S.; Vishwakarma, A. R.; Pathak, T. S.; Kotha, A. M.; Ponrathnam, S. *Bull Mater Sci* 2004, **27**, 529.
126. Germain, J.; Hradil,J; Frechet, J. M. J.; Svec, F. *Chem Mater*, 2006, **18**, 4430.
127. Hebb, A. K.; Senoo, K.; Bhat, R.; Cooper, A. I. *Composites Sci Technol* 2003, **63**, 2379.
128. Santora, B. P.; Gagne, M. R.; Moloy, K. G.; Radu, N. S., *Macromolecules*, 2001, **34**, 658.
129. Deryło-Marczewska, A.; Goworek, J.; Pikus, S.; Kobylas, E. Zgrajka, W. *Langmuir* 2002, **18**, 7538.
130. Deryło-Marczewska, A.; Goworek, J.; Kusak, R.; Zgrajka, W. *Appl Surf Sci* 2002, **195**, 117.
131. Cheng, C. M.; Micale, F. J.; Van der Hoof, J. W. El-Aasser, M. J. *Colloid Interf Sci* 1992, **150**, 549.
132. Sergeyeva, T. A.; Piletsky, S. A.; Piletska, E. V.; Brovko, O. O.; Karabanova, L. V.; Sergeeva, L. M.; Elskaya, A. V.; Turner, A. P. F. *Macromolecules* 2003, **36**, 7352.
133. Dawson, R.; Su, F.; Niu, H.; Wood, C. D.; Jones, J. T. A.; Khimyak, Y. Z.; Cooper, A. I. *Macromolecules*, 2008, **41**, 1591.
134. Seki, K. *Langmuir* 2002, **18**, 2441.
135. Jerábek, K.; Pulko, I.; Soukupova, K.; Stefanec, D.; Krajnc, P. *Macromolecules* 2008, **41**, 3543.
136. Calvino-Casilda, V.; J. Lopez-Peinado, A.; Vaganova, E.; Yitzchaik, S.; Pacios, I. E.; Pierola, I. F. *J Phys Chem B* 2008, **112**, 2809.
137. Rohr, T.; Knaus, S.; Gruber, H.; Sherrington, D. C. *Macromolecules* 2002, **35**, 97.
138. Barret, E. P.; Joyner, L. G.; Halenda, P. H. *J Am Chem Soc* 1951, **73**, 373.
139. Pan, H.; Ritter, J. A.; Balbuena, P. B. *Ind Eng Chem Res* 1998, **37**, 1159.
140. Nguyen, T. X.; Bhatia, S. K. *J Phys Chem B* 2004, **108**, 14302.
141. Dombrokii, R. J.; Hyduke, D. R.; Lastoskie, C. M. *Langmuir* 2000, **16**, 5041.
142. Øye, G.; Roucoules, V.; Cameron, A. M.; Oates, L. J. Cameron, N. R.; Steel, P. G.; Badyal, J. P. S.; Davis, B. G.; Coe, D.; Cox, R. *Langmuir* 2002, **18**, 8996.

143. Zhu, X. X.; Banana,K.; Liu, H. Y.; Krause, M.; Yang, M. *Macromolecules* 1999, **32**, 277.

144. Washburn, E. W. *Phys Rev* 1921, **17**, 273; *Proc Natl Acad Sci USA* 1921, **7**, 115.

145. Rouquerol, J.; Avnir, D.; Fairbridge, C. W.; Everett, D. H.; Haynes, J. H.; Pernicone, N.; Ramsay, J. D. F.; Sing, K. S. W.; Unger, K. K. *Pure Appl Chem* 1994, **66**, 1739.

146. Adamson, A. W.; Gast, A. P. *Physical Chemistry of Surfaces*, 6th ed., Wiley, New York, 1997.

147. Leon, C. A. *Adv Colloid Interf Sci* 1998, **76/77**, 341.

148. Porcheron, F.; Thommes, M.; Ahmad, R.; Monson, P. A. *Langmuir* 2007, **20**, 6482.

149. Porcheron, F.; Monson, P.; Thommes, M.; *Langmuir* 2004, **23**, 3372.

150. Kiseliov, A. V. In *Curso de Fisica Quimica*, Guerasimov, Ya., Ed., Mir, Moscow, 1971, p. 522.

151. Felipe, C.; Rojas, F.; Kornhauser, I.; Thommes, M.; Zgrablich, G. *Adsorpt Sci Technol* 2006, **24**, 623.

152. Egger, C. C.; du Fresne, C.; Raman, V. I.; Schädler, V.; Frechen, T.; Roth, S. V.; Müller-Buschbaum, P. *Langmuir* 2008, **24**, 5877.

153. Shu, J. Master Thesis, Technical University of Delft, Netherlands, 2006.

154. Mulder, M. *Basic Principles of Membrane Technology*, Kluver, Dordrecht, Netherlands, 1996.

155. Baker, R. W. *Membrane Technology and Applications*, Wiley, New York, 2004.

156. Freeman, B. D.; Pinnau, I. Eds. *Polymer Membranes for Gas and Vapor Separation: Chemistry and Materials Science*, American Chemical Society, Washington, DC, 1999.

157. Mauran, S.; Rigaud, L.; Coudeville, O. *Transp Porous Media* 2001, **43**, 355.

158. Roque-Malherbe, R.; del Valle, W.; Marquez, F.; Duconge, J.; Goosen, M. F. A. *Sep Sci Technol* 2006, **41**, 73.

159. Rautenbach, R.; Albrecth, R. *Membrane Processes*, Wiley, New York, 1991.

160. Carman, P. *Flow of Gases Through Porous Media*, Butterworth, London, 1956.

161. Saracco, G.; Specchia, V. *Catal Rev Sci Eng* 1994, **36**, 305.

162. Xiao, J.; Wei, J. *Chem Eng Sci* 1992, **47**, 1123.

163. Roque-Malherbe, R.; Wendelbo, R.; Mifsud, A.; Corma, A. *J Phys Chem* 1995, **99**, 14064.

CHAPTER 6

Nondestructive Evaluation of Critical Properties of Thin Porous Films

MIKHAIL R. BAKLANOV and DENIS SHAMIRYAN
IMEC, Leuven, Belgium

6.1 INTRODUCTION

Thin porous films have received wide applications in various fields, such as catalysis, membranes, chemical sensors, nanotechnology, and microelectronics [1]. They are opening new technological opportunities in different areas. For instance, the application of porous materials in biotechnology allows the recognition of biological molecules such as drugs and proteins, including enzymes, polynucleotides, and polysaccharides. They can also be used as smart membranes for biomedical therapy and as biocompatible materials for biomedical implants and microdevices.

Other recent developments are related to the introduction of porous polymers as interlayer dielectrics (ILDs) with low dielectric constant k for advanced microelectronic technology [2, 3]. Description and analysis of new analytical methods developed for evaluation of these films is the main subject of this chapter. It contains detailed description of different nondestructive techniques most suitable for determination of critical properties of porous films, such as porosity, pore size, and mechanical properties. The content of this chapter is mainly based on our experience in evaluation of porous hybrid polymer films developed for microelectronic applications.

Modern ultra large scale integration (ULSI) devices contain 10^8–10^9 transistors in an area smaller than 1 cm^2 and operate at a clock frequency approaching several gigahertz. As a device's dimensions shrink, the switching speed of its basic elements increases, a consequence of the decreasing carrier transit time across the length of a transistor channel. However, the basic elements (transistors) must be interconnected to provide functionality of ULSI devices. As the functional complexity of devices increase, the number of interconnection levels and metal length continue to increase

Porous Polymers, Edited by Michael S. Silverstein, Neil R. Cameron, and Marc A. Hillmyer

to the extent that an advanced ULSI device may consist of 8 to 10 levels of metal lines. For this reason, the effective speed of the device becomes more dominated by the signal propagation through the horizontal and vertical metal interconnects of components with various functions. Therefore, the resistance R and capacitance C characteristics of the interconnect materials become important. The rapid increase in RC delay time is in fact one of the main bottlenecks for deep submicrometer devices. The RC delay is given by

$$RC = 2\rho k \varepsilon_0 (4L^2/p^2 + L^2/T^2) \tag{6.1}$$

where ρ is the metal resistivity, ε_0 is the vacuum permittivity, k is the relative dielectric constant of the interlayer dielectric, p is the metal line pitch (sum of the line width and the line spacing), T is the metal thickness, and L is the metal line length. This equation demonstrates that RC delay can be reduced using metals with low resistivity and dielectric materials with low dielectric constant. This is why the traditional use of Al has been replaced by that of Cu and technologists are looking for porous polymer films that have a lower dielectric constant than that of the traditional SiO_2.

The dielectric constant of a material is described by the Clausius–Mossotti equation,

$$\frac{(k-1)}{(k+2)} = \frac{4\pi}{3} N\alpha \tag{6.2}$$

where $k = \varepsilon/\varepsilon_0$, with ε and ε_0 the dielectric constants of the material and vacuum, respectively, N is the number of molecules per unit volume (density), and α is the total polarizability, including electronic (α_e), distortion (α_d), and orientation (α_o) polarizabilities. According to Equation 6.2, the dielectric constant of materials can be reduced by decreasing their total polarizability and density. The controllable change of the density is realized by the introduction of porosity.

The relative dielectric constant of porous materials k_r depends on the porosity P and dielectric constant of the film skeleton k_s:

$$\frac{k_r - 1}{k_r + 2} = P \cdot \frac{(k_1 - 1)}{k_1 + 2} + (1 - P) \cdot \frac{(k_s - 1)}{k_s + 2} \tag{6.3}$$

where k_1 is the dielectric constant of the material inside the pores. If the pores are empty, the first term in Equation 6.3 is equal to 0 and we have direct correlation of effective dielectric constant with porosity and skeleton dielectric constant k_s. Materials with relatively small k_s values provide smaller k_r values at lower porosity.

Deposition of a uniform, thin, and porous low-k film is only one of the challenges that have been solved by several industrial companies, research institutes, and universities. The real challenge is integration of these films into ULSI manufacturing processes [4]. Compared to the traditional dense SiO_2, low-k materials are mechanically weak, thermally unstable, penetrable by chemicals, and so on. There are several general requirements for low-k materials to be integrated, including hydrophobicity, mechanical and thermal stability, chemical and physical stability under processing conditions, and compatibility with other materials. There is also a very important challenge for all functional materials: reliability in the user environment.

Porosity affects most film properties. Therefore, it has been necessary to develop new methods or modify existing instrumentation to be able to obtain information related to the critical characteristics of thin porous films. Although many approaches for characterization of porous media are known, most traditional methods are hardly applicable to thin films because of the small amount of matter. Because of this peculiarity (thin films and small amounts of material), the major part of the developed instrumentations is based on highly sensitive techniques. Some of these techniques are based on wave scattering, reflection, and polarization (nonintrusive techniques such as small-angle X-ray scattering [SAXS], neutron scattering [SANS], Brillouin light scattering [BLS] [5, 6], X-ray reflectivity [SXR], ellipsometry, and surface acoustic wave spectroscopy [SAWS] [2]). Ellipsometric and X-ray porosimetry (EP and XRP, respectively) combine nonintrusive and intrusive (adsorption) approaches. Positron annihilation spectroscopy (PAS) and positron annihilation lifetime spectroscopy (PALS) can also be considered as intrusive, but the probes used in these techniques (positrons) have properties fundamentally different from those of the adsorbates used in traditional intrusive techniques. The fundamental difference in the probe properties leads to different physical meaning of the pore interconnectivity measured by adsorption techniques and PALS. All of these methods (see Table 6.1) are analyzed in this chapter.

The materials used as low–dielectric constant films in ULSI technology are both organic and hybrid (organic/inorganic) polymers [7, 8]. Both intrusive and

TABLE 6.1 Methods Used for Characterization of Thin Porous Films

Method	Principle	Information
PAS, PALS	Positron annihilation spectroscopy	Pore size, porosity
SAWS	Surface acoustic wave spectroscopy (measures the velocity of surface acoustic waves)	Young's modulus E, hardness
BLS	Brillouin light scattering based on the inelastic interaction between photons from a laser light source with acoustic phonons of the sample	Young's modulus E, hardness, Poisson ratio
Ellipsometry	Change of polarization of reflected light	Thickness, indices of refraction and absorption
EP	Ellipsometry	Pore size, pore size distribution, and porosity
SXR and XRP	X-ray reflectivity and X-ray porosimetry	Density, thickness (SXR), porosity, pore size distribution (XRP)
SAXS, SANS	Small-angle X-ray and neutron scattering	Pore size distribution

nonintrusive techniques have been successfully used for evaluation of porous low-*k* films during the last 10 years. However, the porosity evaluation in organic low-*k* materials has specific features. For instance, most of the early generations of organic low-*k* materials (Flare, SiLK [2, 3]) swell during the adsorption of organic solvents [9, 10]. For this reason application of intrusive techniques (adsorption porosimetries) might be challenging and limited. However, certain special efforts allow us to solve this problem. For instance, in the case of ellipsometric porosimetry this problem can be solved by separate determination of refractive index and the film thickness. Change of refractive index during the adsorption is more sensitive to the film porosity, while change of the film thickness reflects the degree of swelling. Adsorption porosimetry with organic solvents is sufficiently efficient when the degree of swelling does not exceed 30% and the refractive indices of adsorptive and the polymer matrix are relatively close one to another. In the examples analyzed in Reference 10, the adsorptive and polymer refractive indices were 1.49 and 1.63, respectively, and the degree of swelling was about 20%. A good agreement between the data obtained by PALS and EP has been achieved using EP metrology based on separate determination of refractive index and thickness and further pore size calculation from the change of refractive indices during the toluene adsorption [11]. The recently developed organic low-*k* materials are more resistant to swelling in toluene, which is the most typical adsorbate in ellipsometric porosimetry [12].

The porous materials analyzed in this chapter are mainly porogen-based, porous low-*k* films prepared by dispensing a special liquid precursor on a rotating Si wafer (spin-on deposition) and plasma enhanced chemical vapor deposition (PECVD). In both cases the low-*k* matrix material is codeposited with an additional organic precursor (porogen) that is afterward decomposed and removed, creating the pores. To remove the porogen, thermal annealing combined with ultraviolet (UV) light was used [13]. UV light needs to decompose the porogen, remove volatile fragments, and provide formation of crosslinked low-*k* matrix with sufficiently good mechanical and chemical properties. Most modern low-*k* dielectrics are silica-based materials containing 10%–15% of organic hydrophobic agents. For this reason, they are often termed hybrid polymers. The advantage of these materials is that their chemical properties are similar to traditional SiO_2, making it possible to use traditional microelectronics technology during their integration. Pure organic low-*k* polymers have minimal frequency dispersion and their relatively low k_s value provides low k_r at low porosity. However, they present more problems in integration.

Evaluation of the critical properties of porous low-*k* films (porosity, pore size, mechanical properties, chemical properties, etc.) is extremely important to predicting their behavior during technological processing and the reliability of devices.

6.2 POSITRON ANNIHILATION SPECTROSCOPY

Positronium (Ps) is an "exotic atom" consisting of an electron and a positron. The orbit of the two particles and the set of energy levels are similar to those of the hydrogen atom. The ground state of Ps, like that of hydrogen, has two possible configurations,

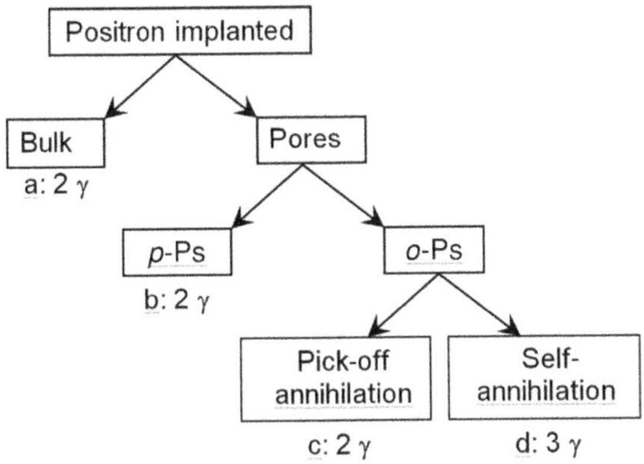

FIGURE 6.1 Channels of positronium annihilation in porous materials.

depending on the relative orientations of the spins of the electron and the positron. The *singlet* state with antiparallel spins is known as *para*-positronium (*p*-Ps) and denoted $^1S_0$. It has a mean lifetime of 125 ps and decays preferentially into two gamma quanta with energy of 511 keV each. Detection of these photons allows reconstruction of the vertex of the decay and is used in positron emission tomography. *para*-Positronium can decay into an even number of photons (2, 4, 6, ...), but the probability quickly decreases with increasing number. The *triplet* state with parallel spins is known as *ortho*-positronium (*o*-Ps) and denoted $^3S_1$. The triplet state in vacuum has a mean lifetime of 142 ns and the leading mode of decay is three γ quanta.

Positronium formed in porous materials annihilates via a number of channels, each producing one or more gamma rays as shown in Fig. 6.1. The most important channel for the pore size evaluation is pick-off annihilation of *o*-Ps (Fig. 6.1c) because it is related to the number of collisions with the pore wall and therefore contains information related to pore size and shape (geometry).

The most probable annihilation channels produce two or three photons, depending on the relative spin configuration of the electron and the positron. The positron implanted in the material is thermalized and has its energy lowered to a few electron volts. Positronium may be formed by the positron capturing a bound molecular electron or recombining with a free electron generated by ionizing collisions. The positronium atoms formed are thus *p*-Ps and *o*-Ps (Fig. 6.1b), and once formed, they begin to diffuse and thermalize in the material. As already mentioned, the lifetime of *p*-Ps is very short (125 ps) because *p*-Ps mainly undergoes 2γ annihilation with its bound electron (Figs. 6.1 and 6.2a, b). The *o*-Ps formed has intrinsically a much longer lifetime than *p*-Ps and therefore is more readily localized in a pore, thermalized, and finally annihilated with molecularly bound electrons through the pick-off process on the pore wall (Fig. 6.2c). Pick-off annihilation is a 2γ process that can decrease the *o*-Ps lifetime to values as low as ~1 ns. The *o*-Ps may also diffuse over long distance if pores are interconnected with the possibility of escaping into the surrounding vacuum and undergoing 3γ annihilation with a natural lifetime

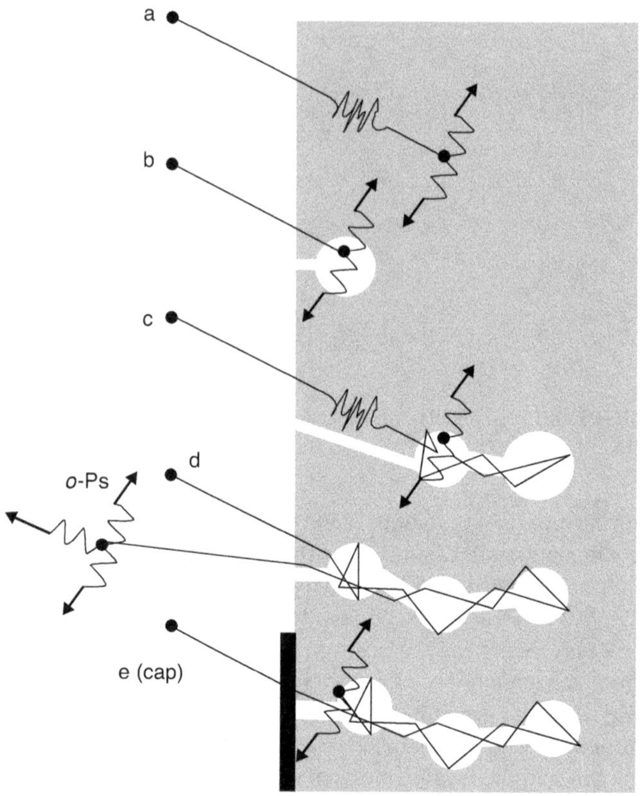

FIGURE 6.2 Mechanisms of positron formation and annihilation in porous media.

of 142 ns (Fig. 6.2d). The analysis of the fate of the positronium can provide valuable information for the characterization of porous materials. Two major techniques have been employed for such analysis, namely the $3\gamma/2\gamma$ ratio technique (PAS) and the positronium lifetime technique (PALS). When films are irradiated with a focused beam of several–kilo electron volt positrons, they penetrate to a certain depth, which depends on the beam energy. Then they either annihilate or form positronium. PALS measures the lifetime of the positronium, whereas PAS extracts information from the Ps annihilation reaction. Reduction of the natural lifetime of *o*-Ps in porous materials yields information about the pore size and geometry. The lifetime depends on the number of collisions with the pore wall (depends on the ratio V/S, similar to the chemistry of radicals, where V and S are pore volume and internal surface area, respectively). This is why Ps lifetime also depends on the pore geometry. In the case of pores with a spherical shape $V/S = r/3$, where r is the pore radius. However, if all pores are open and interconnected, Ps can escape to vacuum and the information related to pore size cannot be obtained (Fig. 6.2d). In this case one needs to deposit a capping layer to measure the pore size [Fig. 6.2e (cap)] [14, 15].

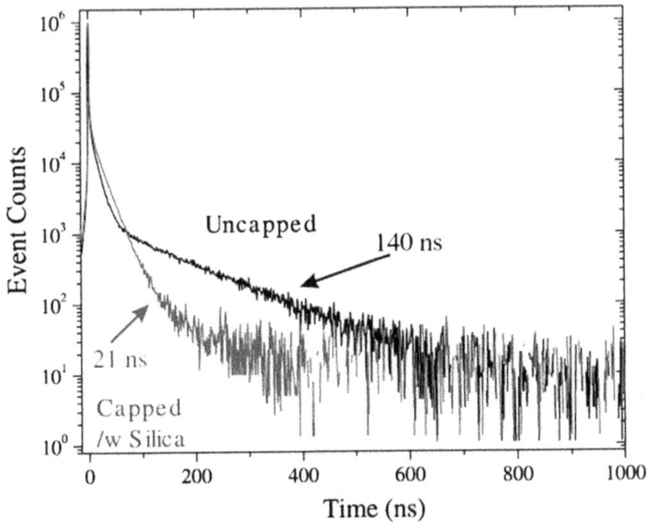

FIGURE 6.3 PALS spectra of an uncapped and capped porous methylsilesquioxane (MSQ) low-*k* film of 22% porosity. Reprinted from reference [23], with permission, from the *Annual Review of Material Research*, Volume 36 © 2006 by Annual Reviews www.annualreviews.org.

To obtain information on the pore size, Ps lifetimes are recorded (Fig. 6.3). As already mentioned, in the case of uncapped film with interconnected pores Ps escapes to vacuum, and shows the natural lifetime 142 ns. If the film has been capped, Ps cannot escape into vacuum, and the lifetime decreases with the decreasing pore size because of higher frequency of collisions with the pore sidewall. In practice, the lifetime distribution curves are obtained with a fitting program developed for this purpose. The distribution curves can then be transformed into pore size data with a certain assumption about the pore geometry. For the calculation it is necessary to perform calibration. For very large pores the calibration is performed using known (standard) high-porosity silica powders. However, when the pores have size larger than 30 nm, the size calculation is challenging because the lifetime is close to the time of self-annihilation [16, 17]. In the subnanometer regime, the quantum mechanical Tao–Eldrup model has been empirically used to calibrate Ps lifetime of several nanoseconds with pore size [18, 19]. The quantum mechanical model has been extended to the classical large-pore limit, and results of the analysis show good agreement with those from other techniques [20, 21].

Depending on the annihilation reaction in PAS, 2γ or 3γ photons are generated. The film porosity can be calculated from a comparison of the measured 2γ and 3γ photon annihilation ratio of Ps atoms (3γ PAS) [22]. This is based on the fact that self-annihilation of *o*-Ps requires at least 3γ photons to convert total momentum, while annihilation with a molecular electron from the pore wall results in the emission of 2γ photons (Fig. 6.1d, c, respectively). As mentioned earlier, the probability of the latter depends on the collision frequency with the pore walls, which affects the

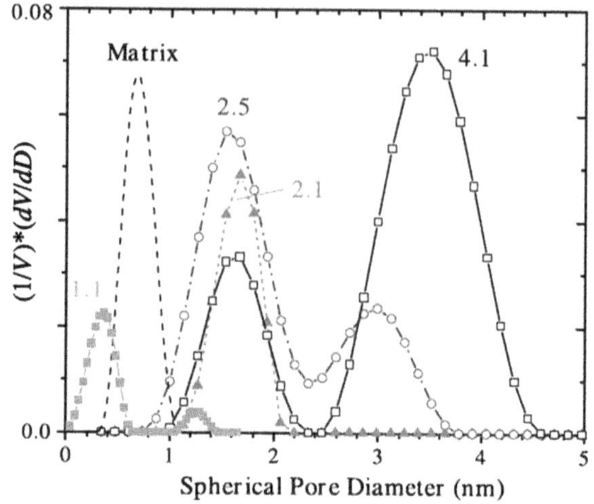

FIGURE 6.4 Pore size distribution in a porous film as a function of beam energy (keV). This film has a complicated depth dependence on porosity. The y axis shows the derivative of the pore volume V with respect to the pore diameter D, normalized to the pore volume. Such results are quite common in porosimetry and reflect comparative pore size distributions. Reprinted from reference [23], with permission, from the *Annual Review of Material Research*, Volume 36 © 2006 by Annual Reviews www.annualreviews.org.

o-Ps lifetime. The rate of pick-off annihilation influences the ratio of the 2γ and 3γ annihilation intensities and can therefore be used to derive the porosity properties. A more detailed description of this procedure can be found elsewhere [22].

PALS is also used for the evaluation of bimodal pore size distributions: They give information related to pore sizes and relative concentration. PALS can also be used for evaluation of pore interconnectivity, as it is clear from the simplified figure, Fig. 6.2. PALS allows analysis of the depth profile of the pore structure using positron beams of different energy (Fig. 6.4). Information related to the pore size at different depths is obtained by changing the energy [23]. The film labeled Matrix in Fig. 6.4 has no engineered pores (deposited without porogen). However, this film shows the presence of micropores with diameter close to 0.6–0.7 nm. The detection of the micropores is related to the presence of methyl groups that lower the density of these films in comparison with SiO_2. It is interesting that the top part of the film with "engineered" pores is quite dense: Only pores with size of 0.25 nm are visible at $E_{beam} = 1.1$ keV. Most probably this phenomenon suggests that the top surface contains fewer CH_3 groups than the matrix material and the bulk of the film. The bottom of the layer ($E_{beam} = 4.1$ keV) has large pores with size close to 3.6 nm.

However, the information related to depth profile and pore interconnectivity (especially in the case of bimodal porosity) obtained by PALS sometimes can be questionable [24]. The coexistence of micropores and mesopores requires careful consideration of the pore connectivity on a subnanometer scale. The transition from a micropore to a mesopore is associated with a gain of kinetic energy, equal to the difference in Ps binding energy [25, 26]. The gained energy is lost rapidly (within a small fraction of the o-Ps lifetime) through collisions with the walls. After several collisions, even an epithermal Ps remains localized to the larger volume (Fig. 6.2c). The reverse transition from a mesopore to a micropore is strongly suppressed because it requires significant energy to overcome the difference in Ps binding energy. PALS experiments detect the final state of the system after diffusion and trapping in mesopores have occurred. A detailed discussion of the Ps trapping kinetics is given in References 21 and 25.

Mogilnikov et al. [24] used a simple quantum mechanical model to show the origin of this irreversible directional transition. The Ps ground level can be estimated as

$$E_0 = \pi^2 h^2 / 2ml^2 \tag{6.4}$$

where h is Planck's constant, m is the Ps mass, and l is the size of the potential well (equal to the pore size in this case). According to Equation 6.4, the o-Ps ground level in a potential well with a diameter of 10 nm is two orders of magnitude lower than that in a well with a diameter of 1 nm ($E_0 = 1.87 \times 10^{-3}$ and 1.87×10^{-1} eV, respectively) (Fig. 6.5). The mean o-Ps energy in large pores at room temperature is close to kT (2.5×10^{-2} eV). Movement of o-Ps between pores with different sizes requires a change in energy, and the probability of the movement is proportional to $\exp[-(E_{02} - E_{01})/kT]$, where E_{02} and E_{01} are o-Ps energies in pores with small and large size, respectively. Therefore, o-Ps has the possibility of easy movement in pores larger than 3–5 nm and from small pores to large pores. If o-Ps is able to escape to vacuum and show a natural lifetime of 142 ns, one can conclude that the large pores are interconnected. The situation is opposite for diffusion from large pores to micropores. The diffusion of o-Ps localized in the large pores (>3 nm) to subnanometer pores requires significant activation energy (of the order of a few

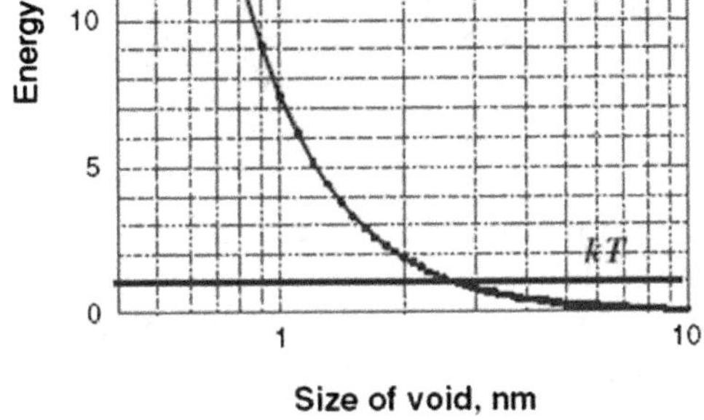

FIGURE 6.5 The o-Ps ground-level energy in pores of different size. (Reprinted from Reference 24 with permission from Japan Society of Applied Physics).

electron volts) that depends on the size of the micropores (Fig. 6.5). This renders the probability for such transition negligible.

Many porous films contain both micropores and mesopores. Although both types generate distinct PALS signals, according to the foregoing analysis, *o*-Ps localized in mesopores virtually cannot escape from the film through the micropores, whereas other species may be able to do so. This allows us to conclude that the PALS percolation threshold is only related to the interconnection of mesopores. Therefore, the practical importance of the PALS percolation threshold might be limited from the point of view of penetration, diffusion, and adsorption of practically important (stable) molecules used for deposition, etching, and cleaning. The quantum mechanical nature of the *o*-Ps transition from mesopores to micropores also suggests that PALS evaluation of the integrity of diffusion barriers deposited on top of porous low-*k* films may also be limited to openings on the mesopore scale. These can still facilitate gas, solvent, and Cu diffusion since the micropores are open for stable molecules but closed for diffusion of *o*-Ps.

6.3 SURFACE ACOUSTIC WAVE SPECTROSCOPY, LASER ULTRASONICS, AND BRILLOUIN LIGHT SCATTERING

6.3.1 Surface Acoustic Wave Spectroscopy

Surface acoustic waves are elastic vibrations propagating along the surface with amplitude decaying exponentially within the material. The energy of the surface acoustic wave is concentrated at the surface due to the exponential characteristics of the amplitude, making the wave sensitive to the surface layers with properties deviating from those of the bulk material. The wave velocity is the essential parameter to be measured, also termed phase velocity. It depends on the elastic properties and the density of the material.

A surface wave can be induced by a pulsed laser beam and, for a homogeneous and isotropic material, has a phase velocity v_{ph} that depends on the material density ρ, Poisson ratio υ, and elastic modulus E as

$$v_{ph} = \frac{0.87 + 1.12\upsilon}{1 + \upsilon} \cdot \sqrt{\frac{E}{2\rho(1 + \upsilon)}} \tag{6.5}$$

A detailed description of the influence of the density and elastic properties of both coating and substrate on the wave dynamics is given by Farnell and Adler [27].

The surface acoustic wave is typically detected at two positions several millimeters from the source (x_1 and x_2), using, for instance, a wideband piezoelectric transducer [28, 29]. From the Fourier transform of the waveform the phase spectra $\phi_1(f)$ and $\phi_2(f)$ can be determined and can be used to calculate the frequency-dependent velocity dispersion curve:

$$v_{ph}(f) = \frac{(x_2 - x_1)\omega}{\phi_2(f) - \phi_1(f)} \tag{6.6}$$

where ω is the angular frequency. The dispersion is a result of the larger penetration depth of the wave for smaller frequencies and the resulting stronger influence of the substrate properties. A fit of the dispersion curve using all known material properties of both substrate and film then yields the elastic modulus of the film [30].

Other systems [31] using lasers to both induce the acoustic wave and measure its propagation have also been developed. Acoustic waves in this case are induced by a localized thermal expansion of the sample through heating from an excitation laser pulse at 532 nm wavelength with duration 0.5 ns. Surface acoustic waves with wavelength varying between 4 and 10 μm can be induced by placing interference grids on the optical path of the excitation beam. The induced surface wave is detected through time-dependent diffraction of a continuous weaker probe beam impinging on the modulated sample surface. A reference beam that coherently interferes with the diffracted signal is used to achieve higher sensitivity (heterodyne detection). The probe laser has a spot size of approximately $25 \times 50\ \mu m^2$. The small spot size makes it applicable for evaluation of patterned structures.

Brillouin light scattering (BLS) also allows measurement of surface wave dispersions. It is based on the inelastic interaction between photons from a laser light source and acoustic phonons of the sample. The small frequency shift Δf of the Brillouin peaks due to scattering of the photons is directly related to the phase velocity of the acoustic phonons v_{ph} (and thus to E) through [32]

$$v_{ph} = \frac{\pi \cdot \Delta f}{k \cdot \sin(\theta_i)} \tag{6.7}$$

for the surface phonon mode and through

$$v_{ph} = \frac{\pi \cdot \Delta f}{n \cdot k} \tag{6.8}$$

for the bulk phonon mode. Here, θ_i is the angle of incidence of the laser beam, k is its wavevector, and n is the refractive index of the material.

The technique was originally developed for investigation of transparent materials, which have a large interaction volume. Since its extension to opaque materials, which have a small penetration depth resulting in surface sensitivity, it strongly depends on the development of high-resolution Fabry–Perot interferometers of the Sandercock type [33], which are used in backscatter mode for higher sensitivity.

In the experiments described later we used the experimental system described by Schneider et al. [30]. The wideband surface acoustic wavepackets are generated from absorption of laser pulse energy. The laser pulse energy (0.5 mJ, 0.5 ns duration, 337 nm wavelength) was focused into a thin line on a sample (10 μm × 6 mm),

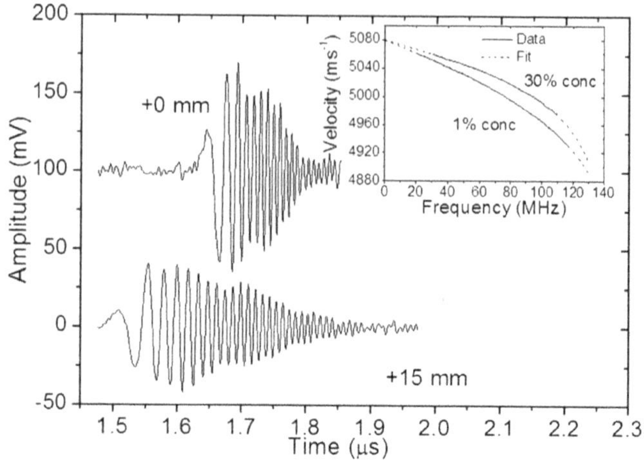

FIGURE 6.6 Dispersive SAW wavepackets on a 1% sample. Inset: Dispersion curves for 1% and 30% with best fits, from which the elastic properties are extracted. (Reprinted from Reference 34 with permission from the IAP.)

generating surface acoustic wavepackets propagating along the sample. The wideband SAW wavepackets are detected by a piezoelectric foil with steel-wedge transducer at different relative propagation distances (here 15 mm) on the sample. Figure 6.6 shows a typical SAW wavepacket at different relative propagation distances for aerogel film deposited on top of Si(001). The broadband SAW wavepacket (approximately 20–100 MHz frequency range) propagates in both layer and substrate and becomes dispersed because waves of different frequency sample a different proportion of layer and substrate, with different net elastic properties, and the wave velocity is therefore frequency dependent. The frequency-dependent velocity dispersion curve is extracted by using a Fourier transform technique. Assuming that thickness and Poisson's ratio υ are known, the density ρ and Young's modulus E of the layer are obtained from the best-fit parameters of the theoretical to the measured dispersion curve. The inset in Fig. 6.6 shows measured dispersion curves for the 1% and 30% porogen materials together with best fits to the data [34, 35].

This technique measures absolute values of the film properties and is independent of changes in chemical makeup of the film, except insofar as density or stiffness are altered. This system requires propagation lengths of ~10 mm. Complementary data were supplied by EP [36] (porosity, pore size, film thickness, and refractive index), SXR (film density) [37], and BLS (Young's modulus). The BLS E values were derived from measurement of the film bulk longitudinal velocity combined with a BLS-measured Poisson ratio of 0.26 for the 1% sample.

Figure 6.7a shows plots of EP porosity and SAWS density of methylsilsesquioxane (MSQ)-type porous materials as a function of porogen concentration. A strong linear dependence is clear. Comparison of the SAWS density to the SXR density (Fig. 6.7b) yields a linear fit with almost ideal slope of 1.02. Therefore, the SAWS technique

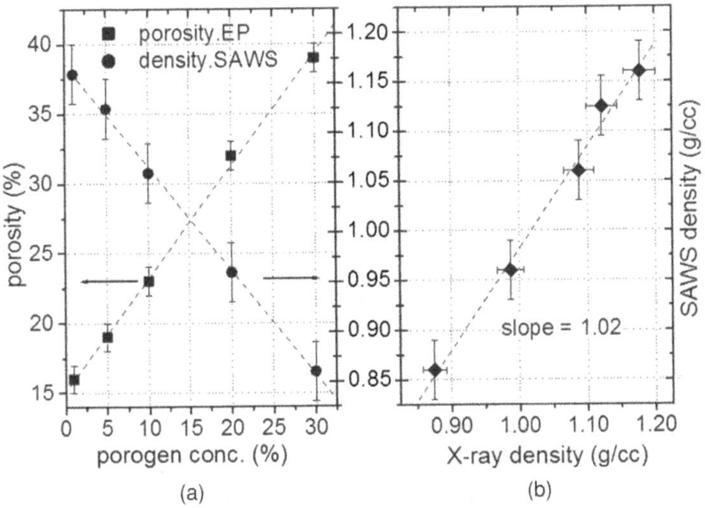

FIGURE 6.7 (**a**) Strong linear dependences of porosity and SAW density on porogen concentration. (**b**) Excellent correlation of SAWS density versus SXR measured density. (Reprinted from Reference 34 with permission from the IAP.)

provides reliable density values with remarkably strong correlations among porogen concentration, porosity, and density. We calculated a skeletal density of ~1.4 g/cm^3, typical for bulk MSQ.

Figure 6.8 plots the E values measured by SAWS, BLS, EP, and nanoindentation (NI) as a function of porosity. The values from all these techniques are, at most, ~20%

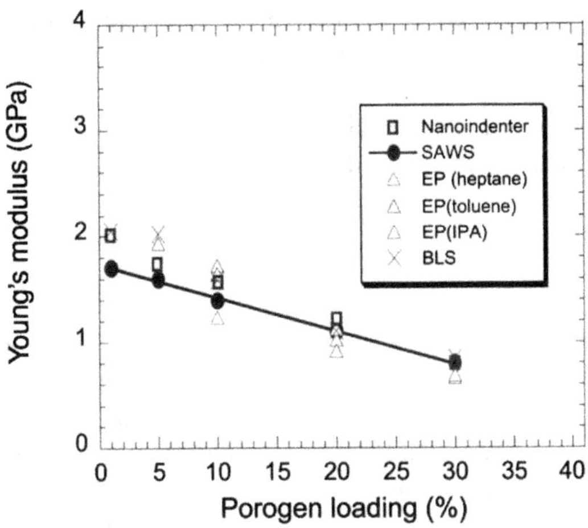

FIGURE 6.8 Comparison of E values measured by SAWS, BLS, EP, and nanoindentation. (Reprinted from Reference 2 with permission from the AIP.)

different. Considering that the BLS E values are derived at ~8 GHz from longitudinal mode and the SAWS values at ~100 MHz from a primarily transverse mode, and that EP and NI use fundamentally different principles, this can be considered an excellent mutual verification of the E values extracted from all these techniques.

Although the results presented in Fig. 6.8 show excellent agreement, such comparison is not always easy. The results of measurements can be affected by different artifacts. For instance, the Young's modulus measured by EP can be affected by surface roughness, and the values measured by SAWS and especially NI become less reliable when the films are too thin. Generally, such comparison is straightforward when the films are thicker than 1 μm [38].

6.3.2 Laser Ultrasonics

A promising application of acoustic techniques for evaluation of thin porous films is based on the generation of acoustic excitations with a pulsed laser [39,40]. The picosecond ultrasonic method was developed primarily to measure thickness of opaque films such as those used for metal interconnect layers. A laser pulse (0.1 ps) heats the film surface, causing it to expand rapidly, and the expansion creates a sound wave that travels down through the film stack, partially reflecting at each interface. The film thickness and various other properties can be calculated by analyzing these echoed sound waves as they return to the surface.

This technique has been extended to the measurement of transparent films such as low-k dielectrics. In this case, most of the laser energy is transmitted through the overlying transparent porous film to the opaque metal deposited on top of silicon substrate, so that the sound wave is generated in the opaque layer. The opaque layers can also be traditional diffusion barrier layers like TaN and TiN. The wave then travels up through the transparent layer, causing a local variation in the refractive index that partially reflects the probe laser. The light reflected from the traveling wavefront interferes with light reflected from the surface of the transparent film, resulting in constructive and destructive interferences that cause the detected signal to oscillate in intensity (Fig. 6.9).

As a result of this interference, the measured signal oscillates with a period τ, from which the sound velocity V in the material can be determined:

$$V = \frac{\lambda}{2n\tau \cos\phi} \tag{6.9}$$

where n is the index of refraction, which can be measured independently with an ellipsometer, λ is the wavelength, and ϕ is the angle of refraction. Combining velocity with density, a known quantity for the material, gives elastic stiffness.

This technique has been applied for the evaluation of mechanical properties of uniform porous low-k films [41,42]. Good correlation of the measured mechanical properties with the results obtained by nanoindentation was established. The Young's modulus dependence on porosity and k value is presented in Fig. 6.10. It was concluded that the picosecond ultrasonic measurements of Young's modulus provide

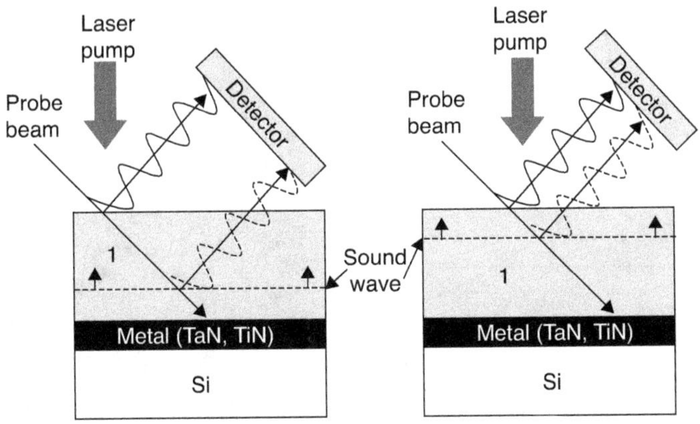

FIGURE 6.9 Picosecond ultrasonic technology can characterize the mechanical properties of transparent films. A subpicosecond laser pulse heats the underlying opaque substrate, which expands rapidly, initiating a sound wave that travels up through the transparent dielectric film. Light from a probe laser partially reflects from the moving wavefront and interferes at the detector with light reflected from the film's surface.

a reliable method for mechanical strength monitoring; they are nondestructive and have sufficient spatial resolution to be used on patterned test structures on production wafers. The measurements require only a few seconds at each site, fast enough to offer timely feedback for process control or to provide multisite full-wafer surveys in process development applications.

Additional application of ultrasonic technique was demonstrated recently [41, 43]. Widely used UV curing generates a gradient of Young's modulus, which depends on the absorption coefficient of low-*k* films [44]. While all the other available

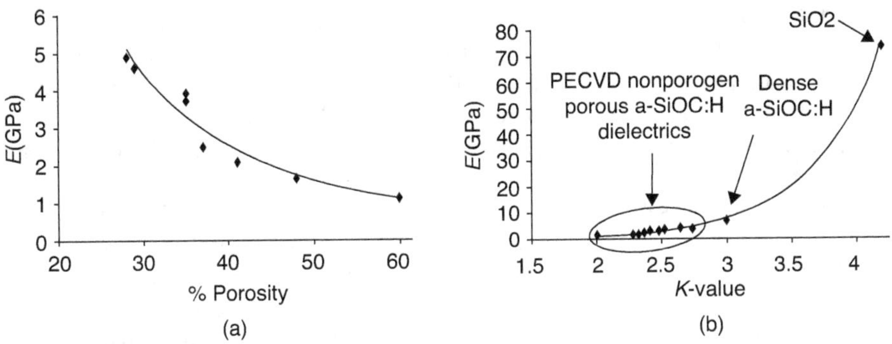

FIGURE 6.10 Experimental values of Young's modulus as measured by picosecond laser ultrasonics of PECVD low-*k* films versus **(a)** EP porosity and **(b)** dielectric constant as measured by a mercury probe. Reprinted from reference 41 with permission from the Elsevier.

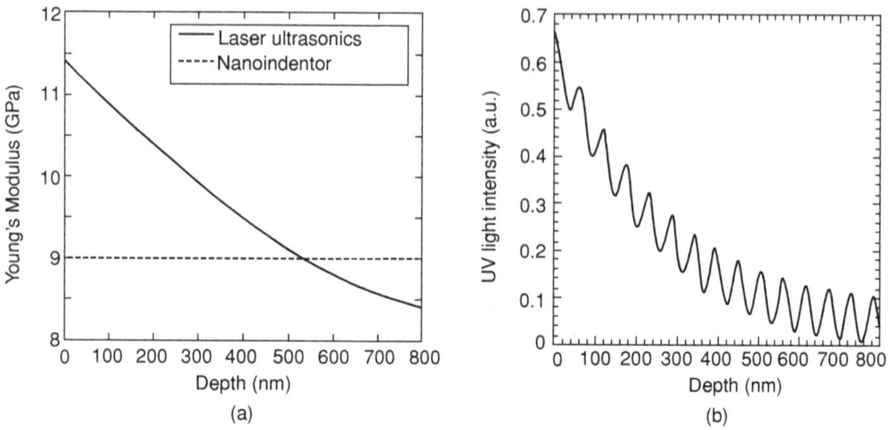

FIGURE 6.11 **(a)** Young's modulus of UV-cured CVD SiCOH films measured by nanoindentation and laser ultrasonics. **(b)** The slope of the *E* curve correlates well with the slope of the 172-nm light intensity calculated with optical characteristics measured by UV ellipsometry. However, the intensity oscillations related to the formation of standing waves are not visible in the *E* curve.

techniques allow measurements of only a mean value of Young's modulus, laser ultrasonics gives a unique opportunity to measure the gradient of the modulus across the film thickness. Such information is extremely important from a UV curing optimization point of view, as well as for reliability of the final devices (Fig. 6.11).

6.4 ELLIPSOMETRY AND ELLIPSOMETRIC POROSIMETRY

6.4.1 Ellipsometry

6.4.1.1 Fundamentals. Ellipsometry is an optical technique widely used for monitoring and investigating the dielectric properties (thickness, complex refractive index, or dielectric function) of thin films. It has applications in many different fields, from microelectronics to bionanotechnology. Ellipsometry is a very sensitive measurement technique and provides unequaled capabilities for thin-film metrology. It is nondestructive, contactless, and fast.

Upon the analysis of the change in light polarization reflected off a sample, ellipsometry can yield information about layers that are much thinner than the wavelength of the probing light. Ellipsometry can probe the complex refractive index or dielectric function tensor, which gives access to fundamental physical parameters and is related to a variety of sample properties, including morphology, crystal quality, chemical composition, and electrical conductivity. It is commonly used to characterize film thickness for single layers or complex multilayer stacks ranging from a few angstroms or tenths of a nanometer to several micrometers with excellent accuracy.

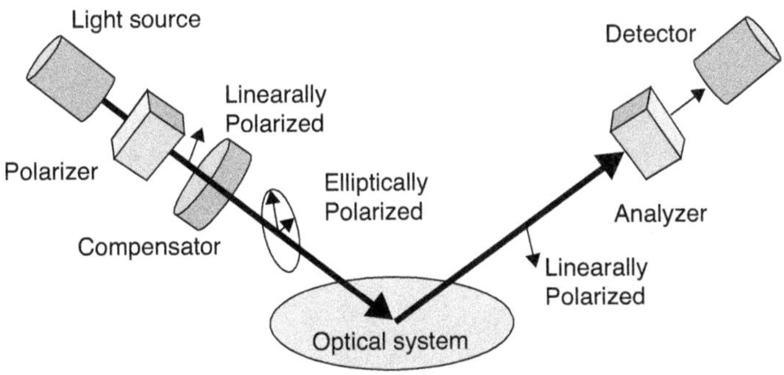

FIGURE 6.12 The classic scheme of zero ellipsometry.

Ellipsometry starts with a light source and ends with a light detector (Fig. 6.12). Two polarizing elements, the polarizer and the analyzer, measure the input and output polarization states. The sample (optical system) to be measured is located in between them. The most traditional form of ellipsometry uses the so-called zero method (*zero ellipsometry*) and is based on finding the parameters of the polarizer and the analyzer at which the intensity of the light registered by the photodetector is equal to zero or minimal (condition of extinction). Zero ellipsometry is used to study samples described by a Johnson matrix (nondepolarizing) systems. In the classical zero ellipsometer, the polarizer represents a combination of a linear polarizing element (a simple polarizer) and a phase-shifting device (compensator) and the analyzer is a simple polarizing element (Fig. 6.12) [45].

The compensator converts linearly polarized light into elliptically polarized light. An ideal compensator is an optical retarder with a retardation exactly equal to 90° (quarter wave). A rotatable compensator combined with a rotatable polarizer can convert unpolarized light into any elliptical polarization. The appropriate choice of the rotation angles of polarizer and compensator creates a polarization that provides linear polarization of the light reflected from the sample. This light can be extinguished by appropriate orientation of the analyzer. The condition of extinction is described by the equation

$$R_p \cos(A) - R_s \sin(A) \exp(i \cdot 2(P + \pi/4)) = 0$$

where A and P are the angles of rotation of the analyzer and polarizer, respectively. This equation can be written as

$$\rho_s \equiv R_p/R_s = \tan(\Psi)\exp(i\Delta) \tag{6.10}$$

where Ψ and Δ are so-called ellipsometric angles that describe the modulus and phase of the relative reflection coefficient ρ_s and are connected with the angles A and P. Here R_p and R_s are the total reflectance coefficients of the waves polarized parallel (p) and perpendicular (s) to the plane of incidence. Equation (6.10) is known as the

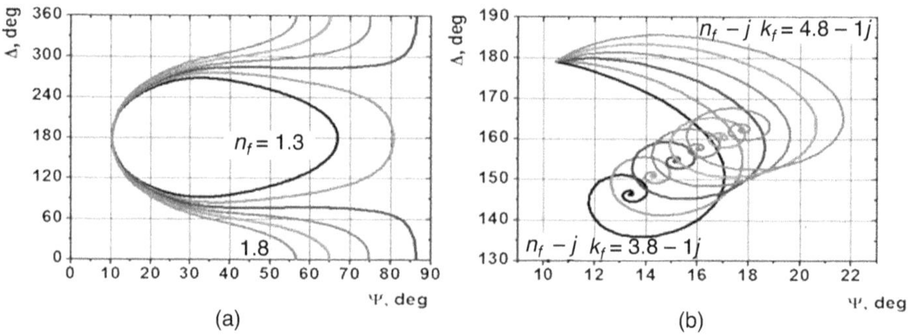

FIGURE 6.13 Calculated Δ–Ψ trajectories for films on top of a silicon substrate. (**a**) Transparent films with refractive indices $n_f = 1.3$–1.8 (step 0.1). (**b**) Nontransparent films with absorption coefficient equal to 1.0 and various indices of refraction from 3.8 to 4.8 (step 0.2).

fundamental equation of ellipsometry and correlates the measured quantities Ψ and Δ with the sample characteristics, such as the film thickness and optical characteristics. Examples of such correlations are presented in Fig. 6.13.

Other versions of ellipsometry have also been developed:

Photometric ellipsometry is based on measurement of the light intensity at the detector versus *P* and/or *A*. This dependence contains information related to the sample characteristics. A different realization of this method allows determining all of the elements of the Jones matrix. In this case the method is called *generalized ellipsometry* [46], in contrast to *conventional ellipsometry,* when only the ratio of diagonal elements is determined. Photometric ellipsometry can also be applied for depolarizing samples.

In the presence of a sufficient number of variable parameters of the polarizer and/or the analyzer it is possible to determine all 16 elements of the Mueller matrix. This method is termed *Mueller ellipsometry* [47]. The photometric methods still have limited application despite the very high accuracy of the measurements. The reason is their strict technical requirements for the detectors and the measurement systems. Progress in electronics and optoelectronics is increasing the application of these systems.

Ellipsometry does not measure thickness or refractive and absorption coefficients of the films directly. To get these values, it is necessary to solve equations that express measured quantities by model parameters, that is, to solve inverse ellipsometric problem. This problem might be complicated and is not always unambiguous [48].

6.4.1.2 Ellipsometry in Visible, Ultraviolet, and Infrared Spectral Ranges. Ellipsometry in the visible spectral range (VIS) is mainly used for most traditional applications (thickness and refractive index measurements of transparent films such as SiO_2 and Si_3N_4). Spectroscopic ellipsometry (SE) is more efficient for the characterization of new materials. In situ ellipsometry enables the study of the kinetics of surface reactions, for instance, during adsorption, film deposition, etching, and curing.

Figure 6.13 shows typical Δ–Ψ trajectories at 633 nm for films deposited on top of an Si substrate. Each Δ–Ψ point gives two numerical values. With this, one can calculate two unknowns about the film, such as thickness and refractive index for a transparent film. It is also clear that there are Δ–Ψ regions that provide much better information than others. For example, near the period point (or the film free point: $\Delta \approx 179^\circ$ and $\Psi \approx 10^\circ$ for the silicon substrate) the curves are not particularly well separated. The Δ–Ψ values in this region are not so efficient for determining the index of refraction of the film. Additional uncertainty might be related to the ellipsometric period equal to $d_0 = \lambda/2\sqrt{n^2 - \sin^2 \varphi}$, where λ and φ are wavelength and incident angle, respectively. Therefore, films with thicknesses $d_m = d + md_0$ ($m = 0, \pm1, \pm2, \ldots$) have the same values of Δ and Ψ. This uncertainty is excluded by using variable wavelength (SE) and/or variable angle of incidence φ.

A Δ–Ψ trajectory for nontransparent films is not periodic (Fig. 6.13b). Initially, at low thicknesses, the curves are well separated. As the thickness increases, the points become closer together and the trajectory begins to spiral into the target value. The Δ–Ψ value at the target point corresponds to the optical characteristics of the bulk materials having the same composition as the film. One can see that determination of three unknown parameters (d, n, and k) is not straightforward because certain points might correspond to films with different characteristics. In this case the use of variable wavelength and variable incident angle is important.

Ellipsometry in the vacuum UV (<190 nm) allows the analysis of materials for advanced lithography (photoresists, antireflective coatings). The short wavelength also increases the sensitivity of ellipsometric measurements to ultra thin films (<10 nm). Many materials absorb UV light, which enables surface-sensitive measurements like surface roughness, native oxide covering, material composition, and structural properties.

An interesting and very important application is related to the measurement of UV optical characteristics of porogen-based low-k materials. Such measurements enable the analysis of the change in the films during UV curing and, as a result, allow optimization of the curing processes [44,49,50]. Figure 6.14a shows the change in optical characteristics during porogen removal. The measured spectra are modeled using the propagation matrix approach in the substrate–film–ambient system. Although some inhomogeneity might be expected in the film thickness due to the depth-dependent effect of the UV cure caused by light absorption and interference, the data are fitted well with the single-layer model; only a small increase of the fitting error is observed for longer curing times. The ellipsometric spectra for all low-k samples were fitted using seven peaks: a Gaussian band at 4.1 eV, three Lorentzians at 4.7, 5.5, and 6.4 eV creating the pronounced peaks, Gaussian bands at 8.1 and 9.9 eV to model the absorption edge with a shoulder, and one δ-shaped peak at 11 eV that represents the contribution of absorption bands above the measured range to the dispersion of the real part of the dielectric function. The modeled dielectric functions are recalculated to refractive index and extinction coefficients; the results are plotted in Fig. 6.14, where the positions of modeled peaks are marked by arrows [44].

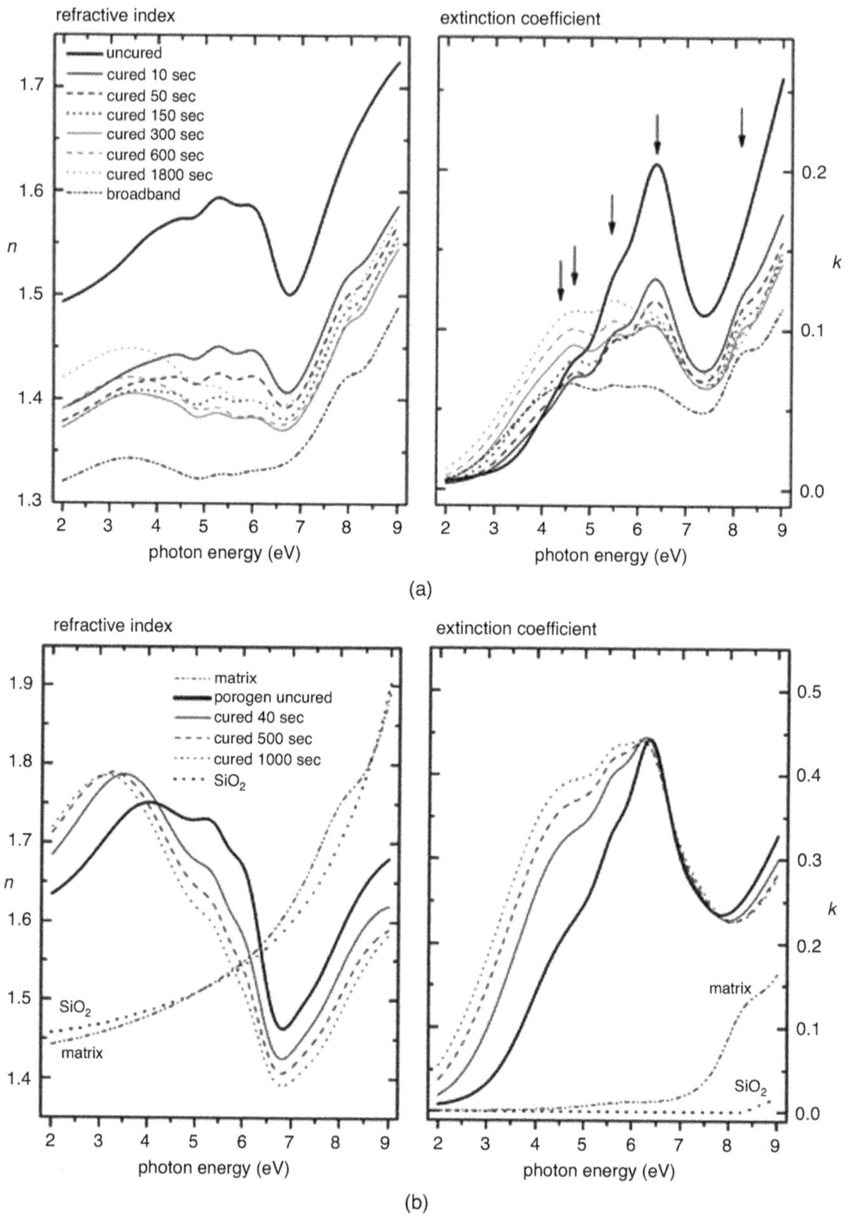

FIGURE 6.14 **(a)** Refractive index and extinction coefficient of UV-cured, low-k dielectric films. The curing time varies from 0 s (*solid thick black line*) to 1800 s (*dotted thin gray line*). The optical function of the sample cured by a broadband lamp with $\lambda > 200$ nm is plotted for comparison (*dash-dot-dot line*). The arrows show the positions of the generalized Gauss–Lorentz (G-L) peaks. **(b)** Optical functions of porogen films UV cured for various curing times from 0 s (*solid thick black line*) to 1000 s (*dotted line*) and the optical functions of the matrix material (*dash-dot-dot line*). The optical functions of SiO_2 are plotted for comparison (*thicker dotted line*).

The refractive index at 2–4 eV decreases with curing time due to the increase in porosity (porogen removal). The organic porogen has strong absorption near 6.5 eV that decreases during the UV curing. However, UV light–assisted structural changes in the deposited porogen material are associated with the effusion of light, hydrogen-rich components and thermal graphitization. A nonvolatile graphitized phase remains inside the low-*k* film, is referred to as "porogen residue," and has an impact on dielectric and mechanical properties of the low-*k* material. The changes in UV spectra are in good agreement with Fourier transform infrared spectroscopy (FTIR) and Raman spectra. Moreover, it was shown that UV optical characteristics are much more sensitive to the amount of porogen residue than IR absorption as measured by FTIR. It has also been shown that the amount of porogen residue can be controlled by the choice of an appropriate wavelength of the UV curing light [44, 51]. Figure 6.14b also shows that the low-*k* matrix is different from SiO_2 and has an absorption edge at 9 eV. The shoulder between 7.5 and 9 eV is related to Si–CH_3 groups that are added to keep the low-*k* matrix hydrophobic.

The near-infrared spectral range (NIR) is mainly used to measure the thickness of single films and layer stacks. Many materials become transparent in this region, which makes this range convenient for measurements. Such applications of ellipsometry are quite traditional, and detailed descriptions of the measurement and calculation procedures are well documented [52].

Most of the materials used in microelectronics are transparent in the infrared range, so the depth of the light penetration into the structure (and hence the sensitivity of the ellipsometry) can be extended up to 15–20 μm [53]. The numerous bands in the infrared spectrum constitute the highly individual fingerprint of a given material, and therefore a sample with a thin surface layer will produce very different contrasts within the infrared spectrum [54]. Moreover, IR ellipsometry can probe the vibrational properties of surface and thin layers [55]. Therefore IR ellipsometry can simultaneously probe the film morphology and the microstructure from the vibrational absorption. Progress in measurement sensitivity in the IR range was obtained from the systematic combination of spectroscopic ellipsometry and Fourier transform spectroscopy [56]. In contrast to many surface analysis methods, the geometric thickness is derived. Generally, IR ellipsometry is sensitive to chemical composition and to the local structural order of the thin-film material [57]. IR spectroscopic ellipsometry (IRSE) is also able to provide important information during the characterization of patterned materials after processing modification and plasma etching [58]. Figure 6.15 illustrates the capabilities of IR ellipsometry for sidewall characterization of Si vias. There are several separate peaks on the Tan Ψ spectrum—one at 3700 cm^{-1} and some at 3000–3500 cm^{-1}. When an area without patterning (only resist is present) is measured, only the peak at 3700 cm^{-1} is observed, so one can conclude that this peak is related to the photoresist. The other peaks start to increase as the via pattern density increases (from 1/5 to 1/1 via/space). At the same time the resist peak decreases as less resist is present due to increased pattern density. Therefore, the other peaks are related to polymers deposited on the via walls.

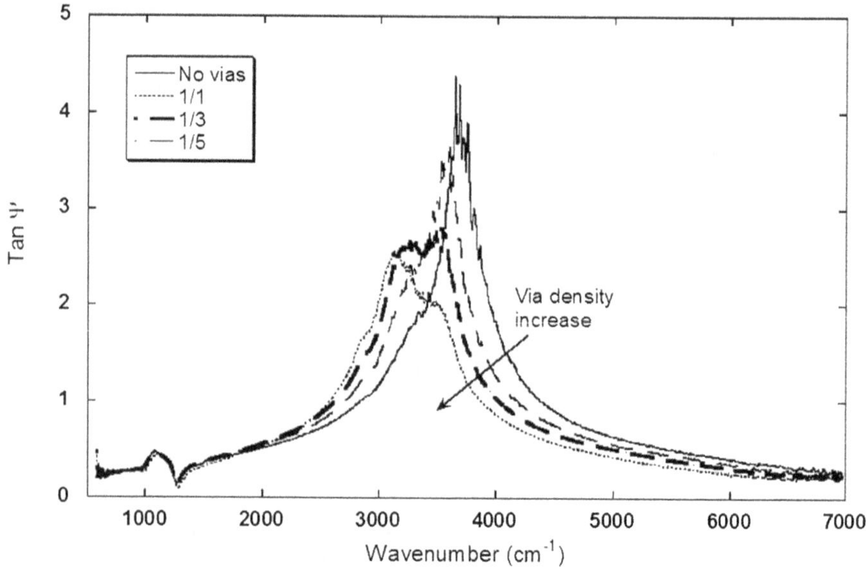

FIGURE 6.15 IRSE measurement of patterned wafers (Tan Ψ versus wavenumber). Plasma polymer deposited on the via sidewall is easily distinguishable from photoresist, which remains in the top surface.

6.4.2 Ellipsometric Porosimetry

6.4.2.1 Porosity and Pore Size Distribution. Ellipsometric porosimetry is a new application of in situ ellipsometry developed for evaluation of porosity and pore size distribution in thin, transparent films [36, 59–61]. Porosity of the films is calculated from the volume polarizability of materials with empty pores and pores filled by an appropriate liquid adsorptive (e.g., toluene).

According to the Lorentz–Lorenz equation the volume polarizability B of a two-component system depends on the refractive indices of each component (indices 1 and 2) and their fractions:

$$B_2 = \frac{n_{re}^2 - 1}{n_{re}^2 + 2} = V\frac{n_1^2 - 1}{n_1^2 + 2} + (1 - V)\frac{n_2^2 - 1}{n_2^2 + 2} \tag{6.11}$$

where n_{re} is the measured refractive index, n_1 and n_2 are refractive indices of the material inside the pores and of the film skeleton, respectively, and V is pore volume. When the pores are filled by a liquid adsorptive,

$$B_3 = \frac{n_{rl}^2 - 1}{n_{rl}^2 + 2} = V\frac{n_{ads}^2 - 1}{n_{ads}^2 + 2} + (1 - V)\frac{n_2^2 - 1}{n_2^2 + 2} \tag{6.12}$$

where n_{ads} is refractive index of the liquid adsorptive. Subtracting B_2 from B_3 gives

$$V = \left(\frac{n_{rl}^2 - 1}{n_{rl}^2 + 2} - \frac{n_{re}^2 - 1}{n_{re}^2 + 2}\right) \Big/ \left(\frac{n_{abs}^2 - 1}{n_{abs}^2 + 2}\right) \tag{6.13}$$

This equation shows that the porosity calculation needs the refractive indices of the film with empty pores n_{re} and that of pores filled by adsorptive n_{rl}, as well as the refractive index of the liquid adsorptive n_{ads}. The refractive index of the film skeleton is generally not needed for the porosity calculation [62]. The porosity calculated by Equation 6.13 includes both the artificial porosity created by the porogen and the constitutive porosity (free volume) of the film matrix. The presence of closed pores (pores not accessible to the probe molecules) can be estimated by analysis of the calculated skeleton refractive index.

Calculation of the pore size distribution uses the phenomenon of progressive emptying of a porous system initially filled at $P = P_0$, where P and P_0 are current and saturated toluene pressures, respectively. The pore size calculation in mesoporous films is based on an analysis of a hysteresis loop that appears during the adsorption and desorption [63]. The hysteresis loops appear because the effective radius of curvature of the condensed liquid meniscus is different during adsorption and desorption. The adsorptive vapor condenses in pores at a vapor pressure P that is lower than the equilibrium pressure of a flat liquid surface P_0. Dependence of the relative pressure (P/P_0) on the meniscus curvature is described by the Kelvin equation:

$$\ln\left(\frac{P}{P_0}\right) = -\frac{f \cdot \gamma \cdot V_L \cos\theta}{r_K \cdot RT} \tag{6.14}$$

where γ and V_L are surface tension and molar volume of the liquid adsorptive, respectively, θ is the wetting angle of the adsorptive, and $f = 1$ for slit-shaped pores and $f = 2$ for cylindrical pores. If the radius of a cylindrical pore is r_p, then $r_p = r_k + t$, where t is the thickness of the layer adsorbed on the pore walls. Values of t are obtained from the adsorption of the same adsorptive on a nonporous sample having a chemically similar surface and are defined by the BET equation [63].

Porosity measurements are performed using in situ ellipsometry combined with a high-vacuum system. The standard EP setup consists of an in situ spectroscopic ellipsometer (λ = 350–850 nm) mounted on a vacuum chamber, a dry pumping system, a pressure gauge, and sources of liquid adsorptives. The vacuum chamber is pumped down before an adsorptive is introduced into the chamber. Vapor of an organic adsorptive (toluene is standard) is introduced into the vacuum chamber using a controllable valve. This step must be done as slowly as necessary to provide equilibrium between the adsorbed phase and the gas phase. The same requirement is applied during desorption. The measurement time depends on the pore size and the film thickness. For most porous low-k films with pore size larger than 1 nm and thickness less than 1 μm, the measurement time is about 1 h.

Initial experimental data for the porosity and PSD calculation are the ellipsometric characteristics Δ and Ψ measured during the adsorption and desorption cycles

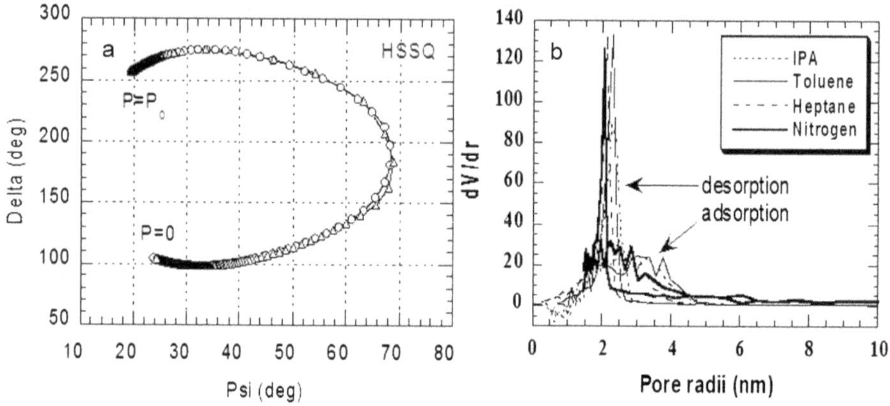

FIGURE 6.16 (**a**) Typical change of Δ–Ψ during toluene adsorption in a mesoporous low-*k* film and (**b**) calculated pore radius distribution. The pore radius distribution demonstrates good agreement between the data obtained with different adsorptives and nitrogen porosimetry.

(Fig. 6.16a). Special EP software calculates the change of refractive indices and thickness of the film during the adsorption and desorption, pore size distribution (PSD), and specific surface area. The change of the adsorbate volume (open porosity) is calculated using Equation 6.13. The dependence of the adsorbate volume on the relative pressure P/P_0 is used to calculate the PSD (Eq. 6.14, Fig. 6.16b). The specific surface area of each small group of pores, δA_i, is calculated from the corresponding pore volume and pore radius as $\delta A_i = \delta V_i/r_i$. By summing the values of δA_i over the whole pore system a value of the cumulative surface area is obtained.

In micropores of a few molecular diameters, the Kelvin equation is no longer valid. Not only would the values of the surface tension and the molar volume deviate from those of the bulk liquid adsorptive, but also the concept of a meniscus would eventually become meaningless. To analyze microporous films, a method based on a theory developed by Dubinin and Radushkevitch (DR) is used. The DR theory takes into account change of adsorption potential when the distance between pore walls is comparable with the size of adsorptive molecules.

Figure 6.17 shows typical experimental data obtained for different types of low-*k* films. One can see that each type of film has its own characteristic shape of the isotherm. Figure 6.17a shows results of porosity evaluation in a CVD low-*k* film with open porosity equal to 30%. The adsorption/desorption isotherms do not have a hysteresis loop, which is a characteristic feature of microporous films. CVD carbon-doped silica films (SiCOH) are typically microporous, and the Dubinin-Radushkevitch method is used for the pore radius calculation. It should be noted that direct observation of micropores is impossible, as they are not visible in the scanning electron microscope (SEM).

Figure 6.17b shows results obtained for porous films with ordered cylindrical pores. The pore geometry significantly affects the thermodynamic properties of fluids

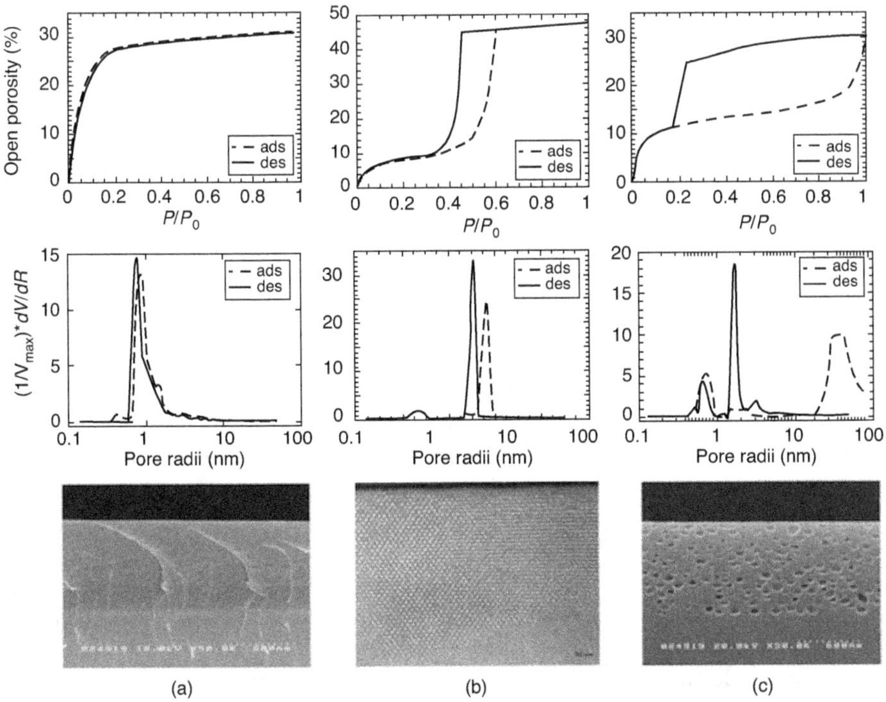

FIGURE 6.17 Adsorption–desorption isotherms, pore radius distribution, and SEM pictures of three different types of low-*k* films. (**a**) CVD low-*k* film with open porosity equal to 30%. (**b**) Porous spin-on-glass (SOG) low-*k* films with ordered cylindrical pores and porosity 47%. (**c**) Porous SOG low-*k* film with large quasi-closed cavities and 30% porosity.

and their adsorption, desorption, and diffusion behavior. In the case of semi-infinite cylindrical pores, a cylindrical interface is formed between the adsorbed layer and the vapor during the adsorption, while evaporation (desorption) is related to the formation of a hemispherical meniscus between the condensed fluid and the vapor. Therefore, according to the Kelvin equation, in the case of infinite cylindrical pores the pore radius calculated from the adsorption curve is twice the radius calculated from the desorption curve. The reason for this difference is as follows. In the general case of noncylindrical pores Equation 6.14 can be written as

$$\frac{1}{r_1}+\frac{1}{r_2}=-\frac{RT}{f\cdot\gamma\cdot V_L\cos\theta}\ln\left(\frac{P}{P_0}\right)$$

where r_1 and r_2 the principal radii of curvature. In a spherical meniscus $r_1 = r_2$ and $1/r_1 + 1/r_2 = 2/r_K$, while in a cylindrical meniscus $r_1 =$ width, $r_2 = \infty$ and $1/r_1 + 1/r_2 = 1/r_K$. In most publications related to adsorption porosimetry the pore size equivalent to cylindrical pores is reported. However, this often contains an uncertainty related to insufficient knowledge of the pore geometry. In some cases, careful analysis of both adsorption and desorption branches can be extremely useful. A typical example is porous films with quasi-closed cavities (Fig. 6.17c). This film contains small micropores (necks) interconnecting the larger cavities. The necks are not visible in the SEM. Evaporation from a pore that is connected to the bulk phase by narrower pores is delayed and occurs spontaneously after the emptying of one of the adjacent pores. This effect of *pore blocking–controlled desorption* is commonly called the ink-bottle effect. More detailed description of analysis of such films can be found elsewhere [63, 64].

In addition to porosity and PSD evaluation, ellipsometric porosimetry has demonstrated its efficiency for other applications. Measurement results can be used for evaluation of mechanical properties of thin films [65] and the integrity of diffusion barriers deposited on top of low-k dielectrics [66]. For instance, a semiautomatic tool developed by SOPRA (EP-5) is equipped with a video camera allowing visualization of the specific behavior of pores in diffusion barriers ("killer" voids) during adsorption. The toluene vapor penetrates through the open holes and locally changes the refractive index of the low-k film. This change is detected by the camera as colored spots. Replacing toluene (the standard adsorptive) with water, EP can be used for evaluation of the internal (bulk) hydrophobicity of low-k materials. This application is especially important for evaluation of processing damage of low-k materials [67]. Evaluation of porosity of low-k films and monitoring of Cu surface cleaning are typical examples demonstrating the importance of ellipsometry in advanced microelectronic technology.

6.4.2.2 *Mechanical Properties of Porous Films.* Ellipsometry allows separate determination of the refractive index and thickness of the film. Both characteristics are measured with high resolution. If the change of refractive index is related to pore filling and enables the calculation of porosity, the change of thickness can be related to the swelling or miniscule reversible film shrinkage during the capillary condensation of a liquid in the pores. In the case of miniscule reversible shrinkage, which is typical for inorganic and hybrid films, the degree of shrinkage depends on capillary forces and the elastic modulus of the film. Therefore, EP can be used to extract the elastic modulus of porous materials from the same data that are used for the evaluation of the pore structure [62, 65, 68].

By measuring the film thickness d as a function of the relative pressure P/P_0 and fitting to

$$d = d_0 - k_{ep} \ln(P/P_0) \qquad (6.15)$$

one can extract the constant k_{ep}, from which the modulus E can be calculated using

$$E = \frac{d_0 RT}{k_{ep} V_L} \qquad (6.16)$$

Here the molecular volume V_L of the adsorbate is the only molecular characteristic needed for determining the elastic modulus. The difference in the film thickness at

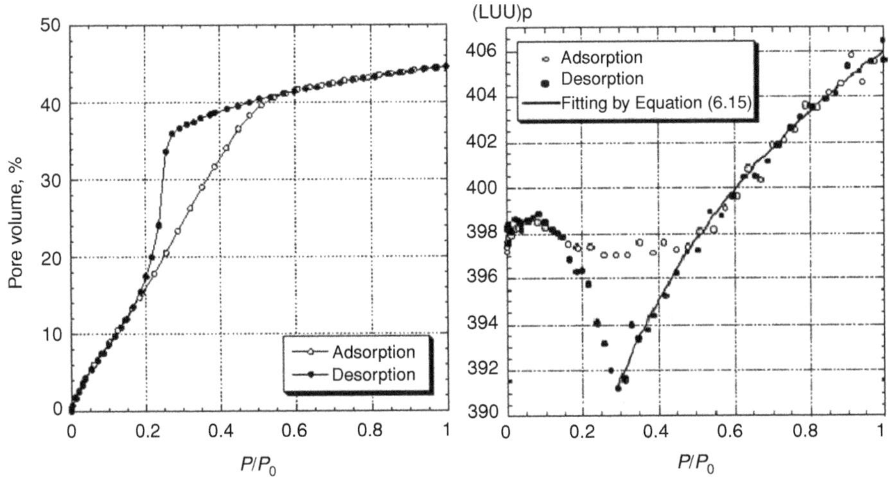

FIGURE 6.18 Adsorption–desorption isotherm and change of the film thickness during toluene adsorption and desorption. The change of the film thickness is fitted by Equation 6.15.

$P/P_0 = 0$ and $P/P_0 = 1$ is related to the change of the film stress during toluene adsorption. More detailed analysis and theory of adsorption deformation of porous films can be found elsewhere [68].

Figure 6.18b shows the typical change of the film thickness during toluene adsorption and desorption. The minimum thickness corresponds to the relative pressure when almost all pores are filled by toluene but the menisci still exist (maximal capillary pressure). This P/P_0 value corresponds to the maximum on the PSD curve. The final film thickness corresponds to the complete filling of all pores at $P/P_0 = 1$ when the liquid surface does not form a concave meniscus. Fitting of Equation 6.15 to the experimental curve allows one to determine the coefficient k_{ep} that can be used for calculation of Young's modulus. If the films are thick enough (>1 μm), the Young's modulus values measured by EP with different solvents, surface acoustic wave spectroscopy, Brillouin light scattering, and nanoindentation [2] are in good agreement (Fig. 6.18).

A possible error in the Young's modulus determination by EP can be related to the surface roughness. In this case, adsorptive condensation at $P/P_0 \leq 1$ gives an additional change of thickness and the E value is underestimated. As mentioned earlier, this error becomes almost negligible when the film thickness is larger than 1 μm.

6.4.2.3 Evaluation of Diffusion Barriers and Interconnectivity. EP also can be used as a pore capping probe [66]. A continuous barrier deposited on top of a porous film prevents solvent vapor from penetrating into the porous film. As a result, no change of ellipsometric characteristics of a porous film is registered when the film is exposed to the vapor. On the other hand, a noncontinuous barrier allows vapor to penetrate into the film and change its ellipsometric characteristics. If the barrier is

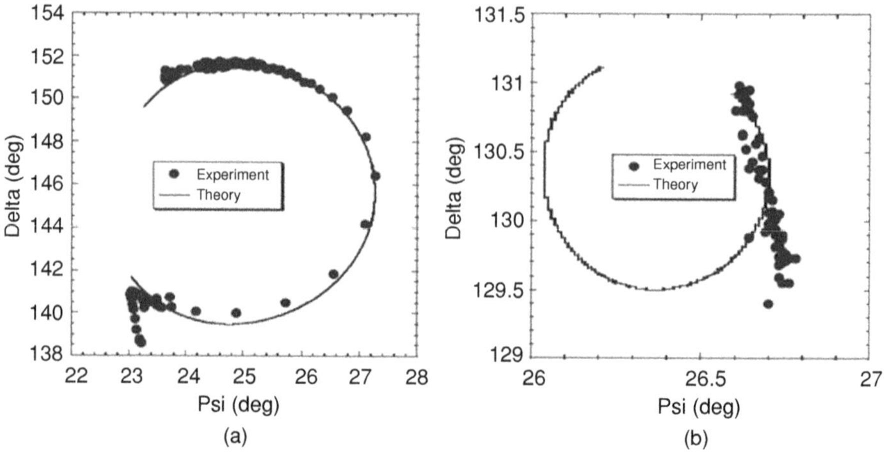

FIGURE 6.19 Ellipsometric angles Δ and Ψ recorded during exposure of a porous HSQ film with (**a**) 30 nm TaN and (**b**) 60 nm TaN. Continuous curves represent calculated theoretical changes of Δ and Ψ assuming that the barrier is porous. (Reprinted from Reference 66 with permission from the AIP.)

not continuous, the change of ellipsometric characteristics enables the calculation of pore size in the bulk film under the barrier and also the size of voids in the barrier. An example of the ellipsometric response recorded during exposure of a 50% porous HSQ-based film with a PVD-deposited TaN barrier is shown in Fig. 6.19.

The noncontinuous barrier of 30 nm (Fig. 6.19a) does not stop solvent from penetrating, and the experimental points follow the theoretical curve. When the thin film is continuous (Fig. 6.19b) the experimental points do not follow the theoretical curve, indicating that there is no solvent adsorption inside the porous film. The small change in Ψ and Δ is attributed to solvent condensation on top of the TaN barrier. Similar results were obtained for barriers deposited on top of organic low-*k* polymers [69].

The EP technique is also able to reveal low-density defects (killer voids). If the barrier has a limited number of large voids, solvent is trapped beneath such defects. If the distance between these voids is larger than the diffusion length of solvent inside the low-*k* film, the change in optical characteristics is local and these voids can be quantified using optical microscopy and a special digital camera installed on top of the EP chamber.

6.5 EVALUATION OF POROUS MATERIALS BY REFLECTIVITY AND SCATTERING

X-ray methods are widely used for characterization of porous low-*k* materials. X-ray techniques are more readily available than positronium (PALS) or neutron beam (SANS) techniques and do not require modeling of the optical properties (as in EP). There are different X-ray–based characterization methods: X-ray reflectivity (SXR), X-ray adsorption porosimetry (XRP), small-angle X-ray scattering (SAXS), and grazing-incidence small-angle X-ray scattering (GISAXS). It is also important that these techniques (except XRP) are nonintrusive and do not have the problems related to swelling.

SXR is able to determine the film thickness and density. Combined with adsorption techniques (when a porous film is measured as it is filled with adsorbate) it enables determination of open porosity (a pore network that is connected to the surface) and pore size distribution. SAXS and GISAXS can be used for measuring pore size distribution without introduction of adsorbate and therefore is able to characterize all pores (not only those open to the surface).

SANS provides information related to the pore size because the SANS intensity is sensitive to density fluctuation correlations. However, neutron scattering is not very practical due to the limited number of neutron sources.

SAXS with a wavelength less than 0.1 nm is also used to get the same structural information. The short wavelength is needed because the X-ray energy is such that transmission experiments through silicon wafers are possible. The physics describing the scattering from SANS and SAXS is identical, but there are differences in the details. In particular, silicon is virtually transparent to neutrons, and a stack of more than 10 low-k films supported on silicon substrates is often used in SANS in order to increase the scattering volume and the neutron scattering signal. In contrast, X-rays have limited penetration power through silicon due to significant absorbance. The small scattering volume, weak scattering of thin porous dielectric films, and high absorbance of the silicon necessitates a high-intensity X-ray source. At present, a synchrotron source is needed to obtain a large enough flux for transmission SAXS measurements with sufficient signal. Another X-ray scattering alternative is GISAXS, which increases the scattering volume by two orders of magnitude over the normal incident geometry.

6.5.1 X-Ray Reflectivity

6.5.1.1 Density and Thickness Measurement. Specular X-ray reflectivity is a well-known technique that enables measurement of the thickness, density, and roughness of a thin film [70, 71]. It is based on the fact that the refractive index of most materials is less than 1 in the X-ray range and X-rays undergo total external reflection under the critical angle Θ_c. The critical angle depends on electron density ρ_e and, therefore, film density:

$$\Theta_c = \lambda(\rho_e r_e/\pi)^{0.5} \tag{6.17}$$

where λ is the X-ray wavelength and r_e is the classical electron radius. Above the critical angle, oscillations of the reflected X-ray intensity related to film thickness are observed. It is possible to calculate the reflected intensity using a one-dimensional Schrödinger equation [72]. Since SXR is sensitive to density, it is possible to detect variations of density across the film thickness—for example, densification (sealing) of the top surface of a porous film. It should be noted that the problem is inverse and

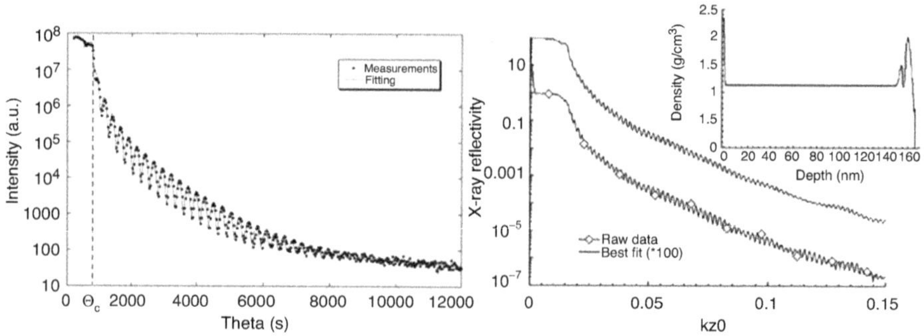

FIGURE 6.20 **(a)** SXR measurement results (*dots*) and fitting results (*solid curve*). The sample is a SiOCH low-*k* film with 35% porosity. The fitting parameters are thickness (60 nm) and density (1.1 g/cm^3). **(b)** Results of SXR analysis of a He plasma–treated (40 s), 160-nm-thick, low-*k* film. This graph demonstrates the possibility of analyzing the formation of a dense capping (sealing) layer by plasma treatment. Reprinted from reference 73 with permission from Wiley.

requires fitting of a model to the measurement data, which might be quite complex in the case of density variation. An example of SXR measurement (relative intensity versus incident angle) is shown in Fig. 6.20.

The main limitation of SXR is that it is based on differences in density, and, therefore, SXR is not able to clearly distinguish materials of similar densities—for example, Si (2.33 g/cm^3) and SiO_2 (2.22 g/cm^3). It should be noted, however, that porous low-*k* films are generally less dense than Si. Another limitation is the difficulty in measuring the thickness of thicker films, since the reflection oscillation frequency increases with an increase in film thickness. Thicker films require higher angular resolution and make fitting more complicated. In general, films thinner than 500 nm are measurable with some amount of effort.

6.5.1.2 Porosity Evaluation.

For porous films, the measured average film density ρ_{avg} can be used for the porosity calculation as

$$\rho_{avg} = \rho_w(1 - P) \tag{6.18}$$

where ρ_w and P are wall (skeleton) density and porosity, respectively. The calculated porosity reflects the wall density and porosity. The wall density is not necessarily equal to the density of a dense prototype, and, therefore, the calculated values can be different from the values obtained by other techniques (EP, XRP, and PAS). It has been reported that SXR can measure the wall density in combination with SANS as applied in transmission to thin porous dielectric films [37, 73]. With two measurements that depend on ρ_w and P in different functional forms from SXR and SANS, the values of ρ_w and P were obtained.

6.5.1.3 *X-Ray Porosimetry.* Since SXR is able to measure film mass and density, it can be employed for adsorption porosimetry, where the density of a porous film increases during adsorptive condensation. Adsorption porosimetry is based on measuring the adsorbate volume condensed inside the pores as a function of the partial pressure of the adsorbate. The volume can be measured by weight (gas adsorption porosimetry) for bulk materials, but it is not applicable for thin films, as their weight is significantly lower than that of the Si carrier wafer. Basically, X-ray porosimetry is similar to ellipsometric porosimetry, and the only difference is in the method of measurement of the amount of condensed adsorbate. The calculation procedure is the same as for EP [74, 75].

The U.S. National Institute of Standards and Technology (NIST) developed XRP porosimetry techniques that are conducted in one of two ways [76]. In the "pressure variation" (PV) method, a stream of saturated toluene in dry air (at constant partial pressure P_0) from a bubbler held at 25°C is combined with a second stream of pure air to control solvent partial pressure. The relative pressure of solvent is increased in increments from $P/P_0 = 0$ (pure air) to $P/P_0 = 1$ (saturated toluene) by controlling the flow rates of the air and saturated toluene streams. In the "temperature variation" (TV) method, a stream of saturated toluene vapor from a bubbler held at 30°C is passed through the sample chamber and P/P_0 is varied by changing the sample temperature, effectively changing P_0 inside the pores. Therefore, the technical difference between existing XRP and EP systems is that XRP operates at atmospheric pressure and EP uses a vacuum chamber. Both XRP and EP use toluene as standard adsorptive.

An example of the evolution of critical angle and, therefore, density upon adsorption of toluene in a porous HSQ film is shown in Fig. 6.21. Since the density of

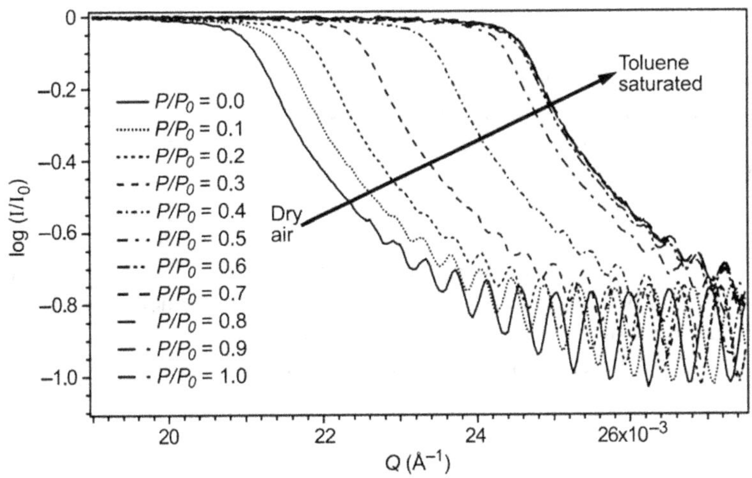

FIGURE 6.21 Critical angle changes for a porous HSQ film as P/P_0 increases systematically from 0 (dry air) to 1 (toluene-saturated air). Condensation of the toluene inside the pores results in appreciable and measurable changes in the critical angle. Reprinted from reference 73 with permission from Wiley.

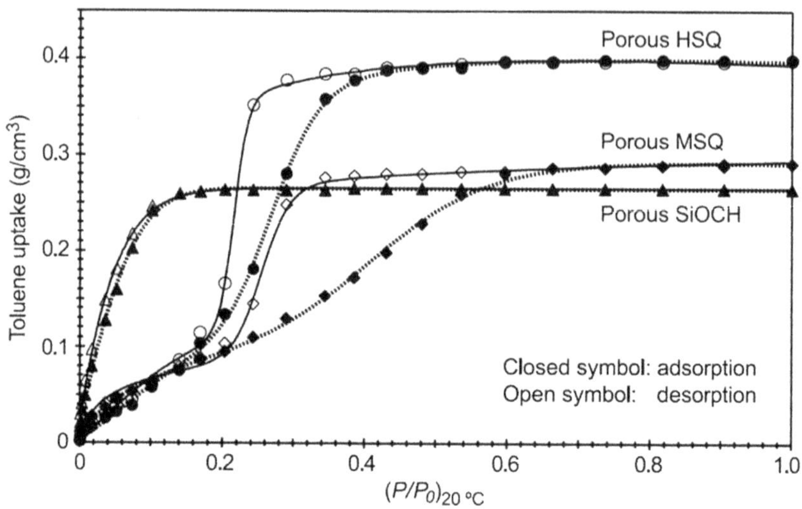

FIGURE 6.22 Physisorption isotherms for porous HSQ, MSQ, and SiCOH films. The lines are smooth fits using the cumulative sum of a sigmoidal and a log-normal function for porous HSQ and MSQ films and the sum of a Gaussian and a sigmoidal function for a porous SiCOH film. Estimated standard uncertainties are comparable to the size of the data markers. Reprinted from reference 73 with permission from Wiley.

toluene is known, it is possible to calculate the percentage of adsorbed toluene and, therefore, the porosity of the film. As soon as adsorption–desorption isotherms are obtained (see Fig. 6.22), one can calculate the pore size distribution using the Kelvin equation in exactly the same way as for EP. The pore size distributions for the films presented in Fig. 6.22 are shown in Fig. 6.23.

The limitations of XRP are similar to those of EP and nitrogen porosimetry: Only pores that are connected and open to the surface can be detected. Both methods are intrusive and require introduction of adsorbate into the porous structure of the films in question.

XRP potentially can provide a depth profile of the density and porosity through the thickness of the film. A multilayer model could be fitted to the reflectivity data. In the preceding examples, the low-*k* films were modeled with a single layer of uniform density both under vacuum and in the presence of toluene. The depth profile of porosity and pore size can also be obtained by PALS (Fig. 6.4) and EP [77]. However, each of these three techniques has its challenges and limitations. The challenge of interpreting PALS results is related to complicated Ps diffusion behavior in pores with different shapes. Both XRP and EP extract this information by fitting models to experimental curves, which may be not sufficiently reliable in certain cases.

6.5.2 Small-Angle X-Ray Scattering

6.5.2.1 SAXS and GISAXS. Small-angle X-ray scattering and a variation, grazing incidence SAXS, were developed to probe pore structure in a nonintrusive way. They are based upon the following principles, as shown in Fig. 6.24: If an X-ray is

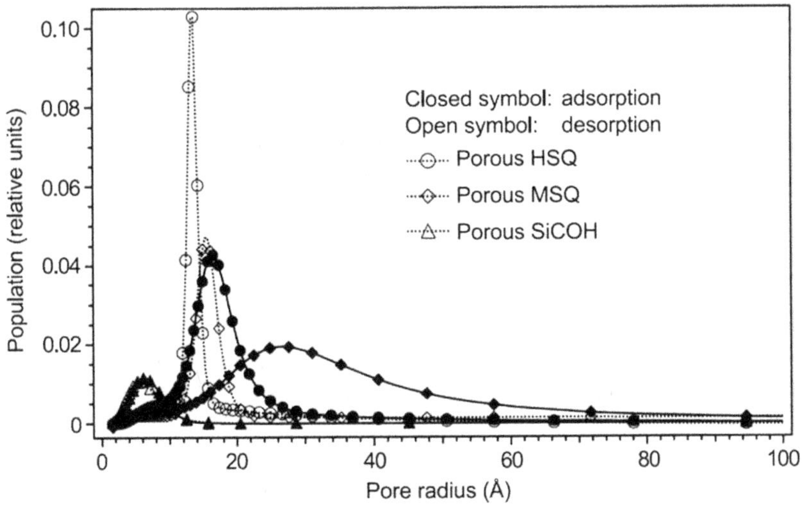

FIGURE 6.23 Pore size distributions from the fits through the physisorption isotherms in Fig. 22, using the Kelvin equation to convert P/P_0 into a pore size. The distributions from the adsorption branch (*solid lines*) can be significantly broader and shifted to larger pore sizes than those from the corresponding desorption branch, especially in those materials (like the MSQ film) with a large distribution of mesopore sizes. Reprinted from reference 73 with permission from Wiley.

incident onto a thin film on a substrate at an angle Θ, then part of the beam is reflected at the specular angle Θ and part is refracted into the film. The refracted beam can be scattered inside the film at a defect that has different electron density from the rest of the film (a pore in our case). The scattered beam will exit the film at a different angle $(\Theta + \delta\Theta)$.

The main difference between SAXS and GISAXS is the incidence angle Θ: SAXS uses normal incidence ($\Theta = 90°$) [78], while GISAXS uses angles close to zero. Application of normal-incidence SAXS is limited by a small scattering volume, weak scattering in thin porous films, and high absorbance of the silicon, and this necessitates a high-intensity X-ray source. At present, a synchrotron source is needed

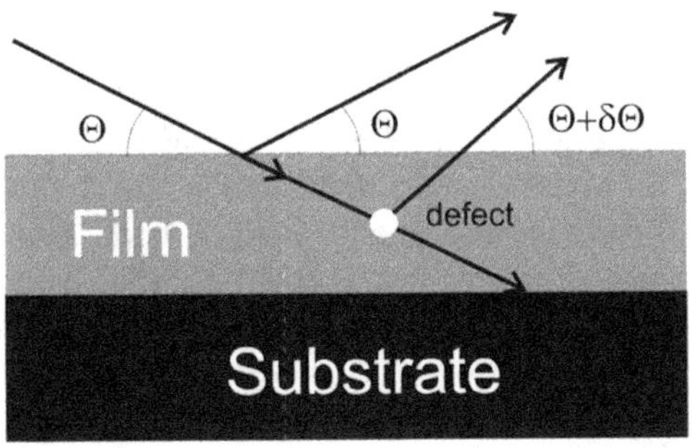

FIGURE 6.24 A scheme of X-ray scattering in a film by a defect (a pore).

to obtain a large enough flux for transmission SAXS measurements with sufficient signal. Unlike transmission SAXS, the GISAXS geometry, in which the scattering signal is collected at the thin-film side, can increase the scattering volume by two orders of magnitude. The X-ray beam does not penetrate the silicon substrate but is reflected from the surface. Experimental techniques have been developed and have demonstrated the capability of collecting GISAXS data within a time period less than 60 s. GISAXS is a powerful tool for studying nanostructured surfaces and thin films, combining the accessible length scales of SAXS and the surface sensitivity of grazing incidence diffraction (GID) [79] and diffuse X-ray reflectivity. The technique was originally introduced by Levine and coworkers [80] but has only recently flourished for the study of nanostructured systems.

In order to make X-ray scattering surface sensitive, a grazing incidence angle α is chosen between about half the critical angle of the film material, α_c, and several times α_c. The actual choice depends on the system to be studied. For free-standing quantum dots, an incident angle below α_c may be chosen to make the scattering exclusively surface sensitive. The largest scattering cross sections are achieved when the incident angle is between the critical angles of the film and the substrate.

The intensity of the scattered light I depends on the form factor F of the defects (pores) [81]:

$$I(q) = |F(q)|^2$$

where $q = \frac{4\pi}{\lambda}\sin\theta$ is the scattering vector, θ is the scattering angle, and λ is the X-ray wavelength. Assuming that pores have a spherical shape of diameter D with the average diameter D_0 and the distribution described by a Γ-function, we have

$$P(D, D_0, \sigma) = \frac{1}{\Gamma\left(\frac{1}{\sigma^2}\right)}\left(\frac{1}{\sigma^2 D_0}\right)^{\frac{1}{\sigma^2}} D^{-1+\frac{1}{\sigma^2}} \exp\left[-\frac{D}{\sigma^2 D_0}\right] \tag{6.19}$$

where σ is the normalized variance. The scattering intensity then can be written as follows:

$$|F(q, D_0, \sigma)|^2 = |\Delta\rho|^2 \int_0^\infty \left|\frac{4\pi}{q^3}\left(\sin\left[\frac{qD}{2}\right] - \frac{qD}{2}\cos\left[\frac{qD}{2}\right]\right)\right|^2 P(D, D_0, \sigma)\frac{D_0^3}{D^3}dD \tag{6.20}$$

where $\Delta\rho$ is the density difference between the film and the pore.

The measurements are done in the following manner [82,83]: A detector is slightly shifted (by about 0.1°) from the angle Θ to avoid strong specular reflection, and the scattering intensity is recorded as a function of the incident angle Θ. Equation 6.20 is then fit to the experimental curve using D_0 and σ as fitting parameters. The resulting pore size distribution is then plotted using Equation 6.19. Examples of such fitting and the resulting pore size distribution for three different low-k films are shown in Fig. 6.25.

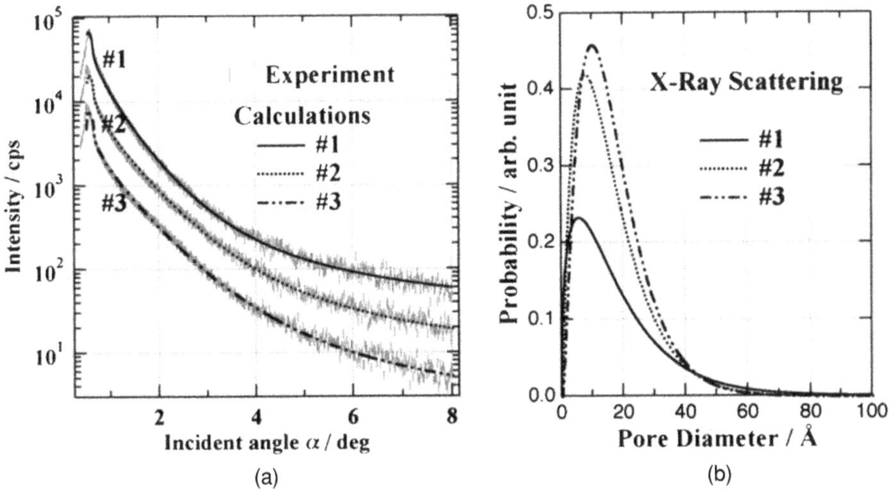

FIGURE 6.25 (**a**) Experimental and calculated X-ray small-angle scattering patterns for the three porous MSQ films. (**b**) Pore-size distribution of the films obtained by X-ray scattering. (Reprinted from Reference 82 with permission from the AIP.)

Diffuse scattering from the film surface can sometimes interfere with the scattering from the pore structure. In that case the contribution of the surface roughness should be estimated. That can be done by measuring rocking scan patterns at several 2Θ positions in addition to the offset scan. Omote et al. [82] reported that the contribution of surface roughness is dominant only at very low incident and exit angles. The main advantage of SAXS is that it can probe all pores regardless their connection to the surface. The main drawback of the method is the fact that the pore size distribution is valid only for spherical pore shapes that are assumed to be described by a Γ-function. Other pore shapes and more complicated distributions would require a more complicated mathematical approach, and multiple scattering effects would have to be taken into account to properly model the data. It is also important to emphasize that scattering techniques are sensitive to density gradients. Therefore, sometimes it is difficult to distinguish between pores and dense inclusions.

6.5.2.2 SANS. As has already been mentioned, SANS is less popular than X-ray techniques due to the limited availability of neutron sources. However, in certain cases, the application of SANS might be very useful. SANS provides information related to the pore size because the SANS intensity is sensitive to density fluctuation correlations. The physics describing the scattering from SANS and SAXS is identical, but there are differences in the details. In particular, silicon is virtually transparent to neutrons, and a stack of more than 10 low-*k* films supported on silicon substrates can be used in SANS in order to increase the scattering volume and the neutron scattering signal. In contrast, X-rays have limited penetration power through silicon due to a significant absorbance, so only a single film is used for SAXS.

Pore size distribution in SANS can be measured and based on scattering inside porous films. A model to describe the scattered intensity was introduced by Debye et al. [84]:

$$\gamma(r) = \exp(-r/\xi) \tag{6.21}$$

where ξ is the correlation length. The average dimension, or the chord length, of the pores is $\xi/(1 - P)$, and the average dimension of the wall between the pores is ξ/P. The scattering intensity based on this model is

$$I(q) = \frac{8\pi P(1 - P)\Delta\rho_n^2\xi^3}{(1 + q^2\xi^2)^2} \tag{6.22}$$

where $\Delta\rho_n$ is the neutron scattering length contrast or the electron density difference between the two phases. For a uniform, porous film, this value is the neutron scattering length or the electron density of the matrix material between the pores because the scattering length of the pores is zero. This quantity is linearly related to the mass density of the matrix, and the proportionality constant can be determined if the atomic composition of the film is known. A more detailed description of the analysis procedure can be found elsewhere [37]. The porosity analysis done using this approach demonstrates good agreement with EP and PALS [20].

A similar approach is also used in SAXS. However, besides the Debye random two-phase model presented by Equation 6.22, there are many other structural models that depend on the sample morphology. Depending upon the type of porous material, SAXS and SANS data may not necessarily yield identical structural information from the same scattering model. The SAXS and SANS results are equivalent only under the condition that the porous material can be modeled with a two-phase structure, that is, a void phase and a matrix phase. When the matrix is not homogenous—for example, if hydrogen is preferentially located at the surface of the void—the contrast factors for X-rays and neutrons can change and yield different structure parameters. Hydrogen is a special case because it has a very different scattering cross section for neutrons. In this case, a three-phase model was developed to accommodate this type of structure [85].

Another SANS application is contrast-match SANS, also known as neutron porosimetry [86]. The idea behind neutron porosimetry is a "contrast match point" determination. The pores are filled with solvent mixtures of variable neutron scattering length density (SLD). The solvent SLD is varied by mixing hydrogen- and deuterium-containing analogs of the solvent. The scattered intensity $I(q)$ from the solvent-filled porous material depends on the difference in SLD or "contrast" between the solvent and the matrix. The solvent SLD is varied systematically, and the scattered intensity $I(q)$ is measured at each composition. If the material is homogeneous in atomic composition and all of the pores are filled with solvent, a contrast-match solvent composition exists for which $I(q)$ becomes zero over all q. If the material has inhomogeneities in its atomic composition or if some of the pores are not filled by solvent ("closed pores"), a contrast match point will not be observed.

The second stage of the measurement is porosimetry using solvent vapor of the contrast-match composition. SANS data are collected as the partial pressure of the probe solvent is increased (sorption) and decreased (desorption). When the pores are filled with the contrast-match liquid, their SLD matches the sample matrix, so the SANS measurement only yields the size of the "empty" pores. Thus, as the solvent partial pressure increases, the pore size is measured for subpopulations of pores of increasingly large size.

6.6 CONCLUSIONS

This chapter has given an overview of recent development of new techniques oriented to the evaluation of critical properties of thin porous films. Although these developments were mainly stimulated by implementation of porous, low–dielectric constant films for advanced microelectronics and demonstrated their efficiency, they certainly can be used for other applications of thin porous films [87]. It is important that despite the different physicochemical principles they demonstrate sufficiently good agreement in the obtained results [20, 21]. However, some observed differences between different instrumentations depend on the principles involved and the nature of the probe used. For practical applications, the chosen method must assess a parameter that is related to the phenomena involved in the application of the porous material as directly as possible [6].

ABBREVIATIONS

BLS	Brillouin light scattering
C	capacitance
E	Young's modulus
E_{beam}	positron beam energy
EP	Ellipsometric porosimetry
HSQ and MSQ	hydrogen- and methylsilsesquioxanes
ILD	interlayer dielectric
IRSE	IR spectroscopic ellipsometery
L	metal line length
n	refractive index
N	number of molecules per unit volume
NI	nanoidentor
p	metal line pitch
$P = V/V_0$	the material's porosity
Ps	Positronium
PALS	Positron annihilation spectroscopy
PECVD	plasma enhanced chemical vapor deposition
r	pore radius
R	resistance
SANS	Small angle neutron scattering

SAWS	Surface acoustic wave spectroscopy
SAXS	Small angle X-ray scattering
SXR	specular X-ray reflectivity
T	metal thickness
ULSI	Ultra Large Scale Integration
V and V_0	pore's volume and the film's volume
XRP	X-ray porosimetry
α	molecular polarisability
ε and ε_0	dielectric constants of the material and vacuum
$k = \varepsilon/\varepsilon_0$	relative dielectric constant of ILD
Ψ and Δ	ellipsometric angles
ρ	metal resistivity

REFERENCES

1. Heo, K.; Yoon, J.; Jin, K. S.; Jin, S.; Ree, M. *IEE Proc Nanobiotechnol* 2006, **153**, 121.
2. Maex, K.; Baklanov, M. R.; Shamiryan, D.; Iacopi, F.; Brongersma, S.; Yanovitskaya, Z. Sh. *J Appl Phys* 2003, **93**, 8793.
3. Baklanov, M. R.; Maex, K. *Philos Trans R Soc A* 2006, **364**, 201.
4. Hoffman, R. J. O. M.; Verheijden, G. J. A. M.; Michelon, J.; Iacopi, F.; Travaly, Y.; Baklanov, M. R.; Tokei, Zs.; Beyer, G. *Microelectron Eng* 2005, **80**, 337.
5. Ramsay, J. D. F. *MRS Bull* 1999, **24**, 36.
6. Rouquerol, J.; Avnir, D.; Fairbridge, C. W.; Everett, D. H.; Haynes, J. H.; Pernicone, N.; Ramsay, J. D. F.; Sing, K. S. W.; Unger, K. K. *Pure Appl Chem* 1994, **66**, 1739.
7. Dubois, G.; Miller, R. D.; Volksen, W. In *Dielectric Films for Advanced Microelectronics*, Baklanov, M.; Green, M.; Maex, K., Eds. Wiley, New York, 2008, p. 33.
8. Plawsky, J. L.; Achanta, R.; Cho W., Rodrieges, O.; Saxena, R.; Gill, W. N. In *Dielectric Films for Advanced Microelectronics*, Baklanov, M.; Green, M.; Maex, K., Eds. Wiley, New York, 2008, p. 137.
9. Silverstein, M. S.; Shach-Caplan, M.; Bauer, B. J.; Hedden, R. C.; Lee H.-J.; Landes, B. G. *Macromolecules*, 2005, **38**, 4301.
10. Baklanov, M. R.; Mogilnikov, K. P. *Microelectron Eng* 2002, **64**, 335.
11. Baklanov, M. R.; Mogilnikov, K. P. SEMATECH Workshop, June 2003.
12. Ferchichi, A.; Travaly, Y.; Carbonell, L.; Vanstreels, K.; Beyer, G.; Baklanov, M.; Asakuma, S.; Nakajima, M. In Advanced Metallization Conference 2008, Materials Research Society, Warrendale, PA, 2009, p. 587.
13. Nakao, S. I.; Ushio, J.; Ohno, T.; Hamada, T.; Kamigaki, Y.; Kato, M.; Yoneda, K.; Kondo, S.; Kobayashi, N. In *Proceedings of the Interconnect Technology Conference*, IEEE, Los Alamitos, CA, 2006, p. 66.
14. Gidley, D. W.; Frieze, W. E.; Dull, T. L.; Sun, J.-N.; Yee, A. F. *Mater Res Soc Symp Proc* 2000, **612**, D4.3.1.
15. Sun, J.-N.; Gidley, D. W.; Dull, T. L.; Frieze, W. E.; Yee, A. E.; Ryan, E. T.; Lin, S.; Wetzel, J. *J Appl Phys* 2001, **89**, 5138.

16. Sun, J.-N.; Gidley, D. W.; Hu, Y. F.; Frieze, W. E.; Yang, S. *Mater Res Soc Symp Proc* 2002, **726**, Q10.5.1.
17. Eslava, S.; Baklanov, M. R.; Neimark, A. V.; Iacopi, F.; Kirschhock, C. E. A.; Maex, K.; Martens, J. A. *Adv Mater* 2008, **20**, 310.
18. Gidley, D. W.; Frieze, W. E.; Dull, T. L.; Yee, A. F.; Ryan, E. T.; Ho, H. M. *Phys Rev B* 1999, **60**, 5157.
19. Gidley, D. W.; Frieze, W. E.; Dull, T. L.; Sun, J.; Yee, A. F.; Nguen, C. V.; Yoon, D. Y. *Appl Phys Lett* 2006, **76**, 1282.
20. Kondoh, E.; Baklanov, M. R.; Lin, E.; Gidley, D.; Nakashima, A. *Jpn J Appl Phys* 2001, **40**, L323.
21. Grill, A.; Patel, V.; Rodbell, K. P.; Huang, E.; Baklanov, M. R.; Mogilnikov, K. P.; Toney, M.; Kim, H.-C. *J Appl Phys*, 2003, **94**, 3427.
22. Petkov, M.; Weber, M. H.; Lynn, K. G.; Rodbell, K. P. *Appl Phys Lett* 2001, **79**, 3884.
23. Gidley, D. W.; Peng, H.-G.; Vallery, R. S. *Annu Rev Mater Res* 2006, **36**, 49.
24. Mogilnikov, K. P.; Baklanov, M. R.; Shamiryan, D.; Petkov, M. P. *Jpn J Appl Phys* 2004, **43**, 247.
25. Goworek, T. J. *Nucl Radiochem Sci* 2000, **1**, 11.
26. Petkov, M. P.; Wang, C. L.; Weber, M.; Lynn, K. G.; Rodbell, K. P. J. *Phys Chem B* 2003, **107**, 2725.
27. Farnell, G. W.; Adler, E. L. In *Physical Acoustics*, Vol. IX, Mason, W. P. R.; Thurston R. N., Eds. Academic Press, New York, 1972, p. 35.
28. Schneider, D.; Schwarz, T.; Buchkremer, H.-P.; Stover, D. *Thin Solid Films* 1993, **224**, 177.
29. Leonhardt, M.; Schneider, D.; Kaspar, J.; Schenk S. *Surface Coatings Technol* 2004, **2/3**, 292.
30. Schneider, D.; Frühauf, S.; Schulz, S. E.; Gessner, T. *Microelectron Eng* 2005, **82**, 393.
31. Nelsen, D.; Gostein, M.; Maznev, A. In *Proceedings of the Symposium on Polymers for Microelectronics*, Wilmington, Delaware, 2000.
32. Carlotti, G.; Doucet, L.; Dupeux, M. *J Vac Sci Technol B* 1996, 3460.
33. Hillebrands, B. *Rev Sci Instrum* 1999, **70**, 1589.
34. Flannery, C. M.; Wittkowski, T.; Jung, K.; Hillebrands, B.; Baklanov, M. R. *Appl Phys Lett* 2002, **24**, 594.
35. Flannery, C.M.; Baklanov, M.R. In *Proceedings of the Interconnect Technology Conference*, IEEE, Los Alamitos, CA, 2002, p. 233.
36. Baklanov, M. R.; Mogilnikov, K. P.; Polovinkin, V. G.; Dultsev, F. N. *J Vac Sci Technol B* 2000, **18**, 1385.
37. Wu, W.-L.; Wallace, W. E.; Lin, E. K.; Lynn, G. W.; Glinka, C. J.; Ryan, E. T.; Ho, H.-M. *J Appl Phys* 2004, **87**, 1193.
38. Abell T.; Iacopi F.; Prokopowicz G.; Sun B.; Mazurenko A.; Travaly Y.; Baklanov M. R.; Sullivan C.; Brongersma S.; Liou H.-C.; Tower J.; Gostein M.; Gallagher M.; Calvert J.; Moinpour M.; Maex K. In: *Advanced Metallization Conference 2004*, Materials Research Society, Warrendale, PA, 2005.
39. Thomsen, C.; Grahn, H. T.; Maris, H. J.; Tauc, J. *Phys Rev B* 1986, **34**, 4129.
40. Antonelli, G. A.; Perrin, B.; Daly, B. C.; Cahill, D. G. *MRS Bull* 2006, **31**, 607.

41. Chapelon, L. L.; Vitiello, J.; Neira, D.; Torres, J.; Royer, J. C.; Barbier, D.; Naudin, F.; Tas, G.; Mukundhan, P.; Clerico, J. *Microelectron Eng* 2006, **83**, 2346.
42. Mechri, C.; Ruello, P.; Mounier, D.; Breteau, J-M.; Edely, M.; Minhao, Y.; Gibaud, A.; Povey, I.; Pemble, M.; Romanov, S. G.; Baklanov, M. R.; Verdonck, P.; Gusev, V. Proceedings of "Acoustics'08", Paris, 2008.
43. Mechri, C.; Ruello, P.; Breteau, J. M.; Baklanov, M. R.; Verdonck, P.; Gusev, V. *Appl Phys Lett* 209, **95**, 091907.
44. Marsik, P.; Urbanowicz, A.; Verdonck, P.; Ferchichi K.; DeRoest D.; Prager L.; Baklanov M. R. In *Advanced Metallization Conference 2008*, Materials Research Society, Warrendale, PA, 2009, p. 543.
45. Polovinkin V. G.; Baklanov M. R. In *Materials for Information Technology*, Zschech, E.; Whelan, C.; Mikolajick, T., Eds., Springer, London, 2005, p. 461.
46. Schubert, M. *Thin Solid Films*, 1998, **313/314**, 33.
47. Compain, E.; Drevillon, B.; Hue, J.; Parey, J. Y.; Bourse, J. E. *Thin Solid Films*, 1998, **313/314**, 47.
48. Jellison, G. E., Jr. *Thin Solid Films*, 1998, **313/314**, 33.
49. Marsik, P.; Verdonck, P.; Schneider, D.; De Roest, D.; Kaneko, S.; Baklanov, M. R. *Phys Stat Sol (c)* 2008, **5**, 1253.
50. Eslava, S.; Eymery, G.; Marsik, P.; Iacopi, F.; Kirschhok, C. E. A.; Maex, K.; Martens, J. A.; Baklanov, M. R. *J Electrochem Soc* 2008, **155**, G115.
51. Prager, L.; Marsik, P.; Wennrich, L.; Baklanov, M. R.; Naumov, S.; Pistol, L.; Schneider, D.; Gerlach, J. W.; Verdonck, P.; Buchmeiser, M. R. *Microelectron Eng* 2008, **85**, 2094.
52. Azzam, R. M. A.; Bashara, N. M. *Ellipsometry and Polarized Light*, North-Holland, New York, 1977.
53. Drevillon, B. *Thin Solid Films*, 1998, **313/314**, 625.
54. Röseler, A. *Infrared Spectroscopic Ellipsometry*, Akademie-Verlag, Berlin, 1990.
55. Kasic, A.; Schubert, M.; Einfeldt, S.; Hommel D. *Vibrational Spectrosc* 2002, **29**, 121.
56. Korte, E. H.; Röseler, A. *Analyst* 1998, **123**, 647.
57. Korte, E. H.; Hinrichs, K.; Röseler, A. *Spectrochim Acta*, 2002, **B57**, 1625.
58. Shamiryan, D. Internal Report, IMEC, Leuven, Belgium, 2007.
59. Baklanov, M. R.; Mogilnikov, K. P. *Optika Applicata* 2000, **30**, 481.
60. Dultsev, F. N.; Baklanov, M. R. *Electrochem Sol St Lett* 1999, **2**, 192.
61. Bourgeois, A.; Turcant, Y.; Walsh, C.; Defranoux C. *Adsorption* 2008, **14**, 457.
62. Baklanov, M. R. In *Dielectric Films for Advanced Microelectronics*, Baklanov, M; Green, M; Maex, K., Eds. Wiley, New York, 2008, p. 117.
63. Gregg, S. J.; Sing, K. W. *Adsorption, Surface Area, and Porosity*, Academic Press, London, 1982.
64. Ravikovitch, P. I.; Neimark, A. V. *Langmuir*, 2002, **18**, 9830.
65. Mogilnikov, K. P.; Baklanov, M. R. *Electrochem Sol State Lett* 2002, **5**, F29.
66. Shamiryan, D.; Baklanov, M. R.; Maex, K. *J Vac Sci Technol B* 2003, **21**, 220.
67. Baklanov, M. R.; Mogilnikov, K. P.; Le, Q. T. *Microelectron Eng* 2006, **83**, 2287.
68. Ravikovitch, P. I.; Neimark, A. V. *Langmuir* 2006, **22**, 10864.

69. Martin Hoyas, A.; Schuhmacher, J.; Whelan, C. M.; Baklanov, M. R.; Carbonell, L.; Celis, J. P.; Maex, K. *J Vac Sci Technol B* 2005, **23**, 1551.

70. Paratt, J. P. *Phys Rev* 1954, **95**, 359.

71. Lekner J. *Theory of Reflection*, Martinus Nijhof, Dordrecht, Netherlands, 1987.

72. Anker, J. F.; Majkrzak, C. J. *Proc SPIE* 1992, **1738**, 260.

73. Soles, C. L.; Lee, H.-J.; Vogt, B. D.; Lin, E. K.; Wu, W.-L. In *Dielectric Films for Advanced Microelectronics*, Baklanov, M.; Green, M.; Maex, K., Eds., Wiley, New York, 2008, p. 100.

74. Lee, H. J.; Soles, C. L.; Liu, W. W.; Bauer, B. J.; Wu, W.-L. J. *Polym Sci B Polym Phys* 2002, **40**, 2170.

75. Lin, E. K.; Lee, H.-J.; Lynn, G. W.; Wu, W.-L.; O'Neill M. L. *Appl Phys Lett* 2002, **81**, 607.

76. Hedden, R. C.; Lee, H.-J.; Soles, C. L.; Bauer, B. J. *Langmuir* 2004, **20**, 6658.

77. Wongmanerod, C.; Zangooie, S.; Arwin, H. *Appl Surf Sci* 2001, **172**, 117.

78. Huang, E.; Toney, M. F.; Volksen, W.; Mercerreys, D.; Brock, P.; Kim, H.-C.; Hawker, C. J.; Hedrick, J. L.; Lee, V. Y.; Magbitang, T.; Miller, R. D.; Lurio, L. B. *Appl Phys Lett* 2002, **81**, 2232

79. Als-Nielsen, J.; McMorrow, D. *Elements of Modern X-Ray Physics*, Wiley, New York, 2001.

80. Levine, J. R.; Cohen, J. B.; Chung, Y. W.; Georgopoulos, P. *J Appl Cryst* 1989, **22**, 528.

81. Saski, A. *Rigaku J* 2005, **22**, 31.

82. Omote, K.; Ito, Y.; Kawamura, S. *Appl Phys Lett* 2003, **82**, 544.

83. Suzuki, T.; Omote, K.; Ito, Y.; Hirosawa, I.; Nakata, Y.; Sugiura, I.; Shimuzu, N.; Nakamura, T. *Thin Solid Films* 2006, **515**, 2410.

84. Debye, P.; Anderson, H. R.; Brumberger, N. J. *Appl Phys* 1957, **28**, 679.

85. Wu, W.-L.; Lin, E. K.; Jin, C.; Wetzel, J. T. In *Materials, Technology, and Reliability for Advanced Interconnects and Low-k Dielectrics*, Maex, K.; Joo, Y.-C.; Oehrlein, G. S.; Ogawa, S.; Wetzel, J. T., Eds., Materials Research Society, Warrendale, PA, 2000, p. 612.

86. Hedden, R. C.; Lee, H.-J.; Soles, C. L.; Bauer, B. J. *Polym Mater Sci Eng* 2004, **90**, 495.

87. Rouessac, V.; van der lee, A.; Bosc, F.; Durand, J.; Ayral A. *Micropor Mesopor Mater* 2008, **111**, 417.

CHAPTER 7

Microscopy Characterization of Porous Polymer Materials

GREGORY MEYERS, ANAND BADAMI, STEVE ROZEVELD,
BOB CIESLINSKI, CLIFFORD TODD, CHARLIE WOOD,
DEBORAH ROTHE, WILLIAM HEESCHEN, and GARY MITCHELL

The Dow Chemical Company, Midland, Michigan, USA

7.1 INTRODUCTION

The application of microscopy characterization to polymer materials is vast [1,2]. Microscopy techniques are used to determine through direct visualization the size, shape, and connectivity of pores over many length scales, from Angstroms to microns. When used in conjunction with indirect methods, microscopy is a powerful complement to developing correct structural models of porous morphology. Microscopy is often a necessary prelude to indirect methods that require, as inputs, structural characteristics of the porosity in order to refine appropriate analysis models. The choice of microscopy methods, sample preparation, and image analysis is critical to the success of any microscopic investigation of porous materials.

7.1.1 Scope

This chapter is designed to focus on direct space interrogation of porous polymer materials using optical, electron, and scanning probe (mechanical) microscopes. Information on porous structure morphology is also accessible through indirect methods such as X-ray, light, or particle scattering, molecular spectroscopy (e.g., nuclear magnetic resonance), or pore-filling approaches using gas penetration or liquid adsorption, all of which are described in greater detail in other chapters. Each technique section in this chapter is designed to give the reader a basic understanding of the contrast mechanisms, the limits of resolution, and the issues or challenges that are specific to porous polymers. Therefore the sections are more descriptive in nature. This

Porous Polymers, Edited by Michael S. Silverstein, Neil R. Cameron, and Marc A. Hillmyer

discussion necessarily leads to issues of sample preparation to reduce or eliminate artifacts. Since the size, shape, connectivity, and dispersion of pores in a material impact performance, the quantification of these measurements is also of interest and will be described briefly in a section on image analysis. Throughout the chapter practical examples are given both from the literature and the authors' laboratory.

7.2 TECHNIQUES

7.2.1 Optical Microscopy

Optical microscopy is often the first screen in the morphological characterization of polymeric materials, with accessible magnification ranges typically from 2 to 2000 times. In its simplest form the optical microscopic image is formed by the collection of scattered or transmitted visible light through or reflected by a specimen using a system of glass lenses. Key advantages of optical microscopy are the ability to detect color, little sample preparation, and the ability to provide through-focus imaging. Several good texts on optical microscopy are available [3–5]. Fundamentally the limitations of resolution are due to diffraction of light through the media governed by the refractive index of the material in front of the lens, the wavelength of the illumination source, and the properties of the objective lens. This suggests that optical microscopy will have a limitation of about 250 nm at best using a high–numerical aperture lens, oil immersion, and a small-wavelength visible light source. With use of white light sources, the resolution will be practically about 500 nm. For direct pore imaging this will limit optical microscopy to only macroporous systems with pore structures in the micrometer range.

The refractive index difference in materials is particularly important in the investigation of porous polymer materials, as the refractive index of air is 1.00 and that of most polymers is in the range 1.45–1.55, implying good imaging contrast. Too large a difference in refractive index may make it difficult to image small features. However, many templating or blending strategies for production of porous polymers rely on formation of phase-separated domains of one polymer in another prior to pore formation, so imaging such precursors may prove difficult without other measures.

Optical microscopy is often used to study other aspects of mesoporous or even nanoporous polymer systems, such as larger-scale defects or macroscale phenomena associated with pore formation. For example, optical microscopy can be used to study phase separation through spinodal decomposition of a polymer blend system that is a precursor state for pore formation (Fig. 7.1). Li and coworkers reported on the chemically induced phase separation of epoxidized soybean oil (ESO) solvent from epoxy monomers and curing agents via spinodal decomposition and further refinement of ESO droplet size with time at cure temperature [6]. The phase-separated morphology is fixed after 70 min, and then the ESO droplets can be extracted with acetone to form a porous epoxy monolith upon vacuum drying.

In addition to providing color imaging, optical microscopy offers a wealth of imaging modes, such as bright and dark field, phase contrast, interference microscopy,

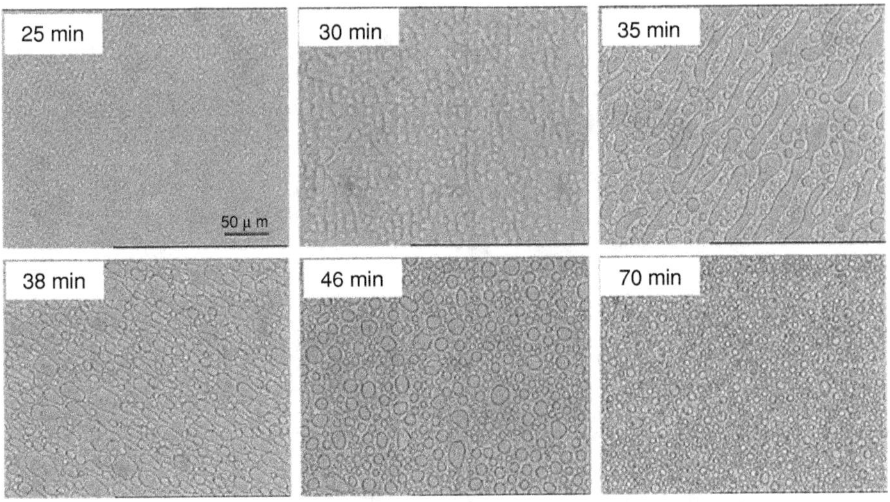

FIGURE 7.1 Time series showing phase separation of epoxidized soy bean oil solvent droplets from diglycidylether of bisphenol A (DGEBA) as a monomer and 4,40-diaminodiphenylmethane (DDM) as a curing agent during isothermal curing at 100°C in an optical microscope with hot stage. (Reprinted from Reference 6 with permission from Elsevier.)

refractive index measurement, and polarization (birefringent) imaging. The spectroscopic imaging modes, which include visible absorption and fluorescence imaging or confocal Raman imaging using additional light sources and detectors, allow mapping of the optical or chemical distribution of components over the morphology.

Newer imaging modes provide three-dimensional (3D) views of materials through various light "sectioning" approaches using white light interferometry or confocal laser scanning microscopy (CLSM) [7]. Both approaches are similar, in that a unique focal plane is scanned at each vertical position of the optics relative to the sample surface. The planes can be reviewed individually or can be used to reconstruct a 3D image. Interferometry relies on constructive interference of reflecting light waves from the surface at each plane and a reference surface, giving vertical resolution on the order of 0.1 nm. Lateral resolution is still diffraction limited. Confocal approaches have slightly poorer vertical resolution but improved lateral resolution compared to interferometry. Confocal imaging of several commercial microfiltration membrane materials representing closed, open, and through-pore systems were described by Charcosset and Bernengo [8] (Fig. 7.2).

A particularly good example demonstrating all the advantages of optical microscopy for characterizing micrometer-sized porosity is in the preparation of breath figures or condensation patterns in polymer thin films [9]. This is a well-studied phenomenon in which the formation of honeycomb-structured, porous films consists of three steps: nucleation and growth of water droplets on a cold surface; arrangement of water droplets into a hexagonal array; and precipitation of the polymer around the water droplets followed by complete evaporation of the solvent and water. The

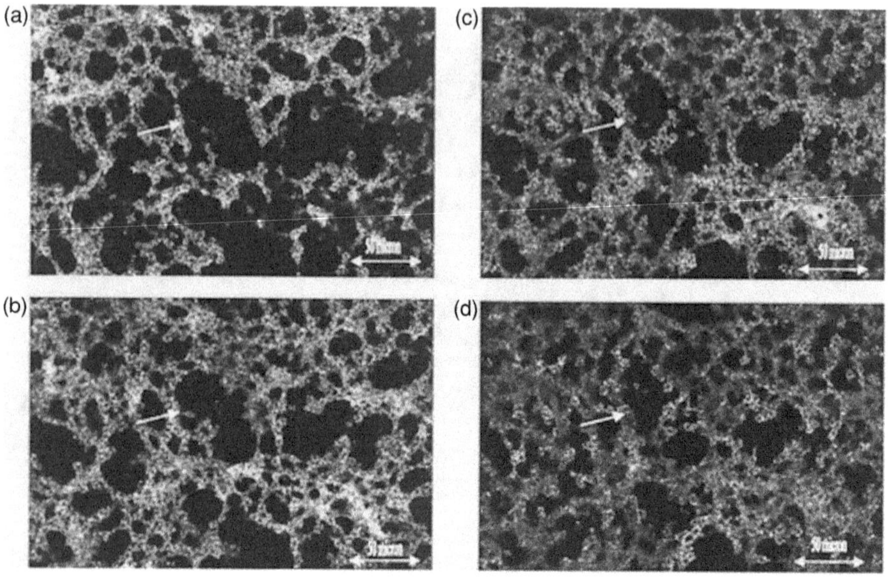

FIGURE 7.2 Confocal laser scanning microscopy image series of a commercial open-cell polymer foam. Images are obtained at the following depths: **(a)** 0 μm or top surface; **(b)** 2 μm below the surface; **(c)** 4 μm below the surface; and **(d)** 8 μm below the surface. Scale bar: 50 μm. The arrow tracks the same location at each depth. (Reprinted from Reference 8 with permission from Elsevier.)

process can be used to introduce pores from 10s of nanometers to 10s of micrometers through control of humidity, temperature, and air flow rate. Three-dimensional patterns of pores can be formed in single-component polymers [10], blends [11], or block copolymers [9]. Figure 7.3 gives an example of controlled introduction of porosity into a polystyrene thin film. The resulting 3D architecture is characterized using reflected light microscopy, bright -field transmission microscopy, and confocal laser scanning microscopy.

Dynamic properties can be studied using fluid or flow cells, hot stages, tensile stages, and even high pressure using specially designed high-pressure cells (e.g., for foaming). High-speed video cameras are also now available that, when coupled to an optical microscope, can capture processes on the submillisecond time scale. Barrow and coworkers [12] used high-speed video microscopy to capture the condensation of water droplets on the surface of an evaporating polymer solution cast as a thin film (Fig. 7.4), which is a precursor to pore formation as shown in Fig. 7.3.

Recently, ultraresolution optical microscopy has been reported [13] and commercialized. The commercial system replaces the illuminator of a conventional microscope with a specialized one that focuses fixed-geometry, highly collimated light at oblique angles on the sample. This serves to dramatically improve contrast and signal-to-noise ratio, which enhances the optimized resolving power and detection capability

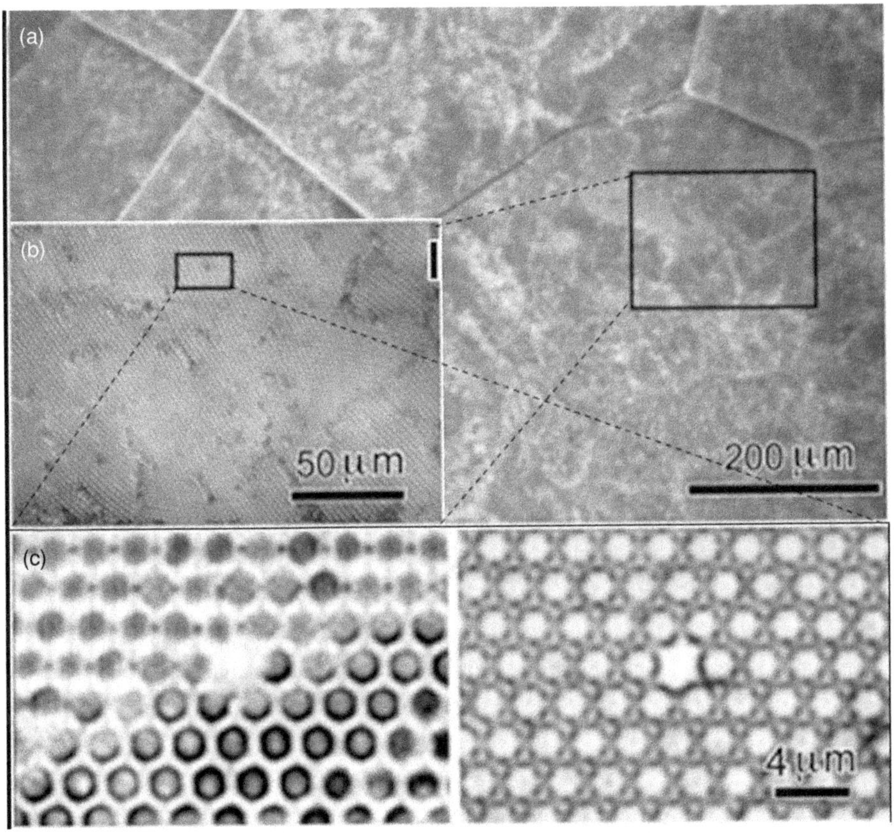

FIGURE 7.3 Various optical microscopy methods used to characterize the 3D architecture of pores in a polystyrene film. **(a)** Low magnification, reflected light showing grayscale variation due to thickness variation. **(b)** Higher magnification using oil immersion from the boxed area in panel (a). **(c)** Confocal image **(left)** and reflected bright-field image **(right)** of the boxed area in panel (b) showing hexagonally packed pores (~2 μm) and a single site defect. (Reprinted from Reference 10 with permission from the American Association for the Advancement of Science.)

for nonfluorescing samples. This system is capable of resolution down to 90 nm and detection to about 10 nm. This technique has been used to size polystyrene particles in water down to 100 nm [14] and may be useful for direct imaging of submicrometer porous polymer materials.

7.2.2 Scanning Electron Microscopy

Scanning electron microscopes (SEMs) are a standard tool used to image the structure of porous polymers at magnifications from 10 to 10,000 thousand times. Electrons are shed at a tip, which in modern instruments is a field emission (FE) gun, conventionally

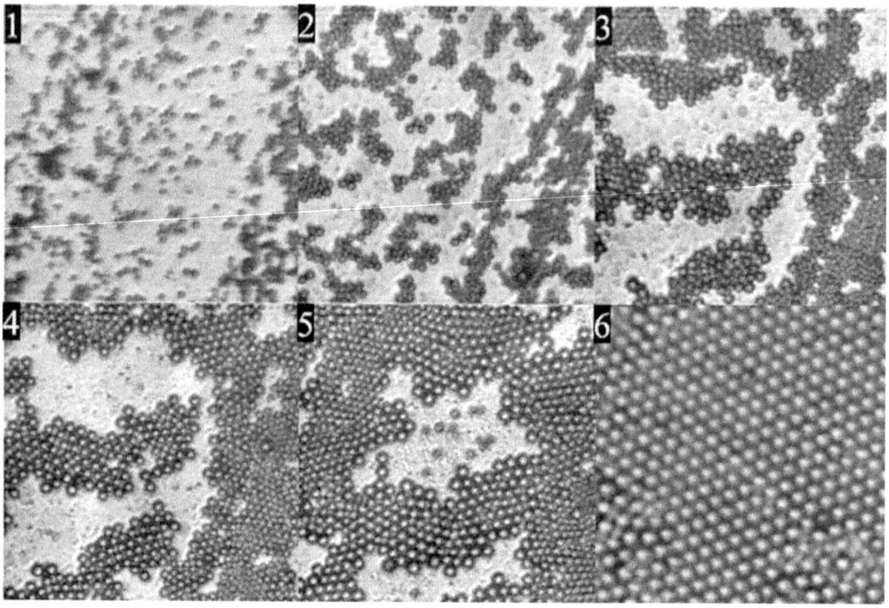

FIGURE 7.4 A sequence of images (180 μm × 180 μm square) depicting the growth and aggregation of water droplets near the surface of an evaporating polymer solution. The time interval between frames 1 and 5 is 50 s. Ultimately, the droplets form an ordered hexagonal structure as shown in frame 6. (Reprinted from Reference 12 with permission from IOS Press.)

referred to as an FESEM. An electrical bias is applied in order to accelerate the electrons. In general, accelerating voltages range from ~1 to 30 keV and currents range from picoamperes to nanoamperes. A series of magnetic lenses is used to focus the beam onto a sample. Depending on the microscope used and the conditions chosen, the beam can be less than 1 nm in diameter upon impact with the sample. When the electron beam interacts with the sample several processes occur. The beam electrons interact with atoms in the sample, leading to changes in the electron trajectory and/or energy. As a result, electrons and X-rays can be generated in the sample. Magnetic coils near the sample are used to raster the electron beam across the surface of the sample. Detectors measure the intensity of the various signals as the beam scans across the surface, and an image can be constructed. Everything from the electron source to the sample is under high vacuum. Several excellent textbooks describe the physics and operational modes of the SEM in detail [15–17].

Several factors impact the quality of the SEM images. First and foremost is the type of signal detected. There are two main types of electrons generated. Secondary electrons (SEs) are usually ejected from the outer orbitals of sample atoms due to interactions with a beam electron. These SEs have energies less than ~50 eV. Therefore only SEs generated within a few nanometers of the sample surface have the energy to make it to the sample surface and escape. SE images are dominated by topographic features and can show very fine surface details. This is the most common kind of image acquired by the SEM. Backscatter electrons (BSEs) are beam electrons that are reflected back due to elastic scattering. They can have energies ranging up to the accelerating voltage of the beam. They can escape from depths up to a few micrometers (and the same in lateral spread), depending on the material properties of the sample and the accelerating voltage used. A major useful aspect of the BSE signal is that it is correlated with the mean atomic number of the material. BSE images can be very useful in identifying inorganic filler particles (higher atomic number) in polymers (lower atomic number) or for visualizing the different components of polymer blends if appropriate heavy metal stains are employed [18]. X-rays generated in the sample can also be used to create images using detectors for energy-dispersive spectroscopy (EDS) or wavelength dispersive spectroscopy (WDS). X-ray signals are usually generated even deeper in the sample than BSE signals.

Lower accelerating voltages lead to shallower penetration of the beam into the sample and therefore a more surface-sensitive analysis [19]. There are fundamental difficulties in controlling the beam at accelerating voltages much below 1 keV. In some SEMs this can be overcome by applying a negative bias to the sample. This decelerates the beam electrons as they near the sample, therefore gaining the benefits of low-voltage imaging while maintaining the advantages of high-voltage focusing optics [20]. This extends high-resolution SEM imaging down to a primary beam energy of a few hundred electron volts [21].

Most polymers are nonconductive; in most cases the beam implants more electrons in the sample than leave by electron ejection processes. There are two main ways to dissipate buildup of this electrical charge. A conductive coating can be applied and connected to ground; carbon or a variety of metals can be evaporated or sputtered onto the sample. The type and thickness of the coating can obscure fine features of the sample at very high magnifications. The other way to compensate charge buildup is to maintain some vapor pressure in the sample chamber. The vapor molecules become ionized by the electron beam and can migrate to the sample, thereby neutralizing the negative charge. Depending on the manufacturer, configuration, and capabilities, SEMs with this function may be referred to as low vacuum (LV-SEM), variable pressure (VP-SEM), or environmental (ESEM). The vapor in the sample chamber can lead to scattering of the primary electron beam and also the detected SE and BSE, possibly degrading the spatial resolution at high magnification.

SEMs are routinely used to characterize the porosity of polymeric membranes (e.g., in water filtration or battery separators), binder porosity (in lithium ion battery cathodes), and novel nanoporous structures made from self-assembled structures (e.g., phase-separated block copolymers). The characterization of closed-cell pore structures in insulating polymer foams provides an opportunity to point out the advantages and challenges associated with SEM imaging. The first such report of cell wall structures of polystyrene and polyurethane foams occurred in 1969 [22]. The authors prepared the cross section using a microtome and had to coat the structures with evaporated gold to mitigate surface charging. BSE imaging makes it possible to image through the skin layer of closed-cell foams by varying the incident beam energy (Fig. 7.5) [23]. The penetration of the beam can be modeled using Monte Carlo simulations, and a model can be developed to estimate the thickness of the skin.

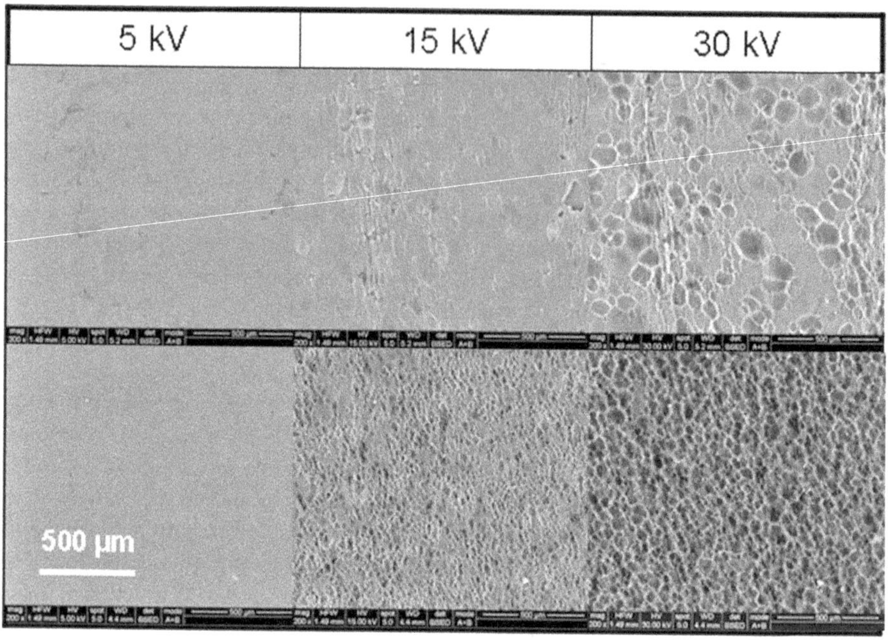

FIGURE 7.5 BSE SEM images of two closed-cell polymer foams as a function of primary beam accelerating voltage. The two images at the left image only the skin layer, while at higher beam voltages the cellular structure below the skin is revealed. (Used with permission from the Dow Chemical Company.)

Recently there has been much effort to improve the insulating values of polymer foams by reduction of the cell size from micrometers to nanometers according to the Knudsen effect [24]. Characterization of nanocellular foams is readily accomplished with modern SEMs equipped with field emission sources and in lens imaging that enables visualization of mesopores (2–50 nm). Most of the approaches to making sub–100 nm pores involve an initial template morphology formed by self-assembly [25], dispersions [26], phase-separated blends [27], or nanocomposites [28]. The pores can be formed using a sacrificial phase that can be removed by solvent extraction [29] or thermal decomposition [30].

Many of the approaches describe the use of supercritical carbon dioxide ($scCO_2$) as a foaming agent and/or a selective solvent for generating porous domains [31]. The use of $scCO_2$ enables nanofoaming in high-T_g polymers without the use of organic solvents. The polymer or targeted domain is preferentially saturated under pressure, effectively plasticizing it (lowering T_g). Upon controlled pressure release, pores nucleate in the more plasticized phase as the T_g of the polymer rapidly increases until it exceeds the processing temperature and the polymer matrix can vitrify around the pores. The conditions controlling the transition from open- to closed-cell morphologies have also been explored [32]. Representative SEM images of nanofoam structures are shown in Fig. 7.6. In all cases the bulk pore morphology can be investigated by cryofracture of the foamed polymer. The resulting fracture surface can be investigated

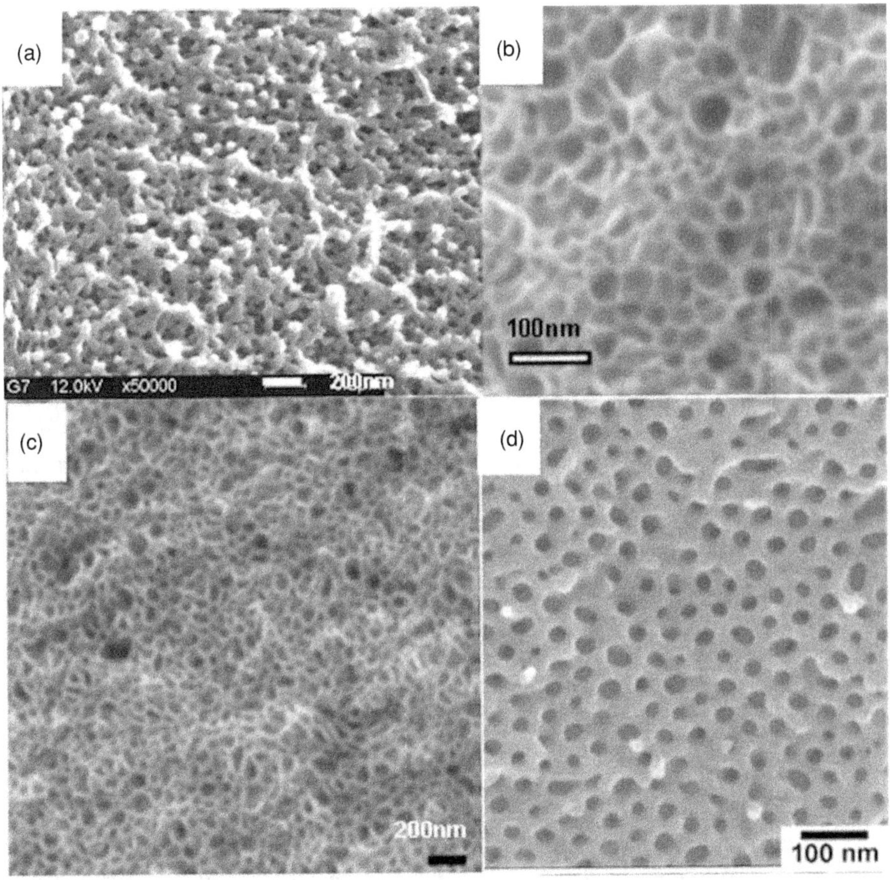

FIGURE 7.6 Internal structure of nanofoams by FESEM. **(a)** Open-cell PEI foam from $scCO_2$ (reprinted from Reference 32 with permission from the American Chemical Society). **(b)** Closed-cell PS/PFMA diblock copolymer foam from $scCO_2$ (reprinted from Reference 33 with permission from John Wiley & Sons). **(c)** Closed-cell PS/PFS diblock copolymer foam from $scCO_2$ (reprinted from Reference 34 with permission from the American Chemical Society). **(d)** Cylindrically aligned PS/PLA diblock extracted with NaOH (reprinted from Reference 29 with permission from the American Chemical Society).

in an FESEM after metal coating or in a variable-pressure mode with water vapor as the imaging gas, where pores as small as a few nanometers may be imaged.

7.2.3 Transmission Electron Microscopy

The technique of transmission electron microscopy (TEM) has a long history of use for polymer morphology characterization. As with other characterization techniques, TEM has its advantages and limitations, but is best used in conjunction with other techniques to solve a materials problem. TEM is typically used for resolving objects ranging from a few micrometers to 1 Å but with a corresponding reduction in the field of view at higher magnifications as well as higher electron doses. In the formation of a TEM image, image contrast is a result of the strong interaction that occurs between the electrons that scatter as they pass through the sample. Several excellent textbooks describe the physics and operational modes of the TEM in detail [35–38]. When electrons pass through the specimen, a number of interactions can take place that can be used to construct the image.

One of the important TEM modes is the technique of dark-field (DF) imaging. The objective aperture is positioned off-axis onto a diffracted spot so that the image is formed by a single diffracted beam. The dark-field image then reveals which region of the sample is contributing to the diffracted spot. The main limitation in DF imaging arises from the spherical aberration of the objective lens. The spherical aberration has the effect of elongating all the image points in proportion to the lens aberration coefficient, and the elongation is greater for higher-order reflections. If a polymeric material contains crystalline domains, then the corresponding electron diffraction spots can be used for DF imaging of these domains.

Bright-field (BF) imaging mode is often used for imaging amorphous polymers and porous materials. In this case, the image is formed by positioning the objective aperture around the transmitted beam, and the contrast depends on the local mass thickness (thickness times density). Electrons will be absorbed depending on the sample thickness and composition (density), and a fraction of the electrons will be scattered outside of the acceptance angle of the objective aperture, resulting in amplitude contrast in the image. Heavy element stains may be used to enhance contrast in bright-field imaging.

As for other materials, TEM of a polymeric sample requires preparation of a 100 nm or less thin section that is electron transparent for 100 to 300 kV beam energies. Figure 7.7 shows several bright-field TEM images of microtomed thin sections of early-stage developmental porous spin on polymer material (p-SiLK low-*k* dielectric film, Trademark of The Dow Chemical Company). Progression from a few pores of a few hundred nanometers to a high loading of sub–10 nm pores has been demonstrated.

In polymer blends, selective and differential heavy metal staining is often incorporated to increase the density of a polymer phase or structure relative to its surrounding to enhance contrast [39]. If staining can be incorporated into a pore, staining can be used to enhance the imaging. For conventional imaging, the incident beam is configured to provide illumination uniformly over the entire sample area of interest. Previous work in nanoporous low-*k* materials indicated that phase contrast, with large defocus values of 600–1000 nm, was better suited to improving the pore contrast compared to using a smaller aperture to increase the amplitude contrast of the pores [6].

For high-resolution or phase contrast imaging, an image is formed from several diffracted beams combined with the unscattered transmitted beam. As a result of the lens properties and focus condition, the scattered beams are phase shifted according to the contrast transfer function of the microscope. This high-resolution mode of TEM (HRTEM) can provide atomic-scale details in crystalline polymer materials (lattice fringes) but is difficult to achieve due to the higher dose sensitivity of many polymer materials [40].

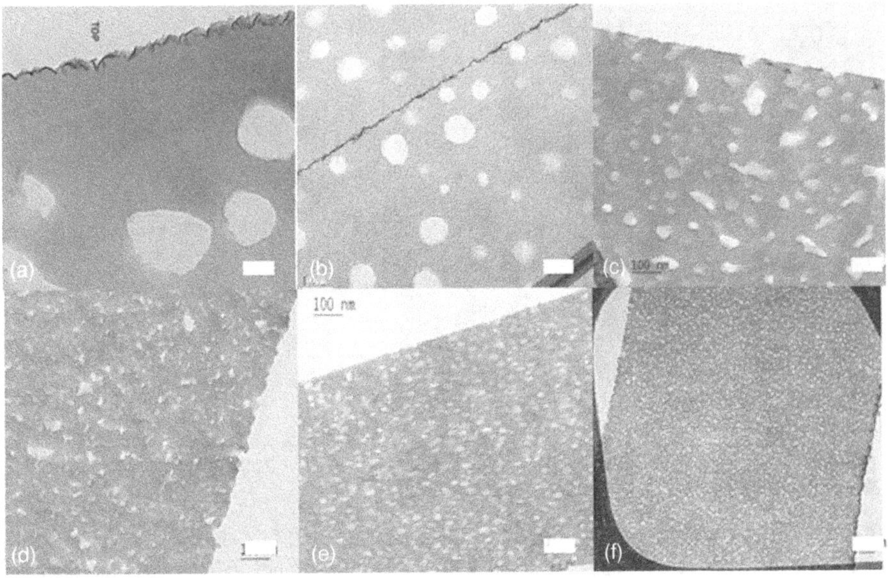

FIGURE 7.7 Bright-field TEM thin sections of porous SiLK low-k dielectric films showing decreasing pore size through the early stages of product development: **(a)** version 1, **(b)** version 3, **(c)** version 5, **(d)** version 7, **(e)** version (8), and **(f)** version 9. The white scale bars correspond to 100nm. (Used with permission from the Dow Chemical Company.)

Phase plates can also be used after the objective lens to cause phase shifting of the scattered beam, leading to observable fringes and enhanced contrast in the image. Phase contrast imaging is usually employed for high resolution, where subtle inhomogeneities in a sample may be amplified without staining. Phase contrast is a complicated function of the defocus, scattering angle, and sample thickness and is often used in conjunction with computer simulations to aid in interpretation.

In analytical TEM instruments, another TEM mode for image formation is scanning–transmission electron microscopy (STEM), in which the electron beam is scanned using a fine probe (e.g., 1 nm). The image is formed as scattered electrons are collected using an annular detector. This method, when coupled with X-ray or energy loss detectors, can be use for elemental mapping [38].

7.2.4 Scanning Transmission X-Ray Microscopy

Scanning transmission X-ray microscopy (STXM) deserves some mention. This technique uses highly coherent, energy-selectable X-ray photons to make a projected image. The high-brightness energy sources required for high spatial resolution are currently only available at a synchrotron source. X-ray optics are necessary to focus the radiation and Fresnel zone plates are used in place of the objective lens for focusing in transmission imaging. The sample can then be raster scanned to generate an image. Resolution in the STXM is currently about 25 nm, comparable to that

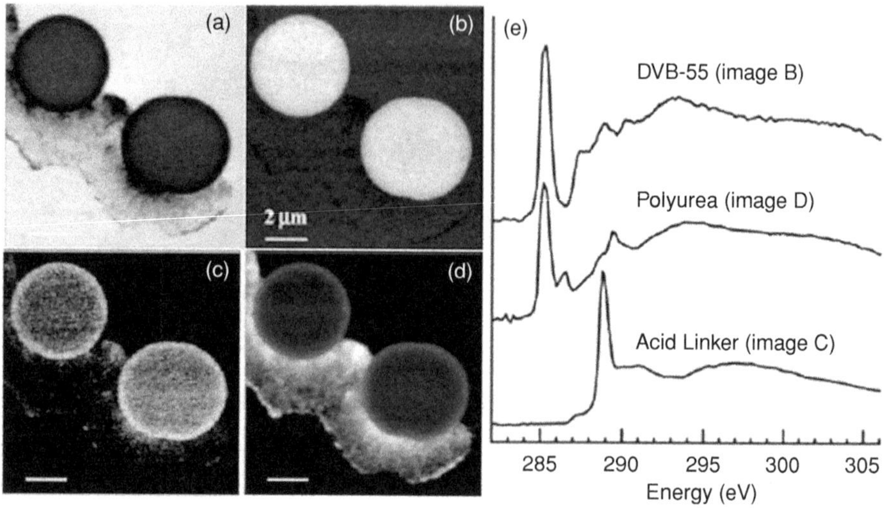

FIGURE 7.8 STXM images obtained from carbon 1*s* NEXAFS spectra of surface-treated DVB-55 microspheres embedded in a polyurea capsule membrane. **(a)** The epoxy (corresponding NEXAFS spectrum is not shown). **(b–d)** Images obtained at selected photon energies where each component is more strongly absorbing in the NEXAFS spectra shown in **(e)**. (Reproduced from Reference 42 with permission from the American Chemical Society.)

accessible by FESEM and AFM. The advantages of STXM are as follows: a tunable X-ray energy source, which allows energy-specific imaging near or at the absorption edges of the elements that make up the material; better energy resolution at the absorption edge compared to electron energy loss spectroscopy in the TEM; and reduced sample damage from the neutral beam. This technique affords spectroscopic contrast without the use of stains. Polymer STXM has been recently reviewed [41]. Image stacks can be obtained at discrete energies and reconstructed to form an image, or spectral data obtained at each pixel can be used to construct images. In Fig. 7.8, STXM is used to visualize the extent of penetration of an acid linker into the surface of crosslinked divinylbenzene microspheres embedded in a polyurea capsule shell. Images were taken at selected photon energies at the C(1*s*) X-ray absorption (NEXAFS) edge (left). The intensity in the images is the thickness (in nanometers) of each component at the given point in the sample. The specific energies were chosen based on strongly absorbing regions in the component spectra (right). The images can then be reconstructed and pseudo colored to form a composite spatiochemical map. The penetration of the acid linker into the microsphere is clearly observed [42]. Additional advantages of STXM include the ability to characterize porous materials in water [43].

7.2.5 Three-Dimensional Imaging (Tomography)

The electron and optical techniques thus far have focused on obtaining 2D images of what are truly 3D objects. One optical technique that can provide 3D images using reconstructions from focused image planes obtained in confocal optical microscopy is, in general, limited to optically transparent materials. A common approach to obtaining a 3D effect in microscopy is the use of stereographic pairs in which two images are obtained from slightly different angles (perspectives) called anaglyphs. Projections of the superimposed images can then be view using optical filters, one for each eye. Anaglyphs are available in any electron or optical microscope with tilt stage and computer software to combine the images. Hawkes and Spence provide a good example of a red-green anaglyph in which FESEM stereo pairs of a mesoporous polymer gel are imaged in the swollen state at room temperature [44]. The stereo pair has been acquired by tilting the sample holder at $\pm 6^\circ$ [44].

Tomography is a method of reconstructing the interior of an object through its projection [45]. By collecting a large number of tilted images, it is possible to perform a true 3D reconstruction of a thin section in a TEM or HAADF-STEM equipped with an appropriate tilt stage. By combining a large number of projections acquired over a wide tilt range (e.g., $\pm 70^\circ$), one can obtain a digital representation of the 3D structure of the material [46]. Because of the high resolution that is achieved in TEM or STEM projections with a thin section, it is possible to image feature sizes in the range of 1–2 nm in three dimensions. In some cases particles of known shapes or other highly scattering (electron-dense) particles may be mixed with or dispersed on top of the thin section to act as references to improve the reconstruction. In other cases, particles are purposely dispersed into the channels of templated nanoporous polymer structures as model catalyst supports, as shown in Fig. 7.9 [47]. Not only has electron tomography been shown to be a good qualitative tool for visualizing nanostructures, but also in recent results the method has been used to obtain quantitative measurements in three dimensions [48].

Mechanical sectioning approaches to 3D imaging include reconstruction of images from serial sections of embedded materials that have been sectioned or polished repeatedly, as has been demonstrated for an atomic force microscope integrated into a room temperature microtome [49]. Tomographic resolution of a few nanometers of an SBS block copolymer thin film has been demonstrated with an AFM using radiofrequency (rf) plasma etching to do the sectioning [50]. Again, the appropriate computer software and hardware are necessary to align, stack, and render the composite 3D image.

Microscale and nanoscale computer-aided X-ray tomography (CT) has recently become available in laboratory-scale systems [51]. In both cases, a collimated X-ray beam is focused onto a sample and the transmitted beam is focused onto a detector. The sample may be translated (2D imaging) or rotated (3D tomography). These are not spectroscopic images, as the X-ray energy is fixed, and contrast is determined by the mass absorption coefficient of the components in the material. The current resolution claim for laboratory-scale X-ray nanotomography is about 50 nm.

7.2.6 Scanning Probe Microscopy

Atomic force microscopy (AFM) is a form of scanning probe microscopy (SPM) in which a nanoscale probe tip is scanned over a surface to obtain a topographical

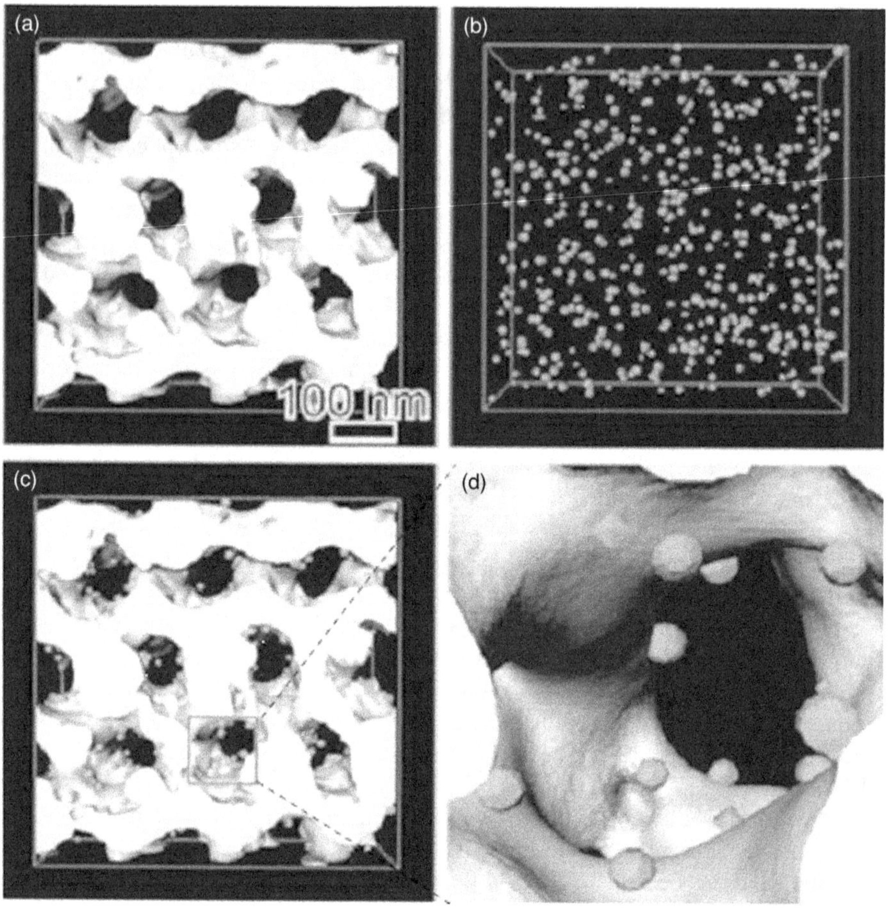

FIGURE 7.9 Three-dimensional reconstruction of the binarized transmission electron microtomography images. **(a)** Poly(2-vinylpyridine) domains. **(b)** Pd nanoparticles. **(c)** Three-dimensional superposition of (a) and (b). **(d)** Close-up view of the region marked by the rectangle in (c), where the entire image field is 100 nm. (Reproduced from Reference 47 with permission from John Wiley & Sons.)

image. The scanning volume accessible in most commercial systems is about 100 μm × 100 μm × 10 μm. As with all scanning probe microscopes, the imaging is accomplished using a feedback mechanism which controls some measurable quantity of the tip–surface interaction. In AFM, the feedback is based on force, which is conveniently measured by deflection of a micrometer-scale cantilever onto which the scanning tip is attached. In most commercial systems, the cantilever deflection is measured by an optical lever; a laser beam is reflected off the back of the lever to a position-sensitive detector. During raster scanning, the feedback acts to control the deflection of the lever at a set value. The feedback loop can operate at rates that

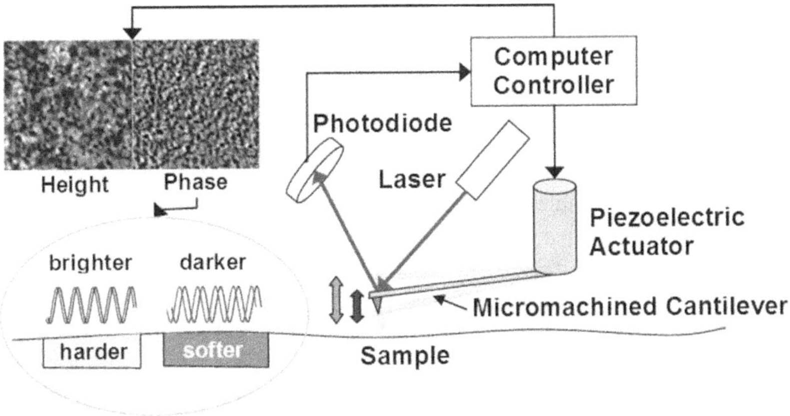

FIGURE 7.10 Schematic of tapping mode AFM operating in air, where height and phase imaging are obtained simultaneously (see text). The height image is obtained under feedback control of the average amplitude of the oscillating lever, while the phase shift is measured directly from the differential phase lag of the tip in contact with dissimilar materials at the surface. (Used with permission from the Dow Chemical Company.)

are orders of magnitude faster than the scanning rates. The positioning and scanning elements are piezoelectric ceramics that have Angstrom-level displacement control. This is the situation for contact mode AFM, where the static deflection is monitored.

Dynamic AFM is now more commonly employed. Using intermittent contact or tapping mode [52], a stiffer cantilever is driven at resonance using a second piezo element at its base. This causes the free end to vibrate, with the tip oscillating at a well-controlled amplitude. In this case, the feedback operates in much the same manner as previously described for contact mode, with the exception that it is the average amplitude that is used for force control. In this mode, the tip literally taps the surface with each oscillation cycle (Fig. 7.10). This has several advantages. First, the surface is tapped at high frequency, on the order of 100 kHz (or 10 μs per cycle). This means that softer polymers can be studied, where mechanical properties are enhanced at high rates (they may effectively "stiffen up" when probed at high frequency). Second, since the scan rates are on the order of 1 s or less per line, the contribution of shear or lateral forces is negligible due to the much higher rate of tapping. This is also a benefit in scanning the surface of mesoporous or macroporous materials, where the tip might get "hung up" while scanning in contact mode. Third, in addition to the amplitude feedback imaging (for height or topography), it is possible to simultaneously monitor the phase lag of the driven tip as a second signal. This signal can provide valuable information about the material composition in heterogeneous systems, where the oscillation delay may change depending on elasticity, adhesion, and/or dissipation within the tip contact zone [53, 54]. Tapping mode phase imaging is a particularly powerful technique for studying the morphology of block copolymer templates and for mesoporous thin films [55]. Thorough reviews of the AFM characterization of polymeric materials can be found in the literature [56, 57], as well as in a recent textbook on AFM of synthetic polymer membranes [58].

One of the consequences of imaging with a physical probe is mixing the geometry of the tip shape with the sample surface to produce the final topography image. Artifacts in AFM images can often be caused by the imaging tip [59, 60]. This is pronounced in the scanning of porous structures, where the tip cannot access undercut features or may not be able to penetrate fully into a pore. The latter issue may confound image analysis when attempting to use AFM to size pores, especially when the tip diameter and pore diameter are close in size. The inability of the tip to penetrate a pore will result in an artificial skewing of the pore size to smaller values. There is a wide selection of probe types that can be used to address these situations (e.g., carbon nanotube tips [61]). Special tips were fabricated using FIB milling to make nanoelectrodes for electrochemical imaging of a model polymer membrane [62].

Additional advantages of AFM include the ability to study morphology in a variety of environments—air, water, vapor, and vacuum. Environmental imaging is one reason AFM has been so well utilized in the study of porous membranes for water filtration [63–65]. Temperature stages are also available in most commercial systems enabling imaging over the temperature range of –30°C to +250°C. AFM is often used in conjunction with other direct imaging techniques (SEM and TEM) as well as indirect techniques (SAXS) to provide complementary information. Tapping mode AFM was used to characterize the morphology of sequential layers of a polyelectrolyte membrane (PEM) made by layer-by-layer self-assembly of alternating DNA and polyferrocenylsilane (PFS) bilayers. The buildup of such layers was shown to generate a hierarchical macroporous polymer network electrostatically bound to a positively charged polyethyleneimine (PEI) base layer (Fig. 7.11). The morphologies of membranes imaged in both the wet and the dry states were compared, with little observable change; they also did not depend on which layer was applied last [66].

7.3 SAMPLE PREPARATION

7.3.1 General Considerations

The quality of the final image is directly related to the quality of the sample preparation, so this must be carefully considered in any investigation. Chapter 4 in Reference 1 and Chapters 9–11 in Reference 2 and references therein provide detailed overviews of preparation methods for polymeric materials. Other useful references include a textbook devoted to sample preparation for SEM [67], a website for TEM sample preparation [68], and a short general review on preparing polymer samples for microscopy characterization [69]. There are many approaches to exposing the internal morphology of a monolithic porous polymer material for microscopy interrogation, and some experimentation may be necessary to find the one that gives the best result. The more commonly employed methods include mechanical polishing routes such as metallography, microtomy, or cryomicrotomy; ion sputtering or milling; gas plasma etching; and freeze fracture. For large-cell polymer foams, simple razor blade

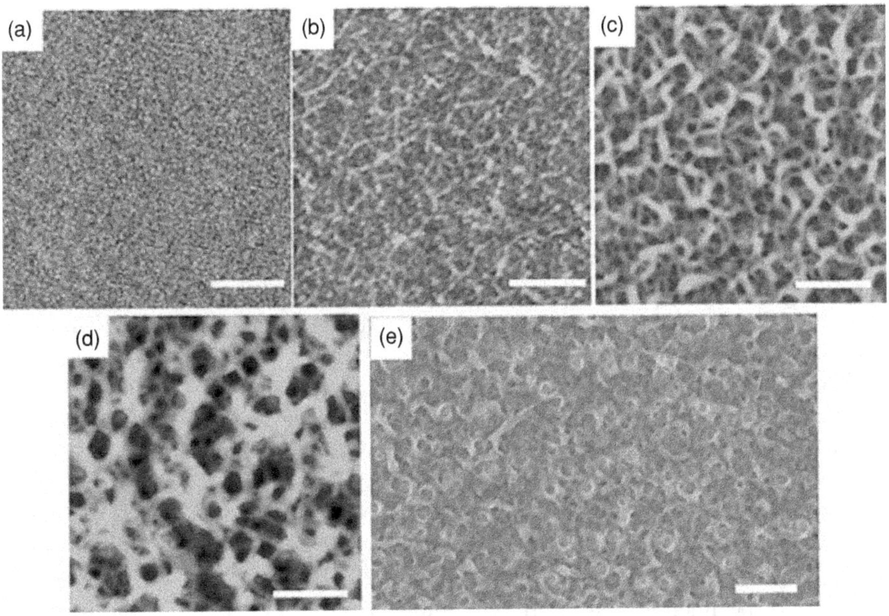

FIGURE 7.11 Tapping mode AFM height images of silicon wafer with **(a)** initial PEI layer, **(b)** PEI + 1 layer (DNA/PFS), **(c)** PEI + 5 layer (DNA/PFS), and **(d)** PEI + 10 layer (DNA/PFS). The z range is (a) 3.0, (b) 20, (c) 80, and (d) 100 nm. The scale bar is 500 nm for (a)–(d). **(e)** Top-view FESEM image of a PEI+ 10 layer (DNA/PFS) film on a silicon surface; scale bar is 1 μm. (Reproduced from Reference 66 with permission from John Wiley & Sons.)

sectioning may suffice. The choice of method will require consideration of the material properties of the polymer (thermal, mechanical, chemical). In addition, the introduction of a porous structure may influence the mechanics of deformation during the fracture, grinding, or cutting processes that could lead to ductility and smearing, even when the polymer is cooled below its glass transition temperature T_g [70]. Sample preparation of mesoporous material as thin films (<5 μm thick) on rigid substrates like silicon wafers will introduce the complexity of dissimilar material properties in the substrate and film. In these cases, milling approaches can be considered, but these processes may introduce sufficient thermal energy to raise the treated region above T_g and cause softening and distortion of the porous structure. The topography of porous structures will lead to surface charging in the SEM, and so metal coating is needed to compensate (although an environmental SEM may be used, as noted earlier). Fine-grained coatings using DC magnetron sputtering of Au/Pd, Cr, Os, or Ir metal targets are available in bench top units for this purpose; slightly thicker coatings may be necessary on such highly textured surfaces. Oxygen plasma etching is another technique that can be used to expose subsurface porosity, but with the caveat that etch rates can be anisotropic due to inherent stresses in the sidewalls and struts of cellular materials or due to compositional variation in heterogeneous materials.

7.3.2 Cryomicrotomy

Specimen preparation for TEM involves the formation of a thin film that is transparent to the electron beam. Typically, the thickness must be less than 100 nm. In the case of thick or bulk polymer specimens, microtomy is the general technique used to produce the thin film. Microtomy is method of cutting thin sections developed by biologists that is now routinely used for polymers [1]. Ultramicrotomes are available from several manufactures for this purpose. Sectioning is done slightly below the glass transition temperature of the polymer (or lowest-T_g component if it is a heterogeneous system) so that the entire sample is rigid. This may require cryogenic accessories, in which case the technique is referred to as cryoultramicrotomy or simply cryomicrotomy.

Highly porous systems often undergo compression during the sectioning process. To make the specimen more rigid and prevent the pore structure from collapsing, embedment and curing procedures are used [1]. Vacuum impregnation [71,72] with epoxy, polyester, and acrylic resin can be used to make the structure stable for sectioning, provided there are channels to the pores. When the pores are isolated from the surface and each other it may be necessary to microtome the bulk specimen first to expose the pore structure, followed by embedment to fill the pores. The embedded specimen is then carefully sectioned to collect the thin sections that have been impregnated.

One useful method employed for soft porous polymeric structures is to first polish (microtome) the face of a bulk specimen using a cryoultramicrotome. Next, a drop of 5 wt% methyl cellulose (Methocel, Dow Chemical Company) is placed on the polished face to fill the pores. The specimen with the filled pores can then be sectioned at −85°C and collected onto a TEM support grid. The methyl cellulose can then be rinsed away with water if so desired. As with any microtomy and embedment procedure, the practitioner should be aware of sectioning or reaction artifacts that may result.

7.3.3 Focused Ion Beam

Focused ion beam scanning electron microscopes (FIB-SEMs) are specialized SEMs that combine the utility of scanning electron microscopy with a focused Ga ion beam that can be used to selectively sputter away precise portions of the sample. These are sometimes also called dual-beam FIB-SEMs since both electron and ion beam optics are employed. The Ga ion beam is used in two ways (Fig. 7.12): to prepare cross sections for direct examination of the bulk sample (FIB-SEM) or to prepare thin section samples for analysis by transmission electron microscopy (FIB-TEM). The instrumentation is described in an early review [73], and more recent applications to nanomaterials characterization have been reviewed by Langford [74].

The primary advantage of FIB-SEM as a characterization or sample preparation tool is the ability to locate specific microscopic features of interest with the SEM and cross section these features with the ion beam. For example, Fig. 7.13 shows high-resolution images of hollow-sphere latex particles, revealing internal polymeric struts [75]. Another important benefit is the ion beam's ability to thin sections of both hard

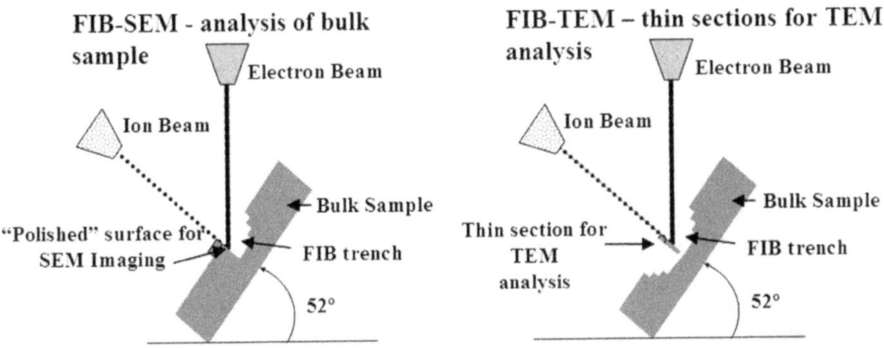

FIGURE 7.12 Schematic of the two primary modes of FIB operation: FIB-SEM **(left)** and FIB-TEM **(right)**. For FIB-TEM the thin section is lifted and remounted on a grid for examination in a TEM. (Used with permission from the Dow Chemical Company.)

and soft materials that cannot be readily microtomed. This unique combination of capabilities accounts for the best-known application of this technology to nanoporous polymers, namely the analysis of low-*k* dielectric materials as interlayer dielectrics for advanced microelectronics technologies [76]. Specific microscopic structures in the circuit need to be characterized, and no other technique is capable of cross sectioning a sample comprised of nanoporous polymer surrounded by hard materials such as silicon, silicon oxide, silicon nitride, and tantalum.

The FIB-TEM image in Fig. 7.14 shows the improvement in the ability to measure the pore size and distribution in a p-SiLK low-*k* dielectric material as the sample is thinned from ~100 nm thick to ~60 nm thick using a low-kV ion-beam thinning method. The degree of pore overlap is significant in the thicker sample, which is using

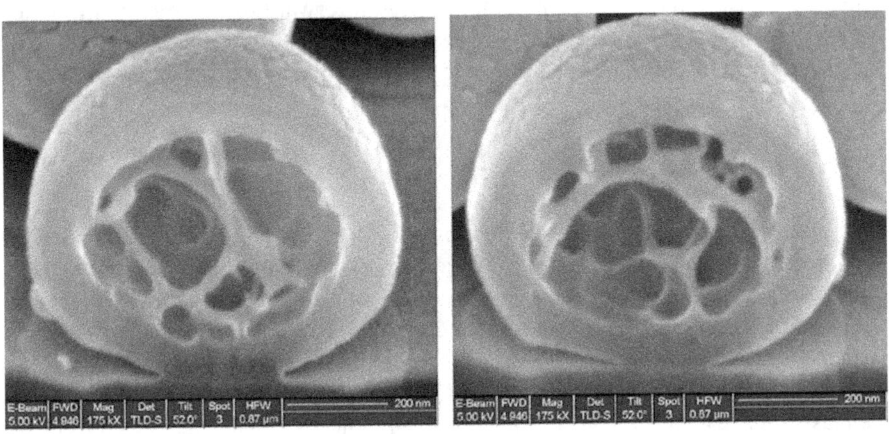

FIGURE 7.13 High-resolution FIB-SEM images of internal polymer structures remaining inside hollow latex particles. White scale bars are 200 nm. (Reproduced from Reference 75 with permission from Elsevier.)

FIGURE 7.14 FIB-TEM of a single-layer damascene structure built using a p-SiLK low-*k* dielectric (version 7, ~20 nm pores). On the left is the section obtained using 30 kV Ga^+ milling (100 nm thick section). On the right is the section after low-angle, low-energy postmilling (10 kV) resulting in a ~60 nm thick section. The overlap of the pores is virtually eliminated. (Used with permission from the Dow Chemical Company.)

an early-generation material with ~20 nm pores (version 7, Fig. 7.7d). For optimum pore sizing, the preferred sample thickness should be 2–10 times the feature size. This will avoid excessive overlap of the pores but not skew the pore size measurement from the true size.

TEM imaging is ideal for understanding the pore distribution near interfaces, for example, at the low-*k* polymer/Ta interface. The ability to detect small pores in the low-*k* material by TEM is a strong function of the sample thickness, and generally pores of less than ~2 nm are difficult to differentiate from the "speckle" contrast typically observed in amorphous materials in a conventional image. However, the image contrast in STEM imaging is more straightforward to interpret and provides improved contrast for automated pore size measurement [39]. The low-kV thinning has added benefits, such as enhancing the imaging of barrier layers.

7.4 IMAGE ANALYSIS

The classic references in this area are those of John Russ [77, 78]. Image analysis is the process of extracting data from images. It is something that one naturally does whenever one sees a scene (live or in a picture). The challenge is getting quantitative descriptions that correspond with what one sees. Even more challenging is the case in which we need objective quantification that seems to disagree with a subjective visual assessment. For this purpose we rely on some type of computer-based routine for unbiased image analysis. The advantages of computerized image analysis are consistent analyses and the ability to run through many images without fatigue.

The key to good image analysis is being able to differentiate the items of interest from the remainder of the scene. If the two are sufficiently segregated, then the automated analyzer has a high probability of success. It is best to do as much of this job as possible when the image is collected. The human eye works in conjunction with the brain and does an excellent job of differentiating items from background. One may stop trying to obtain better images once the human eye is able to discern objects from background. There is a tendency to create an image that utilizes the nuances of shading and framing to make a specific point about a subject. Unfortunately, a typical automated image analyzer is more likely to misinterpret these aesthetically pleasing aspects of the image, even though they help the eye/brain system understand what is being viewed. From a different perspective, an image that is good for image analysis will be fairly high contrast with rather stark features, so that it is often not a particularly attractive image for human review!

Images can come from any source but should be handled in a manner such that little or no loss of information occurs. This can become a major issue when lossy compression algorithms—such as JPEG—are used to store the images. It is generally good practice to keep an "original" file in a safe place and perform processing and analysis on copies since these actions may alter the image.

Image analysis is typically broken into two components: image processing (to get the image to the point at which it can be analyzed) and image measurement. The actions that occur during image analysis may be iterative; that is, one may perform some processing and then some measurement in order to determine what additional processing is necessary to get to the final analysis. Typical image processing actions will include correction of artifacts such as background variation, instrument deficiencies (e.g., scan line artifacts in AFM), filtering, smoothing, and contrast enhancements. This will be followed by application of a threshold to decide what portion of the image is a feature (i.e., to be measured) and what is background. In the case of a two-level or bilevel threshold, the processed image is made binary such that feature pixels are assigned black and background pixels are assigned white. Image measurement can be obtained manually or by using any number of commercial or public domain software packages, such as NIH Image developed at the U.S. National Institutes of Health and available at http://rsb.info.nih.gov/nih-image. In addition to good sample preparation and artifact-free imaging, the image must be calibrated to the correct scale (pixels/unit distance). This assumes that the instrument used to generate the image is suitably calibrated. An important consideration when sizing features from digital images is the determination of the minimum number of pixels that constitutes the feature. This depends on the pixel density in the data. In a similar fashion, it is important to consider, during a comparison between images of different magnifications that have the same pixel density, that the size of each pixel is different and, therefore, the minimum detectable feature size is different. Image processing and analysis of 3D images require more sophisticated computer algorithms since image stack alignment is part of the processing and three-dimensional special relationships must be computed in the measurement.

7.4.1 Pore Size and Shape Measurement

Some of the key metrics that are desired from 2D images of porous polymers include the pore diameter, pore volume, pore size distribution, pore shape factors (e.g., aspect ratio), interpore distance, connectivity, and total porosity. For a foamed material, additional metrics such as cell wall thickness and density and type of struts may be required. Clearly, some metrics will suffer from a stereological effect, that is, the fact that the 2D image is actually a projection of 3D objects. For example, the distribution of circles in a random 2D slice of a material containing either monodispersed pores or a distribution of pore sizes may appear to be similar. However, there are statistical approaches that enable extraction of the correct 3D size distribution in both cases by analysis of the size distribution of the pore cross sections [79].

We have used tapping mode AFM to measure the pore sizes at the surfaces of porous SiLK low-*k* dielectric film materials [80]. Surface porosity is of interest for two reasons. First, it may replicate the interior pore structure and could be useful as a direct microscopy method to evaluate pore size distribution. Second, there is the practical consideration for device integration, where surface porosity could allow chemical intrusion or migration due to poor barrier metal coverage on the side walls during processing (if the pores are interconnected). An approach to sizing pores at the surface using light tapping conditions in the AFM has been developed [80]. The resulting height images are postprocessed with an inverse fast Fourier transform (FFT) filter that is optimized to remove the lower-frequency texture. This creates a background height image which can be subtracted from the original to provide an image that is more forgiving with respect to bilevel thresholding in order to make a binary image for image analysis. Figure 7.15 shows examples of the analysis of early-stage developmental p-SiLK film surfaces that correspond to the cross-sectional images shown in Fig. 7.7d–f.

The binary images can be analyzed for a variety of metrics, including equivalent pore diameter and roundness (Fig. 7.16). The size reduction in the porosity at the surface is consistent with the reduction in size of internal porosity, as shown in Fig.7.7, with the 50% probability distribution showing surface pore diameters of about 18 nm for versions 7 and 8 and of about 11 nm for version 9. The probability distribution for version 7 is broad and consistent with the internal morphology. The roundness is maximized in version 8, consistent with the synthetic strategy for this material.

It is not surprising that much effort has been devoted to the sizing of pores in nanofiltration (NF) and microfiltration (MF) membranes using direct microscopy investigations. Zeman described methods to quantify the pore size in cellulose MF membranes using SEM [81] and correlated these results with performance [82]. Masselin et al. used FESEM to characterize porosity in PES, PVDF, and PAN UF membranes and correlated this with rejection performance [83]. More recently, Otero et al. compared the pore sizes of nanofiltration membranes obtained by AFM with indirect methods [84]. For AFM, the authors used special FIB-sharpened tips and special image filtering to quantify the sub–3 nm diameter pores. Finally, Ziel et al. described methods to quantify pore connectivity in asymmetric polyethersulfone MF membranes [85].

Since indirect methods can provide information on pore sizes and distributions, it is important to consider the physics of the various methods when comparing data from a variety of techniques. For instance, small-angle X-ray scattering may effectively not see pores above a few hundred nanometers even though they may be present in the specimen and can be observed by direct imaging methods such as SEM or AFM. The

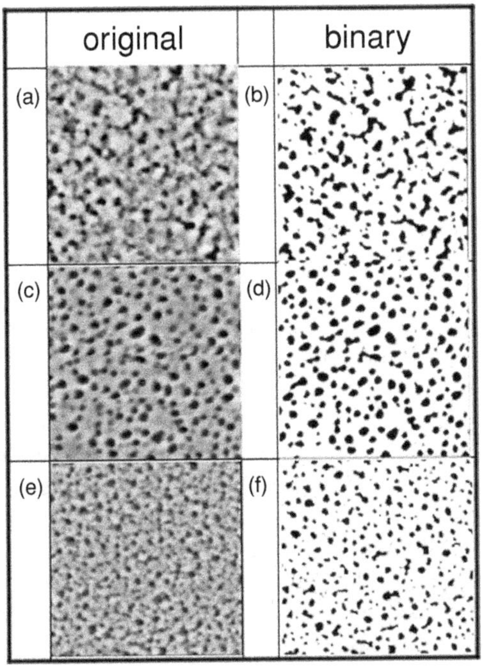

FIGURE 7.15 Tapping mode AFM images of early-stage developmental p-SiLK low-*k* dielectric film surfaces. The original and binary images obtained by bilevel thresholding are shown. **(a, b)** version 7, **(c, d)** version 8, and **(e, f)** version 9 type. Images are 500 nm × 500 nm. (Used with permission from the Dow Chemical Company).

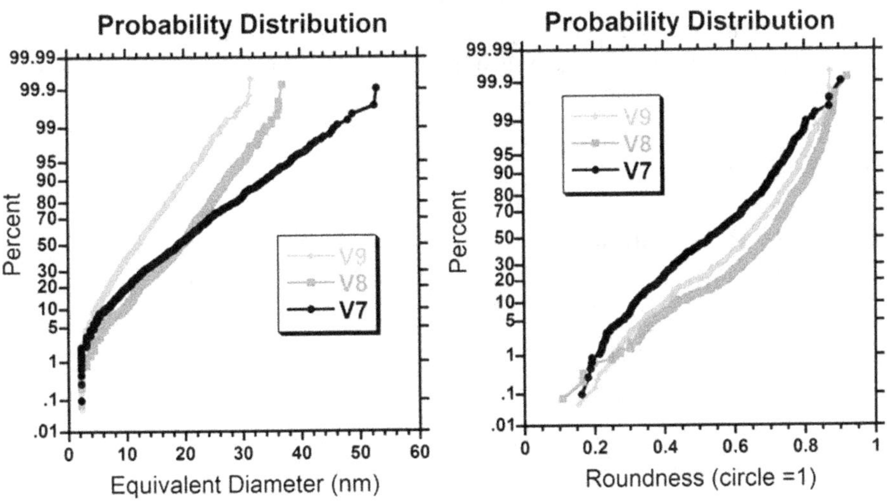

FIGURE 7.16 Probability plots of **(left)** equivalent pore diameter and **(right)** pore roundness from the images shown in Fig. 7.14. A reduction in pore diameter is evident. (Used with permission from the Dow Chemical Company.)

methods may not be in conflict; they just have different limitations on the sizes that can be characterized. This caution is especially noteworthy for the study of materials during the early stages of development, when the pore uniformity may not be well controlled.

7.5 CONCLUSIONS

This chapter has attempted to review the application of optical, electron, X-ray, and probe-based imaging methods to the study of porous polymer materials. It can be hoped that the reader has appreciated that the entire microscopy workflow must be considered—from sample preparation to instrument calibration and data acquisition to postprocessing and analysis—in order to achieve success. Consideration must also be given to the rendering of the image to convey the desired information. Clearly, these techniques need to be used in conjunction with each other where appropriate, but also as complements to indirect methods such as light, X-ray, and neutron scattering, positron annihilation spectroscopy, size exclusion methods, thermal porosimetry, and others. In all cases, careful thought as to the limitations of each technique must be given in order to develop a consistent picture of pore structure, morphology, and chemistry of the material across the various characterization methods. The techniques one chooses in the early development of a new porous material may be those more amenable to rapid screening, while a different suite of methods may be necessary for ongoing quality control or performance monitoring.

The future of porous polymer imaging will grow rapidly into full quantitative, three-dimensional imaging accessible at most length scales of interest. For many applications two-dimensional views are not sufficient to explain performance. Descriptions of the spatial relationships between isolated pores and imaging techniques for quantifying the interconnectivity of pores and through channels will be needed.

ABBREVIATIONS

Polymers and Materials

PAN	Polyacrylonitrile
PEI	Polyethyleneimine
PES	Polyethersulfone
PFMA	Poly(fluorooctylethyl methacrylate)
PFS	Polyferrocenylsilane
PLA	Polylactic acid
PMMA	Poly(methyl methacrylate)
PS	Polystyrene
PVDF	Poly(vinylidene fluoride)
SBS	Styrene–butadiene–styrene block copolymer

Microscopy and Image Analysis Techniques

AFM	Atomic force microscopy
BSE	Backscatter electron
CLSM	Confocal laser scanning microscopy
CT	Computed tomography
EDS	Energy dispersive spectroscopy
FESEM	Field emission scanning electron microscopy
FFT	Fast Fourier transform
FIB-SEM	Focused ion beam scanning electron microscopy
HAADF	High-angle annular dark field
HRTEM	High-resolution transmission electron microscopy
NEXAFS	Near-edge X-ray absorption fine structure
PEELS	Parallel electron energy loss spectroscopy
SAXS	Small-angle X-ray scattering
SEM	Scanning electron microscopy
SPM	Scanning probe microscopy
STEM	Scanning transmission electron microscopy
STXM	Scanning transmission X-ray microscopy
TEM	Transmission electron microscopy
WDS	Wavelength dispersive spectroscopy

REFERENCES

1. Sawyer, L. C.; Grubb, D. T.; Meyers, G. F. *Polymer Microscopy*, 3rd ed., Springer, New York, 2008.
2. Michler, G. H. *Electron Microscopy of Polymers*, 1st ed., Springer, New York, 2008.
3. Murphy, D. B. *Fundamentals of Light Microscopy and Electronic Imaging*, Wiley-Liss, New York, 2001.
4. Hemsley, D. A., Ed. *Applied Polymer Light Microscopy*, Elsevier Applied Science, London, 1989.
5. McCrone, W. C.; Delly, J. G. *Particle Atlas*, Ann Arbor Science, Ann Arbor, MI, 1973.
6. Li, J.; Du, Z.; Li, H.; Zhang, C. *Polymer* 2009, **50**, 1526.
7. Kino, G. S.; Corley, T. R. *Confocal Scanning Optical Microscopy and Related Imaging Systems*, Academic Press, New York, 1996.
8. Charcosset, C.; Bernengo, J.-C. *J Membr Sci* 2000, **168**, 53.
9. Stenzel, M. H.; Barner-Kowollik, C.; Davis, T. P. *J Polym Sci A Polym Chem* 2006, **44**, 2363.
10. Srinivasarao, M.; Collings, D.; Philips, A.; Patel, S. *Science* 2001, **292**, 79.
11. Gliemann, H.; Almeida, A. T.; Petri, D. F. S.; Schimmel, T. *Surf Interf Anal* 2007, **39**, 1.
12. Barrow, M. S.; Jones, R. L.; Park, J. O.; Srinivasarao, M.; Williams, P. R.; Wright, C. J. *Spectroscopy (Amst)* 2004, **18**, 577.
13. Vainrub, A.; Pustovyy, O.; Vodyanov, V. *Opt Lett* 2006, **31**, 2855.

14. Rothe, D. R.; Heeschen, W. A.; Wood, S. P. *Microsc Microanal* 2007, **13** (Suppl 2), 85.

15. Goldstein, J.; Newbury, D.; Echlin, P.; Lyman, C.; Joy, D.; Lifshin, E.; Sawyer L.; Michael, J. *Scanning Electron Microscopy and X-Ray Microanalysis*, Plenum Press, New York, 2003.

16. Flegler, S. L.; Heckman, J. W.; Klomparens, K. L. *Scanning and Transmission Electron Microscopy : An Introduction*, Oxford University Press, Oxford, 1993.

17. Reimer, L.; Hawkes, P. W. Eds., *Scanning Electron Microscopy, Physics of Image Formation and Microanalysis*, Springer, Berlin, 1998.

18. Todd, C. S.; Blackson, J.; Bar, G.; Garcia-Meitin, E.; Reuschle, D.; Janus, M.; Darus, M.; Nickles, A. *Microsc Today* 2008, **16**, 24.

19. Reimer, L. *Image Formation in Low-Voltage Scanning Electron Microscopy*. SPIE, Bellingham, WA, 1993.

20. Phifer, D.; Tuma, L.; Vystavel, T.; Wandrol, P.; Young, R. J. *Microsc Today* 2009, **17**, 40.

21. Young, R.; Templeton, T.; Roussel, L.; Gestmann, I.; Veen, G. v.; Dingle, T.; Henstra, S. *Microsc Today* 2008, **16**, 24.

22. Brumfield, H. L.; Estill, W. B. *J Cell Plast* 1969, **5**, 212.

23. Todd, C. S. *Microsc Microanal*, 2008, **14** (Suppl 2), 698.

24. Reichenauer, G.; Heinemann, U.; Ebert, H.-P. *Colloids Surf A* 2007, **300**, 204.

25. Hillmyer, M. A. *Adv Polym Sci* 2005, **190**, 137.

26. Cooper, A. I. *J Mater Chem* 2000, **10**, 207.

27. Otsuka, T.; Taki, K.; Ohshima, M. *Macromol Mater Eng* 2008, **293**, 78.

28. Ema, Y.; Ikeya, M.; Okamoto, M. *Polymer* 2006, **47**, 5350.

29. Zalusky, A. S.; Olayo-Valles, R.; Wolf, J. H.; Hillmyer, M. A. *J Am Chem Soc* 2002, **124**, 12761.

30. Merlet, S.; Marestin, C.; Schiets, F.; Romeyer, O.; Mercier, R. *Macromolecules* 2007, **40**, 2070.

31. Cooper, A. I. *Adv Mat* 2003, **15**, 1049.

32. Krause, B.; Sijbesma, H. J. P.; Munuklu, P.; Vegt, N. F. A. v. d.; Wessling, M. *Macromolecules* 2001, **34**, 8792.

33. Yokoyama, H.; Li, L.; Nemoto, T.; Sugiyama, K. *Adv Mater* 2004, **16**, 1542.

34. Yokoyama, H.; Sugiyama, K. *Macromolecules* 2005, **38**, 10516.

35. Joy, D. C.; Romig, A. D., Jr; Goldstein, J. I., Eds. *Principles of Analytical Electron Microscopy*, Plenum Press, New York, 1989.

36. Hirsch, P.; Howie, A.; Nicholson, R. B.; Pashley, D. W. *Electron Microscopy of Thin Crystals*, Krieger, New York, 1977.

37. Thomas, G.; Goring, M. J. *Transmission Electron Microscopy of Materials*, Wiley, New York, 1987.

38. Williams, D. B.; Carter, C. B. *Transmission Electron Microscopy: A Textbook for Materials Science*, 2nd ed., Springer, New York, 2009.

39. Foran, B.; Kastenmeier, B.; Bright, D. S. In *Characterization and Metrology for ULSI Technology: 2003*, Seiler, D. G.; Diebold, A .C.; Shaffner, T. J.; McDonald, R.; Zollner, S.; Khosla, R. P.; Secula, E. M., Eds., AIP Press, College Park, MD, 2003, p. 556.

40. Martin, D. C.; Chen, J.; Yang, J.; Drummy, L. F.; Kübel, C. *J Polym Sci B Polym Phys* 2005, **43**, 1749.

41. Hitchcock, A. P.; Stover, H. D. H.; Croll, L.; Childs, R. F. *Aust J Chem* 2005, **58**, 423.
42. Croll, L .M.; Stoever, H. D. H.; Hitchcock, A. P. *Macromolecules* 2005, **38**, 2903.
43. Johansson, G. A.; Tyliszczak, T.; Mitchell, G. E.; Keefe, M. H.; Hitchcock, A. P. *J Synchrotron Radiat* 2007, **14**, 395.
44. Hawkes, P. W.; Spence, J. C. H., Eds., *Science of Microscopy*, Springer, New York, 2007.
45. Frank, J., Ed. *Electron Tomography, Methods for Three-Dimensional Visualization of Structures in the Cell*, Springer, New York, 2007.
46. Banhart, J., Ed. *Advanced Tomographic Methods in Materials Research and Engineering*, Oxford University Press, New York, 2008.
47. Jinnai, H.; Kaneko, T.; Nishioka, H.; Hasegawa, H.; Nishi, T. *Chem Rec* 2006, **6**, 267.
48. Kubel, C.; Niemeyer, D.; Cieslinski, R.; Rozeveld, S. *Mater Sci Forum* 2010, **638–642**, 2517.
49. Efimov, A. E.; Tonevitsky, A. G.; Dittrich, M.; Matsko, N. B. *J Microsc* 2007, **226**, 207.
50. Konrad, M.; Knoll, A.; Krausch, G.; Magerle, R. *Macromolecules* 2000, **33**, 5518.
51. Lau, S. H.; Chiu, W. K. S.; Garzon, F.; Chang, H.; Tkachuk, A.; Feser, M.; Yun, W. *J Phys Conf Ser* 2009, **152**, 012059.
52. Zhong, Q.; Innis, D.; Kjoller, K.; Elings, V. *Surf Sci Lett* 1993, **290**, L688.
53. Garcia, R.; Perez, R. *Surf Sci Rep* 2002, **47**, 197.
54. Schirmeisen, A.; Anczykowski, B.; Fuchs, H. In *Applied Scanning Probe Methods*, Bushan, B.; Fuchs, H.; Hosaka, S., Eds., Springer-Verlag, Berlin, 2003, p. 3.
55. Hamley, I. W. *Prog Polym Sci* 2009, **34**, 1161.
56. Magonov, S. *In Applied Scanning Probe Methods*, Bushan, B.; Fuchs, H.; Hosaka, S., Eds., Springer-Verlag, Berlin, 2003, p. 207.
57. Magonov, S. N.; Yerina, N. A. *Handbook of Microscopy for Nanotechnology*, Yao, N; Wang, Z. L., Eds., Kluwer Academic, New York, 2005, p. 113.
58. Khulbe, K. C.; Feng C. Y.; Matsuura, T. *Synthetic Polymer Membranes: Characterization by Atomic Force Microscopy*, Springer, New York, 2008.
59. Villarrubia, J. S. *In Applied Scanning Probe Methods*, Bushan, B.; Fuchs, H.; Hosaka, S., Eds., Springer-Verlag, Berlin, 2003, p. 147.
60. *E2382-04 Guide to Scanner and Tip Related Artifacts in Scanning Tunneling Microscopy and Atomic Force Microscopy, ASTM Annual Books of Standards*, Vol. 03.06, ASTM International, West Conshohocken, PA, 2004.
61. Nandhakumar, I. S.; Gordon-Smith, T. J.; Attard, G. S.; Smith, D. C. *Small* 2005, **1**, 406.
62. Lugstein, A.; Bertagnoli, E.; Kranz, C.; Kueng, A.; Mizaikoff, B. *Appl Phys Lett* 2002, **81**, 349.
63. Dietz, P.; Hansma, P. K.; Herrmann, K. H.; Inacker. O.; Lehmann, H. D. *Ultramicroscopy* 1991, **35**, 155.
64. Dietz, P.; Herrmann, K. H.; Inacker, O.; Lehmann, H. D.; Hansma, P. K. *Proc SPIE Int Soc Opt Eng* 1992, **1639**, 186.
65. Freger, V. *Environ Sci Technol* 2004, **38**, 3168.
66. Ma, Y.; Dong, W.-F.; Hempenius, M. A.; Moehwald, H.; Vancso, G. J. *Angew Chem Int Ed* 2007, **46**, 1702.
67. Echlin, P. *Handbook of Sample Preparation for Scanning Electron Microscopy and X-Ray Microanalysis*, 1st Ed., Springer, New York, 2009.

68. Ayache, J.; Beaunier, L.; Boumendil, J.; Ehret, G.; Laub, D. *Microsc Today* 2009, **17**, 38.
69. Schrand, A. M. Presented at *Microscopy and Microanalysis 2005*, Honolulu, Hawaii, p. 702.
70. Orelovitch, O. L.; Apel, P. Y.; Sartowska, B. *Mater Chem Phys* 2003, **81**, 349.
71. Glauert, A. M.; Lewis, P. R. *Biological Specimen Preparation for Transmission Electron Microscopy*, Princeton University Press, Princeton, NJ, 1998.
72. Bigelow, W. In *Vacuum Methods in EM: Practical Methods in Electron Microscopy: Vol 15*, Glauert, A. M., Ed., Ashgate, Farnham, Surrey, UK, 1999, p. 512.
73. Meingailis, J. *J Vac Sci Technol B* 1987, **5**, 469.
74. Langford, R. M. *Microsc Res Tech* 2006, **69**, 538.
75. Beach, E.; Keefe, M.; Heeschen, W.; Rothe, D. *Polymer* 2005, **46**, 11195.
76. Hawker, C. J.; Hedrick, J. L.; Miller, R. D.; Volksen, W. *MRS Bull* 2000, **4**, 54.
77. Russ, J. C. *Computer-Assisted Microscopy: The Measurement and Analysis of Images*, Plenum Press, New York, 1990.
78. Russ, J. C. *The Image Processing Handbook*, 5th ed., CRC Press, Boca Raton, FL, 2006.
79. Russ, J. C.; Dehoff, R. T. *Practical Stereology*, 2nd ed., Springer, New York, 2000.
80. Mohler, C. E.; Landes, B. G.; Meyers, G. F.; Kern, B. J.; Ouellette, K. B.; Magonov, S. *AIP Conf Proc* 2003, **683**, 562.
81. Zeman, L.; Denault, L. *J Membr Sci* 1992, **71**, 221.
82. Zeman, L. *J Membr Sci* 1992, **71**, 233.
83. Masselin, I.; Durand-Bourlier, L;. Laine, J. M.; Sizaret, P. Y.; Chasseray, X.; Lemordant, D. *J Membr Sci* 2001, **186**, 85.
84. Otero, J. A.; Mazarrasa, O.; Villasante, J.; Silva, V.; Pradanos, P.; Calvo, J. I.; Hernandez, A. *J Membr Sci* 2008, **309**, 17.
85. Ziel, R.; Haus, A.; Tulke, A. *J Membr Sci* 2008, **323**, 241.

SECTION III

APPLICATIONS

CHAPTER 8

Separation Membranes

MATHIAS ULBRICHT
Universität Duisburg-Essen, Essen, Germany

8.1 MEMBRANES AND MEMBRANE-BASED PROCESSES

A membrane is an interphase between two adjacent phases acting as a selective barrier while at the same time organizing a system into compartments and regulating the transport between the two compartments. This well-known principle from biology has in recent decades been translated into industrial processes with synthetic membranes. The main advantages of membrane technology as compared to other unit operations in (bio)chemical engineering are related to the unique separation principle, that is, the transport selectivity of the membrane. Furthermore, separations with membranes do not require additives, and they can be performed isothermally and at a very competitive energy consumption. Finally, both upscaling and downscaling of membrane processes, as well as their integration into other separation or reaction processes, are easy. The success of membrane technology has already been impressively demonstrated by the first applied examples, blood detoxification by dialysis ("artificial kidney") and water desalination by reverse osmosis [1]. In addition, many other large-scale applications have been realized in the last two decades, especially in the food and pharmaceutical industries, as well as in water treatment. Currently, applications of membrane technology are also emerging in the chemical industry [2]. Serving these needs requires innovative research and development toward improved membrane materials and processes. As will be discussed, most industrial membranes are either entirely porous or at least contain porous parts or components that are essential for their performance. However, it is crucial that the barrier structure of membranes, and hence their function, can be classified according to their porous character.

For *nonporous membranes*, the interactions between permeand and membrane material dominate flux and selectivity; the transport mechanism can be described by

Porous Polymers, Edited by Michael S. Silverstein, Neil R. Cameron, and Marc A. Hillmyer

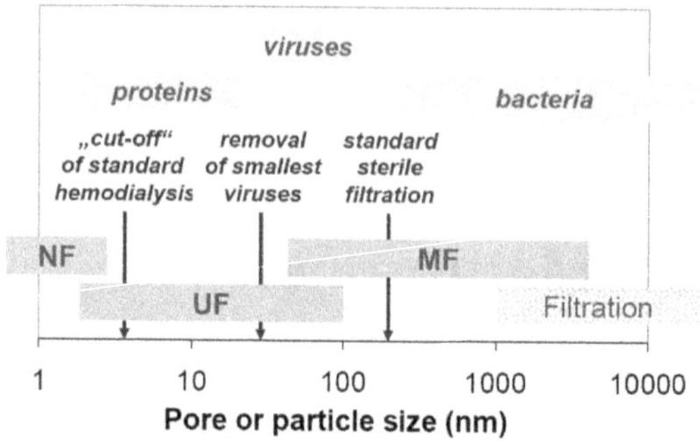

FIGURE 8.1 Ranges for pore sizes of porous membranes and characteristic dimensions of components separated by membrane processes. MF, microfiltration; NF, nanofiltration; UF, ultrafiltration.

the solution/diffusion model. Several important membrane processes belong to this class and will not be discussed further in this chapter:

1. Reverse osmosis.
2. Nanofiltration, pervaporation, and gas separation using polymeric membranes with nonporous polymeric barriers.
3. Electrodialysis, membrane electrolysis, or fuel cell systems using nonporous polymeric ion-exchange membranes.

For *porous membranes* as perm-selective barriers, flux and selectivity are mainly influenced by viscous flow and sieving or size exclusion by the membrane pores.[1] Membrane-based separations should be distinguished from the mechanical separation of particles by filtration. Microfiltration and ultrafiltration are used for size-based removal of colloidal impurities or hazardous components from fluid streams—the membranes provide a perfect mechanical barrier against larger particles (e.g., sterilization and virus removal from gas or aqueous streams)—and for many different concentration and fractionation processes (Fig. 8.1).

Nanofiltration is located at the boundary of the "filtration" spectrum (effective pore size <2 nm), and typical polymeric membranes for such separations have a barrier with no permanent pores (i.e., they function according to the solution/diffusion mechanism). However, it should be noted that in recent years, the formerly rather clear "boundary" between porous and nonporous barrier structures made from polymer membranes has become less distinct. Two different developments have largely contributed to this. First, the development of polymers with a very large fraction of free volume has led to so-called polymers with inherent microporosity (PIMs; see Chapter 1). PIMs have shown superior performance, that is, selectivity to

[1] Transport and separation mechanisms for small pores such as Knudsen and surface diffusion are not further discussed here because porous polymeric membranes with barriers containing permanent micropores or mesopores at a very narrow size distribution are not yet available (see Section 8.2.5.2).

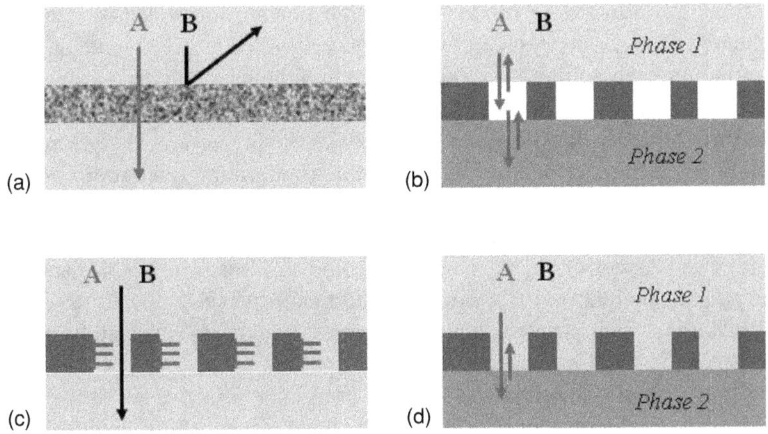

FIGURE 8.2 Schematic overview of separation processes using porous membranes; in all cases, there is a selectivity for A over B. **(a)** Perm-selective porous membrane; the membrane is a separation agent. **(b)** Porous membrane stabilizing an interphase between two phases; the system is a membrane contactor. **(c)** Porous membrane with binding sites on the pore surface; the membrane is a separation agent and often also a membrane contactor. **(d)** Porous membrane stabilizing the interface between two phases; the system is a membrane contactor.

permeability ratios beyond the so-called "upper bound" for the separation of many gas mixtures [3]. However, membranes from these polymers are also very interesting for other membrane separations, such as pervaporation [4]. Second, membranes from polymers with very pronounced microphase-separated morphologies—for example, hydrophilic, water-filled domains ("pores") dispersed in a hydrophobic matrix—have recently exhibited competitive or even superior performance in reverse osmosis [5] or nanofiltration [6].[2] More examples for such "nanoporous" membranes based on controlled microphase-separated polymers are discussed in Section 8.2.5.

Membranes with ion-exchange groups in the barrier are also important; for small barrier pores the interactions of solutes with the charged groups can completely control flux and selectivity (e.g., via Donnan exclusion or ionic binding). More specific noncovalent chemical interactions with substances in the feed mixture are used in affinity membranes. Porous membranes have already found many more applications in membrane contactors (see later discussion) and catalytic membrane reactors [7]. In order to apply the membranes in separation or integrated processes, different driving forces for transmembrane transport can be used; the most important one is a pressure difference. A concentration difference, an electrical potential, or a temperature difference across the membrane are also frequently used (see Section 8.3).

Figure 8.2 gives an overview of the most important separation processes involving porous membranes (due to the porous nature of the barrier, most of the relevant

[2] Analogous microphase-separated structures are known for ion-exchange membranes—for example, those made from perfluorinated polysulfonic acids, which have long been used in electromembrane processes [9].

separations are devoted to mixtures in liquid phase); more details can be found elsewhere [1, 8, 9]. Some details of the processes will be outlined in what follows.

The membrane is acting as a separation agent if the selectivity according to sieving or size exclusion is used in combination with a transmembrane flux by convection (microfiltration and ultrafiltration; pressure as driving force), diffusion (dialysis; concentration difference as driving force), electromigration (electrodialysis; electrical potential as driving force), or combinations thereof (Fig. 8.2a). The selectivity of a size-selective membrane is typically expressed by the "rejection," that is, the percentage of substances or particles with a specific size in a solution or dispersion that cannot pass the membrane. "Tailoring" the barrier properties—ideally, a narrow pore size distribution, high porosity, and a low barrier thickness—is of highest relevance for the separation performance.

The pores of a membrane can also be used as a selective phase between two adjacent phases (phases 1 and 2 in Fig. 8.2b). With a pore-filling liquid containing a selective carrier, the (composite) membrane is a separation agent. The concentration difference of the selectively transported species (here species A) and the mobility of the complex in the membrane will determine the flux. For long-term stability it is critical that the liquid will not be extracted or expelled from the membrane. If the membrane is placed between two liquid phases with similar composition and the pores are filled with vapor, the system is a membrane contactor. The most relevant case is membrane distillation, where a temperature difference is used as driving force for the transport of the more volatile component from the side with the higher temperature through the membrane to the side with the lower temperature. This process has been intensively investigated for water desalination; for long-term stability it is critical that water does not wet the pores.

When the pore surface of the membrane is functionalized with ionic or affinity binding sites, two different separation mechanisms are possible in principle (Fig. 8.2c). Only for very small pores—that is, for substantially reduced nonselective diffusion—can facilitated, fixed carrier transport via solute binding/dissociation to/from affinity sites on pore walls lead to transport-selective membranes for continuous separations [10]. Much more often, pore-functionalized mesoporous and macroporous materials are used for membrane adsorption (of species A in Fig. 8.2c). By convective flow through the membrane pores, mass transfer resistances that are typical for porous adsorbents are significantly reduced; that is, in this case, the membrane could be considered a separation agent and a membrane contactor at the same time. Different from all other cases discussed here, selective separation will only be possible until the binding capacity of the membrane is saturated; thereafter, breakthrough will occur, and the membrane adsorbent must be regenerated by elution of the bound substances.

If the membrane is placed between two different immiscible phases (either liquid/liquid or liquid/gas) and the pores are filled with one of the two phases, the membrane is just stabilizing an interface (Fig. 8.2d). This membrane contactor concept is used for membrane extraction or membrane absorption/stripping processes. The advantage as compared with the conventional thermal separation processes (extraction, absorption, stripping) is that the two-phase systems can be handled in a much more controlled and efficient way because the interfacial area is precisely known and no additional efforts for mixing and/or phase separation are necessary.

Considering all these different process engineering options, membranes have become a key element for so-called process intensification [7].

Furthermore, porous membranes are also frequently used in other applications, in particular for cell culture or as tissue engineering scaffolds [11], in sensor systems and bioanalysis systems [12], for controlled release [13], and as packaging and other barrier materials [14]. In some of these cases, the barrier selectivity of the membrane can be crucial for the function (e.g., porous hydrophobic membranes for "breathable," waterproof clothing [15]); in other cases, the porous membrane is just used as a special kind of membrane contactor or even just a tailored solid support (e.g., macroporous membranes in blotting analyses of biomolecules [16]).

Concentration polarization between bulk fluid and membrane surface, an intrinsic effect caused by the separation of a mixture via convective flow through a membrane, can dominate the transmembrane flux in ultrafiltration, and this can be described by boundary layer models [1]. Because the fluxes are lower than in ultrafiltration, concentration polarization effects are less important in nanofiltration, dialysis, carrier-mediated separation (e.g., with immobilized liquid membranes), and separation processes using membrane contactors (see Fig. 8.2). Interactions between substances in the feed and the membrane surface (solute adsorption and/or particle adhesion; fouling and biofouling) may also significantly influence the separation performance; fouling and biofouling are especially strong with aqueous feeds [1,9].

The membrane cross section can be isotropic ("symmetric"), integrally anisotropic ("asymmetric"), bilayer or multilayer, thin-film composite, or mixed matrix composite. Resistance to mass transfer in all membranes is inversely proportional to barrier thickness. While macroporous membranes for microfiltration or porous membranes for some contactor processes can be relatively thick, separation membranes with small barrier pores (or a nonporous barrier) should have a minimum barrier thickness. This is achieved with integrally "asymmetric" membranes from one material that have a pronounced gradient in porosity and a very thin (<1 μm) layer with the smallest pores or by thin-film composite membranes in which the selective layer is supported by a porous support from a different material; in both cases the layer topology ensures high mechanical stability at minimum overall resistance.

Furthermore, three membrane geometries are possible: flat sheet; tubular, capillary, or hollow fiber[3]; and capsule membranes. The selection is dictated by the specific process (phases involved, driving forces, and separation mechanisms). Often the main criterion is the required membrane area: Modules with capillary or hollow fiber membranes can contain a much higher membrane area per volume than those equipped with flat-sheet membranes. Processes with membrane (micro)capsules still do not play a large role in industrial applications, but they may have a significant potential

[3]The three membrane types that have a hollow tube shape can be roughly distinguished according to their dimensions; tubular membranes have the largest diameter (>5 mm) and are typically cast on a porous tube as mechanical support, while capillary (diameter $\sim$0.2 to $\sim$5 mm) and hollow-fiber membranes (diameter <200 μm) are self-supporting.

for dedicated processes, for example, controlled drug release or delivery [13]. More details about different module and process concepts can be found elsewhere [1].

Membranes can be made from (organic) polymers, inorganic materials, or metals; other special cases are liquid membranes (or immobilized liquid composite membranes; see Fig. 8.2b) and membranes from self-assembled smaller molecules (in analogy to biological membranes). Polymeric membranes dominate a very broad range of industrial membrane applications. This is due to the following advantages: (1) many different types of polymeric materials are commercially available, (2) a large variety of different selective barriers—porous, nonporous, charged, and affinity—can be prepared by versatile and robust methods, (3) production of large membrane areas with consistent quality is possible on an industrial scale at a reasonable cost based on reliable manufacturing processes, and (4) various membrane shapes (flat sheet, tubular, hollow fiber, capillary, capsule) and formats, including membrane modules with high packing densities, can be produced (see Section 8.2). However, membrane polymers also have some limitations. A very well defined regular pore structure and very small pores (in the micropore range) are difficult to achieve, and the mechanical strength, the thermal stability, and the chemical resistance (e.g., at extreme pH values or in organic solvents) are rather low for many organic polymers. In that regard, inorganic materials (porous metal oxides or carbon materials) can offer some advantages, such as high mechanical strength, excellent thermal and chemical stabilities, and, in some cases, a highly uniform pore shape and size (e.g., in zeolites). However, many inorganic materials are very brittle, and due to complicated preparation methods and manufacturing technologies, the prices for most inorganic membranes are still very high. In addition, the variability of chemistries available for inorganic materials is generally much smaller than for organic polymers.

Overall, membrane preparation and modification by/with functional polymers has become a very active and diverse field of research, development, and application on an industrial scale, with a large potential future impact on separation and reaction engineering, as well as on many other crucial technologies [9].

8.2 PREPARATION AND MANUFACTURE OF POROUS POLYMERIC MEMBRANES

An analysis of the state-of-the-art in separation with polymeric membranes reveals that about 20 major membrane polymers are currently used on the industrial scale; most of them had originally been developed for different applications [9]. Of course, many more polymers are being investigated and exhibit promising or superior properties. However, it is very important to note that the intrinsic properties of a polymer are only of decisive importance for membranes with nonporous barriers (working according to the solution/diffusion mechanism) and much less crucial for porous barrier structures. In the latter case, it is of highest relevance whether a desired pore structure can be obtained and whether that structure is stable under application conditions. Therefore, the selection of a polymer for a porous membrane is based on the requirements of the manufacturing process (depending on the preparation process: polymer solid-state structure and morphology, properties of the polymer melt, polymer solubility, precursors for or reactivity of the polymer) and the behavior and performance under application conditions. The following material properties should be considered.

1. *Film-forming properties* indicate the ability of a polymer to form a cohesive film, and the macromolecular structure, especially molar mass and attractive interactions between chain segments, is crucial in this regard. Poly(ether sulfones) (PES), polysulfones (PSf), polyamides (PA), and polyimides (PI) are examples of excellent film-forming materials.
2. *Mechanical properties* involve film strength, film flexibility, and compaction stability (especially of a porous structure). The latter is most important for high-pressure processes (e.g., for the porous substructure of an integrally anisotropic or a composite reverse osmosis membrane). Because hollow-fiber membranes are self-supporting, the mechanical stability will be especially relevant. Many commercial flat-sheet membranes are prepared on a nonwoven support material, typically from robust polymers such as polyolefins or polyesters.
3. *Thermal stability* requirements depend very much on the application. In order to ensure the integrity of a pore structure of nanometer dimensions, the glass transition temperature T_g of the polymer should be higher than the process temperature. For semicrystalline polymers such as polyolefins this is not relevant.
4. *Chemical stability* requirements include the resistance of the polymer at extreme pH values and other chemical conditions. Cleaning agents such as strong acids or bases or oxidizing agents are usually used to clean a fouled membrane. The stability in special solvents is also important in selected cases, such as when processes with nonaqueous mixtures are considered. Examples for very stable membrane polymers are fluorinated polymers, for example, polyvinylidene fluoride (PVDF) or poly(tetrafluoroethylene) (PTFE).
5. *Hydrophilicity–hydrophobicity balance* correlates with the wettability of the material. Complete wetting is important in order to use all pores in ultrafiltration. In contrast, when a porous membrane is applied as a contactor between a liquid and a gas phase, the phase boundary is often stabilized because the liquid will not wet the dry pores of the membrane (see Fig. 8.2). With aqueous liquid phases, ultrafiltration will require a more hydrophilic (e.g., polyacrylonitrile [PAN]) and a liquid/gas contactor a hydrophobic membrane polymer (e.g., polypropylene [PP]). Surface wettability is also critical for fouling; cellulose is an excellent example of a hydrophilic polymer for low-fouling ultrafiltration membranes. Nevertheless, relatively hydrophobic polymers, for instance, PVDF and PES, show much better chemical and thermal stability.

In the following, the various preparation methods can be classified according to the method of processing the membrane polymer, that is, from the solid state (by sintering, Section 8.2.1), from the melt (extrusion, Section 8.2.2; electrospinning, Section 8.2.3), from solutions (electrospinning, Section 8.2.3; phase separation, Section 8.2.4), or via *in situ* polymerization methods (Section 8.2.6). In addition, combinations of these processes with mechanical processing (e.g., stretching) or chemical reactions (e.g., selective etching) are also important. Furthermore, there is an increasing trend to use polymers with more-controlled structures and architectures as membrane materials, typically in conjunction with controlled phase separation methods (Section 8.2.5), and/or to employ templating and self-assembly processes that are linked to the progress in microtechnology and nanotechnology (Sections 8.2.4–8.2.6). Mixed matrix membranes are only briefly covered (Section 8.2.7). Surface functionalization of porous membranes can add another dimension to their functionality for special separation and integrated processes (Section 8.2.8). The current trends toward porous membranes with advanced functionality by highly controlled porosity (even down to the subnanometer range) and/or by the combination of tailored pore structures with special functional groups or architectures will be emphasized. Reference is made to important examples of membranes that are produced and applied on the industrial scale (Section 8.3).

8.2.1 Sintering Polymeric Powders

Porous polymeric materials can be obtained by sintering polymer particles. A powder of a thermoplastic polymer is pressed into the desired shape and then heated to a temperature that is close to the melting temperature of the material so that the particles are fused (sintered) together. This process has the advantage that no solvents are required. However, it is limited to the preparation of flat-sheet or tubular membranes, and the control over pore structure is limited. Examples of polymers used for this process are PTFE and polyolefins. Porosity depends very much on particle size and size distribution; typical ranges of membrane pore size and porosity are 2–20 μm and 20%–50%, respectively. With such characteristics, the resulting membranes can be used for some crude microfiltration processes or as supporting materials for membranes with a finer pore or barrier structures.

8.2.2 Extrusion of Polymer Films

The extrusion of a thermoplastic polymer from its melt is a widely established general way for processing synthetic plastics. Of course, many applications of the resulting nonporous materials, for example, for packing purposes, can be described in the context of membrane (barrier) properties, but this is beyond the scope here. Of interest, however, are processes in which the film formation by extrusion is followed by or combined with (controlled) formation of a porous (barrier) structure.

8.2.2.1 Pore Formation by Stretching. Porous membranes as flat sheets or hollow fibers can be obtained if, during or after the solidification upon cooling, stretching of the nonporous precursor film or hollow fiber is done in one or two directions. Typically, the resulting materials have to undergo a thermal posttreatment in order to reduce internal stress in the porous membrane. This process is limited to semicrystalline polymers because the crystalline domains ensure the integrity of the

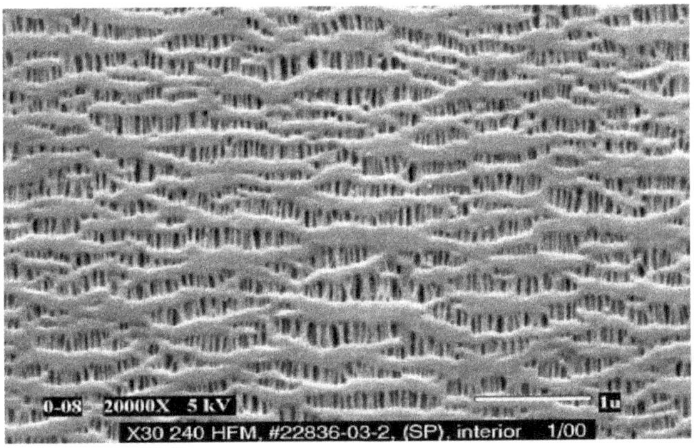

FIGURE 8.3 Inner surface of a polypropylene (PP) hollow-fiber microfiltration membrane with pore size 0.03 × 0.6 μm² prepared by extrusion/stretching (Celgard X30-240, Membrana GmbH, Wuppertal, Germany). The scale bar is 1 μm.

matrix between the opened amorphous domains. Depending on the microstructure of the polymer, a relatively regular pore structure and morphology can be achieved; typical features are slit-like pores for membranes where the effects of uniaxial stretching dominate (Fig. 8.3). This process is used for attractive barrier polymers that can hardly or cannot be processed into a porous structure by phase separation methods (see Section 8.2.4). As for sintering (see Section 8.2.1), this is an intrinsically solvent-free manufacturing process. Additives such as softening agents or particles, respectively, can be used to promote the formation of desired pore structures via "tailored" microstructures of the polymer or its function as a "pore template" [8].

Typical examples of membrane materials with great industrial relevance are PP (see Fig. 8.3), conventional or ultrahigh-molecular-weight polyethylene (PE), and PTFE. Depending on the degree of stretching, membranes with average pore sizes between 0.02 and 10 μm can be produced. For some applications, it is attractive that the membranes can be made relatively thin (down to about 10 μm). The resulting membranes are used in some special ultrafiltration and microfiltration applications (e.g., for aggressive streams in the semiconductor industry), as membrane contactors (e.g., in blood oxygenation), as support materials for composite membranes (e.g., as battery separators), and as barrier materials in textiles (see Section 8.3).

8.2.2.2 Pore Formation by Track Etching. Membranes with very regular pores of sizes down to around 10 nm can be prepared by track etching [17]. A relatively thin [<35 μm] polymer film (typically from poly(ethylene terephthalate) [PET] or aromatic polycarbonate [PC]) is first bombarded with fission particles from a high-energy source. These particles pass through the film, breaking polymer chains and creating damaged "tracks". Thereafter, the film is immersed into an etching bath (strong acid or alkaline), so that the film is preferentially etched along the tracks,

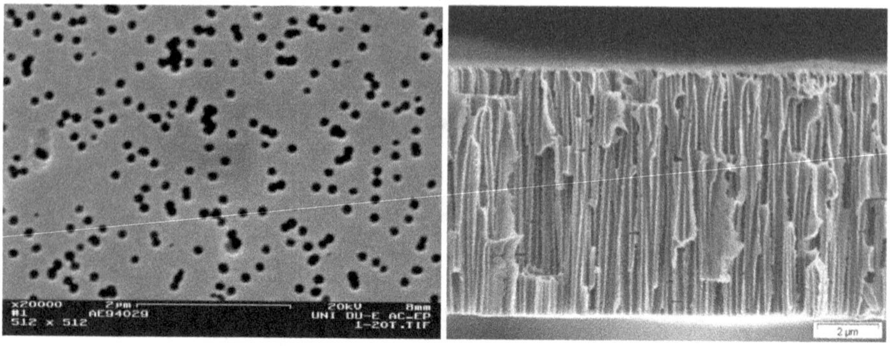

FIGURE 8.4 PET track-etched membranes. **(Left)** top view and **(Right)** cross section of a membrane with a nominal pore diameter of 100 nm (Rotrac, Oxyphen GmbH, Dresden, Germany). The scale bars are 2 μm.

thereby forming pores. The pore density is determined by irradiation intensity and exposure time, whereas etching time determines the pore size. However, the microstructure, degree of crystallinity, and film orientation also have a significant effect on the quality of the pore structure. The advantage of this technique is that uniform and cylindrical pores with very narrow pore size distributions can be achieved (Fig. 8.4). Other pore shapes—in particular conical or double conical—have also been produced by using special etching conditions [18]. In order to avoid the formation of double or multiple pores, which are produced when two nuclear tracks are too close together, the membrane porosity is usually kept relatively low, typically less than 10%.

Track-etched membranes are commercially available with pore sizes from 10 μm to 10 nm and thicknesses between 35 and 6 μm. Such membranes with an isoporous cylindrical structure (confirmed by independent characterization) are also frequently used as model systems for fundamental investigations of the interplay among pore size, functionalization of the pore surface, and barrier or transport properties [9,18]. Recently, PET track-etched membranes have been functionalized with controlled, surface-initiated graft copolymerizations to obtain advanced stimulus-responsive macroporous materials (see Section 8.2.8) [19,20]. Those membranes are the only commercial polymeric membranes with a very narrow pore size distribution. However, industrial applications in microfiltration or ultrafiltration—for example, for the fractionation of macromolecules in solution—are hindered by the low porosity and isotropic pore shape (low permeability), significant fouling (mainly pore blocking), and relatively high price of the material. Nevertheless, track-etched membranes are used in some special analytical applications (e.g., for size-based fractionation and quantitative analysis of (bio)particles; see Section 8.3).

8.2.2.3 Pore Formation by Foaming. Manufacturing of foams from polymers, mainly for insulation materials, is an important means of plastic processing. Various

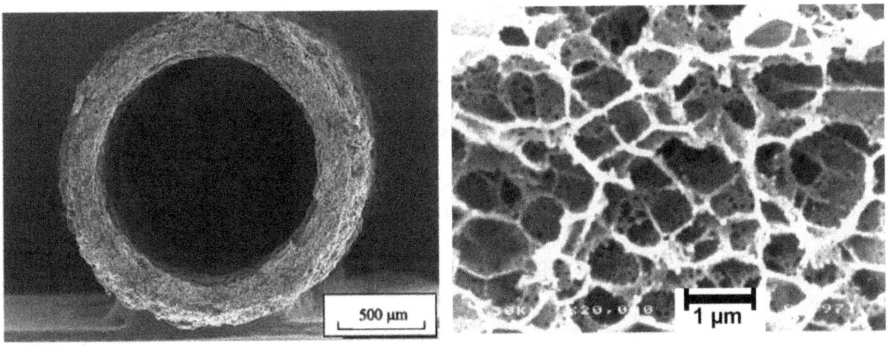

FIGURE 8.5 Porous membranes prepared by extrusion from a pressurized polymer melt in combination with foaming by dissolved carbon dioxide. **(Left)** Overview of the cross section of a polycarbonate hollow fiber. (Reprinted from Reference 21 with permission. Copyright 1999 Elsevier Ltd.) **(Right)** Detail from the cross section of a polysulfone membrane obtained using a small fraction of tetrahydrofuran (THF) as a "pore opener". (Reprinted from Reference 22 with permission. Copyright 1999 Elsevier Ltd.)

physical and chemical "blowing" agents are used, depending on the structure of the polymers and final application. For the preparation of porous membranes, such methods have gained larger interest because solvent-free manufacturing processes could ultimately be developed. The main challenges with respect to useful porous membranes are to achieve high porosities and adjustable pore sizes and, furthermore, to obtain "foams" with interconnected pores. Important achievements have been made, demonstrating that porous polymeric membranes can indeed be produced by extrusion of the pressurized melt with supercritical carbon dioxide as a "blowing" agent. Hollow-fiber membranes, which are in principle suited for microfiltration, have been prepared from polycarbonate in a continuous process (Fig. 8.5 [21]). Flat-sheet membranes from polysulfone with highly interconnected cells were obtained by batch extrusion using carbon dioxide with traces of organic solvents as a "pore opener" [22].

Overall, the feasibility of the technology was demonstrated for a number of interesting thermoplastic membrane polymers and this could also be extended to other foaming agents. However, more work toward a theoretical understanding of all the effects involved in controlling the resulting pore structures and with respect to industrial implementation has to be done before this method could become relevant for industrial membrane manufacturing. The resulting membranes could be of interest for microfiltration and as membrane contactors or supports for composite membranes.

8.2.3 Electrospinning

Technologies for spinning fibers from synthetic polymers have been the basis for preparing woven or nonwoven materials for traditional and novel ("industrial") textile applications. Some polymeric nonwoven materials are also important for polymeric membranes because they are used as supports to increase the mechanical strength of reverse osmosis, nanofiltration, and ultrafiltration membranes. Electrospinning is a radically novel technology that has been developed very rapidly and investigated during the last decade. The process can be summarized as follows. An electric charge is used to draw a fiber from a reservoir containing a polymer melt or solution. As a jet of charged fluid polymer sprays out the bottom of a nozzle, an electric field forces the stream to whip back and forth, stretching the fibers so that their diameters—several 10s of micrometers—shrink by several orders of magnitude. The fiber forms a thin porous membrane as it hits the surface below the nozzle. Many synthetic or biopolymers can be used. Such electrospun "membranes" have a unique combination of stretchiness and strength and are easy to handle, making them suitable for a wide range of applications. Due to the small fiber diameter—down to around 10–100 nm— the resulting nonwoven-like structures could also have filter properties in the range of microfiltration and, potentially, ultrafiltration. Because such "membranes" can be prepared with a very high porosity (up to 90%), they are already used as effective particle filters for treatment of gaseous streams. Electrospun polysulfone membranes with a maximum effective pore diameter of ~5 μm were evaluated as prefilters for water treatment [23]. Electrospun PET membranes were evaluated for the clarification of juice, a typical microfiltration process, and shown to yield higher fluxes at the same rejection [24]. An example of a composite membrane based on an advanced nanofiber layer architecture is shown in Fig. 8.6 [25]. This membrane exhibited a much higher flux than conventional nanofiltration membranes at the same rejection, but it should be noted that the membranes were characterized using an oil-in-water emulsion and that for such (and other nanofiltration and ultrafiltration) separations, the thin coating

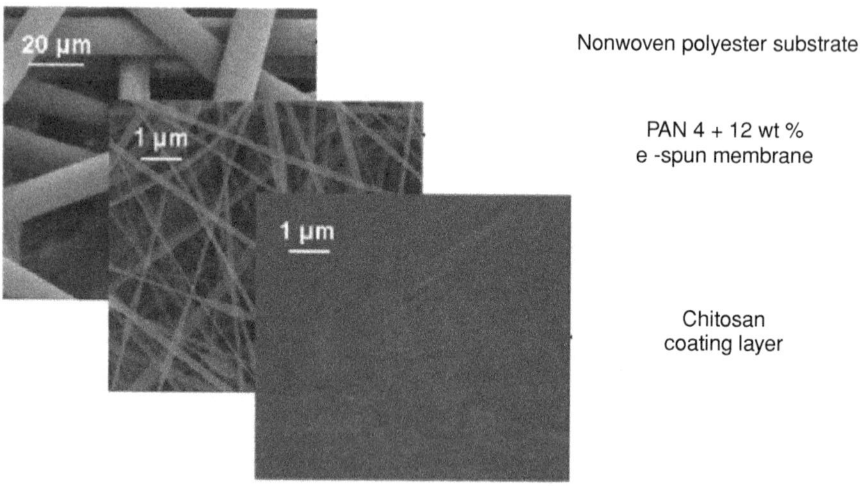

FIGURE 8.6 Three-tier composite membrane for ultrafiltration with a conventional nonwoven material as a mechanical support, a highly porous and permeable two-layer electrospun PAN nanofiber layer, and a thin protective chitosan top layer. (Reprinted from Reference 25 with permission. Copyright 2006 Elsevier Ltd.)

layer (here chitosan) is the actual size-selective barrier and the effective protective layer ensuring low membrane fouling (which would dramatically reduce membrane performance by blocking the inner fibrous filter structure).

This technology holds even more promise because electrospinning from polymer solutions can be combined with phase separation, leading to hierarchical pore structures with porosity in the nanofibers superimposed on the nonwoven macropore structure. Molecularly imprinted nanofiber "membranes" are among the first examples [26] (see Section 8.2.4.5).

The first commercial products were coarse microfiltration membranes for particle removal from gas streams (see Section 8.3). It is expected that many more microfiltration and ultrafiltration membranes for special applications with gaseous and liquid feed streams will follow. Other emerging membrane-related applications are the use as scaffolds for tissue engineering and as affinity membrane adsorbers for bioseparations.

8.2.4 Phase Separation of Polymer Solutions

Most of the industrially relevant and emerging polymer membranes are prepared by phase separation (PS). The method is often called "phase inversion," but it should be described as a phase separation process: A single-phase solution containing the membrane polymer is transformed by a precipitation/solidification process into two separate phases (a polymer-rich solid and a polymer-lean liquid phase). Before the solidification, a transition of the homogeneous liquid into two liquids (liquid–liquid demixing) typically occurs. The "protomembrane" is formed from the solution of the membrane polymer by casting or extruding a film on a suitable substrate or by extruding ("spinning") a liquid hollow fiber together with a bore fluid through a spinneret. For industrial implementation on a large sale, it is important that continuous processes be relatively easy. In particular, the spinning of rather sophisticated hollow fibers has strongly benefited from the comprehensive knowledge of fiber spinning for textile applications. Depending on the way the polymer solution is solidified, the following techniques can be distinguished:

1. *Nonsolvent-induced phase separation (NIPS).* The polymer solution is immersed in a nonsolvent coagulation bath (typically water); demixing and precipitation occur due to the exchange of solvent (from the polymer solution) and nonsolvent (from the coagulation bath), that is, the solvent and nonsolvent must be miscible.
2. *Vapor-induced phase separation (VIPS).* The polymer solution is exposed to an atmosphere containing a nonsolvent (typically water); absorption of nonsolvent causes demixing/precipitation.
3. *Evaporation-induced phase separation (EIPS).* The polymer solution is made in a solvent or in a mixture of a volatile solvent and a less volatile nonsolvent, and solvent is allowed to evaporate, leading to precipitation or demixing/precipitation.

4. *Thermally induced phase separation (TIPS).* A system of polymer and solvent is used that has an upper critical solution temperature; the solution is cast or spun at high temperature, and cooling leads to demixing/precipitation.

Some important principles and ways to tailor membrane pore structures via PS methods will be briefly discussed in the following sections. Only the NIPS and TIPS processes are also compatible with the preparation of (self-supporting) capillary or hollow-fiber membranes; the formation of a solid membrane in the other two processes (VIPS and EIPS) is too slow. With typical industrial polymers, all the approaches yield membranes with relatively broad barrier pore size distributions.

8.2.4.1 *Nonsolvent-Induced Phase Separation.* The cross-sectional structure of an anisotropic membrane is crucial in order to combine the desired selectivity (by a barrier with pores in the lower-nanometer range or by a nonporous polymer) with high fluxes: The top layer acts as a thin selective barrier, and a porous sublayer provides high mechanical strength. Integrally "asymmetric" membranes were discovered by Loeb and Sourirajan [27]. This finding was the first breakthrough for commercial membrane technology, that is, such reverse osmosis membranes from cellulose acetate (CA) showed much higher fluxes than the previously produced ones from the same polymer. This method involves (1) polymer dissolution in single or mixed solvent, (2) casting of the polymer solution as a film ("protomembrane") on a suitable substrate (or spinning as a free liquid film for hollow fibers), (3) precipitation by immersion in a nonsolvent coagulation bath, and (4) posttreatments such as rinsing, annealing, and drying (Fig. 8.7). The membranes resulting from this process have, typically, a very thin (<1 μm, often even <100 nm) top skin layer (selective barrier), which is either nonporous or porous (Fig. 8.8a). However, porous membranes with a more isotropic cross-section morphology can also be obtained (Fig. 8.8b).

The selection of the materials and the discussion of mechanisms for phase separation are based on ternary phase diagrams of the three main components (polymer, solvent, and nonsolvent); a pronounced miscibility gap (unstable region) is an

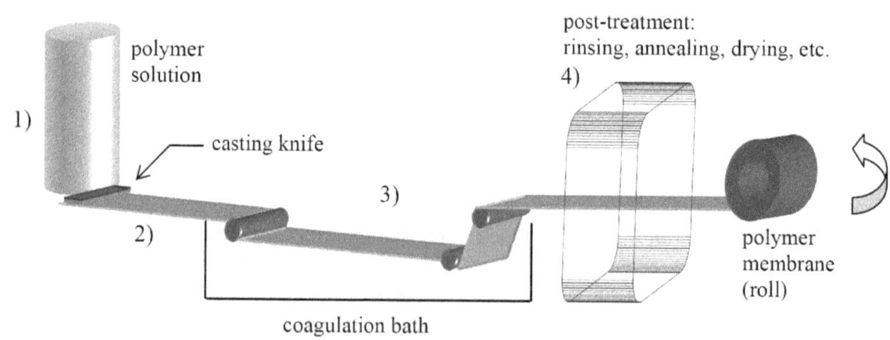

FIGURE 8.7 Schematic depiction of the continuous manufacturing process of polymeric membranes by the NIPS process.

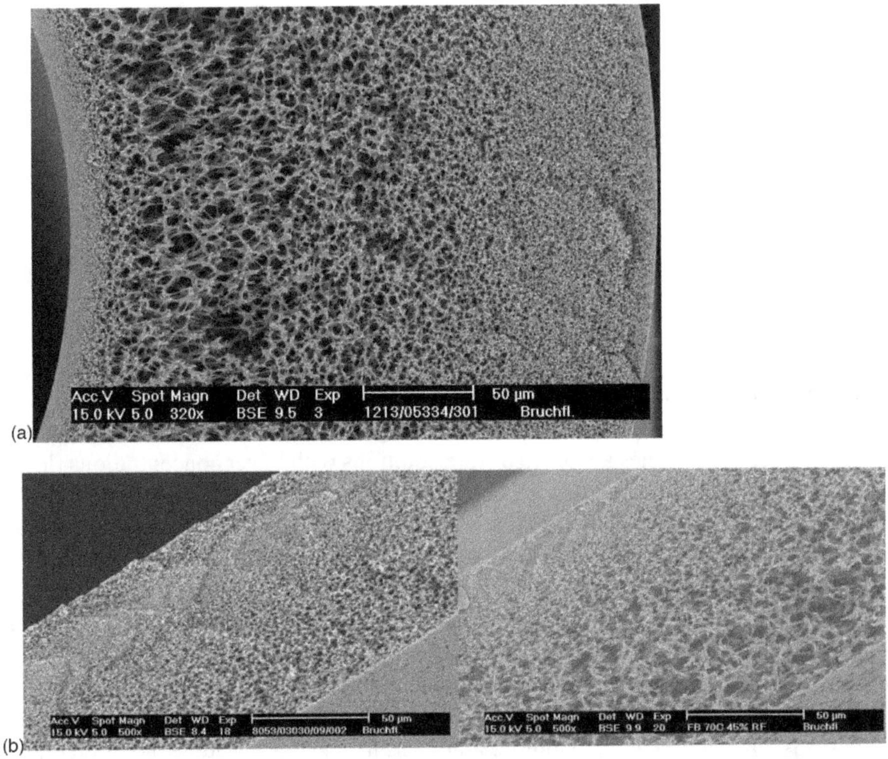

FIGURE 8.8 Cross sections of three different membranes prepared from PES via NIPS. (**a**) A capillary ultrafiltration membrane (UltraPES). (**b**) Flat-sheet microfiltration membranes (**left**: MicroPES; **right**: DuraPES) with a nominal pore size of 0.2 μm (all from Membrana GmbH, Wuppertal, Germany). All the scale bars are 50 μm.

essential precondition. Besides thermodynamic aspects, the onset and rate of precipitation in the liquid film (both are different, depending on the distance to the plane of first contact with the coagulation bath) are also important; the mass transfer (nonsolvent inflow and solvent outflow) can have tremendous influence. Two mechanisms are distinguished: (1) instantaneous liquid–liquid demixing, which will result in a porous membrane, and (2) delayed onset of liquid–liquid demixing, which can result in a membrane with a nonporous barrier skin layer [28]. The rate of precipitation decreases from the top surface (in most cases, the plane of first contact with the coagulation bath will be the barrier in the final membrane) to the bottom surface of the cast film. As precipitation slows down, the resulting pore size increases because the two phases have more time to separate. In practice, most systems for membrane preparation contain more than three components (e.g., polymer blends as materials and solvent mixtures for casting solutions and coagulation baths). Consequently, the mechanisms can be very complex and are still under intense scientific investigation and discussion [29, 30]. In industrial manufacturing, the NIPS process is most frequently applied for membrane polymers such as PES, PSf, PVDF, PAN, PA, PI, and CA, and membranes with either anisotropic or isotropic cross sections are obtained. Important variables to consider in the control of membrane characteristics are as follows.

Characteristics of the casting solution. Most important is the selection of a suitable solvent for the polymer; the strength of mutual interactions is inversely proportional to the ease of precipitation by the nonsolvent. Polymer concentration also plays a vital role in determining membrane porosity. Increasing polymer concentration in the casting solution leads to a higher fraction of polymer and, consequently, decreases the average membrane porosity and pore size. In addition, increasing the polymer concentration could also suppress macrovoid formation and enhance the tendency to form sponge-like structures. However, this can also increase the thickness of the skin layer. Even though details depend on the properties of the membrane polymer, UF membranes can be obtained within a range of polymer concentrations of 12–20 wt%, MF membranes are formed from solutions with lower concentrations (lower limit depending on the polymer), whereas NF/RO membranes are typically prepared from casting solutions with polymer concentrations $\geq$20 wt%.

Solvent/nonsolvent system. The solvent must be miscible with the nonsolvent (here an aqueous system). An aprotic polar solvent like *N*-methyl pyrrolidone, dimethyl formamide, dimethyl acetamide, or dimethylsulfoxide (DMSO) is preferable for rapid precipitation (instantaneous demixing) upon immersion in the nonsolvent water. As a consequence, a high-porosity anisotropic membrane can be achieved. For slow precipitation, yielding low porosity or nonporous membranes, solvents having a relatively low Hildebrand solubility parameter, like THF or acetone, are preferable.

Additives. For certain purposes, an additive or modifier is added to the casting solution. Indeed, this additive can determine the ultimate performance of the membrane and is often not disclosed for commercial membranes. Usually, additives include (1) cosolvents with relatively high solubility parameters (such a solvent can slow down the precipitation rate and higher rejection is achieved); (2) pore-forming agents such as poly(vinyl pyrrolidone) (PVP) or poly(ethylene glycol) (PEG) (these hydrophilic additives can enhance not only membrane pore size, but also membrane hydrophilicity; at least partially, these polymers form stable blends with membrane polymers such as PSf or PES); (3) nonsolvents (should be added only in such amounts that demixing of the casting solution does not occur; promotes formation of more porous structures and could also reduce macrovoid formation); and (4) crosslinking agents in the casting solution (less frequently used, but could also reduce macrovoid formation).

Characteristics of the coagulation bath. The presence of solvent in the coagulation bath can slow down the liquid–liquid demixing rate. Consequently, a less porous barrier structure could be obtained. However, the opposite effect can also occur, that is, addition of solvent could decrease the polymer concentration (in the protomembrane), leading to a more open porous structure. The amount of solvent to be added strongly depends on the solvent–nonsolvent interactions. As the mutual affinity of the solvent and nonsolvent increases, more solvent is required to achieve an effect on membrane structure. For example, in preparing CA membranes, the content of solvent needed in a coagulation bath for a DMSO/water system is higher than for a dioxan/water system. Instantaneous demixing resulting in a porous structure can be obtained by better miscibility between solvent and nonsolvent. In contrast, less miscible solvent/nonsolvent combinations yield a more nonporous structure. Furthermore, addition of solvent to a coagulation bath could also reduce the formation of macrovoids, leading to the desired, more stable sponge-like structure of the supporting layer.

Exposure time of protomembrane before precipitation. The effects of exposure to the atmosphere before immersion depends on the solvent properties (e.g., volatility, water absorption), atmospheric properties (e.g., temperature, humidity), and the exposure ("open") time. This step (i.e., combination of EIPS or VIPS with NIPS) has significant effects on the characteristics of the skin layer and the degree of anisotropy of the resulting membrane [30]. For instance, the cross-sectional pore structure for various commercial microfiltration membranes from the same polymer (PES) with the same nominal pore size can be very different, and the conditions during the "open" time, that is, the structures formed in the "protomembrane" (see Fig. 8.7), are crucial parameters [31].

Overall, the NIPS process is extensively used for the industrial manufacture of many microfiltration and most ultrafiltration membranes and of supports for composite membranes for reverse osmosis, gas separation, and other processes.[4] Furthermore, many membranes for contactors, including the base materials for membrane adsorbers, are also prepared by this process (see Section 8.3).

8.2.4.2 *Evaporation-Induced Phase Separation.* Partial evaporation of solvent from the cast film ("protomembrane") can contribute to the formation of a pronounced "asymmetric" structure (see Section 8.2.4.1). The complete evaporation of the solvent from a polymer solution will result in a nonporous film (e.g., as a selective layer in a composite membrane with a nonporous barrier). This preparation is relevant for polymer membranes for gas separation or pervaporation. Consequently, this is also the method of choice for the preparation of microporous barrier layers from polymers with inherent microporosity (PIMs [4]; see Section 8.1). However, for evaluation of gas separation performance, extraction and conditioning steps are important to obtain PIM membranes with a microstructure that is not significantly influenced by phase separation phenomena [32].

When mixtures of solvents are used, macroscopic phase separation can occur. Often, the resulting "latent" porous structure is "fixed" by quenching (and solvent exchange) with a nonsolvent for the polymer that is miscible with the remaining polymer solvent. Because of the rather slow process (too slow for the preparation of capillary or hollow-fiber membranes with an integral structure), macroporous flat-sheet membranes are usually obtained. Typical examples of polymers processed with this method are cellulose nitrate and cellulose acetate [33]. This leads to membranes

[4]For instance, most state-of-the-art composite membranes for reverse osmosis with a barrier layer from polyamide are prepared on a porous support from polysulfone obtained via NIPS.

for special applications such as blotting or lateral flow/immunochromatographic analyses, as well as to microfiltration membranes (see Section 8.3).

8.2.4.3 Thermally Induced Phase Separation. For polymers with an upper critical solution temperature, controlled phase separation can also be achieved by cooling a liquid film from a polymer solution. This process is less frequently used. However for polymers that cannot be processed by NIPS because no suitable combination of solvent and nonsolvent is available, this method is a very valuable alternative. The most important polymers processed by this method are polyolefins, in particular PP (solvents in industrial manufacturing processes are mineral or vegetable oils). The actual mechanism and resulting structures can be more complicated because crystallization can also occur simultaneously with phase separation and solidification [34]. The range of pore sizes that can be achieved is from 0.05 μm to several micrometers. It has also been demonstrated that modifications of the PP structure (configuration or comonomer content) that are accessible by metallocene catalysis ("tailor-made PP") lead to tunable changes of macroporous membrane morphology under identical TIPS conditions [35]. An example of the structure of an important industrial membrane prepared by the TIPS process is shown in Fig. 8.9.

The TIPS process is typically used to prepare membranes with a macroporous barrier, that is, for microfiltration, or as a support for liquid membranes and as a gas–liquid contactor (see Section 8.3).

8.2.4.4 Combination of Phase Separation with Polymer Crosslinking. There is an increasing demand for more-stable porous polymeric membranes, in particular for processes with aggressive aqueous streams or with organic solutions.

FIGURE 8.9 Cross section of a flat-sheet PP MF membrane with nominal pore diameter of 0.2 μm prepared by the TIPS process (Accurel 2E HF, Membrana GmbH, Wuppertal, Germany).

This is less relevant in the area of microfiltration because a number of membranes from stable polymers such as PTFE, PVDF, and polyolefins are available (however, most of these polymers cannot be processed into membranes with smaller barrier pores; see Sections 8.2.2, 8.2.4.1, and 8.2.4.3).

On one hand, this provides opportunities for novel ultrafiltration membranes made from more specialized polymers. One important example is polyetherimides, which were developed for ultrafiltration and nanofiltration applications involving a variety of organic solvents [36, 37]. On the other hand, stabilizing the polymer membrane structure by chemical crosslinking can significantly enhance its solvent stability. However, for ultrafiltration membranes this can be rather complicated because the fine pore structure formed in the processing step (typically NIPS; see Section 8.2.4.1) should be preserved. A very promising strategy for such postformation stabilization of ultrafiltration membranes with a pore structure "tailored" by NIPS was proposed recently [38]. A copolymer of PAN with a relatively small amount of glycidyl methacrylate (GMA) was synthesized such that the membrane formation from the copolymer P(AN-*co*-GMA) was still controlled by the properties of the PAN, which is a highly versatile membrane polymer. The pore structure can be stabilized in a three dimensional network via reaction with ammonia as a bifunctional or trifunctional crosslinking agent. The pore morphology of the membrane is not changed by exposure to an organic solvent because the reaction can be performed in an aqueous solution. The small size of the reactant ensures that a high conversion is also attained in the bulk of the solid polymer. The resulting crosslinked membranes had the same cross-sectional pore structure and the water permeability was only somewhat reduced. However, the chemical stability was greatly enhanced so that these membranes could even be used for UF separations of strongly acidic and alkaline aqueous solutions and could be used with most organic solvents. Properties of the crosslinked P(AN-*co*-GMA) membranes can be adapted to the requirements of various UF or NF processes where both high separation performance (selectivity and flux) and membrane stability are critical.

8.2.4.5 Combination of Phase Separation with Molecular Imprinting of Polymers. Additives are frequently used in the preparation of porous membranes by PS methods (see Section 8.2.4.1), and the effects of some additives have also been discussed in terms of a "porogen effect," that is, templating of pores during solidification and subsequent removal of the template. A very attractive approach to polymeric materials with "tailored nanoporosity" is molecular imprinting [39]: Molecular templates are used to produce their "replica" in polymers during the formation or solidification of a functional polymer. After removal of the template, these "imprints" can be used as affinity binding sites; in some cases impressive affinities and selectivities have been achieved, so that these molecularly imprinted polymers (MIPs) have also been termed "plastic antibodies". Such nanoscale porosity with "concave" bindings sites can be obtained not only by the most often used *in situ* crosslinking polymerization in the presence of a template (see Section 8.2.6.2), but also via solidification of functional polymers from solutions containing a template, that is, by phase separation.

The combination of molecular imprinting and preparation of affinity membranes toward MIP membranes has been investigated for almost two decades [40]. Yoshikawa et al. [26, 41] prepared imprinted membranes with very low overall porosities via EIPS of solutions of functional polymers, for instance, in the presence of single enantiomers of amino acids as templates. However, the (micro)porous structure of these membranes has not been characterized in detail. Some of the obtained MIP membranes exhibited very interesting transport selectivities, which could be explained by a fixed carrier transport mechanism; therefore, a continuous separation of, for instance, enantiomers by such MIP membranes could be possible [40,42]. Pioneering work on macroporous imprinted membranes via NIPS of functional polymers in the presence of template molecules (also named "scaffold imprinting") was done by Kobayashi's group [43,44]. Because "imprints" are formed on the surface of large pores that do not hinder nonselective diffusion, such MIP membranes cannot be used for continuous transport-selective separations (see Fig. 8.2c). However, very interesting applications could be based on the tailored membrane adsorber properties.

The selection of polymers for which molecular imprinting has been achieved has been extended to many of the commonly used membrane materials: CA, PA, PAN, and PSf; exceptions are the hydrophobic and almost nonfunctional polyolefins, PVDF, and PTFE [9]. The use of a polymer blend in order to tailor both pore structure and binding sites was also investigated. Porous membranes were prepared by NIPS of cellulose acetate/sulfonated polysulfone (CA/SPSf) blends with various compositions and the dye rhodamine B (RhB) as a template, all in DMSO, with a water/DMSO mixture as the coagulation bath [45]. Detailed characterizations of the resulting structures and binding properties were performed (e.g., Fig. 8.10). Significant differences in pore structure between MIP and nonimprinted (NIP) membranes as a function of the polymer blend composition were found; the magnitude of these differences correlated clearly with the imprinting efficiency. In particular, for the CA/SPSf 95/5 blend, which yielded the largest imprinting effect, the characteristic polymer nodule size was much smaller for the MIP than for the NIP (cf. Fig. 8.10e vs. Fig. 8.10b). Hence, fixation of "imprints" occurred mainly in small polymer particles which were formed during very fast demixing upon contact with the nonsolvent. Furthermore, the addition of the template to the CA/SPSf blend solution seemed to facilitate the demixing after contact with the precipitation bath water, presumably via a complexation of the RhB with the sulfonic acid groups of SPSf. The most important conclusion from that work was that the detailed morphologies in correlation with the well-studied mechanisms of membrane formation via NIPS (see Section 8.2.4.1) were successfully used to shed light onto the mechanism of molecular imprinting through the solidification of functional macromolecules.

8.2.4.6 Phase Separation in Microstructures (Phase Separation Micromolding). The combination of microfabrication with membrane technology is attractive with respect to the resulting materials and the potential applications, for instance, microfluidic systems [46]. Here, the focus is on novel porous polymeric materials. A very interesting replica technique for polymeric microsieves and membrane-like porous materials with other shapes and features, so-called "phase

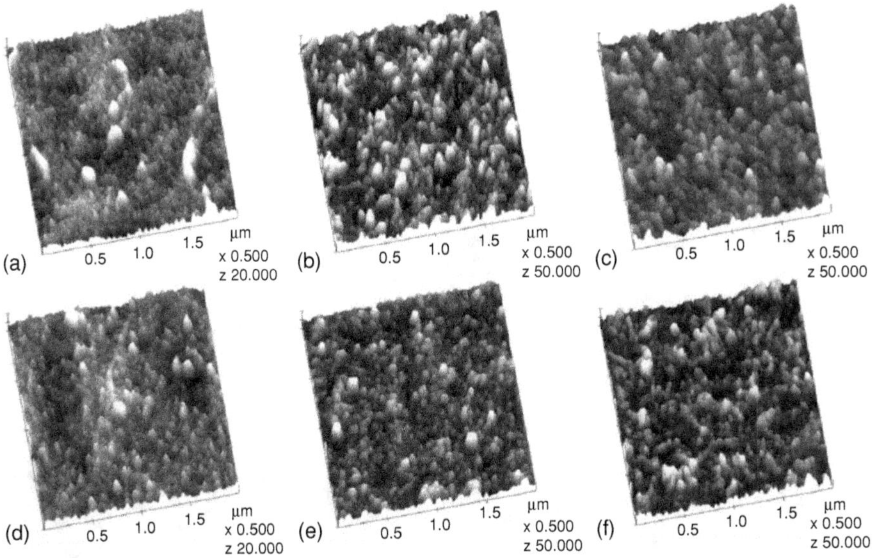

FIGURE 8.10 SFM micrographs of top layers of asymmetric porous membrane adsorbers from CA/SPSf blends of various compositions. **(a)** 100:0 NIP; **(b)** 95:5 NIP; **(c)** 85:15 NIP; **(d)** 100:0 MIP; **(e)** 95:5 MIP; **(f)** 85:15 MIP. (Reprinted from Reference 45 with permission. Copyright 2005 the Royal Society of Chemistry.)

separation micromolding" (PSμM), was introduced by Wessling's group [47]. Solutions of a typical membrane polymer (e.g., PSf) have been cast into microfabricated molds (e.g., for a porous film), phase separated, and, due to some shrinkage, released without major defects from the mold (Fig. 8.11). Very high porosity could be combined with low thickness (a few 10s of micrometers) and, currently, the smallest pore sizes (a few 100s of nanometers) are determined by photolithographic technologies for mold manufacturing. Specific data about membrane properties are rather limited.

A micromolded membrane with a very regular array of pores having a diameter of 1 μm has been shown to achieve a very precise fractionation of microparticles [48]. Another related example was reported recently: An array of silicon nanowires was used as template for preparation, via curing of the polymer precursor in the mold and subsequent etching of the silicon, of PI membranes with very regular pore geometries and diameters down to about 170 nm [49].

When this technology is improved further, including solving the problem of continuous manufacturing, the microfiltration membranes could become very attractive "plastic" counterparts of already established expensive inorganic microsieves for microparticle and colloid separation or fractionation [47]. However, other formats of such micromolded macroporous polymeric materials with inherent microporosity and mesoporosity can be produced and more diverse applications, for instance, in sensor or biomedical systems, can be envisaged [50].

FIGURE 8.11 Fabrication of a microsieve from polyimide by PSμM. (**a**) Mold with a 5-μm-wide and 20-μm-high pillar pattern on its surface. (**b**) PI replica on the mold after phase separation and prior to release; the pillars have perforated the polymer due to the shrinkage of the polymer layer toward the mold during phase separation. (**c**) Completely open porous microstructure following release from the mold. (Reprinted from Reference 47 with permission. Copyright 2005 Wiley-VCH Verlag GmbH & Co. KGaA.)

8.2.5 Novel Approaches Using Self-Assembly of Polymers with Controlled Architectures

Complex macromolecular substances, such as block and graft copolymers with a high degree of control of the macromolecule's chain length, are very attractive nanoscale materials. Many industrial block copolymers have been used as surface-active agents (surfactants) or compatibilizers in blends of engineering polymers for decades. Block copolymers as building blocks for highly ordered three-dimensional polymeric and inorganic structures have gained tremendous attention in the recent decade (recent reviews are available [51, 52]). Here, the focus is on the application of block copolymers and other polymers with controlled structures and architectures in the field of porous membranes. Two routes are followed: their use as (1) surface modifiers and (2) as bulk materials (also in blends with another polymer). The ultimate aim is to employ the self-assembly properties of these tailored building blocks to create novel polymeric barriers with well-defined microporosities and mesoporosities.

8.2.5.1 Block or Graft Copolymers as Surface Modifiers. As already described, polymeric additives are frequently used in NIPS-based industrial manufacturing processes for porous membranes; the most prominent examples are the addition of the hydrophilic homopolymer PVP or PEG to improve the properties of UF or MF membranes from PES or PSf (see Section 8.2.4.1). In addition to the strong influence on pore formation, which is intuitively described as a "porogen" effect, these polymers also enhance the surface hydrophilicity since a fraction of the additive will remain in the membrane. This effect, however, only has a limited stability, and the desired effect is lost due to leaching during application or cleaning of the membrane [9].

In that context, amphiphilic copolymers are very attractive from the membrane preparation and application point of view. If the macromolecular additive would show a pronounced surface segregation along with sufficient surface coverage and a strong anchoring in the polymer bulk, it should be possible to change the membrane surface characteristics with only a minor influence on the bulk (including pore) morphology and properties. In the last decade, "tailored" functional copolymers have been investigated with that motivation. Matsuura's group explored various different syntheses—for example, based on polyurethane chemistry—yielding different "surface-modifying macromolecules" (SMMs) with PEG or fluorinated polymer blocks; they were able to prove surface segregation after membrane formation via NIPS and demonstrate a significant improvement of membrane performance in various UF and MF processes [53]. Copolymers with phosphorylcholine side groups were used as surface-modifying additives in preparation of biocompatible membranes for ultrafiltration or dialysis [54]. Mayes' group prepared interesting copolymers via controlled (ATRP) graft copolymerization of PEG methacrylates (PEGMA) onto the membrane polymer PVDF; PVDF UF membranes with strongly reduced fouling were obtained by NIPS of solutions of PVDF with PVDF-*g*-PPEGMA as an additive [55]. Such PVDF-*g*-PPEGMA materials were also used as a separate barrier layer in an advanced NF composite membrane [6] (see Sections 8.1 and 8.2.5.2).

The effects of commercial block copolymers as additives for UF membranes have also been investigated. In particular, triblock copolymers (PEO-*b*-PPO-*b*-PEO) of polyethylene oxide/polyethylene glycol (PEO/PEG) with polypropylene oxide (PPO), also known under the trade name Pluronic, were studied in detail. With the very important membrane polymer PES, significant surface segregation of PEO (relative to PPO and PES) and a correlated increase in hydrophilicity and reduction in protein fouling tendency were confirmed, but along with an influence on the UF membrane pore structure [56, 57]. It also was demonstrated that for PES UF membranes prepared with the three additives PVP, PEG, and PEO-*b*-PPO-*b*-PEO (all with about same molar mass), the block copolymer Pluronic led to the best overall performance with respect to high hydrophilicity and low protein fouling in combination with the best long-term stability [58]. By adapting the NIPS conditions and identifying the best-suited Pluronic, it was also possible to prepare PES MF membranes with an isotropic pore structure having a high flux and a pronounced pore surface hydrophilicity (imposed by surface-anchored PEO) at the same time [59].

Overall, the use of tailored macromolecular additives consisting of a block for anchoring in the membrane matrix and a block for tuning the surface properties has become a very active field in academic research and in development by membrane companies.

8.2.5.2 Porous Membranes from Tailored Block or Graft Copolymers.

In recent decades, several groups worldwide have been actively pursuing the use of newly synthesized ("tailored") copolymers as membrane materials, with the specific motivation to obtain porous membranes with added functionality and/or improved control over the pore structure. It should be recalled that the search for novel polymers as membrane materials has always been most intense for membranes with nonporous barriers, where the intrinsic properties determine permeability and selectivity (membranes for gas separation are the most prominent example; see Section 8.1). However, that has typically been much less relevant for porous membranes, for which the material properties have been much more important for membrane integrity and stability (see Section 8.2). This last point has the immediate implication that novel "tailored" polymeric materials for porous membranes should also provide suitable mechanical properties, and this can impose significant additional challenges for the synthesis of (completely) suitable copolymers.

Two groups of "tailored" copolymers and two quite different approaches to membrane formation can be distinguished. With respect to polymer synthesis:

(i) Some research groups have been achieving interesting results using polymers with only a moderate control of macromolecular architecture, especially graft copolymers on established membrane polymers.

(ii) Other research groups have been using polymers with full control of structure and architecture, especially diblock copolymers and triblock terpolymers, synthesized using advanced "living" polymerization methods.

With respect to membrane preparation:

(a) Established methods such as film casting followed by EIPS, VIPS, or NIPS (leading to macrophase separation; see Section 8.2.4) have been used with the additional aim of finding conditions under which a controlled microphase separation may lead to improved control over membrane pore structure or additional pore functionality.

(b) Films are prepared under conditions in which the equilibrium bulk or thin-film morphology with fully developed self-assembled order of the macromolecules is obtained. This strategy involves two additional challenges. First, the (hopefully) obtained ordered films contain only "templates" for pores, so that this template polymer must be selectively removed without destroying the ordered morphology (two main routes are chemical etching of one block and selective dissolution of a homopolymer compatible/mixed with the block forming the pore template). Second, not all morphologies will lead to a connected transmembrane pore structure (cylinders perpendicular to the film/membrane surface would be preferred, and bicontinuous gyroid structures would also be acceptable).

It is clear that approach (b) will only lead to the expected results if polymers with precise control over block structures and ratios at minimum polydispersity are used, that is, in combination with method (ii).

Combination of (i) and (a). There are numerous examples of novel functional copolymers with moderate control of architecture that have been processed into MF or UF membranes with added functionalities; the pore morphology can be similar to that of membranes from standard membrane polymers, and rational explanations for the influence of PS parameters onto pore structure can be given. Examples are the work of Wang et al. [60] featuring functional graft copolymers of PVDF or fluorinated polyimides and the work of Xu's group [61] exploring acrylonitrile-based copolymers containing, for instance, phospholipid moieties. The resulting filtration membranes had improved performance due to increased antifouling, antibacterial, and stimulus-responsive properties and increased biocompatibility. Another very promising example for the transformation of the microphase-separated morphology of relatively well defined copolymers into an advanced high performance separation membrane was given by Mayes et al. [62]. Using the graft copolymer PVDF-*g*-PPEGMA synthesized via ATRP (see Section 8.2.5.1), they prepared composite membranes by coating thin films on a support PVDF UF membrane and subsequent phase separation. The structural characterization by high-resolution electron microscopy and by separation performance in nanofiltration experiments suggested that hydrophilic "nanochannels" in a hydrophobic matrix, acting as a transmembrane barrier in the membrane skin layer and a hydrogel-like outer membrane surface, had been obtained. This membrane exhibited a very high flux and molecule selectivity according to size in the range of molar masses <500 g/mol, along with a minimal fouling tendency when used for concentrating oil/water emulsions. For various analogous composite membranes, cut-off values between 3 and 5 nm were determined by fractionation of gold-based nanoparticles in toluene or ethanol [6]. The latter data place the novel composite membranes clearly in the category of ultrafiltration membranes (see Figure 1), but the influences of solvent (organic solvent vs. water) on barrier properties are also well known for other, more conventional polymeric membranes (see Section 8.2.4.4).

Combination of (ii) and (a). An important innovation was reported by Peinemann et al. [63]. They used a well-defined block copolymer, polystyrene-*b*-poly(4-vinylpyridine) (PSt-*b*-P4VP), which was synthesized by sequential anionic polymerization. Membrane preparation was performed via the well-established NIPS process. A mixed solvent system was used, and the residence time in air before the macrophase separation by contact with the nonsolvent water was important (see Fig. 8.7). It had been hypothesized that selective evaporation of one solvent and the subsequent preorganization in the "protomembrane" led to the formation of unique microphase- and macrophase-separated structures, with a relatively thin isoporous layer on top

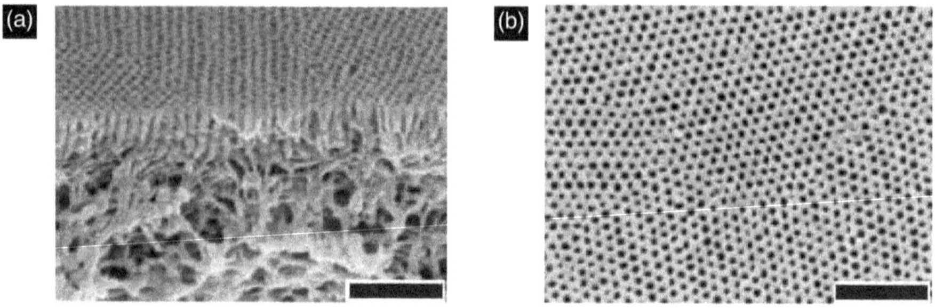

FIGURE 8.12 SEM images of an asymmetric PS-*b*-P4VP diblock copolymer membrane. (**a**) View of the edge between cross section and top layer surface. (**b**) View of the top layer surface. Length of the scale bar is 500 nm. (Reprinted from Reference 63 with permission. Copyright 2007 Nature Publishing Group.)

of a less ordered macroporous sublayer (Fig. 8.12). Water fluxes were, as expected for such an "asymmetric" membrane, in the range of ultrafiltration, but selectivities were not reported. Even though the detailed mechanisms are not yet known, this work indicates that well-defined block copolymers could, indeed, be transformed into membranes with controlled porosity by a manufacturing process that has already been scaled up by industry many times.

Another significant step toward ultrafiltration membranes with tunable permeabilities and selectivities was also based on well-defined block copolymers [64]. Asymmetric membranes were prepared via the NIPS process from polystyrene-*b*-poly(*N*,*N*-dimethylaminoethyl methacrylate) (PS-*b*-PDMAEMA) prepared via sequential living anionic polymerization. Due to the asymmetric cross section with a thin and defect-free porous separation layer and a macroporous volume structure, the membranes were self-supporting. In terms of water flux, the membranes were able to react in a reversible manner on two independently applicable stimuli, pH and temperature. Compared to the conditions under which the lowest water flux was obtained—low temperature and pH—activation of both triggers resulted in a sevenfold permeability increase (Fig. 8.13a). This can be explained with a microphase-separated structure in which polystyrene serves as the membrane matrix, whereas the pH- and temperature-sensitive minority block, PDMAEMA, rendered the pore surface double stimulus responsive (Fig. 8.13b). Even more important, the separation properties were tested through the pH-dependent ultrafiltration of various monodisperse silica particles with sizes of 12–100 nm at ambient temperature, and a reversible shift of the barrier pore cut-off from $>$22 nm at pH $\geq$ 6 to $\geq$15 nm at pH $=$ 2 was confirmed.

In another study, the effect of block copolymer structure (block ratio) and preparation conditions (composition of the solvent mixture, "open time" before immersion in a coagulation bath, cast film thickness) was studied in detail and macrophase separation was interpreted within the framework of theories for the EIPS/NIPS processes [65]. Of interest, ordered, microphase-separated morphologies were also formed (always remote from the skin layer surface) in membranes from a block copolymer with

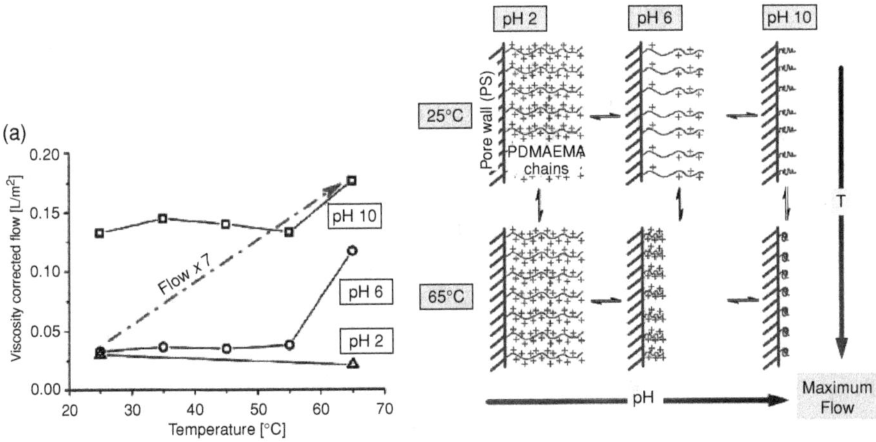

FIGURE 8.13 **(Left)** Water flux (corrected by viscosity as function of temperature) for asymmetric PS-*b*-PDMAEMA diblock copolymer ultrafiltration membranes. **(Right)** Schematic depiction of the different states of the inner part of the membrane pores depending on pH and temperature (idealized illustration, as PDMAEMA also is incorporated into the pore walls). (Reprinted from Reference 64 with permission. Copyright 2009 Wiley-VCH Verlag GmbH & Co. KGaA.)

higher PDMAEMA content, that is, better compatibility with the aqueous coagulation bath and, consequently, slower phase separation.

Combination of (ii) and (b). The ultimate aim is to create a regular high porosity with a tailored pore size and with a very narrow pore size distribution by block copolymer synthesis and processing via self-assembly controlled by block copolymer structure and architecture. The pioneering work toward porous block copolymer films with regular pores, which started in the late 1980s [66, 67], was reviewed recently [9, 68]. In particular, suitable methods and conditions for creating real pores from their polymeric "templates" in the nonporous precursors were explored. Despite the impressive microscopic, spectroscopic, and other characterization studies of the ordered porous morphologies based on diblock and triblock copolymers, a direct demonstration of the membrane function, that is, permeability measurements, or even evidence for selective permeation controlled by the nanoporosity of the polymer film, has only been accomplished recently for a few systems.

Liu et al. [69] reported thin films with self-assembled cylindrical "nanopores" prepared from a triblock copolymer, polyisoprene-*b*-poly(2-cinnamoylethyl methacrylate)-*b*-poly(*tert*-butyl acrylate) (ABC). For film preparation by solution casting and evaporation, the block copolymer was mixed with the homopolymer P*t*BA (homo C). After drying and annealing, the photosensitive block B could be used for UV crosslinking of the AB phase. Pores were obtained by extracting homo C with methylene chloride, and electron microscopy showed a very regular pore morphology. The obtained membranes were permeable for gas but not for liquid, which was related to a discontinuous structure of the pores over the entire film thickness. Hillmyer's group found that polystyrene-*b*-polylactide (PSt-*b*-PL) diblock copolymers can form hexagonally packed nanocylinders of PL in PSt, which can then be converted into pores by selective hydrolysis of the PL [70]. Pore diameters between ~15 and ~45 nm could be controlled by the molar masses and the block ratios of the copolymers. Based on that work, highly ordered monoliths with hydrophilic pores were prepared from PL-*b*-poly(*N*,*N*-dimethylacrylamide)-*b*-PSt triblock copolymers [71]. Alignment of the phase-separated polymer was achieved by using cooling from the melt in a channel die, and, finally, the PL was removed quantitatively, leaving the PSt matrix with the hydrophilic polyacrylamide covering the pore surface. A 2-mm-thick isotropic porous membrane was then characterized by gas (helium, argon, oxygen) and water permeability measurements [72]. Gas fluxes and ideal gas selectivities could be described by Knudsen diffusion through pores with a diameter of 13 nm (at 26% porosity). Water fluxes were in good agreement with this data when assuming Hagen–Poiseuille flow. This work is important because the potentially very promising separation performance of such isoporous polymeric membranes can be based on experimental data. It was postulated that such membranes could yield an improvement in the ultrafiltration separation factor by orders of magnitude at the same permeability relative to existing membranes (with broad pore size distributions). However, the films were still many orders of magnitude too thick, and pore alignment could not be achieved on decreasing the film thickness.

A breakthrough was achieved by Yang et al. [73]: With a polystyrene-*b*-poly(methyl methacrylate) (PS-*b*-PMMA) diblock copolymer blended with additional PMMA, they were able to form a thin (80 nm) film with hexagonally ordered PMMA pore templates on a silicon wafer grafted with a neutral polymer brush; after dissolving the wafer with hydrofluoric acid and transfer of the film to a commercial PSf MF membrane as a support, the PMMA was dissolved with acetic acid. A very regular porosity with a pore diameter of ~20 nm was achieved. Much higher fluxes than through a track-etched membrane with the same pore size, at the same (complete) rejection of a virus with a diameter of 30 nm, confirmed that novel block copolymer-based membranes were, for the first time, superior to state-of-the-art membranes in real separation experiments. More recently, the concept was modified using UV crosslinking of the PS matrix to increase the stability of the membranes by preparing a slightly thicker film (160 nm) that not only had a regular array of 20-nm pores on its outer surfaces, but also had mixed ordered cylinders in the bulk [74]. Both the less regular inner film structure and the crosslinking were crucial for the improved mechanical and chemical stability (Fig. 8.14). It was demonstrated that these membranes could withstand a transmembrane pressure of 2 bar and that they were also stable in toluene, DMF, acetic acid, and a 0.1 N NaOH solution without losing the capability to retain the virus particles completely.

Recently, a series of nanoporous, flexible, free-standing membranes with pore diameters between 5 and 30 nm was prepared from block copolymers PE-*b*-PSt via controlled crystallization of precursor films and subsequent etching of the PSt phase with fuming nitric acid [75]. The membrane with the smallest pore size exhibited

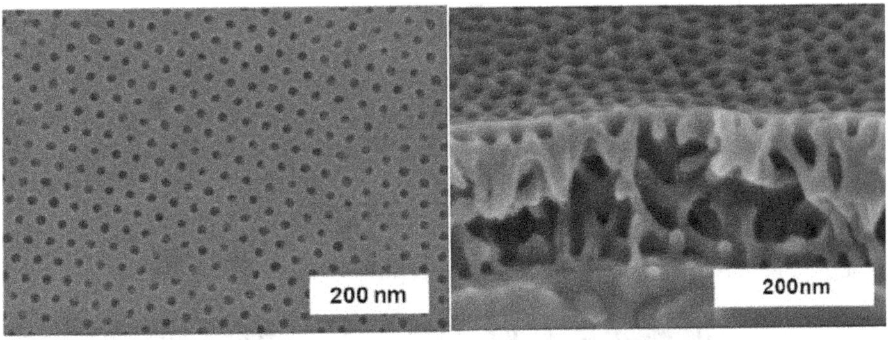

FIGURE 8.14 Selective barrier layer of a dimensionally stable and solvent-resistant PSt-*b*-PMMA/polysulfone composite membrane for high-flux virus filtration. (Reprinted from Reference 74 with permission. Copyright 2008 Wiley-VCH Verlag GmbH & Co. KGaA.)

size-selective diffusion for glucose containing serum albumin and completely rejected the protein.

8.2.6 Porous Membranes via *In Situ* Polymerization and Polymer Crosslinking

Polymerization of many monomers in bulk leads to nonporous polymers, while polymerization in monomer solutions (with the solvent/polymer interactions as a dominant factor) can lead to phase separation at certain monomer conversions; aggregation and agglomeration will then lead to porous polymers as product (see also Chapter 3). With simultaneous crosslinking (with the content of crosslinking comonomer as a dominant factor), *in situ* polymerization of monomer/crosslinker/solvent ("porogen") systems can yield porous monolithic polymers [76]. Such monoliths with bimodal pore size distributions and functional internal surfaces have been developed for various applications, such as for fast chromatographic separations, including those using membrane-like disks [77]. For the preparation of porous membranes, the following *in situ* polymerization approaches have been used or are being explored:

1. Formation of a porous polymer (agglomerate or monolith) by phase separation (induced by solvents or crosslinking).
2. Formation of a porous, molecularly imprinted polymer (MIP [39]; see Section 8.2.4.5); this is similar to the prior approach, but a template is added to the polymerization mixture (porosity is essential for MIPs because the "imprints" should be accessible for extraction of the template and rebinding of the solute during application).
3. Formation of a nonporous polymer around (self-assembled) "templates" for pores; removal of the template yields a porous structure.
4. Fixation of a self-assembled ("nanoporous") structure in a nonporous polymer.

All of these different approaches could, in principle, be performed to yield self-supporting or composite materials (on or in suitable porous supports), and all three membrane geometries (flat sheet, capillary/hollow fiber, capsule) are accessible. Similar to well-established interfacial polymerization toward non-porous polyamide barrier layers of composite membranes for reverse osmosis or nanofiltration [9], other *in situ* polymerization methods can also lead to ultrathin barrier layers (thickness of a few 10s of nanometers). One of the advantages of *in situ* preparation methods is that the desired membranes can be prepared "in place" and this is gaining increasing attention in the field of microfluidic systems [46].

8.2.6.1 Porous Membranes via* In Situ *Polymerization with Phase Separation. Typical polymer systems are based on radical polymerization of vinyl monomers, for example, the crosslinking copolymerization of acrylates or styrene derivatives, or on polycondensation, for example, the synthesis of (crosslinked) polyamides. In order to prepare conventional self-supporting membranes this approach is less attractive because (1) microfiltration membranes from phase separation of engineering polymers are readily available; (2) for membranes with smaller pores (in the mesopore and micropore range) thin barrier layers are desired—such anisotropic ultrafiltration membranes are easily accessible by phase separation methods (see Section 8.2.4); and (3) *in situ* preparation of thin porous layers from monomers *on* porous supports is not straightforward. On the other hand, the controlled preparation of porous, functional polyacrylate-based monoliths with high specific surface areas via *in situ* photoinitiated copolymerization *in* the pores of microfiltration membranes (PET and PP) was demonstrated; crucial for a defect-free pore filling was a premodification of the base membrane with a compatible grafted polyacrylate [78].

An example of the preparation of a separation membrane "in place" in a microfluidic system is the UV-initiated polymerization of a mixture of zwitterionic 2-(*N*-3-sulfopropyl-*N*,*N*-dimethylammonium)ethyl methacrylate and crosslinker methylene bisacrylamide (MBAA) using a focused 355-nm laser [79]. Microdialysis membranes were obtained and, by controlling the course of phase separation during crosslinking polymerization via the ratio between solvent (water) and nonsolvent (2-methoxyethanol), the molecular weight cut-off of the membranes could be engineered for different applications.

Interfacial polymerization reactions in emulsions can conveniently be used to prepare macrocapsules or microcapsules with (porous) membranes as walls. Extensive pioneering work was done with polyamides (e.g., nylon) by Okahata's group [80] and this research has been continued and extended by other groups [81,82]. In combination with the functionalization of the pores by grafting of stimulus-responsive polymers, microcapsules with "gated" walls were obtained; the contents of the capsule could be released as a function of, for example, pH, temperature, or glucose content in the surrounding solution.

8.2.6.2 Molecularly Imprinted Polymer Membranes. The combination of molecular imprinting via *in situ* polymerization with the preparation of porous membranes was promising because the conditions for preparing MIPs and porous polymeric monoliths, typically from acrylate-based monomers, are very similar. These affinity membranes could be used as a membrane adsorber or could be used for continuous (facilitated) transport (see Figure 2c). These membranes could also work as membrane "gates" where the specific binding sites could be used to increase or decrease membrane permeability [40]. Facilitated transport is only effective for relatively small pores, that is, when the nonselective diffusion through the membranes is slow relative to the selective transport via binding/dissociation to/from "imprints" (see Section 8.2.4.5). Overall, only relatively few examples of the preparation of MIP membranes with attractive separation properties via *in situ* polymerization have been reported [9, 40].

Mathew-Krotz and Shea [83] prepared free-standing MIP membranes by thermally initiated crosslinking copolymerization of a mixture of methacrylic acid and ethylene dimethacrylate. From SEM studies, a regular porous structure built up by polymer nodules with 50 to 100 nm diameters was discussed. The properties of the membranes were very interesting because a selective (facilitated) permeation of the template molecule and its derivatives could be observed. Kimaro et al. [84] prepared free-standing MIP membranes by thermally initiated crosslinking copolymerization of styrene monomers followed by leaching of a polyester used as "pore former" at a concentration of a few percent in the reaction mixture. SEM micrographs suggested the presence of isolated pores with diameters of up to 1 μm at a low density ($<2\%$). In line with permeation data showing a very large selectivity for the template uranyl ion, it was speculated that transmembrane channels had been obtained, induced by the presence of a removable macromolecular "pore former". Sergeyeva et al. [85] used oligourethane–acrylate macromonomers in imprinting polymerization mixtures in order to increase membrane flexibility and mechanical stability; self-supporting MIP membranes with a thickness between 60 and 120 μm could be prepared. Such membranes were characterized as barriers in a sensor system, and the response was explained by a "gate" effect, that is, the binding of the template changed the membrane permeability. Macroporous MIP films with higher permeabilities, obtained under various conditions, were very efficient in substance-specific membrane solid-phase extraction (i.e., as membrane adsorbers). Martin's group reported the preparation of a highly selective thin-film composite membrane by photoinitiated crosslinking polymerization of a MIP film on a porous inorganic support [86]. High transport selectivity for the MIP template—theophylline—was attributed to a facilitated transport through a defect-free nanoporous separation layer.

8.2.6.3 Porous Membranes via Templating. A novel strategy toward high-porosity macroporous membranes with uniform pore size distributions has been developed by Xu and Goedel [87] (see also Chapter 4). The so called "particle-assisted-wetting" method was used to prepare porous membranes by spreading a mixture of monodisperse hydrophobized silica particles and a commercially available non-volatile acrylate-based crosslinker monomer (trimethylolpropane trimethacrylate) on a water surface. Depending on the surface properties of the particles and the surface tension of the monomer (here both hydrophobic), the particles could assist wetting of the monomer by the water subphase to form an ultrathin uniform composite layer. UV irradiation was then used for photopolymerization and, after removing the particles by hydrofluoric acid, a thin porous membrane was obtained (Fig. 8.15). The pore size could be tuned by choosing particles of different sizes. To increase the thickness of the membrane, a higher amount of particle–monomer mixture was used and the colloids formed a three-dimensional template embedded in the crosslinker monomer [88]. After UV-initiated polymerization and removal of the particles, a three-dimensional porous structure was obtained. Through this work, novel macroporous membranes with high porosities and uniform pore size distributions were created. Accessible pore diameters were as low as ~200 nm. Other macroporous polymers with an inverted colloidal crystal structure were also obtained via *in situ* polymerization of, for instance, acrylamide-based monomer mixtures around well-ordered 3D particle templates and subsequent template removal; such materials had originally been designed as scaffolds for cell culture [89]. However, the membrane performance parameters must still be explored in more detail to confirm the expectation that the membranes could be used for size-based separations at very high fluxes.

8.2.6.4 Porous Membranes from Fixed Supramolecular Assemblies or Liquid Crystalline Phases. Similar to the fixation of arrays of relatively large particles as templates to yield macropores (see Section 8.2.6.3), supramolecular assemblies that could yield selective transmembrane micropores have also been fixed in films using *in situ* (crosslinking) polymerization. In particular, crosslinked nonlamellar lyotropic liquid crystalline (LLC) assemblies have been proposed for many potential applications, including membrane separations, due to their unique nanoscale porous architectures and tunability. With respect to equilibration of the structures and orientation of the pores, analogous problems as discussed for block copolymers must be solved (see Section 8.2.5.2). Recently, work by Gin, Noble, and coworkers led to innovations with significant potential for the field of industrial membranes (for review see [90]). The resulting "porous" polymer membranes are a "boundary case" with respect to their pore size (the transport through conventional polymeric nanofiltration membranes is typically described by the solution/diffusion mechanism; see Section 8.1). Therefore, only two representative examples will be discussed here.

Pioneering work was done by Beginn et al. [91]. Functional membranes containing ion-selective, matrix-fixed "supramolecular channels" were developed based on the self-assembly of low–molar mass amphiphiles containing crown ethers into long cylindrical aggregates; this effect has been used for the gelation of acrylate-based monomer mixtures. Fixation of the gels was done using *in situ*, UV-initiated polymerization. Because pure monomers, selected to ensure minimal shrinking upon polymerization, were used, the supramolecular structures could be permanently arrested. The resulting membranes, with thicknesses of about 10 μm, exhibited characteristics that were consistent with the concept of pores formed by supramolecular assembly and stacking of the crown ether units, because characteristic differences in ion transport rates were observed (e.g., selectivities for Li^+ and $Na^+ > K^+$ and selectivities for $NO_3^- > Cl^- > ClO_4^-$). Recently, dynamic "self-instructed" membranes based on the self-assembly of heteroditopic crown-ether conjugates in lipid-coated

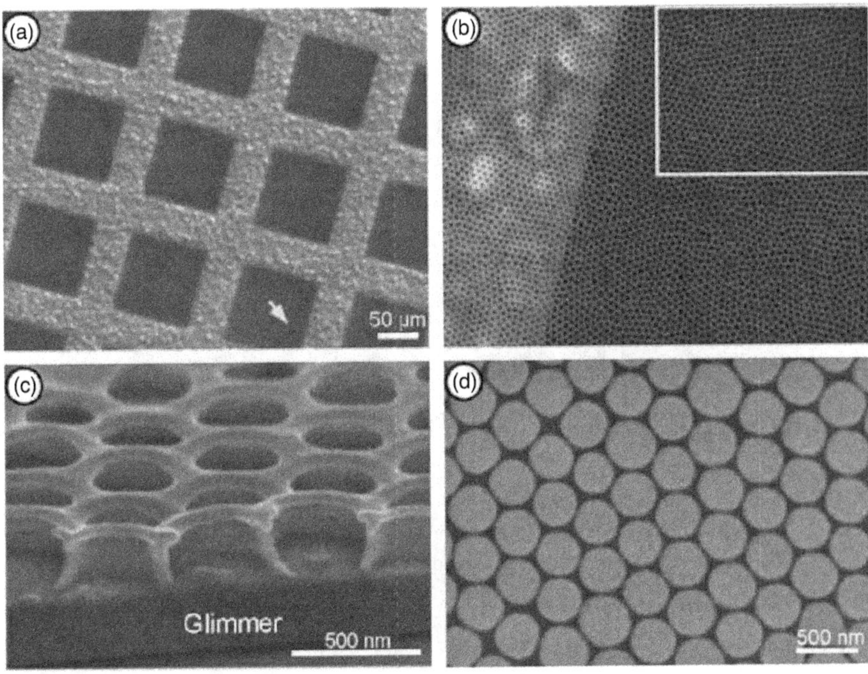

FIGURE 8.15 Thin-film macroporous membrane prepared via particle-assisted wetting of an aqueous subphase by a dispersion of nanoparticles in a monomer mixture, followed by *in situ* crosslinking copolymerization and subsequent removal of the nanoparticle templates. **(a, b)** SEM micrographs of thin, self-supported porous membranes on a metal grid with 100 μm wide windows (two different magnifications). **(c)** High-resolution SEM cross-section micrograph of a porous membrane on a mica support. **(d)** TEM micrograph of a self-supported porous membrane. (Reprinted from Reference 87 with permission. Copyright 2003 Wiley-VCH Verlag GmbH & Co. KGaA.)

polysiloxane-filled cylinder pores were reported to exhibit an increase in flux and selectivity for the respective ion (Na^+ or K^+) when exposed in diffusion experiments to the respective salt [92].

In contrast to the gels just described, a liquid crystalline phase could provide a regular, relatively dense pattern of transport pathways. When separations in aqueous systems are desired, the amphiphilic properties of the building blocks should ensure that connected water-filled domains in a polymer matrix are obtained. A composite membrane with a barrier layer containing 1.2 nm diameter cylindrical pores was obtained by crosslinking an oriented inverted hexagonal LLC phase of a functional monomer, but this membrane had a relatively low water flux, reflecting the low porosity [93]. With a different LLC monomer, a bicontinuous cubic phase (Q_I) was turned into a molecular size-selective water filtration membrane by *in situ* crosslinking polymerization (Fig. 8.16) [94]. This new "nanoporous" material had an effective pore size of 0.75 nm and was capable of high salt rejection, with a water permeability

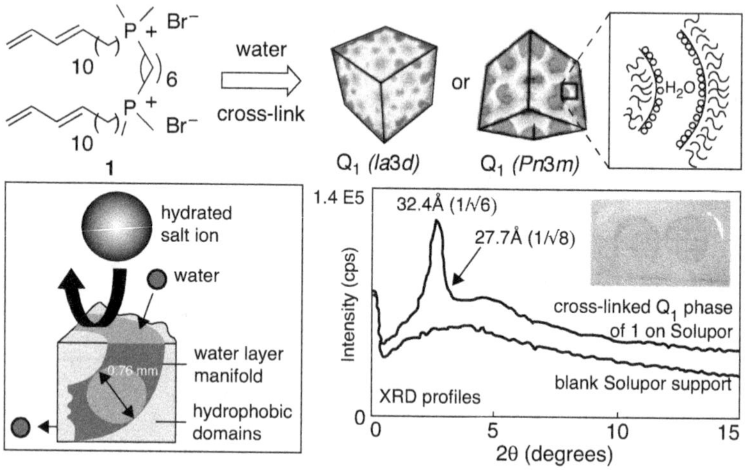

FIGURE 8.16 Structure of the crosslinked Q_I phase of an LLC monomer, **1**. XRD profile and photograph of a Q_I composite membrane (on a PE microfiltration membrane) and the proposed mechanism of water desalination and nanofiltration through the material. (Reprinted from Reference 94 with permission. Copyright 2007 American Chemical Society).

similar to that of commercial reverse osmosis membranes in dead-end filtration. Ongoing work is focused on new processing methods for producing larger areas of the thin-film composite membranes and on tuning the pore size by modifying the structure of the LLC monomer.

8.2.7 Mixed Matrix Membranes

Mixed matrix membranes composed of organic polymers and inorganic or other "fillers" can exhibit largely enhanced separation performance. However, in order to improve the flux and selectivity, this is generally much more important for membranes with a nonporous polymeric barrier. Examples are the attempts to improve gas separation performance of polymers by embedding zeolites. Recently, three interesting examples were reported in which porous materials were incorporated into a nonporous polymeric barrier. First, the pores of the additive contributed significantly to the improvement of membrane performance: The integration of hydrophilic zeolite particles into the ~100 nm thick polyamide barrier of a reverse osmosis membrane led to a significant increase in flux at same salt rejection [95]. Second, the pores of the porous additive completely controlled the membrane properties: An array of carbon nanotubes in a polystyrene matrix enabled studies of convective flow through and sieving properties of these nanomaterials [96]. Third, the incorporation of aquaporin, the water transport channel protein of biological cells, into an ultrathin polymer membrane from a block copolymer was the first proof of concept for an envisaged novel porous bio/synthetic membrane for water desalination [97].

For porous polymer membranes, improvements in mechanical stability can be achieved with inorganic additives (e.g., ZrO_2/polymer composites [98]), but this is rarely done in commercial membranes. Instead, the combination of porous polymeric membranes with functional particles to add specific properties for special applications in combined/integrated processes was proposed. Examples are the incorporation of adsorber particles into a porous membrane to yield a new type of porous membrane adsorber [99], the addition of carbon particles to yield a conductive polymer membrane [100], and the incorporation of silver nanoparticles into an ultrafiltration membrane to obtain antimicrobial properties [101].

8.2.8 Surface Functionalization

The surface properties of materials are of crucial importance in many applications. With porous polymers, a rich diversity of surface properties results from the material used and the membrane preparation method (see Sections 8.2.1–8.2.4 and 8.2.6). Examples of new developments with block copolymers as additives or entirely novel membrane materials have also been discussed (see Section 8.2.5). Nevertheless, postfunctionalization has become a very important field for membrane technology because the requirements with respect to pore structure and membrane stability cannot be combined with all of the desired surface properties. Consequently, many commercial polymeric membranes are surface modified. The intention of membrane surface modification is either to minimize undesired ("secondary") interactions (adsorption or adhesion), which reduce the performance (membrane fouling), or to introduce additional interactions (affinity, responsiveness, biocompatibility, or catalytic properties) for improving the selectivity or creating an entirely novel separation function [9]. A key feature of a successful (i.e., "tailored") surface functionalization is a *synergy* between the useful properties of the base membrane and the novel functional polymer (layer). This is best achieved by a functionalization that essentially preserves the bulk structure of the porous base membrane. Besides coating methods, the most common strategies can be classified into "grafting-to" methods (attachment of presynthesized functional entities) and "grafting-from" methods (*in situ* formation of functional entities via graft copolymerization of functional monomers).

It is beyond the scope of this chapter to give a comprehensive overview of this field. Instead, various relevant examples will help illustrate the diversity and potential of the surface functionalization of porous polymer membranes (Fig. 8.17).

Grafted polymers as a selective barrier (Fig. 8.17a). Examples of pore-filled ultrafiltration membranes with entirely new separation properties are the high-performance pervaporation membranes obtained by photoinitiated "grafting-from" of functional poly(meth)acrylates on a PAN support [103]. Microfiltration membranes prepared by reactive pore filling with cross-linked polyacrylamide hydrogels were evaluated for protein separation via size exclusion under ultrafiltration conditions [104]. A very similar approach but using ion-exchange monomers led to high-performance membranes for removal of multivalent salts from water via nanofiltration [105].

Grafted polymers as an antifouling layer (Fig. 8.17b). Low-fouling ultrafiltration membranes can be prepared under photografting conditions in which the degradation

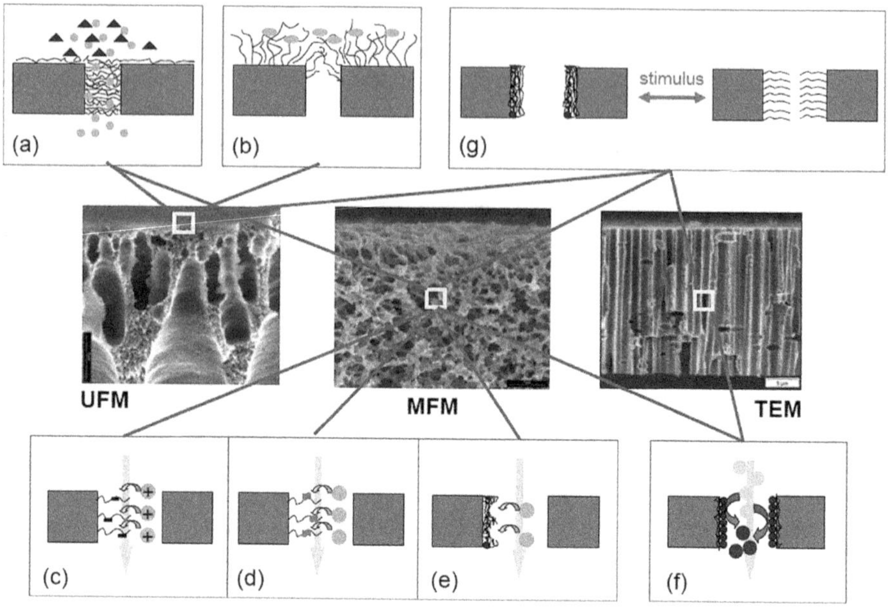

FIGURE 8.17 Representative membrane functionalities achieved by "grafting-from" with various porous polymeric base membranes. **(a)** Pore-filling selective polymer. **(b)** Grafted antifouling layer. **(c)** Ion-exchange membrane adsorber. **(d)** Affinity membrane adsorber. **(e)** Thin-layer MIP membrane adsorber. **(f)** (Bio)catalytic membrane. **(g)** Stimulus-responsive membrane ("valve"). MFM, microfiltration; TEM, track etched; UFM, ultrafiltration. (Modified from Reference 102 with permission. Copyright 2009 Elsevier Ltd.)

of the base membrane pore structure is minimized. The composition, surface coverage, and thickness of the grafted layer are crucial for final membrane performance, that is, low fouling while preserving size exclusion properties at a competitive flux [106].

Grafted polymer layers containing functional groups for reversible binding. Surface-functionalized MF membrane adsorbers for fast protein purification (Fig. 8.17c,d) have been prepared via photografting of two- or three-dimensional layers with suitable functional groups [107, 108]. A novel type of MIP composite membrane (Fig. 8.17e) with high binding specificity at high throughput has been obtained by surface initiated photografting of a very thin crosslinked functional layer [109]. All these cases require sufficient permeability to utilize the main advantage of porous membrane adsorbers, that is, the reduction of mass transfer limitations by directional convective flow though the membrane pores.

Grafted polymer layers for immobilization of (bio)catalyst (Fig. 8.17f). Isoporous track-etched membranes with larger pore diameters (between 100 nm and 3 μm) were functionalized via "grafting-from" reactions in order to prepare enzyme membranes as convective flow microreactors [110]. The surface-selective preparation of catalytic metal nanoparticles on the pore walls of track-etched membranes is an alternative [111].

Grafted stimuli-responsive polymer layers. Using tailored grafted functional polymer layers on the pore walls of membranes, it is possible to reversibly change the permeability and/or selectivity. The most straightforward mechanism is the alteration of the effective pore diameter by changing the conformation of a grafted polymer via solution conditions as a "stimulus" (Fig. 8.17g). For example, reversible switching of permeability was achieved using photografted, pH-responsive poly(acrylic acid) [112] or temperature-responsive chains (polyNIPAAm) [19, 20].

More information on surface functionalization of porous polymer membranes can be found in recent reviews; one discusses the surface engineering of porous PP membranes [113], and others provide a broader overview [9, 102].

8.3 APPLICATIONS OF POROUS POLYMER MEMBRANES

Only a brief overview of the most important industrial applications of porous polymer membranes can be given here (for examples of membranes, processes, and manufacturers see Table 8.1).

The first industrial application of membranes was in the medical field, and blood detoxification by hemodialysis ("artificial kidney") is still the largest segment of the membrane industry by commercial value [114]. The only relevant membrane shape is hollow fiber because, by this means, a large membrane area ($>2\ m^2$) can be incorporated in a relatively compact hemodialysis module ($\sim 0.2\,dm^3$). Other important medical applications are blood oxygenation ("artificial lung"), an example of a gas/liquid membrane contactor [115], and plasmapheresis, in which macroporous membranes are used for blood fractionation [116].

Microfiltration and ultrafiltration have found many diverse applications in the purification of pure water, in waste water treatment, and in the food and chemical industries [1, 2], as well as in pharmaceutical/biotech and medical industries [117, 118]. Tubular and sometimes also capillary membranes are preferred for feed streams with a high fouling potential. Important examples are the removal of colloids (clear filtration) from beverages (juice, beer, wine), often in parallel with sterile filtration. Sterile filtration of all kinds of aqueous streams, including the increasingly rigorous reduction of viruses (with sizes down to 20 nm; see Fig. 8.1), is of crucial importance and is a significant driver for the development of advanced high-flux ultrafiltration membranes [118]. Long-established fractionations using ultrafiltration can be found, for instance, in the milk and dairy industry. Other important processes include the concentration of solutions (typically by ultrafiltration) and the removal of an undesired low–molar mass fraction (e.g., in the manufacturing of alcohol-free beer or wine by dialysis). Membrane bioreactors for water treatment have become a very convincing and successful example of process intensification by using membranes: Biological treatment can be performed at much higher cell densities (and, hence, a lower footprint), and the purified water is obtained directly as the permeate of the membrane [7]. Many systems work with submerged capillary membranes, and aeration is also used to clean the membrane.

TABLE 8.1 Overview of Typical Industrial Membranes for Selected Important Applications

Membrane Process	Membrane Polymer	Membrane Shape	Typical Process Mode, Driving Force	Application	Membrane Manufacturer
Microfiltration	PES, PSf, PA, PVDF, PP, PE, PTFE, cellulose derivatives	Flat sheet	Pressure driven, ≤3 bar	Clarification, concentration, sterile filtration	Asahi, GVS, Koch, Membrana, Millipore, Pall, Sartorius
	PES, PVDF	Tubular			Koch, Siemens
	PES, PVDF, PP, PE	Capillary			Membrana, X-Flow (Norit), Koch
Microfiltration, ultrafiltration	PES	Capillary, hollow fiber	Out-in filtration by vacuum	Membrane bioreactor for waste water	X-Flow (Norit)
Ultrafiltration	PES, PSf, PVDF, PAN, cellulose derivatives	Flat sheet	Pressure driven, ≤6 bar	Clarification, concentration, fractionation, sterile filtration	AlvaLafal, Koch, Membrana, Millipore, Pall, Sartorius
	PES, PVDF	Tubular			Koch
	PES	Capillary	Pressure driven in-out, ≤6 bar	Surface water filtration, filtration prior to desalination	Inge, X-Flow (Norit), Zenon (GE)
Dialysis	PSf, PSf-PA copolymer, PAN, regenerated cellulose	Hollow fiber	Concentration difference	Hemodialysis, fractionation (e.g., ethanol removal from wine or beer)	Asahi, Fresenius, Gambro, Membrana
Gas/liquid contactor	PP, PE, PTFE	Hollow fiber, capillary	Concentration difference	Blood oxygenation, dehumidification of air	GVS, Membrana
Membrane distillation	PP, other polyolefins	Flat sheet, hollow fiber	Temperature difference	Water desalination	Membrana
Membrane adsorber	Regenerated cellulose, PES, PVDF; all with functional layer	Flat sheet, used as stack	Convective flow by transmembrane pressure	Removal of trace impurities ("polishing"), "capturing" of biomolecules or bioparticles	Millipore, Pall, Sartorius

Other applications of microfiltration membranes and porous polymeric depth filters are in air filtration (innovative materials based on electrospun polymer nanofibers as commercial products were introduced by Donaldson; see Section 8.2.3). A well-known example of a porous barrier material that is vapor permeable but not liquid water permeable is based on stretched PTFE (e.g., GORE-TEX; see Section 8.2.2.1).

Membrane contactors with inert polymeric membranes (see Fig. 8.2) can be used for extraction or absorption or to facilitate chemical reactions [7, 8]. Membrane distillation is a special case because a temperature difference serves as the driving force for the transmembrane transport; wetting of the pores (in the submicrometer range) must be prevented [1, 7, 8]. This process is being evaluated intensively as an alternative for water desalination. Porous membrane adsorbers are finding more and more applications as alternatives to particle-base columns for important steps in downstream processing in biotechnology—in particular the manufacture of therapeutic recombinant antibodies. Membrane adsorbers have shown superior performance in polishing steps and are being developed for capturing a target protein [119]. One example of a large-scale industrial process with a catalytic membrane between an aqueous and an organic phase is based on immobilization of lipase in a hollow-fiber ultrafiltration membrane; it is used for the continuous kinetic resolution of a racemate that is an intermediate in the synthesis of a drug [120].

Finally, there are more, very diverse applications of porous polymeric membranes. Very interesting examples can be found in the field of (bio)analytics, as well as in cell culture and tissue engineering; in many—but not all!—cases the porous membrane is just used as support [11]. In very special cases, the membrane pore structure is used in a direction orthogonal to the typical permeation barrier, for instance, in lateral flow or immunochromatographic analyses. One well-known example is a pregnancy test strip in which the analysis is based on lateral flow of the sample through a macroporous cellulose nitrate membrane (see Section 8.2.4.2) containing immobilized antibodies in certain regions where a color is developed upon binding of the specific antigen [33].

8.4 CONCLUSIONS AND OUTLOOK

Separation and integrated processes based on porous polymeric membranes are very well established in many fields of application, including on a very large industrial scale. Due to the intrinsic advantages of membrane-based processes and the well-established manufacturing technologies for many polymeric membranes, this is a highly dynamic field. This provides a perfect basis for the improvement of existing membranes and the development of novel membranes. Needs for improvement result either from undesired secondary processes that limit membrane performance (in particular, membrane fouling) or from a still suboptimal relationship between selectivity and permeability (due to a pore size distribution that is too broad, a porosity that is too low, or a barrier thickness that is too large). Many new advanced porous polymeric membranes will be based on tailored functional macromolecular architectures prepared by state-of-the-art controlled polymerization instead of just "bulk" properties of standard industrial polymers. Examples include the predetermined, regular, "nanoporous" morphologies from phase-separated block or graft copolymers, microporous, mesoporous, or macroporous structures created by using templates during formation or *in situ* synthesis of membranes, functional grafted macromolecular layers to facilitate binding to pore walls or to protect the membrane barriers from unwanted interactions, and affinity binding sites in porous membranes by immobilization through macromolecular linkers or by *in situ* synthesis via molecular imprinting of polymers. For membranes that will ultimately be used in large-scale applications, it must be kept in mind that the current membrane formation processes via phase separation have already been optimized at a large cost, so that one cannot easily deviate very significantly from these without significant economic penalty. However, the existing processes are quite flexible and still offer considerable room for innovative adaptation. Important avenues for such innovations will be the blending of polymers with different functions and the design of polymers for easy and efficient posttreatment. Furthermore, it has been shown that composite membranes can provide very efficient alternatives because smaller amounts of specialty polymers will be required and/or the polymer can be protected from the stress imposed by the process conditions. Advanced membranes of the next generation will also have more functions than just being selective barriers with high performance (flux, stability, etc.). "Smart" membranes with tunable selectivities or adaptive surfaces can be created using the approaches under investigation in research labs. Examples of such stimulus-responsive membranes show that a synergistic interplay of pore structure and tailored functional macromolecular architectures can be used to create "biomimetic" membranes. When this is realized as a composite membrane based on an already established ("industrial") membrane, the novel materials will have a strong potential for future applications because they are already partly "adapted" to the industrial environment. Membrane contactors and the combination of membranes with catalysis are being studied intensively and, in some cases, are already in use on the industrial scale. Considering the limitations of polymeric membranes with respect to stability under process conditions and their advantages in terms of preparing barriers with tailored properties, one can see that membrane-based systems with porous polymeric membranes will also play a very important role in process intensification by combining the conventional unit operations of reaction and separation engineering.

ACKNOWLEDGMENTS

I am very grateful to my former and current students, coworkers, and collaboration partners who have contributed to the work cited in this chapter.

ABBREVIATIONS

ATRP	Atom transfer radical polymerization
CA	Cellulose acetate

DMSO	Dimethyl sulfoxide
EIPS	Evaporation-induced phase separation
EO	Ethyleneoxide
GMA	Glycidyl methacrylate
LLC	Lyotropic liquid crystal
MBAA	*N*,*N*′-methylene bisacrylamide
MF	Microfiltration
MIP	Molecularly imprinted polymer
NIPAAm	*N*-isopropyl acrylamide
NIPS	Nonsolvent phase separation
P4VP	4-Vinyl pyridine
PAA	Poly(acrylic acid)
PAN	Polyacrylonitrile
PC	Polycarbonate
PDMAEMA	Poly(*N*,*N*-dimethylaminoethyl methacrylate)
PE	Polyethylene
PEG	Poly(ethylene glycol)
PEGMA	Poly(ethylene glycol methacrylate)
PEO	Poly(ethylene oxide)
PES	Poly(ether sulfone)
PET	Poly(ethylene terephthalate)
PI	Polyimide
PIM	Polymer with inherent microporosity
PL	Polylactide
PMMA	Poly(methyl methacrylate)
PP	Polypropylene
PPO	Poly(propylene oxide)
PS	Phase separation
PSμM	Phase separation micromolding
PSf	Polysulfone
PSt	Polystyrene
PtBA	Poly(*tert*-butyl acrylate)
PtBMA	Poly(*tert*-butyl methacrylate)
PTFE	Polytetrafluoroethylene
PVDF	Poly(vinylidene fluoride)
PVP	Polyvinylpyrrolidone
RhB	Rhodamine B
SEM	Scanning electron microscopy
SPSf	Sulfonated polysulfone
THF	Tetrahydrofuran
TIPS	Thermally induced phase separation
UF	Ultrafiltration
UV	Ultraviolet
VIPS	Vapor-induced phase separation

REFERENCES

1. Baker, R. W. *Membrane Technology and Applications*, 2nd ed., Wiley, Chichester, UK, 2004.
2. Pereira-Nunes, S.; Peinemann, K. V., Eds. *Membrane Technology in the Chemical Industry*, Wiley-VCH, Weinheim, Germany, 2006.
3. Robeson, L. M. *J Membr Sci* 2008, **320**, 390.
4. McKeown, N. B.; Budd, P. M. *Chem Soc. Rev* 2006, **35**, 675.
5. Park, H. B.; Freeman, B. D.; Zhang, Z. B.; Sankir, M.; McGrath, J. E. *Angew Chem Int Ed* 2008, **47**, 6019.
6. Akthakul, A.; Hochbaum, I., Stellacci, F.; Mayes, A. M. *Adv Mater* 2005, **17**, 532.
7. Drioli, E.; Giorno, L., Eds. *Membrane Operations. Innovative Separations and Transformations*, Wiley-VCH, Weinheim, Germany, 2009.
8. Sirkar, K. K. *Ind Eng Chem Res* 2008, **47**, 5250.
9. Ulbricht, M. *Polymer* 2006, **47**, 2217.
10. Noble, R. D. *J Membr Sci* 1992, **75**, 121.
11. Groth T., Liu Z. M. In *Membranes for the Life Sciences*, Peinemann, K. V.; Pereira-Nunes S., Eds., Wiley-VCH, Weinheim, Germany, 2008, p. 227.
12. Adhikari B.; Majumdar S. *Progr Polym Sci* 2004, **29**, 699.
13. Peppas N. A.; Wood K. M.; Thomas J. B. In *Membranes for the Life Sciences*, Peinemann, K. V.; Pereira-Nunes S., Eds., Wiley-VCH, Weinheim, Germany, 2008, p. 175.
14. Paul D. R., Clarke R. *J Membr Sci* 2002, **208**, 269.
15. GORE-TEX. Available at: http://www.gore-tex.com.
16. Jonathan, M. G.; George, E. P. *Anal Biochem* 1983, **131**, 1.
17. Fleischer, R. L.; Alter, H. W.; Furman, S. C.; Price, P. B.; Walker, R. M. *Science* 1972, **172**, 255.
18. Baker L. A.; Jin, P.; Martin, C. R. *Crit Rev Solid State Mater Sci* 2005, **30**, 183.
19. Geismann, C.; Yaroshchuk, A.; Ulbricht, M. *Langmuir* 2007, **23**, 76.
20. Friebe, A.; Ulbricht, M. *Macromolecules* 2009, **42**, 1838.
21. Huang, Q.; Seibig, B; Paul, D. *J Membr Sci* 1999, **161**, 287.
22. Krause, B.; Boerrigter, M. E.; Van Der Vegt, N. F. A.; Strathmann, H.; Wessling, M. *J Membr Sci* 2001, **187**, 181.
23. Gopal, R.; Kaur, S; Feng, C. Y.; Chand, C.; Ramakrishna, S.; Tabe, S.; Matsuura, T. *J Membr Sci* 2007, **289**, 210.
24. Veleirinho, B.; Lopes da Silva, F. A. *Process Biochem* 2009, **44**, 353.
25. Yoon, K.; Kim, K.; Wang, X.; Fang, D.; Hsiao, B. S.; Chu, B. *Polymer* 2006, **47**, 2434.
26. Yoshikawa M.; Nakai, K.; Matsumoto, H.; Tanioka, A.; Guiver, M. D.; Robertson, G. P. *Macromol Rapid Commun* 2007, **28**, 2100.
27. Loeb, S.; Sourirajan, S. *Adv Chem Ser* 1962, **38**, 117.
28. Mulder, M. *Basic Principles of Membrane Technology*, 2nd ed., Kluwer Academic, Dordrecht, Netherlands, 1996.
29. Van de Witte, P.; Dijkstra, P. J.; van den Berg, J. W. A.; Feijen, J. *J Membr Sci* 1996, **117**, 1.

30. Khare, V. P.; Greenberg, A. R.; Krantz, W. B. *J Membr Sci* 2005, **258**, 140.
31. Ulbricht, M.; Schuster, O.; Ansorge, W.; Ruetering, M.; Steiger, P. *Sep Purif Techn* 2007, **57**, 63.
32. Thomas, S.; Pinnau, I.; Du, N.; Guiver, M. D. *J Membr Sci* 2009, **333**, 125.
33. Wang, J.; Thom, V.; Hollas, M.; Johannsmann, D. *J Membr Sci* 2008, **318**, 280.
34. Hanks, P. L.; Lloyd D. R. *J Membr Sci* 2007, **306**, 125.
35. Yave, W.; Quijada, R.; Lloyd, D. R.; Cerrada, M. L.; Benavente, R.; Ulbricht, M. *Macromol Mater Eng* 2006, **291**, 155
36. Peinemann, K. V.; Maggioni, J. F.; Nunes, S. P. *Polymer* 1998, **39**, 3411.
37. White, L. S.; Nitsch, A. R. *J Membr Sci* 2000, **179**, 267.
38. Hicke, H. G; Lehmann, I.; Malsch G.; Ulbricht, M.; Becker, M. *J Membr Sci* 2002, **198**, 187.
39. Yan, M.; Ramström, O., Eds., *Molecularly Imprinted Materials, Science and Technology*, Marcel Dekker, New York, 2005.
40. Ulbricht, M. *J Chromatogr B* 2004, **804**, 113.
41. Yoshikawa, M.; Izumi, J.; Kitao, T.; Koya, S.; Sakamoto, S. *J Membr Sci* 1995, **108**, 171.
42. Yoshikawa M. *Bioseparation* 2002, **10**, 277.
43. Wang, H. Y.; Kobayashi, T.; Fuji, N. *Langmuir* 1996, **12**, 4850.
44. Faizal, C. K. M.; Kikuchi, Y; Kobayashi, T. *J Membr Sci* 2009, **334**, 110.
45. Ulbricht, M., Malaisamy, R. J Mater Chem 2005, **5**, 1487.
46. de Jong, J.; Lammertink, R. G. H.; Wessling, M. *Lab Chip* 2006, **6**, 1125.
47. Vogelaar, L.; Lammertink, R. G. H; Barsema, J. N.; Nijdam, W., Bolhuis-Versteeg, L. A. M., van Rijn, C. J. M., Wessling, M. *Small* 2005, **1**, 645.
48. Yan, X.; Liu, G.; Dickey, M.; Willson, C. G. *Polymer* **45**, 2004, 8469.
49. Kim, W.; Lee, M. K. *Mater Lett* 2009, **63**, 933.
50. Papenburg, B. J.; Vogelaar, L.; Bolhuis-Versteeg, L. A. M.; Lammertink, R. G. H.; Stamatialis, D.; Wessling, M. *Biomaterials* 2007, **28**, 1998.
51. Park, C.; Yoon, J.; Thomas, E. L. *Polymer* 2003, **44**, 6725.
52. Olson, D. A.; Chen, L.; Hillmyer, M. A. *Chem Mater* 2008, **20**, 869.
53. Rana, D.; Matsuura, T.; Narbaitz, R. M.; Feng, C. *J Membr Sci* 2005, **249**, 103.
54. Ye, S. H.; Watanabe, J.; Iwasaki, Y.; Ishihara, K. *J Membr Sci* 2005, **249**, 133.
55. Hester, J. F.; Mayes, A. M. *J Membr Sci* 2002, **202**, 119.
56. Zhao, W.; Su, Y.; Li, C.; Shi, Q.; Ning, X.; Jiang, Z. *J Membr Sci* 2008, **318**, 405.
57. Shi, Q.; Ye, S.; Kristalyn, C.; Su, Y.; Jiang, Z.; Chen, Z. *Langmuir* 2008, **24**, 7939.
58. Susanto, H.; Ulbricht, M. *J Membr Sci* 2009, **327**, 125.
59. Susanto, H.; Stahra, N.; Ulbricht, M. *J Membr Sci* 2009, **342**, 153.
60. Wang, W. C; Ong, G. T.; Lim, S. L.; Vora, R. H.; Kang, E. T.; Neoh, K. G. *Ind Eng Chem Res* 2003, **42**, 3740.
61. Huang, X. J.; Xu, Z. K.; Wan, L. S.; Wang, Z. G.; Wang, J. L. *Macromol Biosci* 2005, **5**, 322.
62. Akthakul, A.; Salinaro, R. F.; Mayes, A. M. *Macromolecules* 2004, **37**, 7663.
63. Peinemann, K. V.; Simon, P.; Abetz, V. *Nat Mater* 2007, **6**, 992.
64. Schacher, F.; Ulbricht, M.; Müller, A. H. E. *Adv Funct Mater* 2009, **19**, 1040.

65. Schacher, F.; Rudolph, T.; Wieberger, F.; Ulbricht, M.; Müller, A. H. E. *ACS Appl Mater Interf* 2009, **1**, 1492.
66. Lee, J. S.; Hirao, A.; Nakahama, S. *Macromolecules* 1988, **21**, 274.
67. Smith, D. R.; Meier, D. J. *Polymer* 1992, **33**, 3777.
68. Olson, D. A.; Chen, L.; Hillmyer, M. A. *Chem Mater* 2008, **20**, 869.
69. Liu, G.; Ding, J.; Stewart, S. *Angew Chem Int Ed* 1998, **38**, 835.
70. Zalusky, A. S.; Olayo-Valles, R.; Wolf, J. H.; Hillmyer, M. A. *J Am Chem Soc* 2002, **124**, 12761.
71. Rzayev, J.; Hillmyer, M. A. *Macromolecules* 2005, **38**, 3.
72. Phillip, W. A.; Rzayev, J.; Hillmyer, M. A.; Cussler, E. L. *J Membr Sci* 2006, **286**, 144.
73. Yang, S. Y.; Ryu, I.; Kim, H. Y., Kim, J. K.; Jang, S. K.; Russell, T. P. *Adv Mater* 2006, **18**, 709.
74. Yang, S. Y.; Park, J.; Yoon, J.; Ree, M.; Jang, S. K.; Kim, J. K. *Adv Funct Mater* 2008, **18**, 1371.
75. Uehara, H.; Kakiage, M.; Sekiya, M.; Sakuma, D.; Yamonobe, T.; Takano, N.; Barraud, A.; Meurville, E.; Ryser, P. *ACS Nano* 2009, **3**, 924.
76. Svec F. *J Sep Sci* 2004, **27**, 1419.
77. BIA Separations. Available at: http://www.monoliths.com.
78. Salam, A.; Ulbricht, M. *Macromol Mater Eng* 2007, **292**, 310.
79. Song, S.; Singh, A. K.; Shepodd, T. J.; Kirby, B. *J Anal Chem* 2004, **76**, 2367.
80. Okahata, Y.; Noguchi, H.; Seki, T. *Macromolecules*, 1986, **19**, 493.
81. Chu, L. Y.; Liang, Y. J.; Chen, W. M.; Ju, X. J.; Wang H. D. *Colloids Surf B Biointerf* 2004, **37**, 9.
82. Akamatsu, K.; Yamaguchi, T. *Ind Eng Chem Res* 2007, **46**, 124.
83. Mathew-Krotz, J.; Shea, K. J. *J Am Chem Soc* 1996, **118**, 8154.
84. Kimaro, A.; Kelly, L. A.; Murray, G. M. *Chem Commun* 2001, 1282.
85. Sergeyeva, T. A.; Brovko, O. O.; Piletska, E. V.; Piletsky, S. A.; Goncharova, L. A.; Karabanova, L. V.; Sergeyeva, L. M.; El'skaya, A. V. *Analyt Chim Acta* 2007, **582**, 311.
86. Hong, J. M.; Anderson, P. E.; Qian, J., Martin, C. R. *Chem Mater* 1998, **10**, 1029.
87. Xu, H.; Goedel, W. A. *Angew Chem Int Ed* 2003, **42**, 4694.
88. Yan, F.; Goedel, W. A. *Adv Mater* 2004, **16**, 911.
89. Zhang Y.; Wang S.; Eghtedari, M.; Motamedi, M.; Kotov, N. A. *Adv Funct Mater* 2005, **15**, 725.
90. Gin, D. L.; Bara, J. E.; Noble, R. D.; Elliott B. J. *Macromol Rapid Commun* 2008, **29**, 367.
91. Beginn, U.; Zipp, G.; Möller, M. *Adv Mater* 2000, **12**, 510.
92. Cazacu, A.; Legrand, Y. M.; Pasc, A.; Nasr, G.; Van Der Lee, A.; Mahon, E.; Barboiu, M. *Proc Natl Acad Sci USA* 2009, **106**, 8117.
93. Kidd, T. J.; Noble, R. D.; Gin, D. L. *Adv Mater* 2005, **17**, 1850.
94. Zhou, M.; Nemade, P. R.; Lu, X.; Zeng, X.; Hatakeyama, E. S.; Noble, R. D.; Gin, D. L. *J Am Chem Soc* 2007, **129**, 9574.
95. Jeong, B. H.; Hoek, E. M. V.; Yan, Y.; Subramani, A.; Huang, X.; Hurwitz, G.; Ghosh, A. K.; Jawor, A. *J Membr Sci* 2007, **294**, 1.

96. Hinds, B. J.; Chopra, N.; Rantell, T.; Andrews, R.; Gavalas, V.; Bachas, L. G. *Science* 2004, **303**, 62.
97. Kumar, M.; Grzelakowski, M.; Zilles, J.; Clark, M.; Meier, W. *Proc Natl Acad Sci USA* 2007, **104**, 20719.
98. Genné, I.; Kuypers, S.; Leysen, R. *J Membr Sci* 1996, **113**, 343.
99. Avramescu, M. E.; Borneman, Z.; Wessling M. *J Membr Sci* 2008, **322**, 306.
100. Gloukhovski, R.; Oren, Y.; Linder, C.; Freger, V. *J Appl Electrochem* 2008, **38**, 759.
101. Taurozzi, J. S.; Arul, H.; Bosak, V. Z.; Burban, A. F.; Voice, T. C.; Bruening M. L.; Tarabara, V. V. *J Membr Sci* 2008, **325**, 58.
102. He, D. M.; Susanto, H.; Ulbricht, M. *Prog Polym Sci* 2009, **34**, 62
103. Ulbricht, M.; Schwarz, H. H. *J Membr Sci* 1997, **136**, 25.
104. Kapur, V.; Charkoudian, J. C.; Anderson, J. L. *J Membr Sci* 1997, **131**, 143.
105. Mika, A. M.; Childs, R. F. *Ind Eng Chem Res* 2003, **42**, 3111.
106. Susanto, H.; Ulbricht M. *Langmuir* 2007, **23**, 7818.
107. He, D. M.; Ulbricht, M. *J Membr Sci* 2008, **315**, 155
108. Yusof, A. H. M.; Ulbricht, M. *J Membr Sci* 2008, **311**, 294.
109. Sergeyeva, T. A.; Matuschewski, H.; Piletsky, S. A.; Bendig, J.; Schedler, U.; Ulbricht, M. *J Chromatogr A* 2001, **907**, 89.
110. Hicke, H. G.; Becker, M., Paulke, B. R.; Ulbricht, M. *J Membr Sci* 2006, **282**, 413.
111. Dotzauer, D. M.; Dai, J. H.; Sun, L.; Bruening M. L. *Nano Lett* 2006, **6**, 2268.
112. Ulbricht, M.; Yang, H. *Chem Mater* 2005, **17**, 2622.
113. Wan; L. S.; Liu; Z. M.; Xu, Z. K. *Soft Matter* 2009, **5**, 1775.
114. Vienken, J. In *Membranes for the Life Sciences*, Peinemann K. V.; Pereira-Nunes S., Eds., Wiley-VCH, Weinheim, Germany, 2008, p. 1.
115. Wiese, F. In *Membranes for the Life Sciences*, Peinemann K. V.; Pereira-Nunes S., Eds., Wiley-VCH, Weinheim, Germany, 2008, p. 49.
116. Wiese F. In *Membranes for the Life Sciences*, Peinemann K. V.; Pereira-Nunes S., Eds., Wiley-VCH, Weinheim, Germany, 2008, p. 69.
117. Van Reis, R.; Zydney, A. *J Membr Sci* 2007, **297**, 16.
118. Allegrezza, A.; Ireland, T.; Kools, W.; Phillips, M.; Raghunath, B.; Wilkins, R.; Xenopolus, A. In *Membranes for the Life Sciences*, Peinemann K. V.; Pereira-Nunes S., Eds., Wiley-VCH, Weinheim, Germany, 2008, p. 91.
119. Gottschalk, U. *Biotechnol Prog* 2008, **24**, 496.
120. Lopez, J. L.; Matson, S. L. *J Membr Sci* 1997, **125**, 189.

CHAPTER 9

Biomedical Devices

YVONNE REINWALD, KEVIN SHAKESHEFF, and STEVEN HOWDLE
University of Nottingham, Nottingham, United Kingdom

9.1 SOLID POROUS SCAFFOLDS IN TISSUE ENGINEERING

Tissue engineering has been described as a discipline that aims to replace, regenerate, maintain, and/or improve the function of tissue [1]. It involves the utilization of scaffolds [2] onto which cells are then seeded and cultured *in vitro* [3] and/or *in vivo*. Tissue engineering scaffolds act as three-dimensional artificial matrices for cells to adhere onto, enabling them to proliferate, differentiate, and migrate. Cells will then start to produce extracellular matrix (ECM), which is necessary for tissue formation [2]. To meet tissue engineering requirements, scaffolds should be biodegradable, nontoxic, and biocompatible and should not cause an immunological response. Moreover, they should possess large surface area-to-volume ratios, high porosities [2,4,5], pore interconnectivities [6–9], and adequate mechanical strengths.

Since different cell types require different architectures for adhesion, proliferation, and differentiation [10], the choice of scaffold type is essential to provide the adequate environment for tissue formation. There are different types of porous solid scaffolds, such as foam scaffolds, fibrous scaffolds, polymer beds, and scaffolds produced by solid freeform (SFF) technologies. In the following sections various scaffold fabrication methods and their application in tissue engineering are described. Figure 9.1 gives an overview of currently applied fabrication techniques, and Table 9.1 summarizes their advantages and disadvantages. In addition, Tables 9.2–9.7 summarize experimental approaches for the fabrication of different types of scaffolds and their resulting mechanical strengths and applications.

Mechanical strength (compressive, tensile, or shear strength) is the limit state of mechanical stress. Compressive, tensile, and shear strength are distinguished by the type of mechanical stress (compressive, tensile, or shear stress) that is applied. The compressive strength (CS) of a scaffold is its ability to resist axially directed pushing forces. It is determined by compressive testing, in which a uniaxial compressive load

Porous Polymers, Edited by Michael S. Silverstein, Neil R. Cameron, and Marc A. Hillmyer

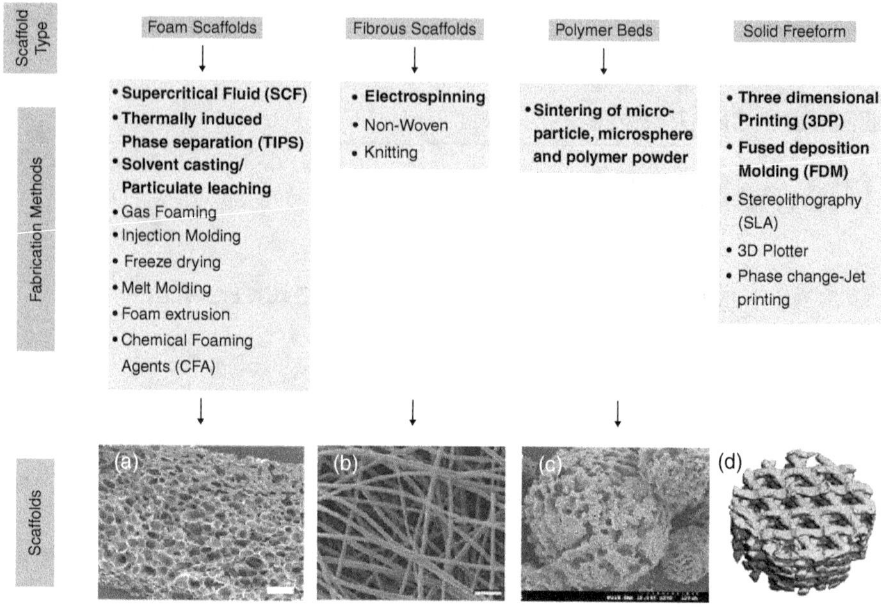

FIGURE 9.1 Overview of current methods for the fabrication of solid porous scaffolds. Foam scaffolds, fibrous scaffolds, polymer beds, and scaffolds produced by solid freeform are regarded as solid porous scaffolds and can be produced by various methods. Examples are shown for **(a)** foam scaffolds produced by supercritical foam technology [25], **(b)** fibrous scaffolds produced by electrospinning [141], **(c)** polymer beds produced by microsphere sintering [94], and **(d)** solid freeform produced by reverse rapid prototyping [105]. Methods in bold are explained in the following sections. (From References 25, 94, and 141 with permission from Elsevier and from Reference 105 with permission from Springer Science and Business Media.)

(compressive stress) is applied to the scaffold. The scaffold breaks when the limit of the compressive strength is reached. The capability of a scaffold to withstand its elongation due to the application of a uniaxial tensile load (tensile stress) is called tensile strength. Tensile strength is determined by tensile testing and can be given as either engineering or true tensile stress. The shear strength of a scaffold is the limit state of a pair of opponent forces applied to a scaffold along parallel lines (shear stress) through the material. An indication for the mechanical strength of a material is the yield strength (YS). It is a point on the engineering stress–strain curve at which the material begins to deform permanently. The stress–strain curve is a graphical representation of the stress applied to a material and its strain (reduced deformation). Stress and strain are directly proportional, resulting in a straight line. The slope of the line is called Young's modulus or modulus of elasticity (E) (also called compressive modulus (CM) for compressive testing, shear modulus (SM), for shear testing, and tensile modulus (TM) for tensile testing).

TABLE 9.1 Summary of Advantages and Disadvantages of Methods Applied for the Fabrication of Porous Scaffolds

	Fabrication Method	Advantages	Disadvantages	Ref.
Foam scaffolds	Supercritical CO_2 ($scCO_2$)	• Control of pore size and porosity by variation of particle/polymer ratio and process conditions • No use of organic solvent	• Low pore interconnectivity • Limited range of materials • Restriction in porosity and pore interconnectivity	3, 13, 18, 27, 33, 34, 36
	Solvent casting/ particulate leaching	• Pore size and morphology controlled by porogen • Controlled porosity	• Limited control of pore interconnectivity, pore shape, and interpore opening • Use of organic solvents	3, 10, 13, 36, 42, 54
	Thermally induced phase separation (TIPS)	• Control of pore size and shape • Highly porous	• Low control over pore size, Distribution, and range • Long fabrication times • Use of organic solvents • User, material, and technique sensitive	3, 10, 33, 34, 41, 42, 106
Fibrous scaffolds	Electrospinning (textile techniques)	• Large surface area to volume ratio • Interconnections between fibers	• Lack of structural stability • Restricted by mechanical properties, pore size, porosity, degradation rate • No defined pore shape or size • Randomly oriented fibers • Limited fiber diameters	3, 10, 33, 34, 36
Polymer beds	Particle sintering	• Graded porosity structures • Controlled porosity • Fabricated into complex shapes	• Use of organic solvents for microsphere production • Closing off pores due to sintering	42, 93, 108
solid freeform (SFF)		• Highly reproducible architecture • Compositional variation of scaffolds • Control of pore size, pore size distribution, porosity • Defined internal and external structure • Highly interconnected porous structure	• Restriction due to resolution of the computer-aided model • Some methods use organic solvents • Scaffold biocompatibility influenced by removal of excess powder (3dp) • Discontinuous material lowers mechanical properties	3, 11, 33, 34, 36, 42

Tissue engineering offers various concepts for the regeneration and repair of lost or injured tissue. One strategy involves the injection of cell suspensions or the transplantation of cell sheets into an injured tissue or defect. Another strategy is the seeding and culturing of cells onto three-dimensional matrices (e.g., foam scaffolds, fibrous scaffolds, and polymer beds). The incorporation of biologically active molecules such as growth factors, cell fractions, and peptides into tissue-engineered scaffolds is another commonly used concept. The fourth method involves the seeding and culturing of cells in/onto scaffolds containing bioactive molecules [11].

Cell-scaffold constructs are then investigated *in vitro* and/or *in vivo* to determine their applicability for the replacement or/and treatment of bone, cartilage, and soft tissue [11, 12]. In the following sections a few of theses studies are summarized to demonstrate the use of scaffold fabrication methods for tissue engineering.

9.2 FOAM SCAFFOLDS

9.2.1 Supercritical Fluid Technology

A supercritical fluid (SCF) is a substance that has exceeded its critical temperature T_c and pressure p_c, the so-called critical point. When this occurs the fluid possesses the density and solvent properties of a liquid and the viscosity and diffusivity of a gas. Further compression will not lead to a phase transition in the fluid [13]. A widely used fluid is supercritical carbon dioxide ($scCO_2$). Its low cost, nonflammability, low toxicity, and stability [14] make it an ideal medium for scaffold production.

A fabrication cycle for supercritical fluid–foamed scaffolds consists of fill, soak, and vent phases. During the fill, CO_2 enters the chamber under high pressure until the operating temperature and pressure, which are higher than T_c and p_c, are reached. Once CO_2 reaches its critical point at 31°C and 73.8 bar, it starts to lower the glass transition temperature T_g of the polymer. The T_g of a polymer is the temperature at which it changes from a glassy (hard, rigid) state to a rubbery (soft, flexible) state [15]. Once the T_g is lower than the operating temperature, the polymer will liquefy [13]. $scCO_2$ is applied under pressure until the polymer is dissolved and a gas-saturated solution has been formed (soak). In order to minimize free energy, gas molecules nucleate within the polymer solution. During depressurization (vent), CO_2 leaves the polymer solution, which causes the T_g to rise, leading to polymer solidification and hence pore generation [16]. The foam porosity is generated by a thermodynamic instability resulting from the depressurization around the dissolved polymer [16, 17]. Porosity and pore size can be controlled by altering the amount of CO_2 dissolved in the polymer and the depressurization rate [17]. Tai et al. investigated the effects of processing parameters such as temperature, pressure, soak/vent time, and polymer characteristics on scaffold properties [18] (Table 9.2). The possibility of tailoring the structure and pore size of poly(lactic-*co*-glycolic) (PLGA) and poly(DL-lactic acid) (PDLLA) scaffolds was demonstrated. Due to an increase in diffusion rate at higher temperatures, larger pores were observed. On the other hand, higher pressures and

TABLE 9.2 Foam Scaffolds Fabricated Using Supercritical Carbon Dioxide

Scaffold Material	Preprocessing	Pore Size, μm	Porosity, %	Mechanical Properties, MPa	Application	Ref.
PDLLA, PLGA, PCL	Polymer powders	0.05–0.1; 0.5–5			Protein release, bone engineering	24
PEMA/THFM, PDLLA, SIS/THFMA	PEMA/THFMA, SIS/THFMA blends	50–900	81.7–88		Cell attachment and proliferation, cartilage and bone tissue engineering	109–111
PDLLA, PLGA		39–580	34–78		Controlled release of growth factors for tissue engineering	18
PLGA	Emulsion with aqueous protein phase				bFGF delivery	112
PLA, PLA + 10% HA, PLA + 10% TCP		200–400	78–92	CM: 10–250	Seeding of human primary bone cells	19
PDLLA	VEGF/PLA lyophilized overnight				VEGF delivery, angiogenesis, bone tissue engineering	25, 114
PDLLA		440 ± 170	85	CS: 0.85 CM: 7.0	Cell penetration and tissue infiltration; microstructure and macrostructure for any scaffold type	115

longer soaking times resulted in higher nucleation densities and the formation of smaller pores. Higher depressurization rates led to the formation of smaller pores due to the shorter times allowed for pore growth [18].

The effects of depressurization rate, cooling rate, saturation pressure, and polymer/ceramic ratio on the three-dimensional (3D) structures and mechanical strengths of polymer and polymer–composite scaffolds were investigated by Mathieu et al. [19]. For polymer scaffolds, a high depressurization rate resulted in ~75% porosity and pore diameters larger than 400 μm; strut and pore sizes were in the range of those of trabecular bone. Lower pore numbers and porosities were observed for slower venting times. Cooling rate and nucleation density influenced the anisotropy in pore morphology; slow cooling allowed pores to elongate. To examine the influence of filler content (β-tricalcium phosphate [β-TCP] and hydroxyapatite [HA]) on foam morphology, processing conditions remained constant. Higher concentrations of ceramics resulted in an increase in solution viscosity and, therefore, caused a reduction in wall ruptures, pore interconnection, pore size, and porosity. Strut sizes, however, increased. Morphological analysis of polymer and polymer–composite scaffolds has shown that structural parameters such as pore size, strut size, and porosity are within the range of those of trabecular bone, and that scaffolds fabricated by $scCO_2$ are, therefore, suitable for its replacement [19].

Supercritical fluid processing strategies were further applied for the fabrication of micrometer-sized particles for use in the fabrication of pharmaceutics [16]. One strategy is the rapid expansion of supercritical solutions (RESS) into low-pressure and low-temperature surroundings, which results in a rapid solidification of the polymer [20, 21]. Another strategy is the gas antisolvent technique (GAS). Here, the SCF acts as an antisolvent precipitating a solute from an organic solvent and thus rendering the solidified polymer in the shape of microparticles [22]. The third method is the formation of particles from gas-saturated solutions (PGSS) [23].

The fabrication of 3D matrixes by supercritical fluid technology enables incorporation of heat- and solvent-sensitive pharmaceuticals and biological agents [24, 25]. However, the range of materials that can be used for this technique is limited. Amorphous polymers or polymer fractions are required since crystalline polymers do not form porous structures due to their inability to dissolve in gases [26]. Furthermore, it was observed that the rapid diffusion of CO_2 from the scaffold during the vent leads to the formation of a nonporous polymer skin at the scaffold surface [17, 27–29].

9.2.1.1 The Application of Foam Scaffolds Produced by Supercritical Carbon Dioxide. Supercritical fluid technology has been utilized for various applications, such as the incorporation of growth factors [24, 30], pharmaceuticals [31], and plasmids [32] and the fabrication of tissue engineering scaffolds.

Porous biodegradable PDLLA scaffolds were fabricated by supercritical fluid technology [25]. Scanning electron microscopic images are shown in Fig. 9.2. Since this technique does not require organic solvents [33, 34], researchers were able to incorporate the thermally sensitive and solvent-sensitive protein vascular endothelial growth factor (VEGF) [25].

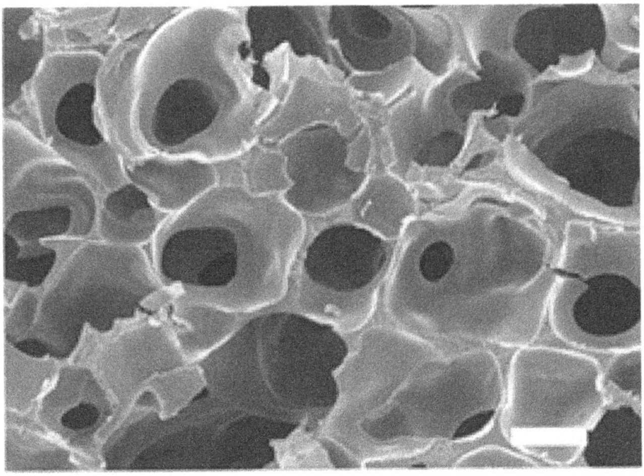

FIGURE 9.2 Foam scaffold fabricated by supercritical fluid technology. An SEM micrograph was taken of a poly(lactic acid) scaffold into which vascular endothelia growth factor was encapsulated. Scale bar = 200 μm. (From Reference 25 with kind permission from Elsevier.)

To investigate the applicability of these scaffolds for bone tissue engineering, control scaffolds (PDLLA only) and human bone marrow stromal cell (hBMSC)–seeded VEGF-containing PDLLA scaffolds were implanted into 5 mm large mouse femur segmental defects for 4 weeks. Results taken from two-dimensional *in vivo* digital X-ray images of these femur defects at 0 and 28 days are shown in Fig. 9.3. Significant bone ingrowth and repair were observed for the VEGF/PLA-hBMSC–treated defects after 28 days. These observations were further confirmed by histological staining and micro computed tomographic (microCT) analysis. Three-dimensional microCT images of the control and VEGF/PLA-hBMSC treated defects are shown in Fig. 9.4.

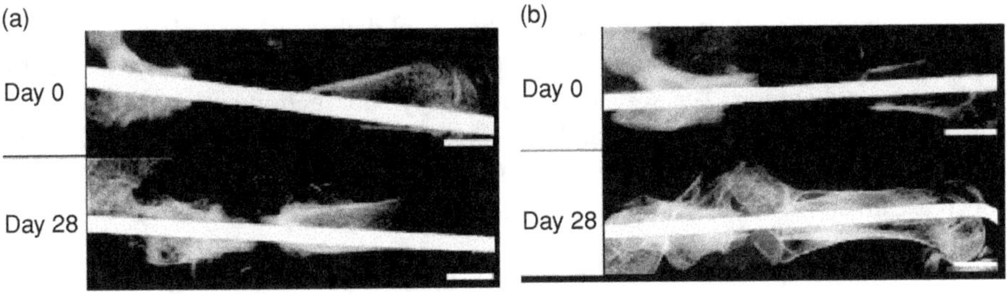

FIGURE 9.3 Digital X-ray images of mouse segmental femur defects. Images of **(a)** control segmental defects (PLA only) and **(b)** defects treated with VEGF-incorporated PLA scaffold seeded with hBMSC were taken at 0 and 28 days postimplantation. Scale bar = 2 mm. (From Reference 25 with kind permission from Elsevier.)

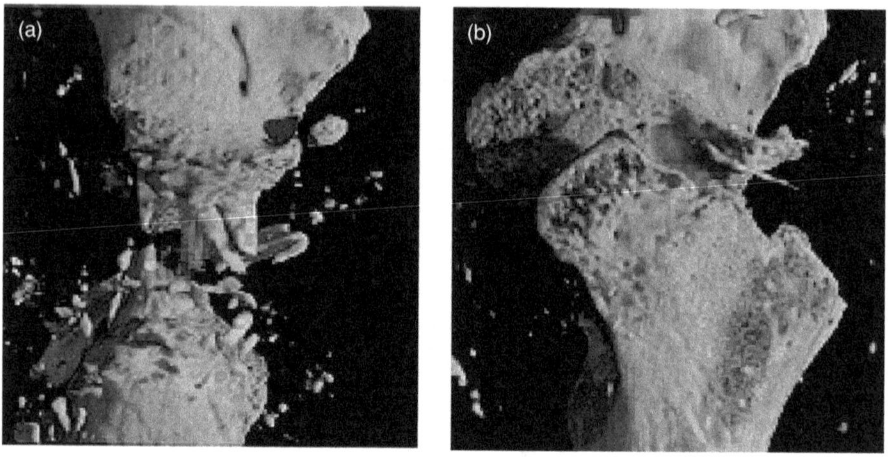

FIGURE 9.4 Micro computed tomographic three-dimensional images of mouse segmental femur defects at 28 days after operation. High-resolution images of **(a)** control defects (PLA only) and **(b)** defects treated with VEGF-containing PLA scaffold seeded with hBMSC. (From Reference 25 with kind permission from Elsevier.)

In this study the local delivery of bioactive growth factors and human bone marrow stromal cells from the biodegradable PDLLA foam scaffolds to the critical-sized bone defects was demonstrated [25], leading to significant bone ingrowth and repair of the injured site after 28 days. These results offer therapeutic treatment in orthopedics.

9.2.2 Thermally Induced Phase Separation

Phase separation is based on the exposure of a homogenous polymer–solvent solution to either immiscible solvents or to lower temperatures (thermally induced phase separation [TIPS]), by which the system then becomes thermodynamically unstable. In order to reduce its free energy, the polymer–solvent mixture separates into a polymer-rich phase and a polymer-lean phase [35].

Solid–liquid phase separation occurs when a solid phase is created within a liquid phase. In this case, the temperature of the polymer solution is lowered to allow solvent crystallization. The solvent is removed by sublimation or solvent exchange, leaving void space (pores) [36] and solidified polymer behind [10]. Liquid–liquid phase separation occurs when continuous liquid–liquid phases are created due to a decrease in temperature, thus leading to the fabrication of scaffolds with an open-pore structure [10].

Different tissues such as nerve, muscle, tendon, and ligament have fibrous or tubular architectures. For the formation of such tissues, highly porous scaffolds with tissue-similar structures are required [10]. By altering the process parameters, TIPS enables control over pore size and shape [38, 39], enabling the fabrication of membranes with micrometer scale pores and microcellular foams [38, 39].

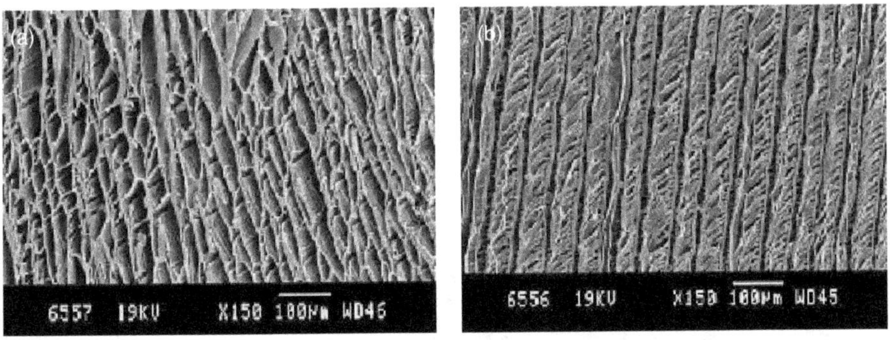

FIGURE 9.5 SEM micrographs of scaffolds fabricated by TIPS. Scaffolds with oriented microtubular architecture were prepared from PLGA/Bioglass. (**a**) Cross section. (**b**) Bottom surface. Scale bars = 100 μm. (From Reference 47 with kind permission from Elsevier.)

Ma and coworkers described the fabrication of parallel-aligned microtubules with mechanical properties similar to those of tubular and fibrillar tissue, which were shown to support the cell arrangement into oriented tissue [40, 41]. First, poly(L-lactic acid) (PLLA) was dissolved in benzene to prepare a 5% (w/v) solution. To induce solid–liquid phase separation, the solution was placed in a freezer. The oriented microtubular scaffold architecture was obtained by applying a uniaxial temperature gradient. The phase-separated solvent/polymer mixture was then freeze-dried at –5°C to –10°C under vacuum.

Figure 9.5 shows scanning electron images of PLGA/Bioglass composite scaffolds with oriented microtubular architecture [47].

Scaffolds fabricated by thermally induced phase separation are characterized by high porosities and highly interconnected pores [42]. However, this method offers only a limited control over pore size distribution and range [10, 36]. Moreover, it is highly dependent on the experimentalist and the equipment [33, 34]. Long fabrication times and the use of organic solvents are further limitations of this technique [33, 34, 42]. Fabrication and characteristics of diverse TIPS scaffolds are summarized in Table 9.3.

9.2.2.1 The Application of Foam Scaffolds Produced by Thermally Induced Phase Separation. Several attempts to engineer bone tissue have been pursued, applying scaffolds in combination with biologically active molecules and cells for the regeneration and replacement of skeletal defects as a major clinical orthopedic requirement [25, 43].

Alongside scaffolds fabricated by SCF, the Joint and Research Group at the University of Southampton also investigated the use of TIPS-fabricated PDLLA scaffolds that contained 45S5 Bioglass (Fig. 9.6a) in different concentrations for skeletal tissue engineering [43]. Bioglass is a bioactive material that consists of 45% (SiO_2), 24.5% sodium oxide (Na_2O), 24.5% calcium oxide (CaO), and 6 wt% phosphorous pentoxide (P_2O_5) and is assumed to be osteoproductive and osteoconductive [44, 45].

TABLE 9.3 Foam Scaffolds Fabricated by Thermally Induced Phase Separation

Scaffold Material	Preprocessing	Pore size, μm	Porosity, %	Mechanical Properties, MPa	Application	Ref.
PLLA, PLGA	Solid–liquid PS	54–113	89–97	CM: 1.5–9 YS :0.1–0.6	Parallel microtubules or random pore structure, various tubular or fibrillar oriented tissues, drug delivery, wound dressing	41
PLGA/ Bioglass		10–100	93	CS: 11.8 (axial) CS: 41.2 (diametric)	Tubular foams for tubular tissues such as trachea, small intestine, esophagus	47
PLGA		50–300			Tubular scaffolds for soft tissue engineering; lumenal intestinal epithelial cell growth	116, 117
PLGA/ Bioglass	Compression molding/ TIPS	100	>90		Soft tissue engineering	115
PDLLA		10–100	>90		Bone tissue engineering	43
PDLLA, PLLA, PLGA	Liquid nitrogen or −15°C	10–170	76–92		Drug delivery	38, 39
PLGA		<300		E: 0.8–0.9		48
PLGA	Injection molding + TIPS		90		Schwann cells, nerve regeneration	118
PLGA	Porous microsphere formation				Drug delivery	119

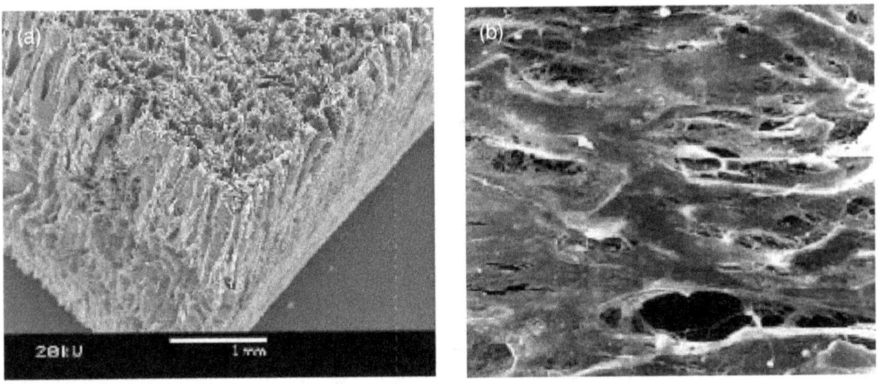

FIGURE 9.6 PDLLA /45S5 Bioglass composite scaffolds fabricated for bone tissue engineering. **(a)** SEM micrograph of a PDLLA /45S5 Bioglass scaffold. **(b)** SEM image of hBMSC adhered to composite scaffold after 6 weeks *in vitro* culture. Images are examples for 5 wt% PDLLA/45S5 Bioglass. Scale bar = 1 mm. (From Reference 43 with kind permission from Elsevier.)

Biologically active materials are defined as substances that induce a particular biological response at the boundary of the material and the tissue, thus leading to bond formation between them [44]. In contact with biological fluids or after implantation, a thin layer of hydroxycarbonate apatite is formed on the 45S5 Bioglass surface, leading to bond formation with the surrounding hard tissue [46,47].

Yang et al. seeded the aforementioned scaffolds with human bone marrow mesenchymal stem cells (hBMSC) and evaluated the *in vitro* and *in vivo* potential of these scaffolds for skeletal tissue engineering [43]. The formation of collagen and proteoglycans *in vitro* and *in vivo* was confirmed by positive Alcian blue and Sirius red staining. Significantly stronger staining was observed after 8 weeks *in vivo*. Viable, proliferating cells and the deposition of extracellular matrix were observed in all scaffolds (Fig. 9.6 b). Even though Yang et al. demonstrated the potential of PDLLA/Bioglass–containing scaffolds for skeletal tissue engineering, further experiments are necessary to investigate the optimum 45S5 Bioglass concentration and to explain the differences between the results obtained *in vitro* and *in vivo* [43].

Another example for the application of foam scaffolds fabricated via thermally induced phase separation has been described for the investigation of tissue ingrowth and vascularization for soft tissue engineering [48] utilizing a previously reported [49] and later adapted [50, 51] arteriovenous loop (AV loop) model. Cassell et al. [50] applied the AV loop model for the formation of blood vessels utilizing poly(lactic-*co*-glycolic acid) (PLGA) scaffolds produced by solvent casting and particulate leaching. However, these scaffolds degraded during histological examinations, which prevented quantitative analysis [50,51].

Therefore, scaffolds manufactured via TIPS were used since these scaffolds were shown to have a slower degradation rate, allowing their morphological and

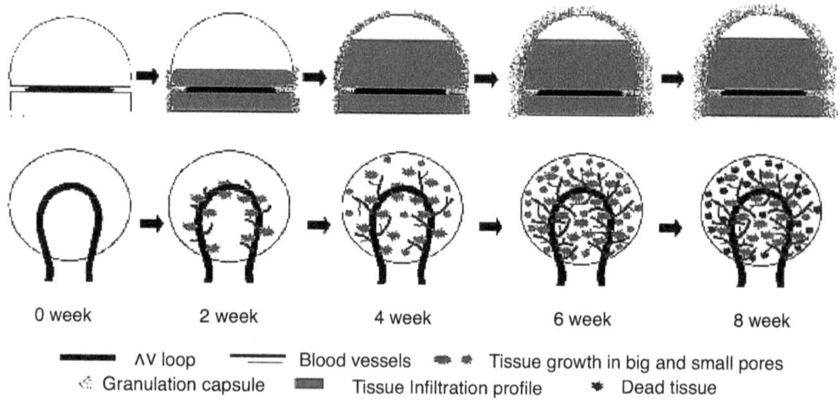

FIGURE 9.7 AV loop model and TIPS scaffolds for the examination of vascularization for soft tissue engineering. Illustration of tissue ingrowth and vascularization into the dome-shaped TIPS scaffold/AV loop construct (longitudinal and horizontal cross sections). (From Reference 48 with kind permission from Elsevier.)

morphometric examination [48]. In this study an AV loop was placed between two TIPS scaffolds (Fig. 9.7), with the lower section being disk shaped and the top section being domed shaped. The scaffold/AV loop construct was transferred into a plastic chamber and implanted beneath the inguinal region of rat skin [48]. It was expected that after implantation blood vessels would spread out from the central AV loop and grow throughout the scaffold [49]. After removing the construct after 2, 4, 6, and 8 weeks (Fig. 9.7) Cao et al. observed continuous tissue ingrowth and blood vessel formation in both the smaller and larger pores of the scaffold, thus leading to a reduction in the polymer/void space ratio [48]. Healthy connective tissue and granulation capsules surrounding the scaffold construct were found. However, tissue death within the construct was also documented. It was assumed that vascularization was negatively influenced by the rapid percolation of foreign body cells, which resulted in the pores smaller than 300 μm being closed, thus preventing further tissue support and vascularization [48].

9.2.3 Solvent Casting/Particulate Leaching

Solvent casting/particulate leaching is an easy and commonly used method for fabricating porous foam scaffolds. Briefly, porogen particles such as salt or sugar [52,53] are sieved into fractions of desired sizes and added into a three-dimensional mold. The polymer is dissolved using an organic solvent and then cast over the porogen particles. After the solvent is allowed to evaporate, porogen particles are embedded in the polymer and subsequently leached out [3, 10, 35, 36, 42, 54]. The shape and size of the porogen particles enable control over pore shape and size, and the polymer/porogen ratio determines the porosity [10, 36] (Fig. 9.8). Increasing the

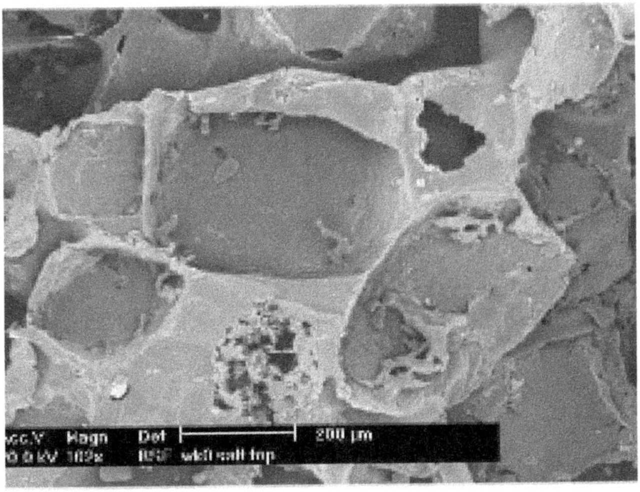

FIGURE 9.8 SEM micrograph of a foam scaffold fabricated by solvent casting/particulate leaching. PLGA disks were prepared using NaCl as a porogen. Pore shape is determined by the rectangular shape of salt particles. Scale bar = 200 μm. (From Reference 48 with kind permission from Elsevier.)

porogen concentration leads to thin 3D scaffolds that are limited in shape and pore wall thickness [33]. If the porogen concentration is too low, the resulting scaffolds will exhibit a nonporous polymer layer on their surface. Solvent casting/particulate leaching has also been applied for the fabrication of scaffolds containing bioactive molecules [55,56]. However, porogen and solvent residues can lead to denaturation and hence reduction of the activity of these molecules [42]. Further limitations of this technique include long fabrication times, no control over window size, and restricted pore interconnectivity [36] (Table 9.1). Interconnected pores play an important role for cell adhesion [9,57,58] and cell migration into polymer matrixes [59,60], as well as for the transport of cellular waste products and nutrients [61,62]. To overcome the limitations of low pore interconnectivity, researchers have modified this fabrication method in a variety of ways. Table 9.4 summarizes several approaches to the fabrication of foam scaffolds by solvent casting/particulate leaching. Thomson et al. modified the salt-leaching technique by including compression molding of gelatin spheres and polymer powders [63]. Polymer foams with a relatively low porosity (~70%) were fabricated, and gelatin residues were observed in the scaffolds. The interconnectivity of pores was not well controlled [40]. Hence, Ma utilized paraffin spheres as a pore-forming compound, also termed a porogen. For the creation of interconnected pores, spheres were bonded together by heat sintering. Porosities greater than 90% were obtained. Porosity and pore size were controlled by changing the concentration of polymer in the solution, the size of the paraffin spheres, and the number of casting steps [40].

TABLE 9.4 Foam Scaffolds Produced by Solvent Casting Particulate Leaching

Scaffold Material	Preprocessing	Pore Size, μm	Porosity, %	Mechanical Properties, MPa	Application	Ref.
PLLA	Sintering		95–96	CM: 0.023–0.082		40
PLGA	Sintering	31 –78	97	CM: 0.06 – 0.13		120
PLCL				E: 0.6 – 5.2	Patient-specific arterial scaffolds	121
PDLLA, PLATMC, PCL	Sintering		>70			122
PGS	Sintering/ nonsintering	75–150; 5–20	>85	EM: >0.004–0.007	Soft tissue engineering	123
Estane	Water as nonsolvent		>72–87	CS: 0.04–0.4	Scaffolds for meniscus	124
Gelatin		350	73 – 88	TT: 0.001–0.008	Re-reepithelialization on skin defects	125
PLLA	Sugar spheres fabrication by emulsion technique		~98	CM: 0.17–0.25	Formation of bone-like apatite layers for bone tissue engineering	53
PLGA, PLGA/PEG		71–154	53–92	SM: 0.91–9.55	Soft tissue engineering, e.g., skin, intestine, vascular grafts	52
PLGA	Ground polymer/ NaCl compression disks	100–300	94		Smooth muscle tissue, regulation of cell phenotype	126
Polyetherurethane		30–450	64			127, 128
PLGA		250–350			Bone tissue engineering, intra-membranous bone formation	55, 56
PLLA	Collagen coated	200–500	91		Neo cartilage formation	104

9.2.3.1 The Application of Foam Scaffolds Produced by Solvent Casting Particulate Leaching. The solvent casting/particulate leaching technique was applied to fabricate disk-like porous poly(D,L-lactide-*co*-glycolide) (PLGA) scaffolds expressing sustained and localized release of active ascorbate-2-phosphate (AsP) and dexamethasone (Dex) [55]. AsP is known to stimulate cell proliferation and collagen synthesis and to influence alkaline phosphatase activity [64, 65]. Dex initiates osteogenic differentiation of human marrow stromal cells at an early stage [66]. These osteogenic scaffolds were investigated for their suitability for *in vivo* bone formation [56]. In this study, hBMSC were seeded onto control (PLGA only) and osteogenic (AsP and Dex) scaffolds. The cell–scaffold complexes were implanted subcutaneously into immune-deficient mice.

Higher calcium deposition and alkaline phosphatase expression were observed in osteogenic scaffolds. Alkaline phosphatase is an enzyme produced by immature bone cells, so-called osteoblasts. It is responsible for the mineralization of bone, and calcium is an inorganic constituent of the bone matrix. Von Kossa stains were prepared to confirm increased mineralized bone formation (Fig. 9.9a, b).

Two different types of bone tissue generation can be distinguished: Intramembranous ossification occurs during fetal development of the skeletal system, namely the healing of bone fractures [67] and the bone formation of the skull [68]. Endochondral ossification, on the other hand, leads to the formation of long bones (e.g., limbs), which are formed from cartilage. To investigate the type of bone formation that occurred, Safranin-O staining was performed. Safranin-O is used to determine the presence of cartilage (cells are stained red/orange) in tissue samples. Negative Safranin-O staining (Fig. 9.9c) suggested that intramembranous ossification occurred.

Thus, this study by Kim et al. presented a successful approach for the intramembranous formation of mineralized bone tissue *in vivo* using osteogenic foam scaffolds and hBMSC [56].

9.3 FIBROUS SCAFFOLDS FABRICATED VIA ELECTROSPINNING

One function of tissue engineering scaffolds is to mimic the extracellular matrix (ECM), the native environment of cells within the body, which consists of nanofibers, ridges, and pores [2, 4]. Electrospinning leads to the formation of a mesh of aligned or random/nonwoven fibers in the submicrometer range suitable to serve as a temporary artificial extracellular matrix for cell adhesion, proliferation, and differentiation.

Table 9.5 summarizes various approaches to the fabrication of electrospun fiber meshes and their characteristics.

Electrospinning is a straightforward, cost-effective, and versatile technique [2, 4]. Its origin can be traced back to 1934 [69, 70]. A basic electrospinning apparatus consists of a syringe pump, a high-voltage power supply with negative or positive polarity, a syringe with a needle containing the polymer solution, and a collector. The collector and syringe needle act as electrodes. Depending on the spinning process, there are different collector geometries (e.g., rotating drums [71–73], collecting

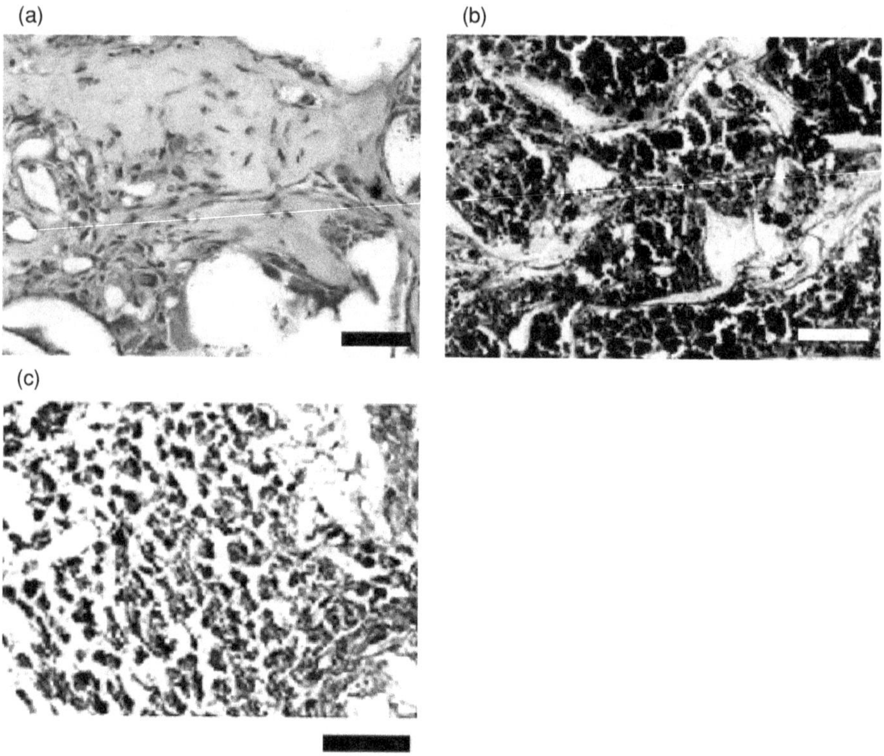

FIGURE 9.9 Representative histological sections of hBMSC and scaffold complexes at week 9 after implantation into athymic mice. Implants of (**a**) control and (**b**) osteogenic scaffolds were excised, fixed, and embedded in paraffin. For the visualization of mineralized bone tissue, the samples were stained with von Kossa. Osteogenic scaffolds were stained stronger [black in (b)] than control scaffolds. (**c**) Negative Safranin-O-staining indicated that no cartilage was present during the ossification. Scale bar = 50 μm. (From Reference 56 with kind permission from Elsevier.)

rings [74], flat metal plates), all of which can influence the alignment of fibers [4] (Fig. 9.10).

To initiate the spinning process, a polymer is dissolved in an organic solvent and then transferred into a syringe needle. Applying high voltage to the polymer solution leads to the accumulation of free ions at the solution surface. An electric field is created between the two electrodes, and a flow rate is applied to the polymer solution [4]. Free charges within the polymer solution travel toward the electrode of opposite polarity. Hence, tensile forces are transferred to the polymer solution [70, 75]. When the electrostatic forces between the two electrodes overcome the surface tension of the solution, a charged polymer jet is ejected from the needle tip [4, 70, 76]. While the polymer jet travels toward the collector, it elongates and thins, and the solvent

TABLE 9.5 Fibrous Scaffolds Fabricated by Electrospinning

Scaffold Material	Preprocessing	Fiber Diameter, μm	Porosity, %	Mechanical Properties, MPa	Application	Ref.
PLA	Addition of organic salt	0.01–2.4				79
PLLA, PLGA/PLA, PLGA/PEG-PLA		~1	71–78		Heart tissue constructs	129
PLGA		0.5–0.8	92	TM: 323 UTS: 23	Skin and cartilage tissue engineering	130
PLA-PEG/PLGA	Fabrication of PLA-PEG copolymer	0.25–5		TM: 35	Plasmid DNA release, gene delivery	80
Collagen, elastin, PLGA					Vessel matrix, vascular bypass craft	85
Gelatin		0.6–3	89–94		Skin tissue engineering, wound healing	90
PLGA		0.15– 6	38–60	TM: 39–79	Skin tissue engineering	140

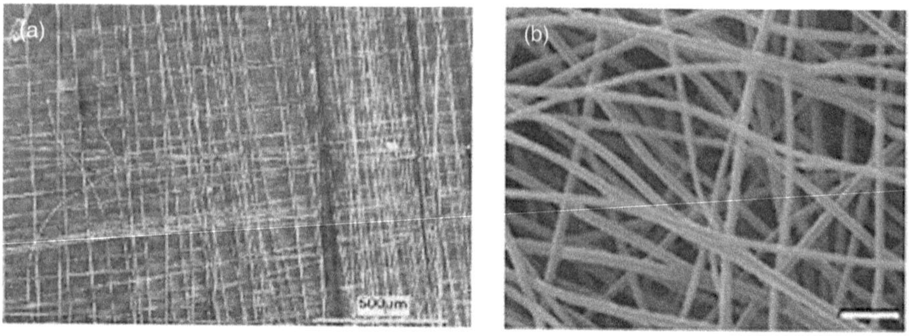

FIGURE 9.10 SEM micrographs of electrospun fibers collected on collectors of different geometries. **(a)** Parallel-aligned polystyrene fibers were collected on a rotating collector (from Reference 72, with kind permission from the American Institute of Physics). **(b)** Randomly oriented PDLLA fibers were collected on a collecting plate. Scale bars = (a) 500 μm and (b) 10 μm. (From Reference 141 with kind permission from Elsevier.)

evaporates, resulting in the collection of solid fibers. The fiber morphology and characteristics are influenced by the process parameters such as flow rate, collecting distance, collector geometry, and applied voltage, as well as polymer concentration and conductivity [70, 77]. Several studies have been carried out to investigate the influence of these parameters and to optimize fiber morphology [78–83].

9.3.1 The Application of Electrospun Fibrous Scaffolds in Tissue Engineering

Due to their properties (dense fiber network, porosity, pore sizes), submicrometer-scaled fibers are applied in various fields, such as pharmacy (drug release), the textile industry, catalysis, filtration, and medicine (scaffolds for tissue engineering) [4, 84].

To reduce the limitations of the currently applied synthetic and autologous grafts for cardiovascular surgery, Stitzel et al. used electrospinning to fabricate scaffolds that could be applied as alternatives to vascular transplants [85]. Biodegradable vascular scaffolds were made of blends of PLGA, collagen type I, and elastin. Optimization of blend composition and electrospinning parameters led to the fabrication of 12 cm long scaffolds with 1 mm thick walls (Fig. 9.11a). The vessel matrices consisted of randomly oriented fibers with an average diameter of 0.720 ± 0.35 μm. To enhance stability and strength, the electrospun scaffold was crosslinked by applying glutaraldehyde. Immunohistochemical staining confirmed a homogeneous distribution of elastin and collagen throughout the scaffold. Biocompatibility of the electrospun matrixes was investigated by seeding endothelial and smooth muscle cells. Cell viability and proliferation were determined by mitochondrial metabolic (MTT) activity and neutral red assay. Here, more than 70% of the seeded cells survived, and double immunostaining revealed the attachment of smooth muscle and endothelial cells

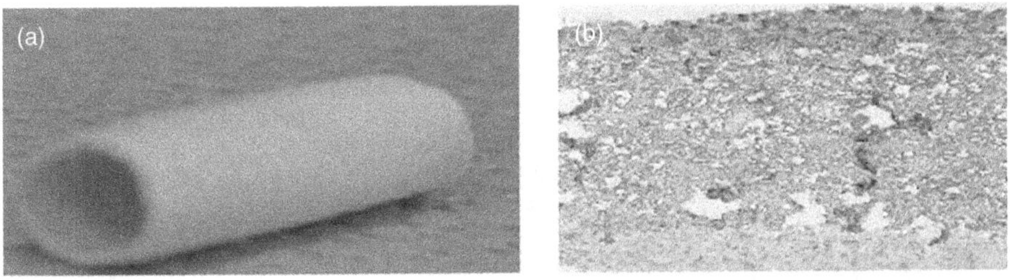

FIGURE 9.11 Electrospun scaffold for soft tissue engineering. PLGA, collagen type I, and elastin were electrospun into a 12 cm long scaffold with a wall thickness of 1 mm. **(a)** A representative 2 cm section of the vascular scaffold before crosslinking. **(b)** Double immunostaining was performed to demonstrate adhesion of endothelia cells (light gray) and smooth muscle cells (dark gray) onto electrospun vascular matrix. (From Reference 85 with kind permission from Elsevier.)

on both sides of the scaffold (Fig. 9.11b). Moreover, no visible changes of physical impairments were observed after implantation of scaffolds subcutaneously into mice. Hence, Stitzel et al. fabricated biocompatible electrospun vascular scaffolds that possessed mechanical properties and material compositions comparable to those of blood vessels and are, therefore, promising for clinical application. Further studies will investigate the applicability of these scaffolds as bypass grafts in sheep [85].

Another application of electrospun fibers is the treatment of severe burns. To overcome the limitations of conventional skin substitutes, tissue-engineered skin substitutes have become more important. The advantages of tissue-engineered skin grafts include, for example, the larger expansion of surface area from donor skin and the need for patients to undergo fewer surgeries [86]. Skin substitutes usually consist of fibroblasts and/or keratinocytes on collagen fabricated by freeze-drying, which can lead to heterogeneous scaffold structures [87–89].

Powell et al. compared skin substitutes fabricated by freeze-drying (FD) and electrospinning (ES), where electrospinning was assumed to yield a more homogeneous scaffold structure [90]. After fabrication, electrospun scaffolds consisted of randomly aligned fibers with an average diameter of 1.3 μm and were made of bovine collagen. For the assessment of cell maturation, organization, distribution, and proliferation, human dermal fibroblasts (hF) were seeded onto electrospun scaffolds and cultured *in vitro*. Immunohistological images (Fig. 9.12) showed cell alignment (keratinized epithelium) and the presence of collagen IV as evidence for the formation of a basement membrane at the junction between dermis and epidermis [90]. The basement membrane is a thin fiber sheet located either beneath the epithelium (covering organs and cavities) or endothelium (covering the interior of blood vessels) anchoring cells to the connective tissue.

Posttransplantation of skin substitutes (electrospun scaffold seeded with hF) into athymic mice, no bovine collagen was observed in the wounds after 8 weeks (Fig. 9.12). Hence, the engraftment rate was 100%, and blood vessel formation

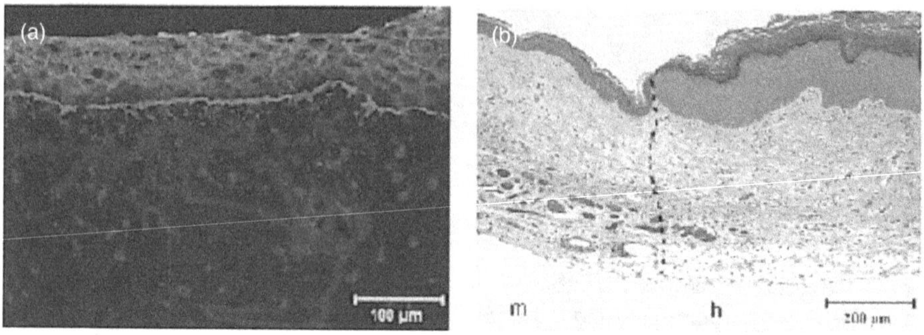

FIGURE 9.12 Electrospun skin graft for full-thickness burns. (**a**) Cell nuclei (dark gray), human involucrin (light gray), and human collagen type IV (line of very light gray) were observed on immunohistological images of an electrospun scaffold seeded with hF and cultured *in vitro*. (**b**) Masson trichrome stains of healed skin 8 weeks postimplantation showed no residuals of ES. The junction between murine (m) and human (h) skin is indicated by the dashed line. Scale bars = (a) 100 µm and (b) 200 µm. (From Reference 90 with kind permission from Elsevier.)

occurred in both human and mouse tissue. In addition, the junctions between human and mouse skin tissue could be detected.

In comparison with freeze-drying, electrospun scaffolds lead to the formation of skin replacements with properties (cell maturation, proliferation, organization) similar to those of the clinically applied skin substitutes [90].

9.4 POLYMER BEDS

Several approaches have been described for the production of 3D porous polymer beds from microparticles or microspheres (Table 9.6).

Polymeric microspheres are most commonly fabricated by an emulsion solvent evaporation technique also called single or oil-in-water emulsion technique. First, the polymer is dissolved in a volatile organic solvent (oil phase), which is immiscible in water. To form an oil-in-water emulsion, this polymer solution is mixed with an aqueous phase containing, for example, poly(vinyl alcohol) (PVA), which acts as a stabilizer and/or viscosity builder. While the emulsion is agitated, the organic solvent evaporates, resulting in solid microspheres which can be collected by filtration or centrifugation [91]. To fabricate porous tissue-engineering scaffolds, microspheres or polymer powders are sintered by applying heat [92–94] or a chemical [95] treatment. Nukavarapu et al. described an approach for the synthesis of microsphere composite scaffolds synthesized from polyphosphazene (PPH) and hydroxyapatite particles utilizing a solvent/nonsolvent sintering method in which polymer dissolution and precipitation were influenced by the hydrogen bonding, polar, and dispersion forces of the solvent (Flory-Huggins solution theory) [96]. Here, tetrahydrofuran (THF) was

TABLE 9.6 Polymer Beds Fabricated by Various Sintering Procedures

Scaffold Material	Preprocessing	Particle Size, μm	Pore size, μm	Porosity, %	Mechanical Properties, MPa	Application	Ref.
HA powder	HA heat sintering	4–50	<1 to >100	30–64		Vascularized bone grafts	98
HA powder	HA heat sintering					Clinical applications of HA particles	131
PLGA/BG composite	Microsphere heat sintering	350–500				Bone tissue engineering, differentiation of MSC into osteoblasts	132
PLGA	Microsphere heat sintering	100–250	100	75		Bone tissue engineering	92
PLGA	Microsphere heat sintering	355–600			E: 0.691–64.7	Bone tissue engineering	93
Calcium phosphate glass, sucrose particles	Sucrose particle heat sintering	<500 sucrose particles, 125 glass powder	100–500	77		Bone tissue engineering	97
PLGA-50/50	Microspheres chemical sintering: EtOH	220	~20–120	>40	EM: 0.14–0.3		95
Polyphosphazene-nano-sized HA particle composite, PLGA	Dynamic solvent/heat sintering	350–500	86–145	15–24	E: 46–81 CS: 6–13	Bone tissue engineering	96
CaP/PLGA composite		>0–200		<60			133
Ti/HA	Heat sintering	<100 nm			CS: 137–276		134
CMP/PLA composite	Pressure and heat sintering, porogen leaching	CMP: 100–300; NaCl: 300–500		90	TS: 0.13	Bone tissue engineering	135
Bioactive macroporous microsphere	Microspheres heat sintering	~240	~100			Bone tissue engineering, cell delivery	94

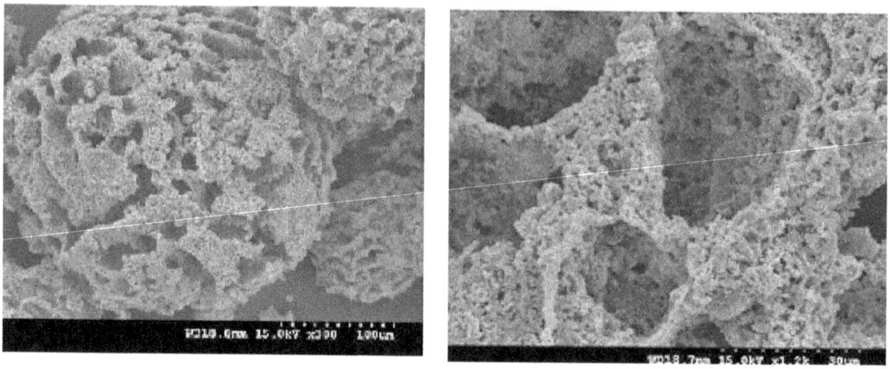

FIGURE 9.13 SEM micrographs of polymer beds fabricated from heat-sintered porous microspheres (microsphere pore size ~20–50 μm). Scale bars = (left) 100 μm and (right) 30 μm. (From Reference 94 with kind permission from Elsevier.)

used as a solvent and *n*-hexane as a nonsolvent. The solvent/nonsolvent solutions were mixed with microspheres, placed into a mold, and allowed to dry [96]. The resulting scaffolds possessed average pore diameters of 86–145 μm and compressive moduli of 46–81 MPa. Adhesion and proliferation of primary rat osteoblasts and their expression of alkaline phosphatase indicated the suitability of these scaffolds for orthopedic applications [96].

Another approach reported the fabrication of bioactive ceramic scaffolds with an interconnected macropore structure by mixing calcium silicate powder and camphene, with camphene acting as a porogen. Microspheres containing camphene and calcium silicate powder were produced by an oil-in-water emulsion technique. After sublimation of the porogen, porous microspheres were heat sintered (Fig. 9.13) [94].

The influence of microsphere and scaffold preparation on mechanical properties was investigated by Khan et al. [93]. Poly(lactic-*co*-glycolic acid) (PLGA) microspheres consisting of various polymer/ceramic ratios were sintered at 4°C and room temperature (RT) for 30, 60, and 90 min. The compressive modulus increased from 691 kPa (microspheres sintered at RT for 30 min, medium polymer/ceramic ratio) to around 65 MPa (microspheres produced at 4°C for 90 min, high polymer/ceramic ratio). Longer sintering times led to enhanced fusion of microspheres and pore closure. Thicker sphere walls, obtained at lower sintering temperatures, and higher polymer/ceramic ratios resulted in higher mechanical strengths of the scaffolds [93]. Moreover, the control over porosity was enabled by altering sintering temperature, particle size, and polymer/porogen ratio [97].

9.4.1 The Application of Particle Beds in Tissue Engineering

Vascularization is required for the transition of a scaffold from a structural backbone into a construct integrated into the host tissue [98]. Two different types of

FIGURE 9.14 Scaffold construct fabricated by 3DP and particle sintering for the formation of vascularized bone tissue. A CAD model **(a)** was designed to fabricate the scaffold by three-dimensional printing and sintering **(b)** of a powder mixture **(c)**. Parallel-printed macroscopic channels had a diameter of 1 mm. Scaffolds investigated in an *in vivo* study had a porosity of 50% and ~18 μm pore size. Scale bar in (c) = 50 μm. (From Reference 98 with kind permission from Elsevier.)

neovascularization can be distinguished, both including cells seeded onto three-dimensional matrixes. The classic model of "extrinsic" vascularization involves the ingrowth of blood vessels from the surrounding area into the scaffold [50]. However, this approach is limited by the death of transplanted cells due to slow angiogenesis and the strong inflammatory responses caused by implanted grafts.

An alternative is so-called "intrinsic" vascularization by which the formation of blood vessels occurs from the core of the implant toward its periphery [99, 100]. The structure of vascularization thus resembles that of an organoid [99, 101].

Will et al. fabricated hydroxyapatite–starch scaffolds by a combination of three-dimensional printing (3DP) and particle sintering for use as intrinsic vascularized bone graft substitutes [98]. The macroscopic shape of the scaffolds was designed by computer-aided design (CAD), and indirect three-dimensional printing was performed for its fabrication. The construct was sintered at temperatures above 1000°C and then cooled to room temperature. Images of the polymer powder and scaffold are shown in Fig. 9.14. To investigate the applicability of these scaffolds for intrinsic vascularization, the constructs were implanted subcutaneously in rats. At 4 weeks postimplantation, the constructs seemed be enclosed by fibrous tissue [98].

Fibroblast ingrowth from the periphery toward the center of the scaffold matrix was observed. Scaffold channels were occupied by mature blood vessels of different sizes, indicating elevated levels of differentiation and organization of blood vessels (revealed by alpha–smooth muscle actin staining) (Fig. 9.15). However, no bone formation was detected in the osseous sections of the scaffold construct. Neovascularization and granulocytic invasion were observed in a similar model [102]. Further *in vivo* experiments involving bone morphogenetic proteins (BMPs) are necessary to investigate whether these scaffolds assist ossification and vascularization within the respective areas of the scaffold, leading to the formation of bone-like structures with myeloid and Haversian sections [98].

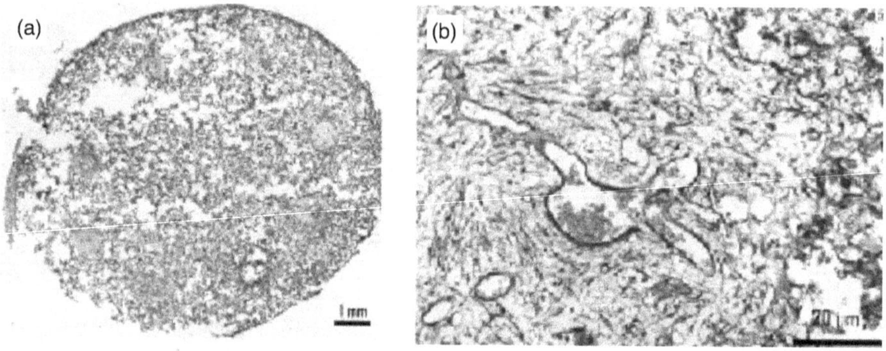

FIGURE 9.15 Histological sections of the vascularized bone graft substitute 4 weeks postimplantation. **(a)** As indicated by hematoxylin-eosin staining (H&E), the scaffold matrix is completely covered with fibrovascular structures. Eosinophilic staining was observed in the channel space, indicating the presence of extracellular matrix and red blood cells in formed blood vessels. **(b)** Positive staining for smooth muscle actin demonstrated the high level of maturation of newly formed blood vessels in the intersection between the scaffold matrix and channels (H&E staining). (From Reference 98 with kind permission from Elsevier.)

9.5 SOLID FREEFORM TECHNIQUES

Solid freeform (SFF), also called rapid prototyping technology (RP), is a technique that aims to produce 3D structures. The process is based on the computer-aided design of an object [33,34], which is expressed as two-dimensional cross sections [3]. These 2D data are then utilized by an RP machine to fabricate the 3D object by repeatedly depositing and processing layers of material from the bottom to the top, with each layer representing a 2D cross section [3].

Commonly applied RP methods to produce tissue engineering scaffolds are 3D printing (3DP), fused deposition modeling (FDM), stereolithography (SLA), 3D plotting, and phase-change jet printing [3]. Fabrication and characterization of scaffolds fabricated by SFF techniques are summarized in Table 9.7. SFF allows the manufacture of scaffolds with predefined reproducible morphology, controlled pore size and porosity [3], and an interconnected porous network [34]. Furthermore, structures that mimic the 3D environment of living tissues are fabricated [33]. However, rapid prototyping techniques are limited in the choice of material, resolution, and structural homogeneity. Reverse prototyping was developed to overcome some of those limitations [10]. In this procedure, a negative replica of the scaffold (Fig. 9.16) is manufactured. The polymer solution is cast into this mold, and solvent evaporation leads to the solidification of the polymer. The mold is then removed, leaving the polymer scaffold behind [36]. Indirect SFF fabrication methods offer more control over scaffold architecture but also result in longer fabrication times [42].

TABLE 9.7 Scaffolds Fabricated by Solid Freeform Techniques

Scaffold Material	Preprocessing	Pore size, μm	Porosity, %	Application	Ref.
PLGA	Solvent casting, particulate leaching	100–150; 300–500		Complex tissues, zygoma	125
OKM-2, PDLLA, PLGA	Addition of HA			Individual implants for maxillofacial surgery and stomatology and matrices for tissue engineering	136
TEOS	Sintered	50–200	48	Mandibular bone	137
Organic inks				Microvascular networks, anatomic healing materials, microfluidic systems	138
Chitosan/ hydroxyapatite		200–1000			139
PLGA/PLLA, PLGA/TCP, PLGA/PLA	NaCl leaching and solvent extraction	>100	55–90	Osteochondral scaffold, articular cartilage repair, joint replacement	103

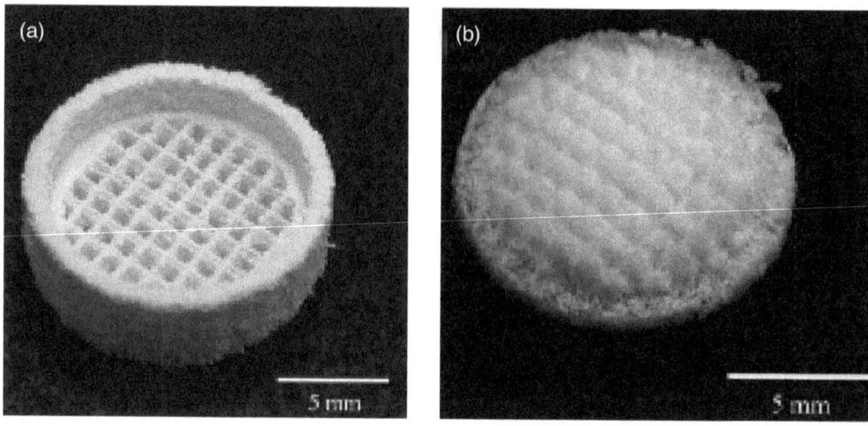

FIGURE 9.16 Reverse rapid prototyping. (**a**) A computer-aided design was created for the fabrication of the negative replica of the scaffold. (**b**) A digital image shows the polymer scaffold. Scale bars = 5 mm. (From Reference 142 with kind permission from Elsevier.)

9.5.1 The Application of Scaffolds Fabricated by Solid Freeform Techniques

Sherwood et al. applied a 3DP technique to develop an osteochondral scaffold consisting of bone and cartilage regions [103]. Each domain was characterized by a different porosity, pore size, macroarchitecture, mechanical properties, and material composition and was optimized for bone and cartilage formation (Fig. 9.17a). The lower, or bone, region was fabricated of a poly(lactic-*co*-glycolic acid)/calcium phosphate tribasic (PLGA/TCP) composite with a porosity of 55% that was designed to supply the required mechanical strength and maximum bone ingrowth. Staggered channels were inserted in the top, or cartilage, section. This region was composed of PLGA/PLA with 90% porosity [103].

In order to prevent delamination, these two areas were separated by the transition region, which consisted of PLGA/PLA and a porosity gradient. To investigate cell attachment onto the cartilage–bone composite scaffold, chondrocytes were seeded and attached preferably to the cartilage region. The staggered channels permitted the transport of nutrients and cellular waste products and allowed cell seeding throughout the entire thickness of the cartilage region, which is important for homogeneous cartilage formation [104].

Sherwood et al. demonstrated that the SFF technique enables the design and fabrication of complex scaffolds composed of different materials and different internal architectures and, thus, enables the seeding of different cell types. Moreover, *in vitro* experiments and mechanical testing suggested that these osteochondral scaffolds could lead to the repair of articular defects *in vivo* and, potentially, to the repair of whole joints for osteoarthritis treatment [103].

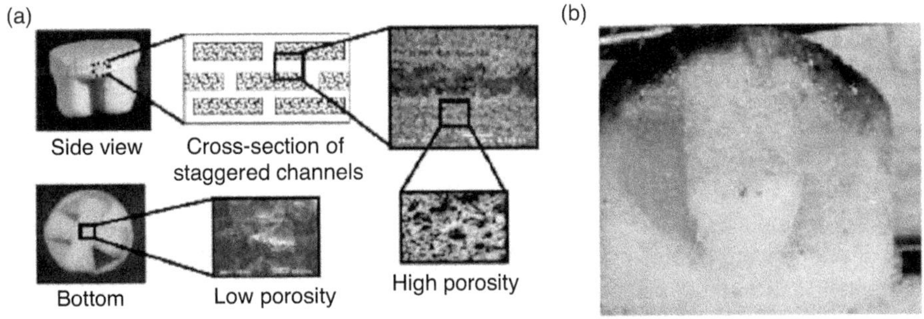

FIGURE 9.17 Osteochondral scaffold fabricated by 3DP. Scaffolds consisted of a bone and a cartilage region, which were separated by a transition section. **(a)** Each domain consisting of different materials and porosities. **(b)** Chondrocytes were seeded onto the cartilage–bone scaffold and MTT stained 24 h after seeding. Cells attached to the highly porous cartilage region. (From Reference 103 with kind permission from Elsevier.)

Even though solid freeform technologies have been widely used to fabricate scaffolds, very few have reached clinical application [11]. Medical-grade polycaprolactone (mPCL) and mPCL composites were produced using FDM by an interdisciplinary group at the National University of Singapore. These scaffolds were used in a pilot study for cranioplasty [105] for the treatment of patients suffering from chronic subdural hematoma. Subdural hematomas were trepanned and evacuated and subsequently filled with moldable, slowly degrading mPCL sheets [105]. Clinical results postoperation were positive, showing that all patients tolerated the implants, and no severe side effects were observed. Computed tomographic scans suggested that the implants were stably integrated in the host tissue and the void space was filled by new bone tissue (Fig. 9.18). Functional cranioplasty could be observed in all patients postoperation [11, 105].

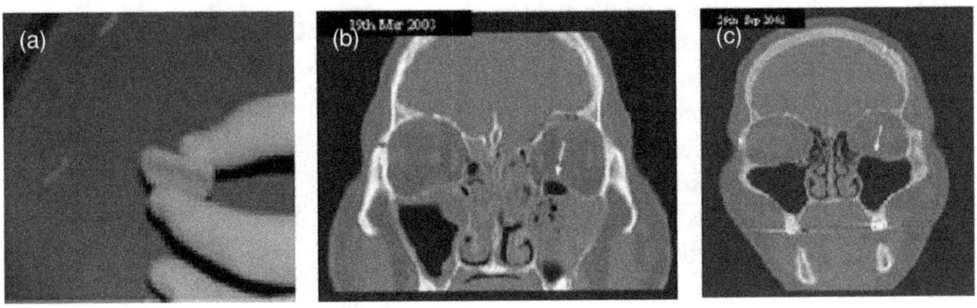

FIGURE 9.18 Biodegradable mPCL scaffolds for craniofacial reconstruction. **(a)** Moldable medical-grade PCL scaffold sheets were implanted into (**b**, *arrow*) trepanned and evacuated subdural hematomas. **(c)** Computed tomographic scans postimplantation show full regeneration of the defect site (*arrow*). (From Reference 11 with kind permission from Blackwell Publishing.)

Hence, biodegradable mPCL scaffolds were successfully applied for the regeneration and/or replacement of posttrephination defects. To date, scaffolds for cranioplasties have been received by more than 200 patients [11].

9.6 SUMMARY

Solid porous scaffolds include foam and textile scaffolds, polymer beds, and scaffolds produced by solid freeform techniques. Over the last decades, various techniques for their fabrication have been developed. Tissue engineering scaffolds are utilized in combination with, or without, cells to regenerate and repair various lost or damaged tissues, for example, bone, cartilage, or soft tissues. The scaffold type and its fabrication method depend on the desired application and target tissue.

ABBREVIATIONS

AsP	Ascorbate-2-phosphate
AV-loop	Arteriovenous loop
bFGF	Basic fibroblast growth factor
BG	Bioactive glass
BMP	Bone morphogenetic protein
CAD	Computer-aided design
CFA	Chemical foaming agent
CM	Compressive modulus
CMP	Calcium metaphosphate
CS	Compressive strength
CT	Computed tomography
Dex	Dexamethasone
DNA	Deoxyribonucleic acid
E	Young's modulus
ECM	Extracellular matrix
EM	Elasticity modulus
ES	Electrospinning
EtOH	Ethanol
FD	Freeze-drying
FDM	Fused deposition modeling
GAS	Gas antisolvent technique
HA	Hydroxyapatite
hBMSC	Human bone marrow stromal cells
H&E staining	Hematoxylin and eosin staining
hF	Human dermal fibroblasts
MicroCT	Micro computed tomography
mPCL	Medical-grade polycaprolactone
MSC	Mesenchymal stromal cells
MTT assay	Mitochondrial metabolic activity assay

OKM-2	Photopolymerizing composite polymer
PEMA/THFMA	Poly(ethyl methacrylate)/tetrahydrofurfuryl methacrylate
p_c	Critical pressure
PCL	Polycaprolactone
PDLLA	poly(D,L-lactic acid)
PGS	Poly(glycerol sebacate)
PGSS	Particles from gas saturated solutions
PLA	Poly(lactic acid)
PLA-PEG	Poly(lactic acid)–poly(ethylene glycol) composite
PLA-PEG/PLGA	Poly(lactic acid)–poly(ethylene glycol) composite/poly(lactic-*co*-glycolic acid)
PLATMC	Poly-(L-lactide-*co*-trimethylene carbonate)
PLCL	Poly(L-lactide-*co*-ε-caprolactone)
PLGA	Poly(lactic-*co*-glycolic acid)
PLGA/PEG	Poly(lactic-*co*-glycolic acid)/poly(ethylene glycol)
PLGA/PLA	Poly(lactic-*co*-glycolic acid)/poly(lactic acid)
PLGA/TCP	Poly(lactic-*co*-glycolic acid)/calcium phosphate tribasic
PLGA-50/50	PLGA with 50:50 lactic acid:glycolic acid
PLLA	Poly(L-lactic acid)
PMMA	Poly(methyl methacrylate)
PPH	Polyphosphazene
PS	Phase separation
PVA	Poly(vinyl alcohol)
RESS	Rapid expansion of supercritical solutions
RP	Rapid prototyping technology
RT	Room temperature
$scCO_2$	Supercritical carbon dioxide
SCF	Supercritical fluid
SEM	Scanning electron microscopy
SFF	solid freeform
SIS	Isoprene–styrene copolymer elastomer
SLA	Stereolithography
SM	Shear modulus
T_c	Critical temperature
TEOS	Tetraethylorthosilicate
T_g	Glass transition temperature
THF	Tetrahydrofuran
Ti/HA	Titanium/hydroxyapatite
TIPS	Thermally induced phase separation
TT	Tensile testing
UTS	Ultimate tensile stress
VEGF	Vascular endothelial growth factor
YS	Yield strength
3DP	Three-dimensional printing
45S5 Bioglass	Bioactive glass

REFERENCES

1. Langer, R., Vacanti, J. P. *Science* 1993, **260**, 920.
2. Murugan, R., Ramakrishna, S. *Tiss Eng* 2006, **12**, 435–447.
3. Sachlos, E.; Czernuszka, J. T. *Eur Cells Mater* 2003, **5**, 29–40.
4. Kumar, C. S. S. R., Ed., *Nanodevices for the Life Sciences*, Wiley-VCH, Weinheim, Germany, 2006.
5. Shi, X. F.; Sitharaman, B.; Pham, Q. P.; Liang, F.; Wu, K.; Billups, W. E.; Wilson, L. J.; Mikos, A. G. *Biomaterials* 2007, **28**, 4078–4090.
6. Karande, T. S.; Ong, J. L.; Agrawal, C. M. *Ann Biomed Eng* 2004, **32**, 1728–1743.
7. Taboas, J. M.; Maddox, R. D.; Krebsbach, P. H.; Hollister, S. J. *Biomaterials* 2003, **24**, 181–194.
8. Moore, M. J.; Jabbari, E.; Ritman, E. L.; Lu, L. C.; Currier, B. L.; Windebank, A. J.; Yaszemski, M. J. *J Biomed Mater Res A* 2004, **71A**, 258–267.
9. Gong, S. J.; Wang, H. J.; Sun, Q. S.; Xue, S. T.; Wang, J. Y. *Biomaterials* 2006, **27**, 3793–3799.
10. Ma, P. X. *Mater Today* 2004, **7**, 30–40.
11. Hutmacher, D. W.; Cool, S. *J Cell Mol Med* 2007, **11**, 654–669.
12. Hutmacher, D. W.; Sittinger, M.; Risbud, M. V. *Trends Biotechnol* 2004, **22**, 354–362.
13. Quirk, R. A.; France, R. M.; Shakesheff, K. M.; Howdle, S. M. *Curr Opin Solid State Mater Science* 2004, **8**, 313–321.
14. Kazarian, S. G. *Polymer Science C Select Top* 2000, **42**, 78–101.
15. Andrews, R. J.; Grulke, E. A. In *Polymer Handbook, Fourth Edition*, Brandrup, J.; Immergut, E. H.; Grulke, E. A., Eds., Wiley-Interscience, New York, 1999, p VI/193.
16. Cooper, A. I. *Adv Mater* 2003, **15**, 1049–1059.
17. Goel, S. K.; Beckman, E. J. *Polym Eng Science* 1994, **34**, 1137–1147.
18. Tai, H.; Mather, M. L.; Howard, D.; Wang, W.; White, L. J.; Crowe, J. A.; Morgan, S. P.; Chandra, A.; Williams, D. J.; Howdle, S. M.; Shakesheff, K. M.; Darr, J.; Layrolle, P.; Buckland, T. *Eur Cells Mater* 2007, **14**, 76–77.
19. Mathieu, L. M.; Montjovent, M. O.; Bourban, P. E.; Pioletti, D. P.; Manson, J. A. E. *J Biomed Mater Res* Part A 2005, **75A**, 89–97.
20. Kim, J. H.; Paxton, T. E.; Tomasko, D. L. *Biotechnol Prog* 1996, **12**, 650–661.
21. Matsuyama, K.; Mishima, K.; Hayashi, K. I.; Ishikawa, H.; Matsuyama, H.; Harada, T. *J Appl Polym Science* 2003, **89**, 742–752.
22. Elvassore, N.; Baggio, M.; Pallado, P.; Bertucco, A. *Biotechnol Bioeng* 2001, **73**, 449–457.
23. SencarBozic, P.; Srcic, S.; Knez, Z.; Kerc, J. *Int J Pharmaceut* 1997, **148**, 123–130.
24. Howdle, S. M.; Watson, M. S.; Whitaker, M. J.; Popov, V. K.; Davies, M. C.; Mandel, F. S.; Wang, J. D.; Shakesheff, K. M. *Chem Commun* 2001, 109–110.
25. Kanczler, J. A.; Ginty, P. J.; Barry, J. J. A.; Clarke, N. M. P.; Howdle, S. M.; Shakesheff, K. M.; Oreffo, R. O. C. *Biomaterials* 2008, **29**, 1892–1900.
26. Sheridan, M. H.; Shea, L. D.; Peters, M. C.; Mooney, D. J. *J Controlled Release* 2000, **64**, 91–102.

27. Mooney, D. J.; Baldwin, D. F.; Suh, N. P.; Vacanti, L. P.; Langer, R. *Biomaterials* 1996, **17**, 1417–1422.
28. Barry, J. J. A.; Silva, M.; Popov, V. K.; Shakesheff, K. M.; Howdle, S. M. *Phil Trans R Soc A Math Phys Eng Sci* 2006, **364**, 249–261.
29. Barry, J. J. A.; Gidda, H. S.; Scotchford, C. A.; Howdle, S. M. *Biomaterials* 2004, **25**, 3559–3568.
30. Murphy, W. L.; Peters, M. C.; Kohn, D. H.; Mooney, D. J. *Biomaterials* 2000, **21**, 2521–2527.
31. Kompella, U. B.; Koushik, K. *Crit Rev Therapeut Drug Carrier Syst* 2001, **18**, 173–199.
32. Jang, J. H.; Shea, L. D. *J Controlled Release* 2003, **86**, 157–168.
33. Hutmacher, D. W. *Biomaterials* 2000, **21**, 2529–2543.
34. Hutmacher, D. W. *J Biomater Sci Polym Ed* 2001, **12**, 107–124.
35. Chung, H. J.; Park, T. G. *Adv Drug Deliv Rev* 2007, **59**, 249–262.
36. Phillips, R. In MIE497P: Introduction to Biomechanics. Final Group Research Proceedings, 2006, pp. 27–34, University of Massachusetts, Amherst MA.
37. Ma, P. X.; Langer, R. *J Biomed Mater Res* 1999, **44**, 217–221.
38. Nam, Y. S.; Park, T. G. *J Biomed Mater Res* 1999, **47**, 8–17.
39. Nam, Y. S.; Park, T. G. *Biomaterials* 1999, **20**, 1783–1790.
40. Ma, P. X.; Choi, J. W. *Tiss Eng* 2001, **7**, 23–33.
41. Ma, P. X.; Zhang, R. Y. *J Biomed Mater Res* 2001, **56**, 469–477.
42. Rezwan, K.; Chen, Q. Z.; Blaker, J. J.; Boccaccini, A. R. *Biomaterials* 2006, **27**, 3413–3431.
43. Yang, X. B. B.; Webb, D.; Blaker, J.; Boccaccini, A. R.; Maquet, V.; Cooper, C.; Oreffo, R. O. C. *Biochem Biophys Res Commun* 2006, **342**, 1098–1107.
44. Hench, L. L. *J Biomed Mater Res* 1998, **41**, 511–518.
45. Hench, L. L.; Xynos, I. D.; Edgar, A. J.; Buttery, L. D. K.; Polak, J. M.; Zhong, J. P.; Liu, X. Y.; Chang, J. *J Inorg Mater* 2002, **17**, 897–909.
46. Roether, J. A.; Boccaccini, A. R.; Hench, L. L.; Maquet, V.; Gautier, S.; Jerome, R. *Biomaterials* 2002, **23**, 3871–3878.
47. Boccaccini, A. R.; Blaker, J. J.; Maquet, V.; Day, R. M.; Jerome, R. *Mater Sci Eng C Biomimet Supramol Syst* 2005, **25**, 23–31.
48. Cao, Y.; Mitchell, G.; Messina, A.; Price, L.; Thompson, E.; Penington, A.; Morrison, W.; O'Connor, A.; Stevens, G.; Cooper-White, J. *Biomaterials* 2006, **27**, 2854–2864.
49. Mian, R. W.; Morrison, W. A.; Hurley, J. V.; Penington, A. J.; Romeo, R.; Tanaka, Y.; Knight, K. R. *Tiss Eng* 2000, **6**, 595–603.
50. Cassell, O. C. S.; Morrison, W. A.; Messina, A.; Penington, A. J.; Thompson, E. W.; Stevens, G. W.; Perera, J. M.; Kleinman, H. K.; Hurley, J. V.; Romeo, R.; Knight, K. R. *Bioartificial Organs III Tiss Sourc Immunoisolat Clin Trials* 2001, **944**, 429–442.
51. Hofer, S. O. P.; Knight, K. M.; Cooper-White, J. J.; O'Connor, A. J.; Perera, J. M.; Romeo-Meeuw, R.; Penington, A. J.; Knight, K. R.; Morrison, W. A.; Messina, A. *Plast Reconstruct Surg* 2003, **111**, 1186–1192.
52. Wake, M. C.; Gupta, P. K.; Mikos, A. G. *Cell Transplant* 1996, **5**, 465–473.
53. Wei, G. B.; Ma, P. X. *J Biomed Mater Res A* 2006, **78A**, 306–315.

54. Jacobs, L. J. M.; Kemmere, M. F.; Keurentjes, J. T. F. *Green Chem* 2008, **10**, 731–738.
55. Kim, H.; Kim, H. W.; Suh H. *Biomaterials* 2003, **24**, 4671–4679.
56. Kim, H.; Suh, H.; Jo, S. A; Kim, H. W.; Lee, J. M.; Kim, E. H.; Reinwald, Y.; Park, S. H.; Min, B. H.; Jo I. *Biochem Biophys Res Commun* 2005, **332**, 1053–1060.
57. Hou, Q. P.; Grijpma, D. W.; Feijen, J. *J Biomed Mater Res B Appl Biomaterials* 2003, **67B**, 732–740.
58. Murphy, W. L.; Dennis, R. G.; Kileny, J. L.; Mooney, D. J. *Tiss Eng* 2002, **8**, 43–52.
59. Kuboki, Y.; Takita, H.; Kobayashi, D.; Tsuruga, E.; Inoue, M.; Murata, M.; Nagai, N.; Dohi, Y.; Ohgushi, H. *J Biomed Mater Res* 1998, **39**, 190–199.
60. Otsuki, B.; Takemoto, M.; Fujibayashi, S.; Neo, M.; Kokubo, T.; Nakamura, T. *Biomaterials* 2006, **27**, 5892–5900.
61. Gross, K. A.; Rodriguez-Lorenzo, L. M. *Biomaterials* 2004, **25**, 4955–4962.
62. Ghosh, S.; Viana, J. C.; Reis, R. L.; Mano, J. F. *J Mater Sci Mater Med* 2007, **18**, 185–193.
63. Thomson, R. C.; Yaszemski, M. J.; Powers, J. M.; Mikos, A. G. *J Biomater Sci Polym Ed* 1995, **7**, 23–38.
64. Geesin, J. C.; Gordon, J. S.; Berg, R. A. *Skin Pharmacol* 1993, **6**, 65–71.
65. Zhang, J. Y.; Doll, B. A.; Beckman, E. J.; Hollinger, J. O. *Tiss Eng* 2003, **9**, 1143–1157.
66. Porter, R. M.; Huckle, W. R.; Goldstein, A. S. *J Cell Biochem* 2003, **90**, 13–22.
67. Netter, F. H. "Musculoskeletal system: Anatomy, physiology, and metabolic disorders", Ciba-Geigy Corporation, Summit, NJ, 1987, p. 129.
68. Brighton, C. T.; Hunt, R. M. *J Bone Joint Surg Am Vol* 1991, **73A**, 832–847.
69. Formhals, A. (1934). US Patent 1975504.
70. Subbiah, T.; Bhat, G. S.; Tock, R. W.; Pararneswaran, S.; Ramkumar, S. S. *J Appl Polym Science* 2005, **96**, 557–569.
71. Matthews, J. A.; Wnek, G. E.; Simpson, D. G.; Bowlin, G. L. *Biomacromolecules* 2002, **3**, 232–238.
72. Sundaray, B.; Subramanian, V.; Natarajan, T. S.; Xiang, R. Z.; Chang, C. C.; Fann, W. S. *Appl Phys Lett* 2004, **84**, 1222–1224.
73. Chew, S. Y.; Wen, J.; Yim, E. K. F.; Leong, K. W. *Biomacromolecules* 2005, **6**, 2017–2024.
74. Dalton, P. D.; Klee, D.; Moller, M. *Polymer* 2005, **46**, 611–614.
75. Reneker, D. H.; Chun, I. *Nanotechnology* 1996, **7**, 216–223.
76. Taylor, G. I. *Proc R Soc Lond* 1969, **A313**, 453.
77. Huang, Z. M.; Zhang, Y. Z.; Kotaki, M.; Ramakrishna, S. *Composites Sci Technol* 2003, **63**, 2223–2253.
78. Zong, X. H.; Kim, K.; Fang, D. F.; Ran, S. F.; Hsiao, B. S.; Chu, B. *Polymer* 2002, **43**, 4403–4412.
79. Jun, Z.; Hou, H. Q.; Schaper, A.; Wendorff, J. H.; Greiner, A. *e-Polymers* 2003, 9.
80. Luu, Y. K.; Kim, K.; Hsiao, B. S.; Chu, B.; Hadjiargyrou, M. *J Controlled Release* 2003, **89**, 341–353.
81. Katti, D. S.; Robinson, K. W.; Ko, F. K.; Laurencin, C. T. *J Biomed Mater Res B Appl Biomaterials* 2004, **70B**, 286–296.
82. Gu, S. Y.; Ren, J. *Macromol Mater Eng* 2005, **290**, 1097–1105
83. Tan, S. H.; Inai, R.; Kotaki, M.; Ramakrishna, S. *Polymer* 2005, **46**, 6128–6134.

84. Greiner, A.; Wendorff, J. H. *Angew Chem Int Ed* 2007, **46**, 5670–5703.

85. Stitzel, J.; Liu, L.; Lee, S. J.; Komura, M.; Berry, J.; Soker, S.; Lim, G.; Van Dyke, M.; Czerw, R.; Yoo, J. J.; Atala, A. *Biomaterials* 2006, **27**, 1088–1094.

86. Boyce, S. T.; Kagan, R. J.; Greenhalgh, D. G.; Warner, P.; Yakuboff, K. P.; Palmieri, T.; Warden, G. D. *J Trauma Injury Infect Crit Care* 2006, **60**, 821–829.

87. Yannas, I. V.; Burke, J. F.; Orgill, D. P.; Skrabut, E. M. *Science* 1982, **215**, 174–176.

88. Hansbrough, J. F.; Boyce, S. T.; Cooper, M. L.; Foreman, T. J. *JAMA* 1989, **262**, 2125–2130.

89. Park, S. N.; Lee, H. J.; Lee, K. H.; Suh, H. *Biomaterials* 2003, **24**, 1631–1641.

90. Powell, H. M.; Supp, D. M.; Boyce, S. T. *Biomaterials* 2008, **29**, 834–843.

91. Chau, D. Y. S.; Agashi, K.; Shakesheff, K. M. *Mater Sci Technol* 2008, **24**, 1031–1044.

92. Ambrosio, A. M. A.; Sahota, J. S.; Khan, Y.; Laurencin, C. T. *J Biomed Mater Res* 2001, **58**, 295–301.

93. Khan, Y. M.; Katti, D. S.; Laurencin, C. T. *J Biomed Mater Res A* 2004, **69A**, 728–737.

94. Hong, S. J.; Yu, H. S.; Kim, H. W. *Acta Biomater* 2009, **5**, 1725–1731.

95. Singh, M.; Morris, C. P.; Ellis, R. J.; Detamore, M. S.; Berkland, C. *Tiss Eng C Methods* 2008, **14**, 299–309.

96. Nukavarapu, S. P.; Kumbar, S. G.; Brown, J. L.; Krogman, N. R.; Weikel, A. L.; Hindenlang, M. D.; Nair, L. S.; Allcock, H. R.; Laurencin, C. T. *Biomacromolecules* 2008, **9**, 1818–1825.

97. Wang, C.; Kasuga, T.; Nogami, M. *J Mater Sci Mater Med* 2005, **16**, 739–744.

98. Will, J.; Melcher, R.; Treul, C.; Travitzky, N.; Kneser, U.; Polykandriotis, E.; Horch, R.; Greil, P. *J Mater Sci Mater Med* 2008, **19**, 2781–2790.

99. Polykandriotis, E.; Arkudas, A.; Horch, R. E.; Sturzl, M.; Kneser, U. *J Cell Mol Med* 2007, **11**, 6–20.

100. Polykandriotis, E.; Tjiawi, J.; Euler, S.; Arkudas, A.; Hess, A.; Brune, K.; Greil, P.; Lametschwandtner, A.; Horch, R. E.; Kneser, U. *Microvasc Res* 2008, **75**, 25–33.

101. Kneser, U.; Schaefer, D. J.; Polykandriotis, E.; Horch, R. E. *J Cell Mol Med* 2006, **10**, 7–19.

102. Polykandriotis, E.; Arkudas, A.; Euler, S.; Beier, J. P.; Horch, R. E.; Kneser, U. *Handchirurg Mikrochirurg Plast Chirurg* 2006, **38**, 217–223.

103. Sherwood, J. K.; Riley, S. L.; Palazzolo, R.; Brown, S. C.; Monkhouse, D. C.; Coates, M.; Griffith, L. G.; Landeen, L. K.; Ratcliffe, A. *Biomaterials* 2002, **23**, 4739–4751.

104. Freed, L. E.; Vunjaknovakovic, G.; Biron, R. J.; Eagles, D. B.; Lesnoy, D. C.; Barlow, S. K.; Langer, R. *BioTechnology* 1994, **12**, 689–693.

105. Schantz, J. T.; Brandwood, A.; Hutmacher, D. W.; Khor, H. L.; Bittner, K. *J Mater Sci Mater Med* 2005, **16**, 807–819.

106. Maquet, V.; Boccaccini, A. R.; Pravata, L.; Notingher, I.; Jerome, R. *J Biomed Mater Res A* 2003, **66A**, 335–346.

107. Mooney, D. J.; Baldwin, D. F.; Suh, N. P.; Vacanti, L. P.; Langer, R. *Biomaterials* 1996, **17**, 1417–1422.

108. Lu, H. H.; El–Amin, S. F.; Scott, K. D.; Laurencin, C. T. *J Biomed Mater Res A* 2003, **64A**, 465–474.

109. Barry, J. J. A.; Gidda, H. S.; Scotchford, C. A.; Howdle, S. M. *Biomaterials* 2004, **25**, 3559–3568.

110. Barry, J. J. A.; Nazhat, S. N.; Rose, F.; Hainsworth, A. H.; Scotchford, C. A.; Howdle, S. M. *J Mater Chem* 2005, **15**, 4881–4888.

111. Barry, J. J. A.; Silva, M.; Popov, V. K.; Shakesheff, K. M.; Howdle, S. M. *Phil Trans R Soc A Math Phys Eng Sci* 2006, **364**, 249–261.

112. Hile, D. D.; Amirpour, M. L.; Akgerman, A.; Pishko, M. V. *J Controlled Release* 2000, **66**, 177–185.

113. Mathieu, L. M.; Montjovent, M. O.; Bourban, P. E.; Pioletti, D. P.; Manson, J. A. E. *J Biomed Mater Res A* 2005, **75A**, 89–97.

114. Kanczler, J.; Barry, J.; Ginty, P.; Sebald, W.; Howdle, S.; Shakesheff, K.; Oreffo, R. O. C. *J Bone Mineral Res* 2006, **21**, 1153–1153.

115. Silva, M.; Cyster, L. A.; Barry, J. J. A.; Yang, X. B.; Oreffo, R. O. C.; Grant, D. M.; Scotchford, C. A.; Howdle, S. M.; Shakesheff, K. M.; Rose, F. *Biomaterials* 2006, **27**, 5909–5917.

116. Day, R. M.; Boccaccin, A. R.; Maquet, V.; Shurey, S.; Forbes, A.; Gabe, S. M.; Jerome, R. *J Mater Sci Mater Med* 2004, **15**, 729–734.

117. Day, R. M.; Maquet, V.; Boccaccini, A. R.; Jerome, R.; Forbes, A. *J Biomed Mater Res A* 2005, **75A**, 778–787.

118. Sundback, C.; Hadlock, T.; Cheney, M.; Vacanti, J. *Biomaterials* 2003, **24**, 819–830.

119. Blaker, J. J.; Knowles, J. C.; Day, R. M. *Acta Biomater* 2008, **4**, 264–272.

120. Murphy, W. L.; Dennis, R. G.; Kileny, J. L.; Mooney, D. J. *Tiss Eng* 2002, **8**, 43–52.

121. Uchida, T.; Ikeda, S.; Oura, H.; Tada, M.; Nakano, T.; Fukuda, T.; Matsuda, T.; Negoro, M.; Arai, F. *J Biotechnol* 2008, **133**, 213–218.

122. Draghi, L.; Resta, S.; Pirozzolo, M. G.; Tanzi, M. C. *J Mater Sci Mater Med* 2005, **16**, 1093–1097.

123. Gao, J.; Crapo, P. M.; Wang, Y. D. Tissue Engineering 2006, 12, 917–925.

124. Heijkants, R.; Van Tienen, T. G.; De Groot, J. H.; Pennings, A. J.; Buma, P.; Veth, R. P. H.; Schouten, A. J. *Journal of Materials Science* 2006, **41**, 2423–2428.

125. Lee, S. B.; Kim, Y. H.; Chong, M. S.; Hong, S. H.; Lee, Y. M. *Biomaterials* 2005, **26**, 1961–1968.

126. Kim, J. S.; Reneker, D. H. *Polym Eng Sci* 1999, **39**, 849–854.

127. Haugen, H.; Aigner, J.; Brunner, M.; Wintermantel, E. *J Biomed Mater Res B Appl Biomater* 2006, **77B**, 73–78.

128. Haugen, H.; Will, J.; Fuchs, W.; Wintermantel, E. *J Biomed Mater Res B Appl Biomater* 2006, **77B**, 65–72.

129. Zong, X. H.; Bien, H.; Chung, C. Y.; Yin, L. H.; Fang, D. F.; Hsiao, B. S.; Chu, B.; Entcheva, E. *Biomaterials* 2005, **26**, 5330–5338.

130. Li, D.; Ouyang, G.; McCann, J. T.; Xia, Y. N. *Nano Letters* 2005, **5**, 913–916.

131. Wang, F.; Shor, L.; Darling, A.; Khalil, S.; Sun, W.; Guceri, S.; Lau, A. *Rapid Prototyping Journal* 2004, **10**, 42–49.

132. Yao, J.; Radin, S.; Leboy, P. S.; Ducheyne, P. *Biomaterials* 2005, **26**, 1935–1943.

133. Link, D. P.; van den Dolder, J.; van den Beucken, J.; Cuijpers, V. M.; Wolke, J. G. C.; Mikos, A. G.; Jansen, J. A. *J Biomed Mater Res A* 2008, **87A**, 760–769.

134. Li, Z. S.; Qu, Y.; Yang, B. C.; Zhang, B.; Kim, H. M.; Zhao, H. B.; Zhang, X. D. *J Biomed Mater Res A* 2008, **86A**, 333–338.

135. Jung, Y.; Kim, S. S.; Kim, Y. H.; Kim, S. H.; Kim, B. S.; Kim, S.; Choi, C. Y.; Kim, S. H. *Biomaterials* 2005, **26**, 6314–6322.

136. Popov, V. K.; Evseev, A. V.; Antonov, E. N.; Bagratashvili, V. N.; Konovalov, A. N.; Panchenko, V. Y.; Barry, J. J. A.; Whitaker, M. J.; Howdle, S. M. *J Opt Technol* 2007, **74**, 636–640.

137. Rainer, A.; Giannitelli, S. M.; Abbruzzese, F.; Traversa, E.; Licoccia, S.; Trombetta, M. *Acta Biomaterialia* 2008, **4**, 362–369.

138. Therriault, D.; Shepherd, R. F.; White, S. R.; Lewis, J. A. *Adv Mater* 2005, **17**, 395–399

139. Ang, T. H.; Sultana, F. S. A.; Hutmacher, D. W.; Wong, Y. S.; Fuh, J. Y. H.; Mo, X. M.; Loh, H. T.; Burdet, E.; Teoh, S. H. *Mater Science Eng C Biomimet Supramol Syst* 2002, **20**, 35–42.

140. Kumbar, S. G.; Nukavarapu, S. P.; James, R.; Nair, L. S.; Laurencin, C. T. *Biomaterials* 2008, **29**, 4100–4107.

141. Li, W. J.; Cooper, J. A., Jr.; Mauck, R. L.; Tuan, R. S. *Acta Biomater* 2006, **2**, 377–385.

142. Lee, M.; Dunn, J. C. Y.; Wu, B. M. *Biomaterials* 2005, **26**, 4281–4289.

CHAPTER 10

High-Performance Microelectronics

CHARLES T. BLACK

Brookhaven National Laboratory, Upton, New York, USA

10.1 INTRODUCTION

The engine fueling more than 40 years of performance improvements in semiconductor integrated circuits (ICs) has been industry's ability to pattern circuit elements at ever-higher resolution and with ever-greater precision. Steady advances in photolithography—the process by which ultraviolet light chemically changes a photosensitive polymer resist material in order to create a latent image—have resulted in scaling of minimum printed feature sizes from 10s of micrometers during the 1980s to sub-50-nm transistor gate lengths in today's state-of-the-art ICs. The history of semiconductor technology scaling as well as future technology requirements are documented in the *International Technology Roadmap for Semiconductors* (ITRS) [1]. The progression of the semiconductor industry to the realm of nanometer-scale sizes has brought enormous challenges to device and circuit fabrication, rendering performance improvements by conventional scaling alone increasingly difficult. Most often this discussion is couched in terms of field-effect transistor (FET) feature sizes such as the gate length or gate oxide thickness; however, these challenges extend to many other aspects of the IC, including interconnect dimensions and pitch, device packing density, power consumption, and heat dissipation. The ITRS forecasts a difficult set of scientific and engineering challenges with no presently known solutions.

The complexities associated with meeting the demands of continued performance improvements in future IC generations have created opportunities for new device architectures, new materials integration, and alternative patterning approaches. For many years, research groups at IBM, as well as other groups around the world, have pursued applications of self-assembling materials in the IC fabrication process. Block copolymers are an attractive self-assembling material choice for applications in microelectronics [2] because, like photoresists, they can act as sacrificial templates for patterning integrated circuit elements. Although block copolymers are similar to

Porous Polymers, Edited by Michael S. Silverstein, Neil R. Cameron, and Marc A. Hillmyer

conventional photoresists in that they can be used to form well-defined patterns, they do so *autonomously* at sublithographic feature sizes (<20 nm) and pitches (<40 nm) and with a high density of features ($\sim 10^{11}/cm^2$) [2–9]. The first proposal for using block copolymer thin films in high-resolution lithography was in 1995 [10], and initial experimental demonstrations employed a polystyrene-*b*-polybutadiene (PS-*b*-PB) block copolymer material [11]. Further initial implementations of "block copolymer lithography" utilized self-assembled patterns as both positive and negative resists for etch transfer to underlying silicon, silicon nitride, and germanium [12].

Diblock copolymer materials spontaneously form patterns by minimizing their free energy through microphase separation [13]. Two chemically immiscible polymer "blocks" are covalently bonded end to end to comprise each copolymer molecule—a situation that drives the system toward a minimum interaction volume between blocks. The resulting equilibrium patterns have molecular-scale dimensions determined by intrinsic polymer properties such as the degree of polymerization N, the Flory–Huggins interaction parameter between polymer blocks χ, and the relative constituent block volume fractions in the copolymer f [13] (Fig. 10.1a shows schematic pattern representations). Highly asymmetric diblock copolymer materials (having volume fractions $f < 0.2$) form patterns of close-packed zero-dimensional spheres. More symmetric materials ($f \sim 0.3$) form one-dimensional columnar patterns, while symmetric diblock copolymers ($f \sim 0.5$) create alternating two-dimensional lamellar sheets. The diblock copolymer compositional phase space is roughly symmetric, with inverse cylinders and inverse spheres also possible. Cylinder and lamellar patterns resemble conventional photoresist structures because nanometer-scale domains run continuously from top to bottom through the entire film thickness. The different diblock copolymer phases can be considered a catalog of achievable patterns—unlike conventional photoresists, each diblock material defines only a specific shape. For example, cylindrical phase diblock copolymers are a natural choice for defining circular elements such as transistor contact holes, while lamellar materials are better suited for linear structures such as wiring or transistor gates.

The primary focus of this chapter is the research performed at IBM on diblock copolymer films composed of polystyrene (PS) and poly(methyl methacrylate) (PMMA) (PS-*b*-PMMA) with total molecular weights M_n in the range of ~60K (g/mol) and polydispersities (*PD*) of ~1.1 [2]. These materials self assemble to form patterns having feature sizes in the range of 15–20 nm. PS-*b*-PMMA was selected as a self-assembling patterning material due to its compatibility with the semiconductor microelectronics manufacturing infrastructure, as well as the significant body of existing research on understanding its material properties [14].

Forming self-assembled PS-*b*-PMMA diblock copolymer thin-film patterns involves spin casting the polymer from a dilute solution (e.g., toluene or propylene glycol methyl ether acetate) to an appropriate film thickness, followed by thermal annealing above the polymer glass transition temperature [14, 15]. Polymer phase separation occurs rapidly, with pattern domain size uniformity and positional order improving more slowly over the course of hours [14–16]. The three steps of film casting, latent image formation, and chemical develop (shown schematically in Fig. 10.1b) are all procedurally similar to a conventional photolithography process.

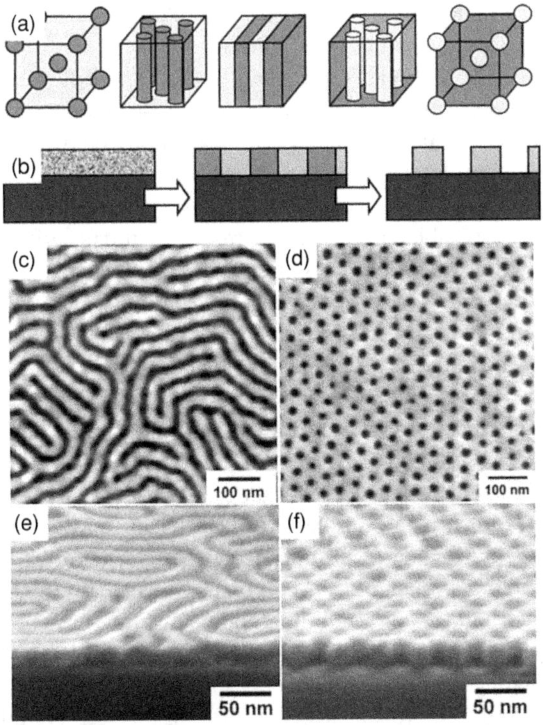

FIGURE 10.1 (**a**) Schematic diblock copolymer phase diagram. (**b**) Schematic process flow for diblock copolymer thin film template formation. Top-down and 70° cross-sectional scanning electron microscope images of (**c, e**) parallel-oriented and (**d, f**) perpendicular-oriented cylindrical-phase PS-*b*-PMMA diblock copolymer films. [Panel (a) reprinted from Reference 99 with permission. Copyright 2006 SPIE.]

Although the self-assembly process is driven by microphase separation of the constituent copolymer blocks, the polymer domain orientation is determined by each block's relative surface affinity. For PS-*b*-PMMA diblock copolymers, PMMA preferentially wets silicon dioxide (SiO_2) and thus drives orientation of the cylindrical PMMA domains parallel to the surface [17]. When viewed from the top surface (SEM image in Fig. 10.1c), a parallel-oriented PS-*b*-PMMA cylindrical film (PS-*b*-PMMA, $f \sim 0.3, M_n = 64K$) shows meandering 20 nm diameter PMMA domains (with 40 nm separation) covering the entire surface.

Preferential surface wetting can be prevented by creating a surface neutral to PS and PMMA using a P(S-*r*-MMA) random copolymer brush [14, 18, 19] prior to diblock copolymer film application and self-assembly. Both random copolymer brushes and thermally crosslinked random copolymers are now widely used as surface neutralization layers to promote preferred domain orientation [14, 18–22]. The same $f \sim 0.3$ PS-*b*-PMMA polymer orients with cylindrical domains perpendicular to the surface, resulting in a close-packed, two-dimensional, 20 nm diameter PMMA cylinder array in a matrix of PS (Fig. 10.1d, f). A significant attractive feature of the PS-*b*-PMMA material is that after pattern formation, the PMMA block can be chemically removed by optionally exposing to UV light followed by immersion in an acetic acid developer [14, 15, 23], thereby creating a nanoporous PS film template. In Fig. 10.1 and all SEM images, the contrast was improved by chemically removing the PMMA block after pattern formation. The lighter-colored regions are PS and the darker regions show the PMMA location prior to its removal.

As-formed, self-assembled polymer patterns have little positional order and a high density of defects. Furthermore, the self-assembly process provides no method for registering self-assembled patterns to other preexisting features on a surface. Nevertheless, porous films such as these have applications in high-performance semiconductor IC fabrication. The examples presented are taken largely from research performed at IBM, although there are certainly many other beautiful demonstrations from groups around the world. In the first examples, the device performance benefits derive from the self-assembled template sublithographic dimensions and high feature density and do not rely on pattern perfection or positional order. The relatively loose demands on perfection for the porous templates mean a lower barrier for process incorporation into future microelectronics technology generations.

A more challenging application for polymer self-assembly within the IC fabrication process is as a high-resolution lithography substitute in which a self-assembly process is used to define individual circuit elements. An area of active research interest is in designing methods for reducing the defectivity of self-assembled polymer films. One popular approach involves forming the self-assembled template on a lithographically prepatterned surface rather than a flat surface, a process referred to by various names, such as self-aligned self assembly, directed self-assembly, topography-induced assembly, or graphoepitaxy. The surface topography provides boundary conditions for the assembly process and, for the appropriate confining dimensions, these boundaries can improve the quality of the self-assembled pattern [24]. For example, an unconfined cylindrical-phase PS-*b*-PMMA material forms a polycrystalline film with a characteristic grain size of order 0.25 μm (Fig. 10.2a). A similar material formed on a prepatterned line/space pattern (Fig. 10.2b) can result in defect-free hexagonal domains extending many micrometers in length (the vertical direction in Fig. 10.2b). Lithography applications require not only high pattern resolution, pattern alignment, and precise pattern registration to other lithography levels, but also a near-zero pattern defectivity and roughness. The technological barriers are extremely high and also fall outside of the present discussion (for further reading on this subject, see, for example, Reference 2).

10.2 BLOCK COPOLYMER TEMPLATES: PORE DENSITY AND POROSITY

Before illustrating different applications of porous block copolymer materials in IC fabrication, the relationship between the polymer material properties and resulting

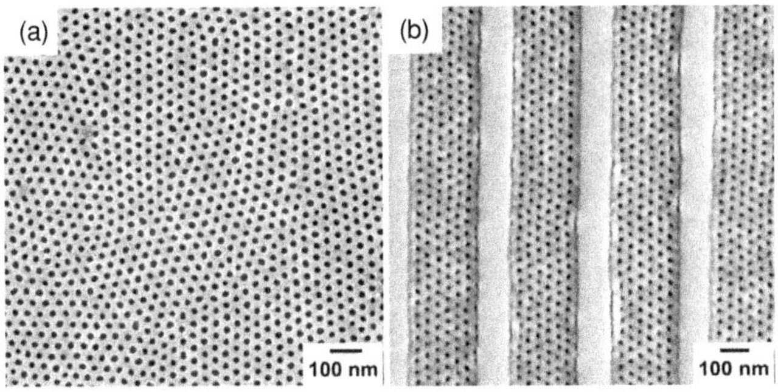

FIGURE 10.2 Top-down SEM images of perpendicularly oriented cylindrical-phase PS-*b*-PMMA template formed on **(a)** a flat surface and **(b)** a lithographically defined topographic pattern of lines/spaces.

self-assembled template dimensions will be discussed. The extent by which template pore density ρ and template porosity ϕ may be adjusted to suit a particular application will be illustrated.

10.2.1 Pore Density

The pore density in a self-assembled polymer film is determined by the relationship between pore size d and separation ℓ. The two-dimensional hexagonal pore arrangement (e.g., Fig. 10.2a) makes estimating pore density a straightforward exercise in geometry. A close-packed, two-dimensional, hexagonal pore array has a unit cell area a given by (Fig. 10.3a)

$$a = \frac{3\sqrt{3}}{2}\ell^2 \tag{10.1}$$

where ℓ is the nearest-neighbor pore separation. Each unit cell contains three pores, so that the pore density ρ (pores per unit area) is

$$\rho = 3\frac{2}{3\sqrt{3}}\frac{1}{\ell^2} = \frac{1.15}{\ell^2} \sim \frac{1}{\ell^2} \tag{10.2}$$

For example, the cylindrical PS-*b*-PMMA film shown in Fig. 10.2 has an average pore separation $\ell \sim 40$ nm, with corresponding pore density $\rho \sim 6 \times 10^{10}$ cm^{-2}.

An attractive feature of diblock copolymer films for patterning microelectronic circuit elements is their ability to scale to smaller dimensions, meaning that the technique can in principle meet the requirements for successive future microelectronics generations. For example, the schematic porous pattern of Fig. 10.3b has a greater

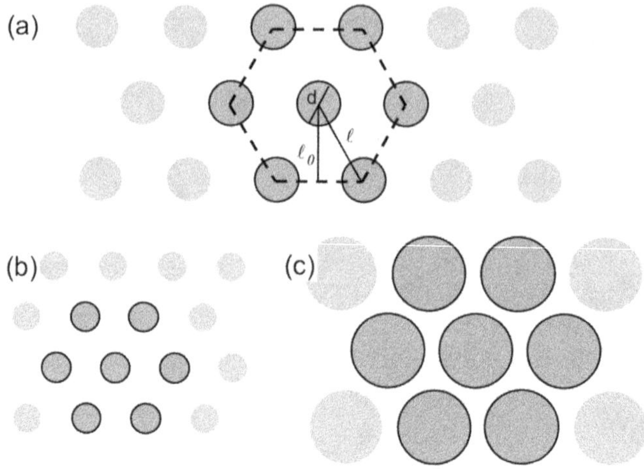

FIGURE 10.3 (**a**) Hexagonal pore array illustrating pore diameter d, pore spacing ℓ, and cylinder row separation ℓ_0. (**b**) Hexagonal pore arrays with having the same porosity and increased pore density [relative to panel (a)]. (**c**) Hexagonal pore array having increased porosity and the same pore density [relative to panel (a)].

pore density than that of Fig. 10.3a because ℓ and d are scaled proportionally. The block copolymer pattern characteristic domain size ℓ_0 scales with the material's total molecular weight or degree of polymerization N [25–27]:

$$\ell_0 \propto N^{\delta} \tag{10.3}$$

In a cylindrical-phase material, ℓ_0 is the distance between cylinder rows and is related to the pore separation ℓ by $\ell_0 = \sqrt{3}\ell/2 \approx 0.87\ell$. Theoretical models all predict a scaling exponent in the range of $0.5 < \delta < 1$, depending on the strength of repulsion between constituent polymer blocks. In the limit of two strongly immiscible polymer blocks, Equation (10.3) becomes [25]

$$\frac{\ell_0}{a_0} = \chi^{1/6} N^{2/3} \sim 0.5 N^{2/3} \tag{10.4}$$

where a_0 is a statistical monomer segment length that is estimated as ~1 nm for clarity of discussion. The last equality of Equation 10.4 assumes $\chi \sim 0.03$ for PS-*b*-PMMA [28], independent of temperature. For a monodisperse cylindrical-phase ($f \sim 0.3$) PS-*b*-PMMA diblock copolymer, the relationship between N and the total polymer molecular weight is

$$N = M_n \left(\frac{0.7}{m_s} + \frac{0.3}{m_{mma}} \right) \approx \frac{M_n}{100} \tag{10.5}$$

where M_n is the polymer number-averaged molecular weight, and m_s and m_{mma} are the styrene and methyl methacrylate monomer molecular weights (each ~100 g/mol), respectively [29]. The relationships 2, 4, and 5 imply that the film pore density ρ can be adjusted by selecting the appropriate polymer molecular weight (keeping the block volume fraction f constant) according to

$$\rho \sim \frac{1}{\ell^2} \sim \frac{1}{\ell_0^2} \sim \frac{4}{a_0^2 N^{4/3}} \sim \frac{4}{a_0^2}\left(\frac{100}{M_n}\right)^{4/3} \tag{10.6}$$

Equation 10.6 reasonably estimates the pore density of an $M_n \sim 67$K PS-*b*-PMMA film (Fig. 10.2a) as $\rho \sim 7 \times 10^{10}$ cm^{-2}.

The maximum pore density for a given material is dictated by the driving force for microphase separation, which is proportional to polymer molecular weight. The product χN parameterizes the block incompatibility responsible for pattern formation. Mean field theory calculations [25, 30] predict an order–disorder transition in the polymer at $\chi N = 10.5$ [25], below which the two polymer blocks are miscible. Typically χ decreases with increasing T according to [25, 31, 32]

$$\chi = c_1 + \frac{c_2}{T} \tag{10.7}$$

where c_1 and c_2 are material constants for specific fractional compositions. For PS-*b*-PMMA diblock copolymers $c_1 = 0.028$ and $c_2 = 3.9$ K [28], so that χ is essentially constant, ~0.03, in the temperature range of interest ($T \sim 300$–500 K).

The criterion $(\chi N)_{ODT} = 10.5$ defines a minimum condition for pattern formation. With $\chi_{PS\text{-}PMMA} \sim 0.03$ this condition defines a lower bound on $N = 10.5/\chi \sim 350$, or $M_n \sim 35$ kg/mol, from which a maximum achievable PS-*b*-PMMA pore density $\rho \sim 2 \times 10^{11}$ cm^{-2} can be calculated. This estimate is not strictly correct because, at $N = 350$, PS-*b*-PMMA is more appropriately in the weak-segregation regime where Equation 10.4 does not apply.

The extendibility of PS-*b*-PMMA diblock copolymer films to higher pore densities has been investigated by systematically synthesizing and evaluating a series of polymers having decreasing molecular weights [2, 15, 33, 34]. Previous experiments have also shown pattern formation in cylindrical-phase PS-*b*-PMMA materials with pore diameters ranging in size from 15 to 50 nm, with corresponding molecular weights of 42–295 kg/mol [15, 34]. Cylindrical-phase PS-*b*-PMMA diblock copolymers with total molecular weights ranging from 38K to 130K and volume fractions $f \sim 0.3$ were synthesized. Top-down SEM images of patterns from two representative materials ($M_w = 67$K and 38K) confirm the scaling of both pore size and separation with molecular weight (Fig. 10.4a, b). The lower–molecular weight synthesized material ($M_w = 38$K) provided a minimum pore separation of 27 nm with corresponding pore density of ~1.3×10^{11} cm^{-2}. The relationship between pore separation ℓ and M_w is well described by the power-law relationship of Equation 10.3 with scaling exponent $\delta \sim 0.77$ [Fig. 10.4c (solid circles)]. Previously reported measurements of scaling in cylindrical-phase films have shown slightly different scaling exponents

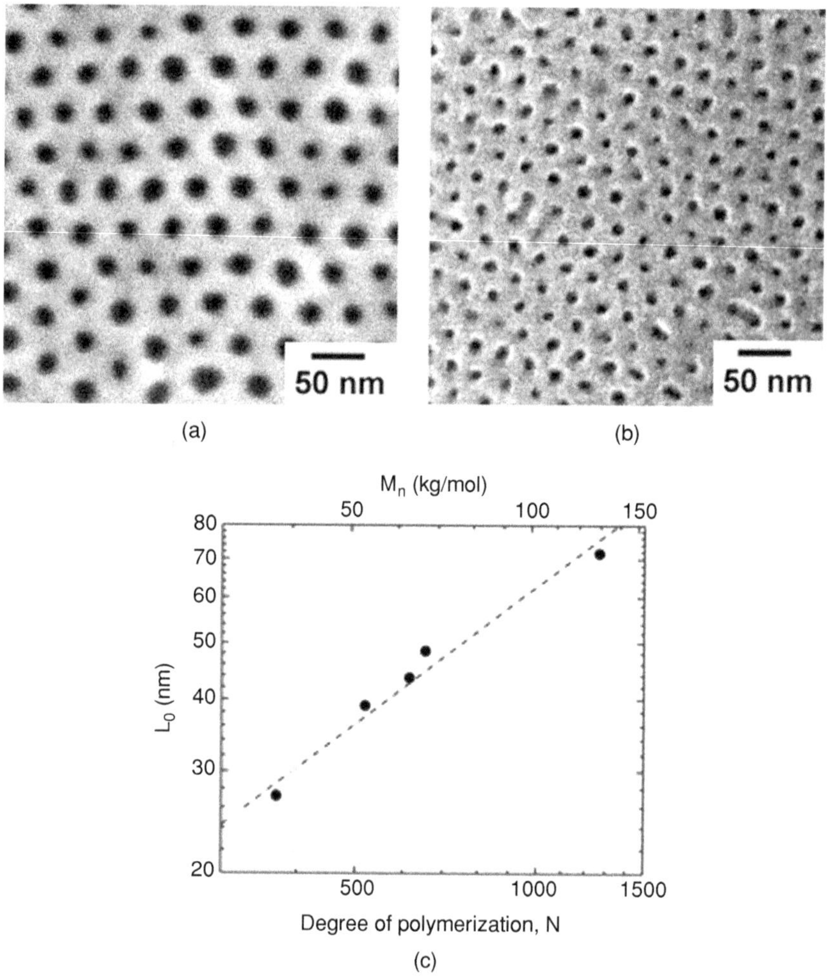

FIGURE 10.4 **(a, b)** SEM images of porous PS films: **(a)** $M_n = 64$ kg/mol. Mean pore diameter is 20 nm and mean center-to-center spacing is 40 nm. **(b)** $M_n = 38$ kg/mol. Mean pore diameter is 12 nm and mean center-to-center spacing is 25 nm. **(c)** ℓ_0 versus polymer degree of polymerization (bottom axis) and molecular weight (top axis). The slope of the dashed line is 0.77. [Panels (a) and (b) reprinted from Reference 33 with permission. Copyright 2006 AVS. Panel (c) reprinted from Reference 2 with permission. Copyright 2007 IBM.]

($\delta \sim 0.58$ [15], $\delta \sim 0.64$ [34], and $\delta \sim 0.65$ in lower-molecular-weight deuterated PS-*b*-PMMA [35]).

10.2.2 Porosity

The diblock copolymer film *porosity* ϕ is the fractional film area occupied by pores. Our previous discussion illustrated the manner by which scaling the total block copolymer molecular weight (or degree of polymerization N) changes the film pore density. If the constituent polymer block volume fraction f remains constant (e.g., ~0.3), then scaling the pore density leaves the film porosity unchanged (e.g., the schematic patterns of Fig. 10.3a, b have the same porosity). Here, the effects of changing block volume fraction on the resulting film structures are described. As an example, the schematic porous pattern of Fig. 10.3c has a higher porosity than the pattern of Fig. 10.3a, even though the pore densities are equal. The film porosity can conveniently be related to intrinsic polymer material properties by observing that perpendicularly aligned, cylindrical-phase films have a uniform density throughout the film thickness (i.e., surface perpendicular) (Fig. 10.1f). Assuming the film phase-separated domain volume fractions are unchanged from constituent block volume fractions, the film porosity ϕ can then be equated to the relative block volume fraction:

$$\phi \approx f \tag{10.8}$$

The block copolymer phase diagram where cylindrical phases exist only for volume fractions in the range of $\sim 0.2 < f < \sim 0.4$ has been described previously. Therefore, cylindrical phase diblock copolymer films are constrained to porosities in the range $20\% < \phi < 40\%$.

The three circular pores of each unit cell in a hexagonal close-packed array occupy an area given by

$$\text{Pore area} = 3\pi \left(\frac{d}{2}\right)^2 = \frac{3}{4}\pi d^2 \tag{10.9}$$

where d is the pore diameter. The film porosity ϕ is the fractional unit cell area occupied by these three pores, given by the ratio of Equations 10.1 and 10.9:

$$\phi = \frac{3}{4}\pi d^2 \frac{1}{\frac{3\sqrt{3}}{2}\ell^2} = \frac{\pi}{2\sqrt{3}}\left(\frac{d}{\ell}\right)^2 \approx 0.9\left(\frac{d}{\ell}\right)^2 \tag{10.10}$$

By equating relations 8 and 10, we find a useful relationship between pore size and separation:

$$d = \sqrt{\frac{2\sqrt{3}\phi}{\pi}}\ell \approx 0.95\sqrt{\phi}\ell \approx \sqrt{f}\ell \tag{10.11}$$

Equation 10.11 demonstrates that cylindrical-phase diblock copolymer films must have pore separations roughly twice their pore diameter, extending over the range $0.45\ell < d < 0.63\ell$ for $0.2 < f < 0.4$.

The foregoing discussion highlights both the flexibility and the limitations for forming porous templates from cylindrical-phase diblock copolymer materials. Template pore density ρ is ultimately limited by the minimum degree of polymerization necessary to facilitate microphase separation and pattern formation, and template porosity ϕ is determined by the polymer block volume fractions necessary to maintain a cylindrical film morphology. Changing intrinsic polymer properties (such as molecular weight and weight ratio) requires both additional polymer synthesis and reoptimization of the experimental conditions for proper self-assembly.

Alternatively, separate processing techniques for adjusting the template pore size *after* polymer film self-assembly can be employed. The advantage here is that different film porosities can be achieved with *the same* initial diblock copolymer template, thus eliminating the need for new polymer synthesis and the need to optimize self-assembly for only a single polymer molecular weight. Unfortunately, there is no similar method for adjusting the template pore density ρ after film assembly.

For example, an oxygen plasma etch process can be used to widen film pore diameters while keeping pore separation constant, thereby increasing template porosity. The oxygen plasma removes polymer material equally from all exposed surfaces, including the pore wall interiors. In this way, the pores are widened as the overall polymer film thickness decreases during the etch process. Exposure of a porous PS film with 20-nm mean pore diameter (Fig. 10.5a) to a brief isotropic O_2 plasma etch successively increases the average pore diameter to 24 nm (Fig. 10.5b), 26 nm (Fig. 10.5c), and finally 30 nm (Fig. 10.5d). The corresponding film porosity ϕ increases from 22% to 50% in Fig. 10.5d. Significantly, the technique achieves a film porosity of $\phi \sim 50\%$ that is not achievable with a cylindrical-phase diblock copolymer material because of limitations on the required block volume fraction f (Eq. 10.8).

A similar type of template post processing to *decrease* film porosity has also been used [33]. In this case, the dimensions of a self-assembled porous PS template were replicated in inorganic silicon material using a hydrogen bromide–based plasma etch [36]. The etch has a high degree of selectivity against the porous PS mask and achieves a vertical etch profile, allowing the etch depth to be controlled by plasma process time. After removal of any remaining PS mask, the patterned silicon reflects the pore dimensions of the initial PS-*b*-PMMA diblock copolymer template (Fig. 10.5e). Subsequent thermal oxidation of the porous silicon consumes surface silicon to form silicon dioxide within the template pores as well as on the top film surface. Because the density of silicon dioxide is 0.44 times the density of silicon, the resulting volume expansion necessarily shrinks the template pore diameters to as small as 12 nm (Fig. 10.5f), corresponding to a decrease in film porosity ϕ from 22% to 8%. Here again, the minimum porosity $\phi \sim 8\%$ falls outside the achievable range for cylindrical-phase diblock copolymer films (Eq. 10.8). Thermal oxidation of a silicon template can also serve as a tool for *increasing* film porosity if the resulting silicon dioxide is subsequently stripped using buffered hydrofluoric acid. This technique has been used to increase template pore diameters from 20 nm to as large as 35 nm (with corresponding porosity $\phi \sim 70\%$) [37].

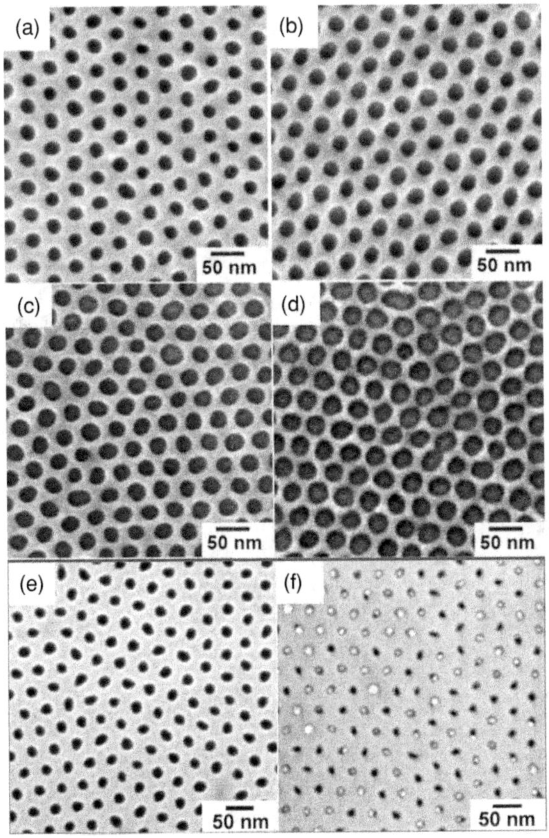

FIGURE 10.5 **(a–d)** SEM images of porous PS templates. O_2 plasma treatment of porous PS templates result in average pore diameter increases from **(a)** 20, **(b)** 24, **(c)** 26, and **(d)** 30 nm. **(e, f)** SEM images of porous silicon templates. **(e)** Pattern transfer into silicon layer by plasma etch through porous PS mask. Mean pore diameter is 20 nm. **(f)** Mean pore diameter of silicon template shrunk by thermal oxidation to 12 nm. [Panels (e) and (f) reprinted from Reference 33 with permission. Copyright 2007 AIP.]

10.3 ENGINEERING DIELECTRIC PERMITTIVITY

An important example of the way in which of self-assembled porous polymer templates can affect IC performance is in engineering the permittivity of dielectric materials. Consider two metal plates separated by a dielectric material with relative permittivity ε_1 (schematic side view in Fig. 10.6a). The capacitance of this planar structure C_0 is proportional to ε_1 and the device footprint A_0 and inversely proportional to the dielectric thickness t:

$$C_0 = \varepsilon_0 \varepsilon_1 A_0 / t \tag{10.12}$$

where ε_0 is the vacuum permittivity.

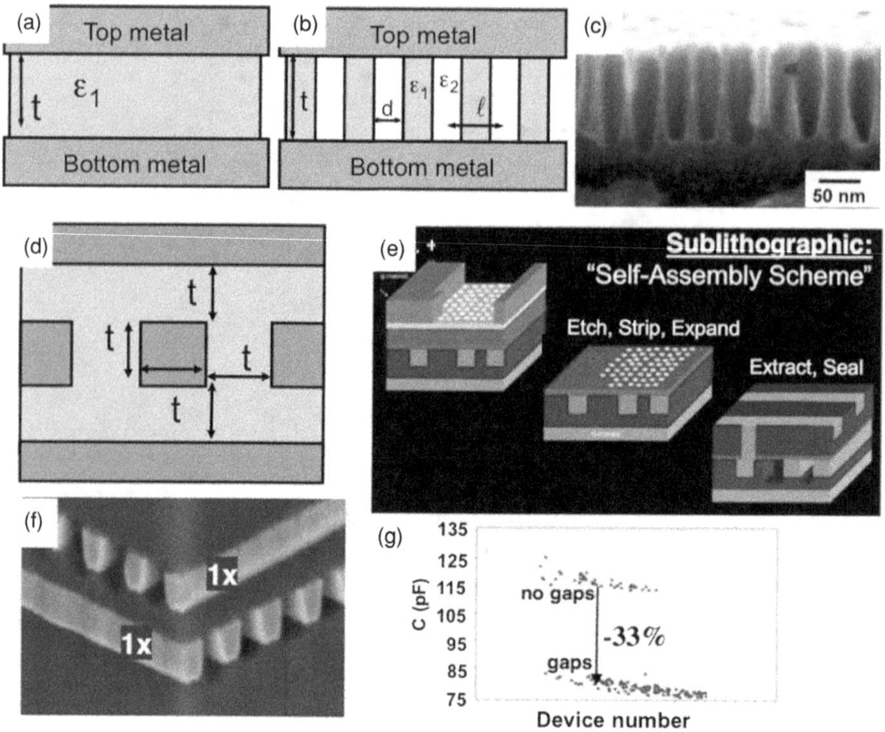

FIGURE 10.6 (**a, b**) Schematic side views of a capacitor device containing (**a**) nonporous and (**b**) porous dielectric materials. (**c**) SEM cross section of a dielectric material made porous by patterning with a diblock copolymer template. (**d**) Schematic circuit cross section. (**e**) Schematic process of IBM self-assembly fabrication of porous airgap interconnect structure. (**f**) SEM cross section of IBM circuit 1X wiring levels containing porous airgaps. (**g**) Capacitance comparison of IBM 1X wiring levels with and without airgaps. [Panel (c) reprinted from Reference 2 with permission. Copyright 2007 IBM. Panels (e)–(g) reprinted from Reference 42 with permission. Copyright 2007 MRS.]

The capacitance of this structure can be reduced by replacing a fraction of the dielectric with a second material of lower relative permittivity ε_2 (schematic side view in Fig. 10.6b). The capacitance of the composite dielectric was estimated by considering its effective permittivity ε_{eff} as a volume-weighted average of the two constituent material permittivities:

$$\varepsilon_{eff} \approx (1-\phi)\,\varepsilon_1 + \phi\varepsilon_2 \tag{10.13}$$

where ϕ is the dielectric material porosity. The total device capacitance becomes

$$C_{porous} \approx \frac{\varepsilon_0\varepsilon_{eff}}{t} = \frac{\varepsilon_0}{t}\left[(1-\phi)\,\varepsilon_1 + \phi\varepsilon_2\right] = C_0\left(1 - \phi\frac{(\varepsilon_1-\varepsilon_2)}{\varepsilon_1}\right) \tag{10.14}$$

and the capacitance change caused by introduction of the porous material is

$$\frac{\Delta C}{C_0} = \left[\frac{C_{porous}}{C_0} - 1\right] = \phi\,(\varepsilon_2 - \varepsilon_1)/\varepsilon_1 \tag{10.15}$$

Cylindrical-phase diblock copolymer templates have proven useful tools for rendering underlying dielectric materials porous when they are used as physical masks for plasma or chemical etch processes [36, 38]. For example, a cylindrical PS-*b*-PMMA template was formed on a thermally oxidized silicon wafer and SiO_2 was selectively etched using a trifluoromethane/argon reactive plasma. The degree of achievable capacitance decrease was determined by considering a dielectric film of permittivity $\varepsilon_1 \sim 3$ rendered porous in such a manner (e.g., Fig. 10.6c). Typical cylindrical-phase polymer templates have porosities $\phi \sim 0.3$ (Eq. 10.8). The etch process replaces a dielectric volume ($\sim\phi t$) with vacuum ($\varepsilon = 1$), so that the resulting capacitance of the porous structure is reduced by (Eq. 10.15) $\Delta C/C_1 = \phi\,(\varepsilon_2 - \varepsilon_1)/\varepsilon_1 = 0.3(1-3)/3 = 20\%$. The maximum achievable capacitance reduction occurs in the limit of maximum template porosity ($\phi \sim 0.9$ when $d \sim \ell$; see Eq. 10.10), so that $\Delta C/C_0 = 0.9(1-3)/3 = 60\%$.

In high-performance integrated circuits, the dielectric permittivity ε plays an important role in determining the maximum frequency at which electrical signals can be transmitted between transistor circuit elements. This cutoff frequency is often described by a quantity called the interconnect delay τ, equal to the product of a wire's resistance R and its capacitance C to the surrounding environment, $\tau = RC$.

An unavoidable consequence of integrated circuit scaling is that metal interconnects become smaller and more densely packed, thereby increasing wire resistance ($R \propto 1/A$, where A is the cross-sectional area) and wire capacitance ($C \propto 1/t$, where t is the thickness of the dielectric separating the wires). The result is that scaling interconnects to smaller sizes results in *poorer wiring performance*, in contrast to transistor device scaling, in which smaller dimensions generally mean *better switching performance*.

The scaling effect on wiring performance can be estimated by considering a model circuit (schematic cross section shown in Fig. 10.6d) [39]. Here a square wire with cross-sectional area A and length L (perpendicular to the printed page) is separated laterally from two similar parallel-running interconnects and vertically from two perpendicular-running interconnects. For simplicity, the wire cross-sectional dimension is made equal to the thickness of the dielectric, t. The total wire capacitance is the sum of capacitances to the two laterally adjacent interconnects C_L and the two vertically adjacent interconnects C_V:

$$C = 2\,(C_L + C_V) = 2C_V\left(1 + \frac{C_L}{C_V}\right) \tag{10.16}$$

$$C = 2\varepsilon_0\varepsilon_V L\left(1 + \frac{\varepsilon_L}{\varepsilon_V}\right) = 4\varepsilon_0\varepsilon_V L \tag{10.17}$$

where ε_v and ε_L are the relative permittivities of the dielectrics vertically and laterally adjacent to the wire, respectively. The interconnect delay is then [39]

$$\tau = RC = \left(\frac{\rho L}{t^2}\right) * 2\varepsilon_0\varepsilon_V L\left(1+\frac{\varepsilon_L}{\varepsilon_V}\right) = \frac{2\rho_r\varepsilon_0\varepsilon_V L^2}{t^2}\left(1+\frac{\varepsilon_L}{\varepsilon_V}\right) = \frac{4\rho_r\varepsilon_0\varepsilon_V L^2}{t^2} \tag{10.18}$$

where the last equalities in relations 10.17 and 10.18 hold true when $\varepsilon_V = \varepsilon_L$. For example, an ~100-nm square copper wire ($\rho_r \sim 17 \times 10^{-7}$ Ω-cm) of length $L \sim 1$ mm embedded in a dielectric with $\varepsilon_v \sim 3$ has a delay $\tau = RC \sim 0.2$ ns, meaning that the wire can support signal frequencies only as high as $1/\tau \sim 5$ GHz. Present-day microprocessor clock speeds operate in the range of ~3 GHz, so that interconnect delays are an important limiter in IC performance.

It is possible to *reduce* the interconnect delay τ by *lowering* the permittivity of the surrounding dielectric material. For example, reducing ε_L by 20% by introducing porosity into only the laterally surrounding dielectric material (leaving the dielectric material between vertical interconnects unchanged) will correspondingly reduce the interconnect delay by ~10% (Eqs. 10.15 and 10.18).

IBM recently demonstrated the promise of self-assembled diblock copolymer films for solving key challenges in scaling interconnect performance [40, 41]. A reliable and potentially manufacturable process for generating porous dielectric materials in successive back-end-of-the-line (BEOL) wiring levels of state-of-the-art microprocessors was successfully implemented. The resulting performance gains meet the targets outlined in the ITRS for BEOL performance [1].

The process involves first depositing capping layers on top of an integrated circuit wiring level (Fig. 10.6e) [42]. A PS-*b*-PMMA diblock copolymer film self-assembles on the capping layer, and select regions are then protected using standard optical lithography. The subsequent etch processes transfer the porous pattern first into the capping layers and subsequently into the dielectric material between copper interconnects. After pattern transfer into the dielectric material, the porous capping layer is "sealed" and "pinched off" during conformal deposition of the dielectric material for the successive wiring level. Key advantages of using self-assembled diblock copolymer films for this process include (1) there is minimal added process complexity involved with block copolymer patterning (compared with lithography processes), (2) the nanometer-scale feature sizes of block copolymer domains allow patterning of the dielectric in the smallest wiring levels (often called the 1X wiring levels), and (3) the block copolymer domain sizes are easily pinched off by subsequent dielectric depositions.

A high-resolution SEM image of part of the completed microprocessor structure illustrates the precision with which this technique can be applied (Fig. 10.6f). Here, the dielectric material between two successive wiring levels (labeled 1X in Fig. 10.6f) can be completely removed using the diblock copolymer patterning scheme described earlier. Capacitance measurements show a reduction of 33% from a dielectric constant of 3.2 in the nonporous material to an effective $\varepsilon = 2.1$ in the microprocessor containing porous airgaps between lateral wires (Fig. 10.6g). This capacitance reduction is described remarkably well by Equation 10.15, which predicts $\Delta C = \left(C_{porous}/C_0 - 1\right) = 0.5 * (1 + \varepsilon_L/\varepsilon_V) - 1 = -0.34 = -34\%$ for $\varepsilon_L = 1$ and $\varepsilon_V = 3.2$.

10.4 INCREASING DEVICE SURFACE AREA

Methods by which unwanted device capacitances were reduced through the introduction of porous dielectric materials were described previously. Other aspects of IC fabrication rely on increased device capacitance for improved performance. Porous polymer patterns can again be employed to solve the scaling challenge of increasing a material's effective dielectric constant by engineering its surface area. The concept is best illustrated by first considering a planar capacitor device structure (Fig. 10.7a). The device capacitance is proportional to the electrode areal footprint A_0 (Eq. 10.12), so that the device performs its charge storage function *less proficiently* as its area A_0 is reduced. For reference, a planar capacitor with dielectric thickness $t \sim 4$ nm and dielectric permittivity $\varepsilon \sim 4$ has a capacitance density of ~0.9 μF/cm^2. This planar device can be compared to a structure having the same lateral footprint but instead a three-dimensional internal geometry (Fig. 10.7b). Here the device electrode area is defined by its total surface area A_{rough}, which can be much larger than the device footprint A_0.

The effect of roughening the device electrode can be calculated by considering a surface patterned using a diblock copolymer material having a hexagonal pore arrangement. In this case, the surface area enhancement is [36, 43]

$$\frac{A_{rough}}{A_0} = 1 + \frac{3\pi dh}{a} \tag{10.19}$$

where a is the hexagonal unit cell area (Eq. 10.1) and h is the depth of surface roughness. Equation 10.19 assumes a cylindrical etch profile for the roughened structure. The roughened device capacitance can be expressed in terms of an effective dielectric constant ε_{eff}:

$$C_{rough} = \frac{\varepsilon_0 \varepsilon A_{rough}}{t} = \frac{\varepsilon_0 \varepsilon_{eff} A_0}{t} \tag{10.20}$$

where

$$\frac{\varepsilon_{eff}}{\varepsilon} = \left(1 + \frac{3\pi dh}{a}\right) = 1 + 4\phi\frac{h}{d} \tag{10.21}$$

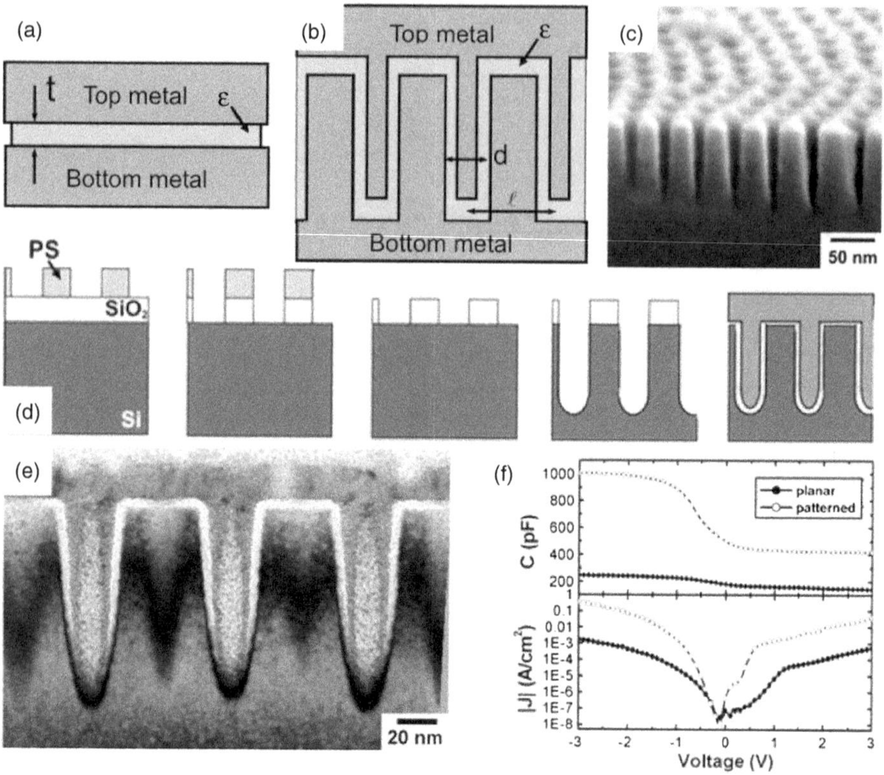

FIGURE 10.7 **(a, b)** Schematic side views of **(a)** planar and **(b)** shallow-trench-array capacitors. **(c)** SEM cross section of a shallow-trench-array silicon counterelectrode. **(d)** Schematic side view of process flow for shallow-trench-array capacitor fabrication. **(e)** TEM cross section of a section of a completed shallow-trench-array capacitor. **(f)** Capacitance–voltage and current–voltage relationships for planar (*filled circles*) and shallow-trench-array (*open circles*) devices. [Panels (c), (d), and (f) reprinted from References 43 with permission. Copyright 2004 IEEE. Panel (e) reprinted from Reference 2 with permission. Copyright 2007 IBM.]

Equations 10.20 and 10.21 illustrate that a porous cylindrical-phase diblock copolymer template (with porosity $\phi \sim 0.3$) can double the device capacitance by merely roughening to a depth on the order of the pore diameter, $h \sim d \sim 20$ nm.

High-performance integrated circuits operate at ever-higher frequencies and powers, and one resulting consequence is noise generation in the IC power-supply grid [43]. Power-supply noise is buffered by decoupling capacitor devices placed within the IC so that they reside close to the logic switching elements. Efficient capacitor design involves occupying minimal valuable chip real estate while maintaining a required charge storage capacity. This challenge is inherent to all charge storage elements within the IC, including on-chip decoupling capacitors [44], dynamic random access memory (DRAM) cells [45], and ferroelectric nonvolatile memories [46]. Fabrication approaches such as deep-trench etching [47], three-dimensional stack capacitors [48], surface roughening [49], and high-*k* dielectrics [50] have all been considered and implemented as ways to maximize charge storage capacity in a minimum amount of lateral device footprint. The importance of capacitance density has been discussed in terms of both on-chip decoupling capacitors [43] and DRAM storage nodes that store an information bit as charge on a capacitor [36]. While both devices share a similar goal of storing a sufficient amount of electric charge in a minimum footprint, the target capacitance densities for decoupling capacitors ($\sim$1 μF/cm^2) are roughly an order of magnitude smaller than for DRAM storage nodes ($\sim$10 μF/cm^2).

Several years ago IBM demonstrated an on-chip decoupling capacitor using self-assembled block copolymer films to increase device capacitance without expanding the device lateral footprint [2, 43, 51]. The shallow trench array device has potential for use in decreasing the amount of required IC chip area occupied by capacitors or, alternatively, in packing a greater decoupling capacitance into the same device footprint. Nanostructuring device electrodes in this way provides a realistic method for achieving up to $\sim$10-fold capacitance increases over equivalent planar area structures using relatively standard fabrication processes and without introducing novel dielectric materials. The large-area devices comprise many shallow trenches, so that their performance tolerates variations in the self-assembly process.

Shallow-trench-array capacitor fabrication is accomplished by first replicating a self-assembled porous polymer pattern into the device Si counterelectrode (schematic process flow shown in cross section in Fig. 10.7d) [36]. The resulting dense arrays of shallow trenches can have height to diameter aspect ratios h/d in excess of 5:1 (Fig. 10.7c). The apparent variations in trench depth (Fig. 10.7c) are an artifact of sample preparation because cleaving through hexagonal pore arrays invariably results in slices through different pore sections.

The shallow-trench-array geometry provides a surface area increase (compared to equivalent planar-area devices) proportional to the trench depth (Eq. 10.19), a dimension controllable through the etch time [36]. Equations 10.20 and 10.21 predict an enhancement $[(C_{rough}/C_0) - 1] = 4\phi h/d$ of $\sim$600% for trench aspect ratio $h/d \sim 5$ [43]. The capacitance increase is determined by the initial template porosity rather than pore density, such that the shallow-trench-array surface area can be controlled by tuning the trench diameter [36, 37]. The process by which the shallow-trench-array device porosity can be increased by widening the original pore diameter through growth of 18-nm sacrificial thermal oxide (consuming $\sim$9 nm silicon) followed by an oxide strip has been demonstrated [37].

A cross-sectional TEM image of a section of a shallow-trench-array device (Fig. 10.7e) shows three pores coated with a 4.5-nm SiO_2 gate dielectric (lighter colored in the image) and filled with a tantalum nitride top electrode. The device accumulation capacitance is $\sim$410% more than that of a planar device of the same lateral footprint (Fig. 10.7f). The deviation between the measured enhancement and the geometric estimate (~600%) is understood as being due to the tapered etch profile of the shallow trench.

A disadvantageous effect of the shallow-trench-array geometry is the corresponding increase in device leakage current compared to a planar structure (Fig. 10.7f). The leakage current per lateral device area (J) for the shallow-trench-array capacitor is roughly ~100 times higher than for the planar device (Fig. 10.7f)—a much greater increase than is explained by the increased device surface area alone (4.1 times). The excess current likely stems from higher charge tunneling rates in the high-curvature trench bottoms, where electric fields are enhanced [52]. The shallow-trench-array structure demonstrates the performance tradeoff often encountered in device design—in this case, the need to strike a balance between enhanced capacitance and appropriate leakage current levels.

10.5 FIELD-EFFECT TRANSISTOR CONTACT HOLES

The previous discussion has focused on semiconductor device applications in which nanometer-scale porosity of block copolymer thin films provides a unique performance advantage. In these examples, the self-assembled block copolymer films are used to improve material properties through controlled nanostructuring with a high degree of nanometer-scale size uniformity. The advantage of self-assembly processes for large-area patterning of nanometer-scale features has been highlighted. In a practical sense, however, it is often necessary to pattern materials only in select regions of the IC. In this case, a degree of control over the location of the self-assembled pattern is also important.

Self-assembled polymer patterns formed on flat surfaces provide a film of uniform porosity with hexagonal pore patterns having little positional order and a high defect density (Fig. 10.2a). Enhancements to the polymer assembly process can improve both pattern order and defectivity. For example, externally applied forces such as electric fields [53–59], mechanical shear [60–65], or shear flow; solvent interactions, including solvent annealing [66–69], solvent evaporation, and solvent crystallization [70, 71]; or nano-imprint [63] and supercritical carbon dioxide methods [72] can impart preferred directionality to self-assembled patterns and reduce defectivity while also controlling pattern orientation. Two additional methods are particularly well suited to semiconductor device fabrication processes—topography induced alignment (i.e., graphoepitaxy) [24, 64, 73–80] and chemical prepatterning techniques [20, 81–87].

The focus here will be on topography-induced methods for controlling the placement of self-assembled block copolymer thin films—a process known as pattern registration. The first demonstration of self-alignment of a block copolymer pattern to a lithographically defined pattern used thin films of sphere-forming polystyrene-*b*-polyvinylpyridine (PS-*b*-PVP) [24]. Since this initial demonstration, similar self-aligned behavior has been shown in other spherical-phase materials as well as both parallel- and perpendicular-oriented cylinder patterns [69, 73, 74, 79, 80, 88–97]. The approach is demonstrated here by first patterning the regions of interest using

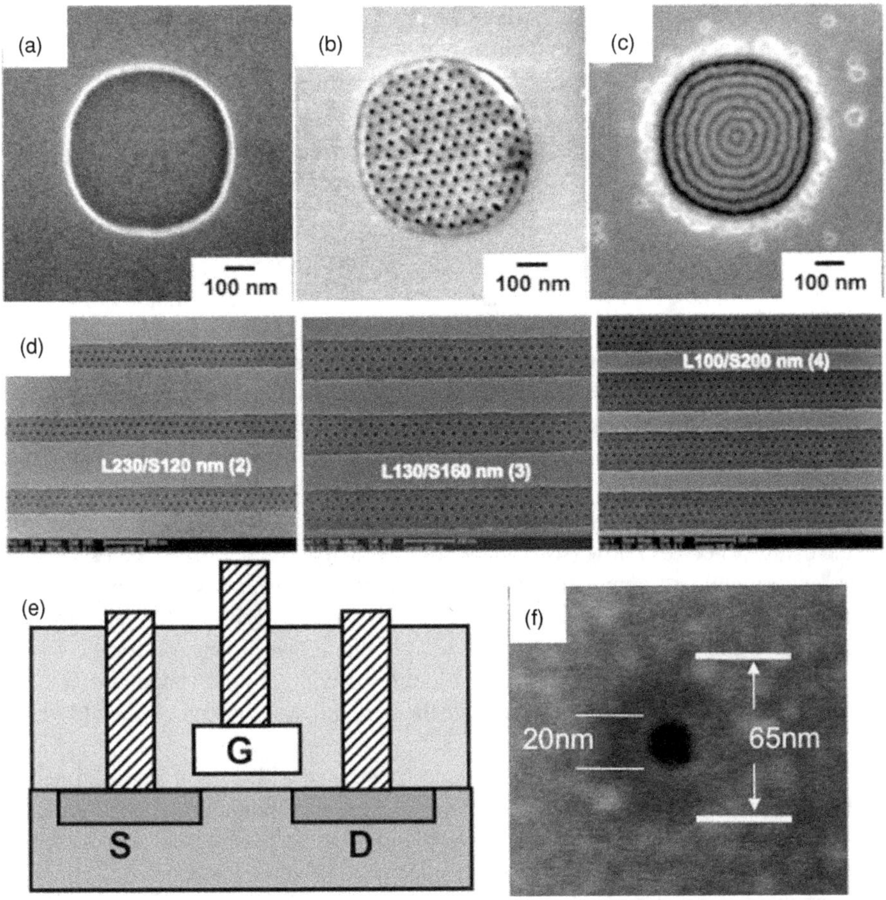

FIGURE 10.8 **(a)** A lithographically defined 0.5-μm circle is subdivided by **(b)** perpendicularly oriented and **(c)** parallel-oriented cylindrical-phase diblock copolymer templates. **(d)** Contact-hole formation in lithographically defined lines of different widths and pitches. **(e)** Schematic side view of contact hole vias (diagonal rectangles) connected to an FET gate, source, and drain. **(f)** A lithographically defined 65-nm contact hole containing a single cylindrical-phase diblock copolymer template pore. [Panel (d) reprinted from Reference 89 with permission. Copyright 2006 SPIE. Panel (f) reprinted from Reference 98 with permission. Copyright 2007 AVS.]

large-area lithography and plasma etching [73]. In the example shown in Fig. 10.8a, a 0.5-μm-diameter circle has been etched to a depth ~30 nm into a silicon dioxide surface. Spin casting a cylindrical-phase polymer film of suitable thickness onto the entire wafer surface results in pattern formation of perpendicular-oriented (Fig. 10.8b) and parallel-oriented (Fig. 10.8c) cylinder domains only within the lithographically defined regions.

Neglecting the effects of the pattern periphery, each lithographically defined area A_0 contains a number of perpendicular cylinder pores determined by

$$n \approx 3\frac{A_0}{a} = \frac{2}{\sqrt{3}\ell^2}A_0 \sim \frac{A_0}{\ell^2} \qquad (10.22)$$

where a is the hexagonal unit cell area. For example, the 0.5-μm-diameter circle ($A_0 \sim 2 \times 10^{-9}$ cm^2) in Fig. 10.8b contains ~150 pores, which the estimate of Equation 10.22 predicts to within ~15% (~125 pores).

Two recent experimental demonstrations of graphoepitaxy of porous cylindrical phase block copolymer films have highlighted its utility in addressing fabrication of transistor contact holes—one of the most difficult lithography challenges of advanced microelectronics processing. Field-effect transistor fabrication involves patterning transistor gate, source, and drain elements at high resolution and further providing metallic connections to each of these three elements. Electrical connections are made using structures known as contact holes, which are vertically etched structures filled with metal and connecting FET elements to the wiring levels above. Contact hole dimensions are scaled nearly as aggressively as the FET gate length with each successive technology generation. For example, the target dimension for the 65-nm technology generation is ~70 nm [1]. Fidelity of contact hole shape and size are paramount, as these dimensions determine the wiring resistivity and ultimately circuit performance.

The FET source and drain contact holes are contained within boundaries defined by the device active area and the gate electrode. These natural boundaries can provide the topographic features necessary for the self-aligned self-assembly process described previously. A first demonstration of the manner in which such a process could be implemented showed multiple contact hole rows contained within lithographically defined line/space patterns [89] (Fig. 10.8d). A schematic contact hole cross section is shown in Fig. 10.8e. In this work, the authors recognized complications presented by the lithography pattern periphery, which facilitates self-assembly of partial hole patterns. Suitable surface preparation can alleviate this integration challenge.

IBM has implemented a self-aligned self-assembly process to demonstrate single 20-nm-diameter contact holes for advanced device applications [98]. Their process is integrated within a state-of-the-art semiconductor fabrication facility processing 300-mm-diameter silicon wafers. The IBM demonstration is essentially an extension of the process shown in Fig. 10.8b to the limit of a single self-assembled pore contained within the lithographic structure. By lithographically defining a "placer" structure with an area less than the area of a single hexagonal unit cell ($a = 3\sqrt{3}\ell^2/2$), the polymer self-assembly process will only allow inclusion of a *single pore* within the confined structure. For a circularly defined placer shape, this condition is met for placer diameter less than $d < \sqrt{6\sqrt{3}\ell^2/\pi} \sim 70$ nm. Fig. 10.8f shows a lithographically defined placer feature with diameter ~65 nm in which a cylindrical PS-*b*-PMMA diblock copolymer material self-assembles for form a single pore. In this way, a method for combining ~65-nm lithography with a self-assembly process in order to produce ~20-nm-diameter contact holes has been demonstrated.

10.6 SUMMARY

This chapter has described the development and integration of fabrication processes based on porous block copolymer materials for use in high-performance semiconductor devices. While the discussion has focused on IBM's research, the research field includes exciting progress from many different groups around the world. Many examples are highlighted in the cited references.

Self-assembled porous patterns provide sublithographic dimensions and high feature densities that facilitate material nanostructuring—a useful technique for controllably improving material properties. An exciting property of block copolymer self-assembly is its ability to scale to successively finer dimensions through adjustment of material parameters, thereby potentially providing patterning solutions for successive future semiconductor technology nodes. Ways to control both the block copolymer template pore density and porosity, as well as applications of these materials in both reducing and increasing device capacitances, have been discussed. These desirable material enhancements arise from having many nanometer-scale elements operating within a single device, such that a lack of *positional control* of the self-assembly process does not degrade performance. A more challenging application of porous polymer materials involves precise control of pore locations on a surface, and the utility of this type of control has been illustrated with an example of patterning high-resolution transistor contact holes. As the challenges of continued improvements in integrated circuit performance become ever greater, new prospects for implementing unconventional fabrication approaches afforded by self-assembled block copolymer materials will be discovered.

ACKNOWLEDGMENTS

Much of the research described was performed at the IBM Thomas J. Watson Research Center during the years 1999–2006. The author gratefully recognizes many meaningful collaborations over the years, including foremost the scientific interactions with K. W. Guarini, R. Ruiz, Y. Zhang, M. C. Colburn, H.-C. Kim, and J. Cheng (IBM). T. P. Russell and M. T. Tuominen (University of Massachusetts-Amherst) provided the author a valuable introduction to polymer self-assembly processes. C. J. Hawker (University of California-Santa Barbara) synthesized several of the PS-*b*-PMMA diblock copolymers and the P(S-*r*-MMA) random copolymer used in the reported experiments and also contributed numerous welcome discussions. R. L. Sandstrom, N. Ruiz, I. V. Babich, E. M. Sikorski, K. R. Milkove, G. Breyta, E. Huang, and B. To (IBM) all contributed in important ways to the materials integration effort at IBM Yorktown. S. Yeung, O. Bezencenet, S. Cowan, and K. Bosworth participated in many of the experiments reported in this chapter.

ABBREVIATIONS

1X	Smallest wiring level in an integrated circuit
a	Unit cell area
a_0	Statistical monomer segment length
A	Cross-sectional area
A_0	Areal footprint
A_{rough}	Surface area
χ	Flory–Huggins interaction parameter
c_1	Material constant parameterizing the Flory–Huggins interaction parameter
c_2	Material constant parameterizing the Flory–Huggins interaction parameter
C	Capacitance
C_L	Capacitance to a laterally adjacent structure
C_V	Capacitance to a vertically adjacent structure
C_0	Planar capacitance
C_{rough}	Capacitance of a device with roughened electrode surface
C_{porous}	Capacitance of planar device containing porous dielectric material
ΔC	Capacitance change
δ	Scaling exponent for block copolymer domain size
d	Pore size
DRAM	Dynamic random access memory
ε_0	Vacuum permittivity
ε_1, ε_2, ε_v, ε_L	Material-dependent relative permittivities
z_{eff}	Effective dielectric permittivity
f	Block copolymer constituent block volume fraction
FET	Field-effect transistor
h	Height of surface roughness
IBM	International Business Machines
IC	Integrated circuit
ITRS	*International Technology Roadmap for Semiconductors*
J	Current per lateral device area
ℓ	Pore separation
ℓ_0	Block copolymer characteristic length scale
L	Wire length
m_s	Styrene monomer molecular weight
m_{mma}	Methyl methacrylate monomer molecular weight
M_n	Number-averaged molecular weight
M_w	Weight-averaged molecular weight
N	Degree of polymerization
ϕ	Porosity
PB	Polybutadiene
PD	Polydispersity
PMMA	Poly(methyl methacrylate)

PS	Polystyrene
PVP	Polyvinylpyridine
PS-*b*-PB	Polystyrene-*b*-polybutadiene block copolymer
PS-*b*-PMMA	Polystyrene-*b*-poly(methyl methacrylate) block copolymer
P(S-*r*-MMA)	Random copolymer of styrene and methyl methacrylate
PS-*b*-PVP	Polystyrene-*b*-polyvinylpyridine block copolymer
ρ	Pore density
ρ_r	Wire resistivity
R	Wire resistance
SEM	Scanning electron microscope
SiO_2	Silicon dioxide
τ	Interconnect delay
t	Dielectric thickness
T	Temperature
TEM	Transmission electron microscope
UV	Ultraviolet

REFERENCES

1. *International Technology Roadmap for Semiconductors: 2005 Edition*, International SEMATECH, Austin, TX, 2005.
2. Black, C. T.; Ruiz, R.; Breyta, G.; Cheng, J. Y.; Colburn, M. C.; Guarini, K. W.; Kim, H. C.; Zhang, Y. *IBM J Res Dev* 2007, **51**, 605–633.
3. Hamley, I. W. *Nanotechnology* 2003, **14**, R39–R54.
4. Harrison, C.; Park, M.; Chaikin, P. M.; Register, R. A.; Adamson, D. H. *J Vac Sci Technol B* 1998, **16**, 544–552.
5. Hawker, C. J.; Russell, T. P. *MRS Bull* 2005, **30**, 952–966.
6. Lazzari, M.; Lopez-Quintela, M. A. *Adv Mater* 2003, **15**, 1583–1594.
7. Li, M.; Coenjarts, A. C.; Ober, K. O. *Patternable Block Copolymers*, Springer-Verlag, Berlin, 2005.
8. Park, C.; Yoon, J.; Thomas, E. L. *Polymer* 2003, **44**, 6725–6760.
9. Segalman, R. A. *Mater Sci Eng Rep* 2005, **48**, 191–226.
10. Mansky, P.; Chaikin, P.; Thomas, E. L. *J Mater Sci* 1995, **30**, 1987–1992.
11. Mansky, P.; Harrison, C. K.; Chaikin, P. M.; Register, R. A.; Yao, N. *Appl Phys Lett* 1996, **68**, 2586–2588.
12. Park, M.; Harrison, C.; Chaikin, P. M.; Register, R. A.; Adamson, D. H. *Science* 1997, **276**, 1401–1404.
13. Bates, F. S.; Fredrickson, G. H. *Phys Today* 1999, **52**, 32–38.
14. Thurn-Albrecht, T.; Steiner, R.; DeRouchey, J.; Stafford, C. M.; Huang, E.; Bal, M.; Tuominen, M.; Hawker, C. J.; Russell, T. *Adv Mater* 2000, **12**, 787–791.
15. Guarini, K. W.; Black, C. T.; Yeuing, S. H. I. *Adv Mater* 2002, **14**, 1290–1294.
16. Black, C. T.; Guarini, K. W. *J Polym Sci A Polym Chem* 2004, **42**, 1970–1975.

17. Menelle, A.; Russell, T. P.; Anastasiadis, S. H.; Satija, S. K.; Majkrzak, C. F. *Phys Rev Lett* 1992, **68**, 67–70.
18. Huang, E.; Russell, T. P.; Harrison, C.; Chaikin, P. M.; Register, R. A.; Hawker, C. J.; Mays, J. *Macromolecules* 1998, **31**, 7641–7650.
19. Mansky, P.; Liu, Y.; Huang, E.; Russell, T. P.; Hawker, C. *Science* 1997, **275**, 1458–1460.
20. Cheng, J. Y.; Rettner, C. T.; Sanders, D. P.; Kim, H. C.; Hinsberg, W. D. *Adv Mater* 2008, **20**, 3155–3158.
21. Ji, S.; Liu, G.; Zheng, F.; Craig, G. S. W.; Himpsel, F. J.; Nealey, P. F. *Adv Mater* 2008, **20**, 3054–3060.
22. Ryu, D. Y.; Shin, K.; Drockenmuller, E.; Hawker, C. J.; Russell, T. P. *Science* 2005, **308**, 236–239.
23. Xu, T.; Stevens, J.; Villa, J. A.; Goldbach, J. T.; Guarim, K. W.; Black, C. T.; Hawker, C. J.; Russell, T. R. *Adv Funct Mater* 2003, **13**, 698–702.
24. Segalman, R. A.; Yokoyama, H.; Kramer, E. J. *Adv Mater* 2001, **13**, 1152–1155.
25. Bates, F. S.; Fredrickson, G. H. *Annu Rev Phys Chem* 1990, **41**, 525–557.
26. Helfand, E.; Tagami, Y. *J Chem Phys* 1972, **57**, 1812–1813.
27. Semenov, A. N. *Sov Phys JETP* 1985, **61**, 733.
28. Russell, T. P.; Hjelm, Jr., R. P.; Seeger, P. A. *Macromolecules* 1990, **23**, 890–893.
29. Ferry, J. D. *Viscoelastic Properties of Polymers*, Wiley, New York, 1980.
30. Leibler, L. *Macromolecules* 1980, **13**, 1602–1617.
31. Hamley, I. W. *The Physics of Block Copolymers*, Oxford University Press, New York, 1998.
32. Kleman, M.; Lavrentovich, O. D. *Soft Matter Physics. An Introduction*, Springer-Verlag, New York, 2003.
33. Black, C. T.; Guarini, K. W.; Breyta, G.; Colburn, M. C.; Ruiz, R.; Sandstrom, R. L.; Sikorski, E. M.; Zhang, Y. *J Vac Sci Technol B* 2006, **24**, 3188–3191.
34. Xu, T.; Kim, H.-C.; Rouchey, J.; Seney, C.; Levesque, C.; Martin, P.; Stafford, C. M.; Russell, T. P. *Polymer* 2001, **42**, 9091–9095.
35. Anastasiadis, S. H.; Russell, T. P.; Satija, S. K.; Majkrzak, C. F. *J Chem Phys* 1990, **92**, 5677–5691.
36. Guarini, K. W.; Black, C. T.; Zhang, Y.; Kim, H.; Sikorski, E. M.; Babich, I. V. *J Vac Sci Technol B* 2002, **20**, 2788–2792.
37. Black, C. T. In *Proceedings of the IEEE Custom Integrated Circuits Conference*, 2005, p. 86.
38. Black, C. T.; Guarini, K. W.; Ruiz, R.; Sikorski, E. M.; Babich, I. V.; Sandstrom, R. L.; Zhang, Y. In *2006 IEEE International Electron Devices Meeting Technical Digest*, 2007, 16.13.
39. Bohr, M. T. In *IEEE International Electron Devices Meeting Technical Digest*, 1995, pp. 241–244.
40. Adee, S. *IEEE Spectrum* 2008, **45**, 39–42.
41. Nitta, S.; Edelstein, D.; Ponoth, S.; Clevenger, L.; Liu, X.; Standaert, T. In *International Interconnect Technology Conference 2008*, IEEE, 2008, pp. 191–192.
42. Nitta, S.; Ponoth, S.; Breyta, G.; Colburn, M.; Clevenger, L.; Horak, D.; Bhushan, M.; Casey, J.; Chan, E.; Cohen, S.; Colt, J.; Flaitz, P.; Fluhr, E.; Fuller, N.; Kniffin, A.;

Huang, E.; Kumar, K.; Landis, H.; Li, B.; Li, W.-K.; Liniger, E.; Lisi, A.; Liu, X.; Lloyd, J. R.; Melville, I.; Muncy, J.; Nogami, T.; Ramachandran, V.; Rath, D. L.; Standaert, T.; Sucharitaves, J.-T.; Turnbull, D.; Crabbe, E.; McCredie, B.; Lane, M.; Purushothaman, S.; Edelstein, D. In *Proceedings of the Advanced Metallization Conference (AMC) 2007*, Materials Research Society, 2008, pp. 329–336.

43. Black, C. T.; Guarini, K. W.; Zhang, Y.; Kim, H. J.; Benedict, J.; Sikorski, E.; Babich, I. V.; Milkove, K. R. *IEEE Electron Device Lett* 2004, **25**, 622–624.
44. Chen, H. H.; Ling, D. D. In *Proceedings of the IEEE Design Automation Conference*, 1997, pp. 639–643.
45. Kasai, N. *NEC Res Dev* 1999, **40**, 267–271.
46. Scott, J. F.; Araujo, C. A. P. D. *Science* 1989, **246**, 1400–1403.
47. Muller, K. P.; Flietner, B.; Hwang, C. L.; Kleinhenz, R. L.; Nakao, T.; Ranade, R.; Tsunashima, Y.; Mii, T. In *International Electron Devices Meeting*, 1996, pp. 507–510.
48. Kotecki, D. E.; Baniecki, J. D.; Shen, H.; Laibowitz, R. B.; Saenger, K. L.; Lian, J. J.; Shaw, T. M.; Athavale, S. D.; Cabral, C.; Duncombe, P. R.; Gutsche, M.; Kunkel, G.; Park, Y. J.; Wang, Y. Y.; Wise, R. *IBM J Res Dev* 1999, **43**, 367–382.
49. Ino, M.; Miyano, J.; Kurogi, H.; Tamura, H.; Nagatomo, Y.; Yoshimaru, M. *J Vac Sci Technol B* 1996, **14**, 751–756.
50. Gusev, E. P.; Cariter, E.; Buchanan, D. A.; Gribelyuk, M.; Copel, M.; Okorn-Schmidt, H.; D'Emic, C. *Microelectron Eng* 2001, **59**, 341–349.
51. Black, C. T.; Guarini, K. W.; Milkove, K. R.; Baker, S. M.; Russell, T. P.; Tuominen, M. T. *Applied Physics Letters* 2001, **79**, 409–411.
52. Ellis, R. K. *IEEE Electron Device Lett* 1982, **EDL-3**, 330–332.
53. Ashok, B.; Muthukumar, M.; Russell, T. P. *J Chem Phys* 2001, **115**, 1559–1564.
54. Boker, A.; Schmidt, K.; Knoll, A.; Zettl, H.; Hansel, H.; Urban, V.; Abetz, V.; Krausch, G. *Polymer* 2006, **47**, 849–857.
55. Morkved, T. L.; Lu, M.; Urbas, A. M.; Ehrichs, E. E.; Jaeger, H. M.; Mansky, P.; Russell, T. P. *Science* 1996, **273**, 931–933.
56. Schmidt, K.; Boker, A.; Zettl, H.; Schubert, F.; Hansel, H.; Fischer, F.; Weiss, T. M.; Abetz, V.; Zvelindovsky, A. V.; Sevink, G. J. A.; Krausch, G. *Langmuir* 2005, **21**, 11974–11980.
57. Thurn-Albrecht, T.; Schotter, J.; Kastle, C. A.; Emley, N.; Shibauchi, T.; Krusin-Elbaum, L.; Guarini, K.; Black, C. T.; Tuominen, M. T.; Russell, T. P. *Science* 2000, **290**, 2126–2129.
58. Wang, J. Y.; Xu, T.; Leiston-Belanger, J. M.; Gupta, S.; Russell, T. P. *Phys Rev Lett* 2006, **96**, 128301.
59. Xu, T.; Zhu, Y. Q.; Gido, S. P.; Russell, T. P. *Macromolecules* 2004, **37**, 2625–2629.
60. Angelescu, D. E.; Waller, J. H.; Adamson, D. H.; Deshpande, P.; Chou, S. Y.; Register, R. A.; Chaikin, P. M. *Adv Mater* 2004, **16**, 1736–1740.
61. Arya, G.; Rottler, J.; Panagiotopoulos, A. Z.; Srolovitz, D. J.; Chaikin, P. M. *Langmuir* 2005, **21**, 11518–11527.
62. Hamley, I. W.; Castelletto, V.; Mykhaylyk, O. O.; Gleeson, A. J. *J Appl Crystallog* 2004, **37**, 341–344.
63. Li, H. W.; Huck, W. T. S. *Nano Lett* 2004, **4**, 1633–1636.
64. Sundrani, D.; Sibener, S. J. *Macromolecules* 2002, **35**, 8531–8539.

65. Wu, M. W.; Register, R. A.; Chaikin, P. M. *Phys Rev E* 2006, **74**, 040801.
66. Corte, L.; Yamauchi, K.; Court, F.; Cloitre, M.; Hashimoto, T.; Leibler, L. *Macromolecules* 2003, **36**, 7695–7706.
67. Freer, E. M.; Krupp, L. E.; Hinsberg, W. D.; Rice, P. M.; Hedrick, J. L.; Cha, J. N.; Miller, R. D.; Kim, H. C. *Nano Lett* 2005, **5**, 2014–2018.
68. Kim, S. H.; Misner, M. J.; Russell, T. P. *Advan Mater* 2004, **16**, 2119–2123.
69. Kim, S. H.; Misner, M. J.; Xu, T.; Kimura, M.; Russell, T. P. *Advan Mater* 2004, **16**, 226–231.
70. Park, C.; De Rosa, C.; Thomas, E. L. *Macromolecules* 2001, **34**, 2602–2606.
71. Reiter, G.; Castelein, G.; Sommer, J.-U.; Rottele, A.; Thurn-Albrecht, T. *Phys Rev Lett* 2001, **87**, 226101.
72. RamachandraRao, V. S.; Gupta, R. R.; Russell, T. P.; Watkins, J. J. *Macromolecules* 2001, **34**, 7923–7925.
73. Black, C. T.; Bezencenet, O. *IEEE Trans Nanotechnol* 2004, **3**, 412–415.
74. Cheng, J. Y.; Ross, C. A.; Thomas, E. L.; Smith, H. I.; Vancso, G. J. *Advan Mater* 2003, **15**, 1599–1602.
75. Hammond, M. R.; Cochran, E.; Fredrickson, G. H.; Kramer, E. J. *Macromolecules* 2005, **38**, 6575–6585.
76. Hammond, M. R.; Kramer, E. J. *Macromolecules* 2006, **39**, 1538–1544.
77. Segalman, R. A.; Hexemer, A.; Kramer, E. J. *Phys Rev Lett* 2003, **91**, 196101.
78. Segalman, R. A.; Hexemer, A.; Kramer, E. J. *Macromolecules* 2003, **36**, 6831–6839.
79. Sundrani, D.; Darling, S. B.; Sibener, S. J. *Langmuir* 2004, **20**, 5091–5099.
80. Sundrani, D.; Darling, S. B.; Sibener, S. J. *Nano Lett* 2004, **4**, 273–276.
81. Edwards, E. W.; Stoykovich, M. P.; Muller, M.; Solak, H. H.; De Pablo, J. J.; Nealey, P. F. *J Polym Sci B Polym Phys* 2005, **43**, 3444–3459.
82. Fasolka, M. J.; Harris, D. J.; Mayes, A. M.; Yoon, M.; Mochrie, S. G. J. *Phys Rev Lett* 1997, **79**, 3018–3021.
83. Kim, S. O.; Solak, H. H.; Stoykovich, M. P.; Ferrier, N. J.; de Pablo, J. J.; Nealey, P. F. *Nature* 2003, **424**, 411–414.
84. Rockford, L.; Liu, Y.; Mansky, P.; Russell, T. P.; Yoon, M.; Mochrie, S. G. *J. Phys Rev Lett* 1999, **82**, 2602.
85. Ruiz, R.; Kang, H.; Detcheverry, F. A.; Dobisz, E.; Kercher, D. S.; Albrecht, T. R.; de Pablo, J. J.; Nealey, P. F. *Science* 2008, **321**, 936.
86. Wang, Q.; Nath, S. K.; Graham, M. D.; Nealey, P. F.; de Pablo, J. J. *J Chem Phys* 2000, **112**, 9996–10010.
87. Yang, X. M.; Peters, R. D.; Nealey, P. F.; Solak, H. H.; Cerrina, F. *Macromolecules* 2000, **33**, 9575–9582.
88. Asakawa, K.; Hiraoka, T.; Hieda, H.; Sakurai, M.; Kamata, Y. *J Photopolym Sci Technol* 2002, **16**, 465–470.
89. Chang, L.-W.; Wong, H. S. P. *SPIE Advan Resist Technol Processing XXIII* 2006, **6156**, 615611–615616.
90. Cheng, J. Y.; Mayes, A. M.; Ross, C. A. *Nat Mater* 2004, **3**, 823–828.
91. Cheng, J. Y.; Ross, C. A.; Thomas, E. L.; Smith, H. I.; Vancso, G. J. *Appl Phys Lett* 2002, **81**, 3657–3659.

92. Li, M. Q.; Douki, K.; Goto, K.; Li, X. F.; Coenjarts, C.; Smilgies, D. M.; Ober, C. K. *Chem Mater* 2004, **16**, 3800–3808.

93. Naito, K.; Hieda, H.; Sakurai, M.; Kamata, Y.; Asakawa, K. *IEEE Trans Magn* 2002, **38**, 1949–1951.

94. Park, S.-M.; Stoykovich, M. P.; Ruiz, R.; Zhang, Y.; Black, C. T.; Nealey, P. F. *Adv Mater* 2007, **19**, 607–611.

95. Xiao, S. G.; Yang, X. M.; Edwards, E. W.; La, Y. H.; Nealey, P. F. *Nanotechnology* 2005, S324–S329.

96. Yamaguchi, T.; Yamaguchi, H. *J Photopolym Sci Technol* 2006, **19**, 385–388.

97. Yang, X. M.; Xiao, S. G.; Liu, C.; Pelhos, K.; Minor, K. *J Vac Sci Technol B* 2004, **22**, 3331–3334.

98. Li, W.; Yang, S. *J Vac Sci Technol B* 2007, **25**, 1982.

99. Black, C. T.; Ruiz, R. *SPIE Adv Resist Technol Processing XXIII* 2006, **6153**, 615302.

CHAPTER 11

Polymer-supported Reagents and Catalysts

JONATHAN BEHRENDT and ANDREW SUTHERLAND
Aston University, Birmingham, United Kingdom

11.1 POLYMER-SUPPORTED REAGENTS IN ORGANIC SYNTHESIS

This review chapter focuses on the utility of catalysts and reagents immobilized on crosslinked polymer supports. Particular emphasis is directed toward the exploitation of reagents and catalysts supported on highly crosslinked porous polymers. The first part of the chapter provides an introduction to the concept of polymer-supported synthesis and reagents, and the second part describes recent advances involving porous polymers. Supported catalysts and reagents that employ ionic liquids and noncrosslinked polymers, such as poly(ethylene) glycol (PEG) and other soluble supports, are not included in the discussion, but the interested reader is directed to a number of recent reviews centered on these matrices [1].

11.1.1 Introduction

In 1963 Bruce Merrifield published the concept of using a polymer support for the solid-phase synthesis of the tetrapetide H_2N-Leu-Ala-Gly-Val-COOH [2]. Shortly afterward (896 journal pages later) in the same year, Robert Letsinger and Milton Kornet published a similar paper independently describing the solid-phase synthesis of a dipeptide, H_2N-Leu-Gly-COOH [3]. Both routes involved the immobilization of the first amino acid residue onto a suitably activated polymer support via its amino terminus (Letsinger and Kornet) or its carboxyl terminus (Merrifield). However, the amino-terminus linkage employed by Letsinger and Kornet complicated the chemistry required to generate peptides; for example, it utilized a strong base, likely to cause epimerization of the chiral centers of the amino acid building blocks. Thus the milder Merrifield procedure became the favored polymer-supported methodology for

Porous Polymers, Edited by Michael S. Silverstein, Neil R. Cameron, and Marc A. Hillmyer

peptides. Letsinger went on to publish the first example of the polymer-supported synthesis of oligonucleotides in 1965 [4] and is quite rightly acknowledged today as the pioneer of polymer-supported polynucleotide synthesis. Polymer-supported peptide synthesis was further developed by Merrifield in an astonishing piece of work published in 1971 [5] in which he successfully undertook the solid-phase peptide synthesis of a synthetic variant of ribonuclease A (an endonuclease secreted by the pancreas that catalyzes the hydrolysis of RNA) and showed that this entirely synthetic enzyme possessed similar (80%) activity to the native enzyme. Ribonuclease A consists of 124 amino acid residues, and Merrifield's on-support synthesis proceeded in an overall yield of 17% (subsequent purification steps led to the isolation of the "pure" enzyme in overall 3%), which for 246 sequential reactions steps on a "home-modified" resin would be extremely impressive even today. At the time it was quite simply a remarkable achievement!

Key to the concept of the solid-phase peptide synthesis that Merrifield accomplished was the recognition that the polymer was not simply an inert material but was a chemically reactive species that could be exploited in synthesis. Until that time the vast majority of researchers simply thought of polymers as being essentially chemically inert materials, albeit often with highly desirable material properties (P. Hodge, personal communication; [6]). Merrifield's pioneering work changed that misperception completely and opened up an entirely new synthetic area. Merrifield was awarded the Nobel Prize for Chemistry in 1984 "for his development of methodology for chemical synthesis on a solid matrix" [7].

Following Merrifield's successful demonstration of solid-phase synthesis, he and others went on to demonstrate that the approach was generally applicable, and over the years many other compound types have been constructed using this methodology. However, solid-phase synthesis of peptides and oligonucleotides are by far and away the two groups of compounds most commonly synthesized on a solid support, even today. There are a number of reasons for this, but an important one is that the chemistry used in the construction of these two classes of biomolecules has been thoroughly investigated and optimized so that it rarely, if ever, fails. This optimization work occurred from 1963 onward as large numbers of scientists jumped on the solid-phase peptide/oligonucleotide synthesis bandwagons. At the same time, however, a relatively small number of chemists, such as Hodge, Sherrington, Grubbs, Frechet, and Leznoff, sought to develop the field of chemically reactive polymers. These pioneers generated examples of polymer-supported reagents and catalysts (both chiral and achiral), systems that we now refer to as "catch and release," and even looked at simple flow systems [6]. At the time this work was difficult to pursue from both a synthetic point of view and a financial one—money to pursue polymer-supported peptide/oligonucleotide synthesis was relatively easy to come by, but funding for work on reactive polymers was more difficult to obtain. This situation remained the *status quo* until the onset of combinatorial chemistry, a synthetic approach first conceived of and published by Arpad Furka [8], who is widely acknowledged as the father of this area of science. The combinatorial approach enables large ensembles of molecules (libraries) to be constructed relatively quickly; for example, in "split and mix" synthesis, the number of library compounds synthesized easily outstrips the number of synthetic steps used in their construction. Consequently combinatorial chemistry was hugely attractive to the pharmaceutical industry, and from the mid-1990s it invested millions of dollars in the technology. One area that was identified as critical was the nature and reactivity of the solid supports used in library construction. As a result of this, people who had been conducting curiosity-driven research in the area of chemically reactive polymers suddenly became very high profile, and a renaissance of interest in chemically reactive polymers resulted. Initially, libraries of compounds were generated on the solid phase, but fairly rapidly it became clear that a number of problems arose with this approach. For example:

- Sequential reactions on the support must proceed in near-quantitative yield, or else byproducts rapidly accrue on the support.
- Solution-phase reactions are not always readily adaptable to the solid phase.
- Multistep organic syntheses on a polymer support are difficult to accomplish because one support may not be amenable to all of the different reaction conditions/solvents employed in a synthesis.
- Only linear and split syntheses may be accomplished, as convergent syntheses are not possible.
- On-support screening is often precluded and even if possible may give erroneous results. Off-support screening requires additional cleavage and purification steps.
- The nature/properties of the support can change significantly as the substrate attached to it changes.

As outlined previously, polymer-supported synthesis of peptide libraries was relatively facile, and many combinatorial libraries of this type were generated to test out screening strategies, automated equipment, and so on. However, it soon became clear, as a result of the problems just listed, that the best way to use polymer supports in synthesis was to utilize them as reagents in solution-phase syntheses of compounds. This approach, also known as polymer-assisted synthesis, was championed by Ley at Cambridge, who rapidly published a number of high-profile exemplar papers in the late 1990s [9] and a large review article that comprised an entire issue of the Royal Society of Chemistry's journal, *Perkin Transactions 1*, in 2000 [10].

This excellent review summarized many of the advantages of conducting polymer-assisted chemistry, which include the following:

- Reactions may be driven to completion by addition of excess supported reagent(s).
- Purification is facile—a simple filtration procedure removes spent and excess supported reagents.
- Linear, split, and convergent syntheses are possible (see Fig. 11.1).
- Analysis of the progress of reactions is facile, as the reaction solution does not contain soluble reagents/reagent byproducts.

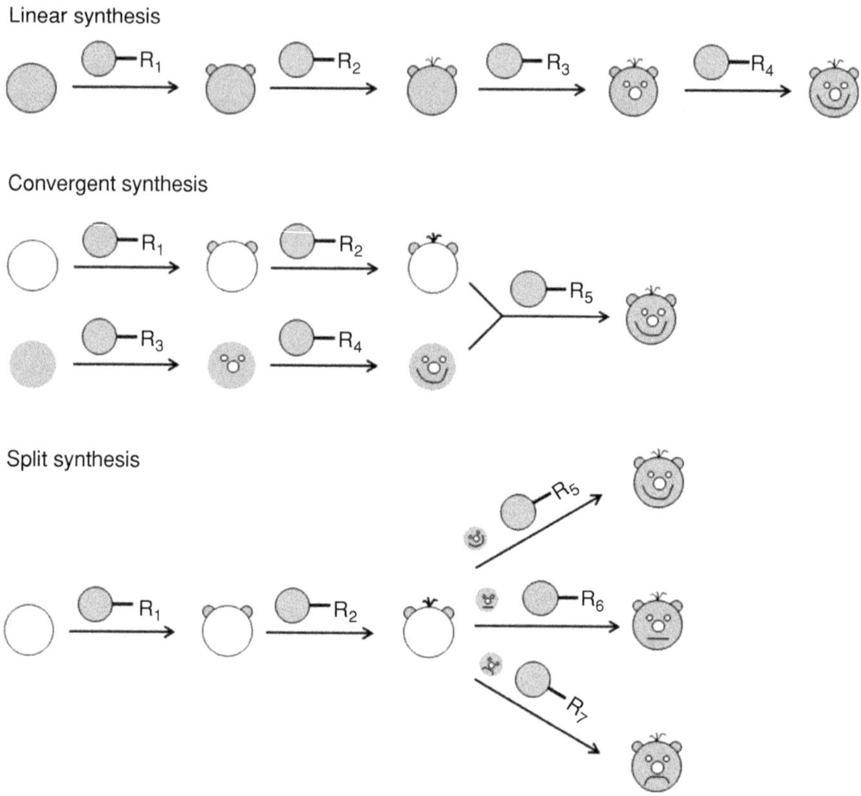

FIGURE 11.1 Schematic representation of linear, convergent, and split synthetic routes. The figure shows how starting chemical substrates are converted into a variety of products through the action of polymer-supported reagents.

- Soluble byproducts may be removed by adding one or more scavenger resins (a scavenger is a polymer-supported reagent that sequesters a known reaction byproduct; see Fig. 11.2).
- A "catch-and-release" strategy may be employed to purify products (in this two-step approach a scavenger is first used in a reversible and selective sequestration of the desired product from a mixture of one or more soluble compounds; after filtration the product is released from the scavenger resin in a second stage; see Fig. 11.2).
- Incompatible reagents may be used in concert.
- Volatile/toxic/noxious reagents/byproducts may be rendered more "user friendly" through attachment to a polymer support.
- Where reaction kinetics allows, columns packed with supported reagents may be employed in flowthrough synthesis systems.
- Automation in either batch or flowthrough format is possible.

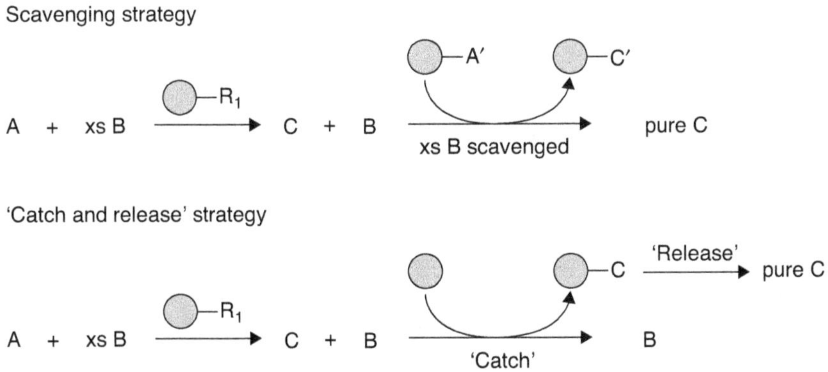

FIGURE 11.2 Schematic representations of scavenging and "catch and release" strategies.

11.1.2 Polymer-Supported Reagents—A Historical Perspective

When embarking on the synthesis of a new polymer-supported reagent it is important to remember that the renaissance of interest in polymer-supported reagents in the middle 1990s was not the start of the area. A thorough search of the literature may well save considerable time reinventing the wheel. With incomplete electronic records and search engines it is very easy to overlook previous work in this area. This is especially the case when searching for polymer-supported structures, which typically are not as straightforward to search for as soluble organic compounds. To help highlight factors worthy of consideration when designing a new reagent or in deciding on what form of the reagent to buy, historical descriptions of an impregnated reagent and two well-known polymer-supported reagents, triphenylphosphine and tetrapropyl ammonium perruthenate, are discussed below.

11.1.3 Impregnated Supports

One of the simplest methods that can be used to generate a polymer-supported reagent is to take a preformed polymeric support and impregnate it with a reagent for a given chemical transformation. The reagent is imbibed into pores within the support by dissolving the reagent in a solvent that is also compatible with the support. Depending on the polymer morphology, the solvent either swells the polymer, in the case of a gel-type resin, creating open spaces within the matrix, or else it penetrates the pores of the support if it is a macroporous type of polymer. The soluble reagent is then able to enter the support, where it becomes trapped either by (1) simple removal of the solvent or (2) addition of a solvent that is not compatible with the reagent and that displaces the original solvent and leaves the reagent precipitated within the polymer matrix comprising the support. Such an approach is often used effectively in the generation of materials for use in biological assays. For example the second impregnation strategy has been used to precipitate 2,5-diphenyloxazole **1**, a scintillant molecule, within the pores of a macroporous polystyrene-based resin. In a two-step process the beads are

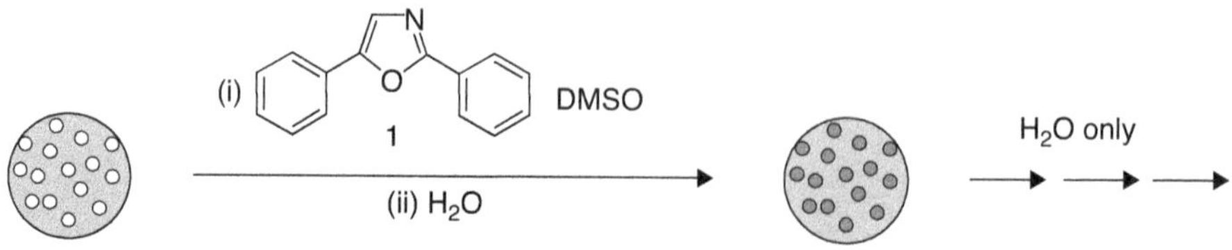

SCHEME 11.1 Impregnation of microspheres with 2,5-diphenyloxazole **1**. The microspheres are first incubated with a DMSO solution of **1** [step (i)]; in a second step the DMSO is removed by the addition of water causing **1** to precipitate in the pores of the microspheres [step (ii)]. After this two-step process the beads can only be used in aqueous media since exposure to organic solvents results in dissolution of encapsulated **1**.

first incubated with a DMSO solution of the scintillant molecules **1**. In a second step, water is added to displace the DMSO, which in turn results in the scintillant molecules being precipitated within the pores of the microspheres (Scheme 11.1). The resultant scintillation proximity assay (SPA) beads may then be used to detect radioactivity in close proximity to the impregnated polymer support (European Patent 0154734 describes an immediate ligand detection assay that is available commercially from GE Healthcare Life Sciences as the scintillation proximity assay; see also Reference 11). It is important to note that assay systems of this type must necessarily be used in aqueous media, as addition of organic solvents to the doped beads results in simple dissolution of the dopant material and they are lost from the support. Bearing this caveat in mind, a similar doping approach can be used to generate impregnated reagents for synthetic applications. A good example of a simple and yet effective impregnated polymeric reagent was developed by Sukata [12]. The reagent, potassium cyanide–impregnated Amberlite XAD resin, was synthesized, in an extension of others' work on inorganic supports [13], by mixing Amberlite XAD resin (a polymeric adsorbent with a highly porous structure) with potassium cyanide in 90% aqueous methanol. The solvent was subsequently removed under reduced pressure to leave the impregnated reagent, which was used directly in the synthesis of a series of aromatic acyl cyanides **3** from the corresponding acyl chlorides **2**. The syntheses all proceeded smoothly simply by adding impregnated resin to a substrate dissolved in benzene containing a trace of water. Notably very little acyl cyanide dimer **4** formation was observed—a side reaction (Scheme 11.2) that occurs to a significant degree when the

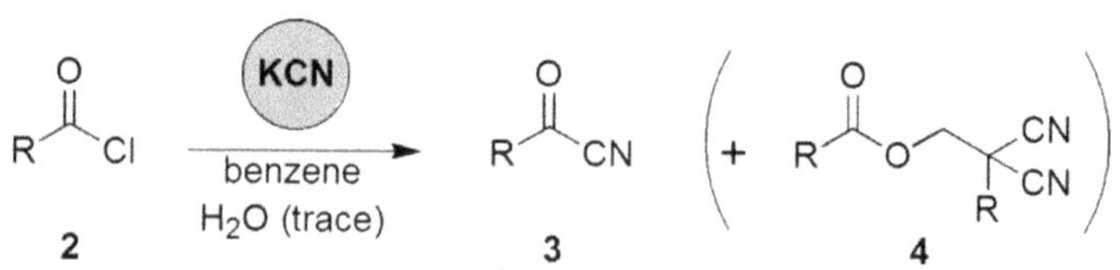

SCHEME 11.2 Synthesis of acyl cyanides using potassium cyanide impregnated Amberlite XAD.

reaction is conducted using more conventional reagents. The use of this impregnated reagent minimized the unwanted dimerization reaction by effectively providing a system with "infinite dilution" of the cyanide anion. The addition of a trace amount of water (10 μL in 40 mL of benzene) was key to driving the reactions to completion, presumably by enabling slight solvation of the cyanide anion, but without generating sufficient free cyanide anion to enable the dimerization reaction to occur.

A related approach to impregnated supports is the immobilization of catalysts via the process of microencapsulation. This approach is the subject of a recent excellent and very comprehensive review [14], and so only a brief treatment will be given here. The methodology was developed by Kobayashi, who published his first paper describing the technique in 1998 [15]. This seminal work describes the immobilization of scandium trifluoromethanesulfonate [scandium triflate, $Sc(OTf)_3$], a water-soluble Lewis acid, onto polystyrene. The procedure for forming the microencapsulated scandium triflate [MC $Sc(OTf)_3$] simply involves stirring a solid sample of $Sc(OTf)_3$ in powder form in a cyclohexane solution of polystyrene. Cooling results in the solid particles of the Lewis acid becoming enveloped by the polystyrene, and upon addition of hexane, hardened particles form that may be collected by filtration and washed easily. The metal-containing species, in this exemplar scandium triflate, is thought to be immobilized by the polystyrene both by being physically enveloped by it and by the interaction between the pi electrons in the aromatic rings within the polymer and the vacant orbitals on the metal atom (akin to the crystal field theory description of electron backdonation by ligands in classic ligand–metal complex formation). Microencapsulation is a generic approach and has been applied to other catalytic systems, such as osmium tetroxide (MC OsO_4) [16] (notably microencapsulation greatly reduced the volatility of the osmium tetroxide, which makes it easier to handle this highly toxic reagent), vanadyl(acetylacetonate) [MC $VO(acac)_2$] [17], and palladium (MC Pd) [18], or, via a different microencapsulation process, involving an *in situ* interfacial polymerization approach, PdEnCat [19] (marketed by Reaxa, Manchester, UK), among others. The interested reader is referred to Akiyama and Kobayashi's excellent review of this area [14].

While microencapsulation does not formally constitute a polymer-supported catalyst, the interface is a gray one; for example, microencapsulation of osmium tetroxide by a crosslinked polystyrene resin (PS resin-MC Os) has also been described [20], and a closely related area involves polymer-incarcerated catalysts, which are microencapsulated systems that undergo a postencapsulation crosslinking procedure [14]. It is likely that microencapsulation using crosslinked supports (either preformed or formed after the encapsulation procedure) will be expanded upon considerably in the future. The next section details supports where there is a formal bonding interaction between the support and the reagent.

11.1.4 Ion Exchange Resin–Based Reagents

In 1935 Adams and Holmes published a description of the first totally synthetic ion exchange resins (see Reference 21). In their approach a phenol functionalized crosslinked polymer could be produced by the polymerization reaction between

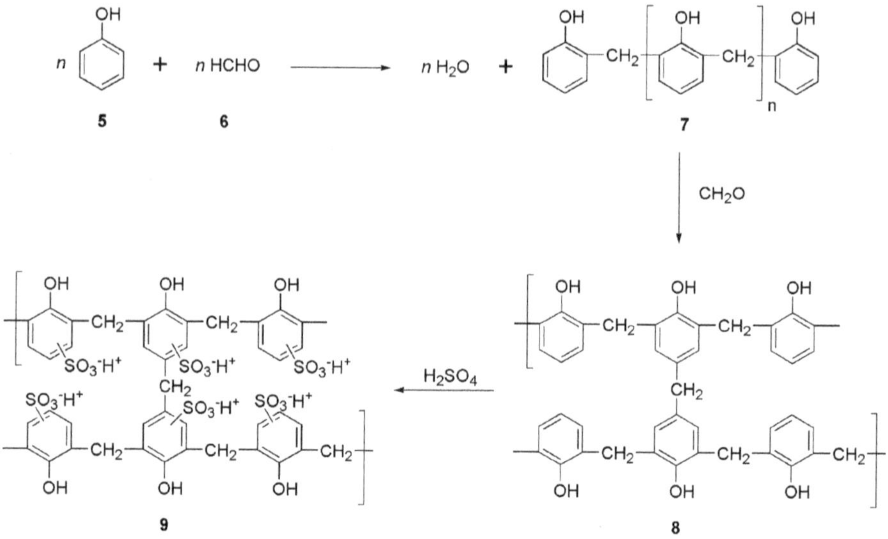

SCHEME 11.3 Summary of the synthesis of the earliest ion-exchange resins developed by Adams and Holmes. (Adapted from Reference 21.)

phenol **5** and formaldehyde **6** (Scheme 11.3). The initial linear condensation polymerization products (represented by structure **7**) were subsequently crosslinked by further reaction with formaldehyde, resulting in a highly crosslinked polymer matrix with representative structure **8**. The phenolic OH groups rendered this a weak cation exchanger, but simply treating the resin with sulfuric acid resulted in a "phenol–formaldehyde" sulfonic acid cation exchange resin with representative structure **9**, which is a much stronger cation exchange resin.

Since the pioneering work of Adams and Holmes, a host of different types of synthetic ion exchange resins have been constructed. The principle of ion exchange is straightforward. In the case of sulfonic acid derivative **9**, the resin is easily ionized to give a negatively charged resin bead with the associated protons free in the solution in which the resin is suspended but held close to the resin by electrostatic charge–charge attraction. Placing this resin in a solution containing a large excess (in molar terms) of a salt results in these readily ionized protons being exchanged for the cationic species (for example, a positively charged reagent, R^+) of the dissolved salt and results in a new ion exchange resin (see Scheme 11.4). This is the principle that underlies many solid-supported reagents used in polymer-assisted synthesis. It is thus possible to use the same base resin to generate a large number of solid-supported reagents and catalysts related by the same underlying resin. The most commonly employed resins in this respect are quaternary ammonium salts (with associated active anions) derived from amino functionalized macroporous resins; some representative examples are shown in Fig. 11.3. For a more comprehensive listing of reagents and catalysts of this type the interested reader is referred to the 2000 review by Ley et al. [10].

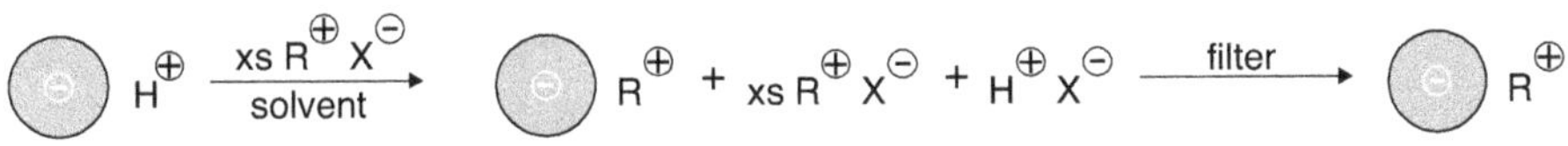

SCHEME 11.4 Principle of ion exchange for generating solid-supported reagents and catalysts.

11.1.4.1 Polymer-Supported Tetrapropylammonium Perruthenate (TPAP). To place the development of polymer-supported reagents on charged supports in context, this subsection gives a brief overview of the development of the solid-supported variant of the versatile designer oxidant tetra-*n*-propylammonium perruthenate (TPAP). This reagent has been chosen because it highlights the key stages necessary to develop a solid-supported reagent (see Table 11.1), which apply both to reagents immobilized on the basis of charge, such as those under discussion here, and to reagents linked covalently to a support—the subject of the following section.

One of the most common transformations in organic synthesis is oxidation. A plethora of reagents for this transformation exists, and reagent choice can sometimes be rather bewildering. A subset of this transformation is the selective oxidation of primary alcohols into the corresponding aldehyde, in which overoxidation can be problematic, and the oxidation of secondary alcohols. This was the situation back in the mid-1980s, despite the large number of available reagents available, and resulted in the development by Griffiths et al. [22] of TPAP, an oxidant that enabled these transformations to proceed smoothly and in high yield with a wide range of substrates. This highly versatile reagent was a perfect choice to develop into a solid-supported reagent (Table 11.1, Stage 1). Initially, the solid-supported reagent, polymer-supported perruthenate (PSP), was constructed simply by treating Amberlyst A-27 chloride with potassium perruthenate in water (the potassium perruthenate was solubilized in water by the action of ultrasound); see Fig. 11.4 (Table 11.1, Stage 2) [23]. Loading of the resultant black polymer resin was established by performing the oxidation of benzyl alcohol to benzaldehyde in the presence of an excess of the alcohol and gave a loading of 0.16 mmol g^{-1} (Table 11.1, Stage 3) [24]. In terms of loading for a reagent this is on the low side; typically it is desirable to achieve a

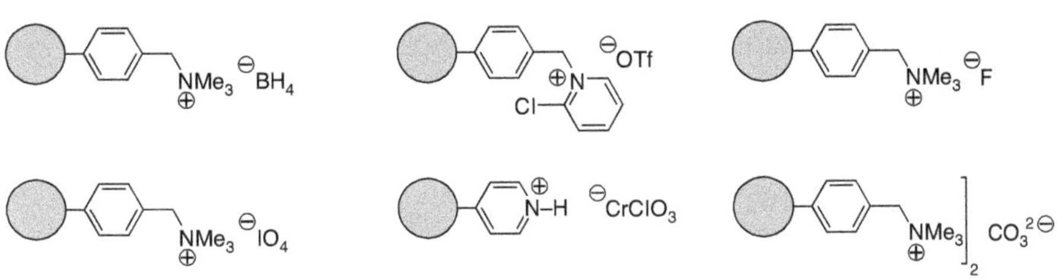

FIGURE 11.3 Representative examples of polymer-supported reagents on positively charged resins.

TABLE 11.1 The Key Steps in Developing a Solid-Supported Reagent

Stage	Task
1	Identify solution-phase reagent
2	Attach to solid support
3	Assess loading
4	Assess reactivity
5	Optimize reaction conditions
6	Recycle?
7	Optimize support (if necessary)

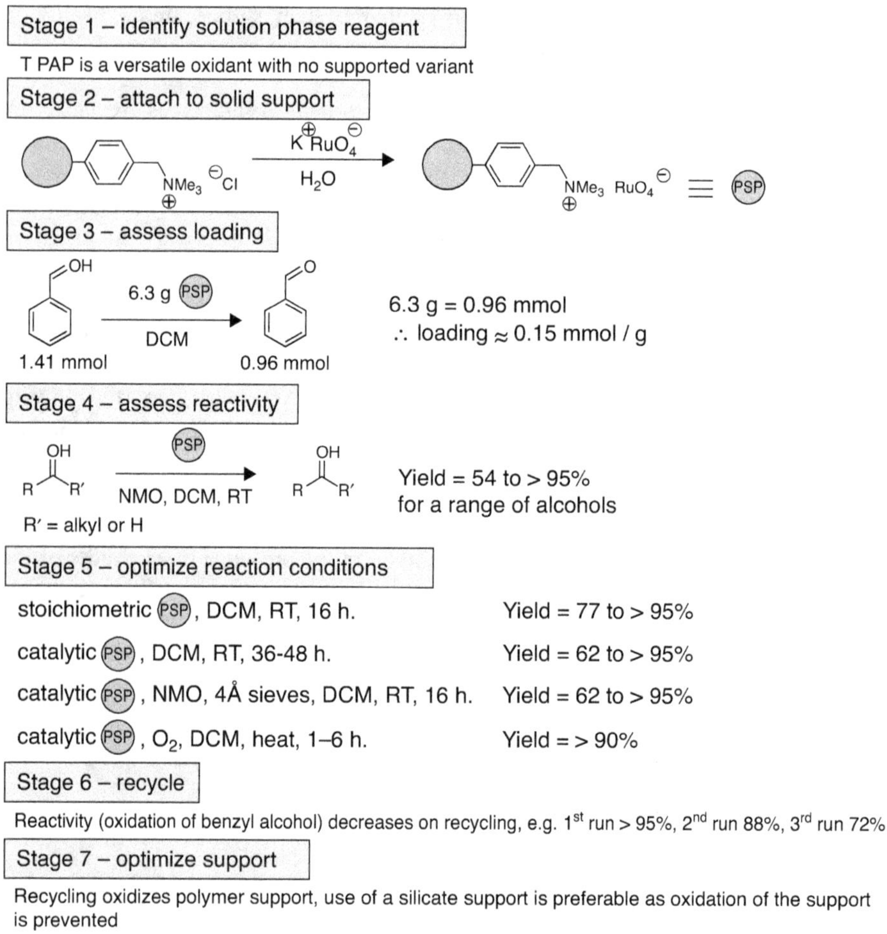

FIGURE 11.4 The development and optimization of polymer-supported perruthenate (PSP).

loading that is as high as possible, and usually this is greater than 1.5 mmol g^{-1}. In the case of polymer-supported TPAP, however, a low loading was acceptable, as the reagent was known to be catalytic in the presence of a cooxidant such as *N*-methyl morpholine *N*-oxide (NMO). The reactivity of the PSP was assessed initially by looking at the oxidation of a range of alcohols, but its efficacy in additional reactions was also assessed subsequently [25] (Table 11.1, Stage 4). As just mentioned, PSP, like TPAP, can be regenerated once it has been reduced by addition of a suitable cooxidant, and, accordingly, a number of cooxidants and reaction conditions were evaluated and the optimal conditions, shown in Fig. 11.4, identified [24] (Table 11.1, Stage 5). The ability to recycle a supported reagent is one of the attractive reasons for using these reagents and is especially relevant, as reducing the environmental impact of synthesis is now a major consideration. In the case of PSP, the efficacy of the reagent after a number of cycles was assessed by monitoring the percentage of product produced (Table 11.1, Stage 6). In this particular instance, the solid support was found to be degraded by the conditions employed to regenerate the PSP, and so an alternative inorganic support, a mesoporous silicate, was developed to circumvent this problem when recycling was required and allowed up to 12 cycles without any observable loss in activity [26]. This last stage is an interesting one, as it is easily overlooked—despite the warnings of polymer chemists! Often the users of polymer-supported reagents are synthetic chemists with little, if any, experience of polymers. They are thus in danger of considering the polymer as an entirely inert support that does nothing more than serve as a tether for the reagent and thus does not participate in any way in the synthesis. This may be unimportant in many instances, but it is important to bear in mind, especially when things do not go according to plan, such as in the recycling process discussed earlier.

11.1.5 Covalently Linked Reagents

In 1958, Leebrick and Ramsden first reported the synthesis of *p*-vinylphenylmagnesium chloride **14** [27], and 3 years later this Grignard reagent was employed successfully in the first reported synthesis of *p*-styryldiphenylphosphine **15** [28]. The availability of this monomer led, 10 years later, to the first reported example of the polymer-supported Wittig reagent, polystyryl-*p*-diphenylphosphine **16** [29]. In this seminal report, the utility of the polymer-supported triphenylphosphine was demonstrated with a successful exemplar synthesis of *cis*- and *trans*-stilbene. The polymer-supported Wittig reagent **16** was made using the chemistry highlighted earlier in this chapter and shown in Scheme 11.5. The support bearing the reagent was a gel-type resin crosslinked with 2% divinylbenzene (DVB) and constructed from a 3:1 monomer mix of styrene:*p*-styryldiphenylphosphine. The following year, polymer-supported Wittig reagents were also reported by Heitz and Michels [30] and by McKinley and Rakshys [31]. Both reports describe the postpolymerization derivatization of preformed polystyrene beads **17** via the routes of either bromination, lithiation, and reaction with chlorodiphenylphosphine or bromination and subsequent reaction with sodium diphenylphosphide; see Scheme 11.6.

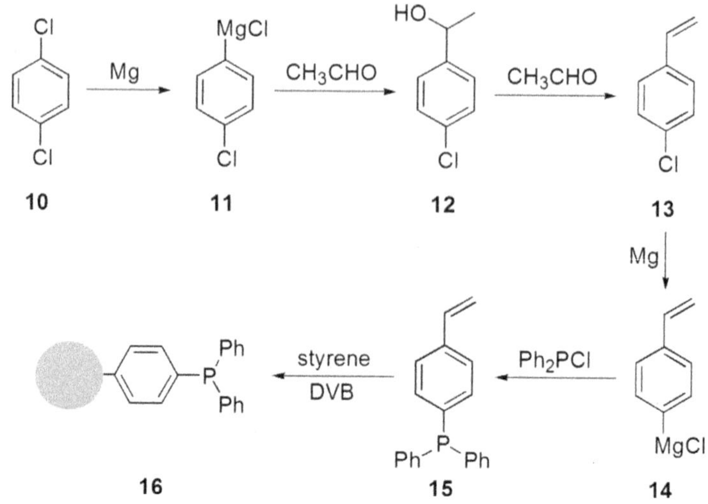

SCHEME 11.5 Synthesis of polymer-supported triphenylphosphine **16.**

These two different methods of constructing polymer-supported triphenyl phosphine **16** led to an interesting dilemma applicable to the design and synthesis of any reagent that is to be covalently linked to a support. Is it better to construct the supported reagent directly by copolymerizing a suitably functionalized monomer (i.e., Scheme 11.5), or is postpolymerization modification of a preformed support preferable (i.e., Scheme 11.6)? The approach preferred by many, including the authors, is to carry out as much synthesis as possible up front before constructing the polymer since solution-phase chemistry is far easier to monitor than reactions carried out on a support. That said, clearly both approaches have merit; for example, the report by McKinley and Rakshys [31] describes the synthesis of the polystyryl-*p*-diphenylphosphine reagent via a suspension polymerization reaction involving styrene, *p*-styryldiphenylphosphine **15** (prepared via the route outlined earlier in this

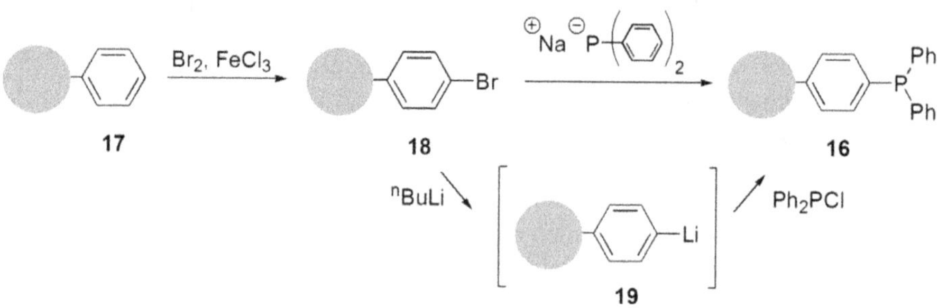

SCHEME 11.6 Two alternative postpolymerization routes to polymer-supported triphenylphosphine **16.**

chapter [28]), and divinylbenzene in addition to the postpolymerization modification of a preformed support methodology referred to earlier. The authors compared the loading of both resins and found that, in theory, while they both should have had similar loadings, the reagent constructed using a number of postpolymerization derivatization steps had a lower loading than the reagent formed directly from the suspension polymerization reaction. They ascribed this observation to the fact that while the lithiation step was assumed to proceed in high yield (as demonstrated by the absence of bromine in the product resin **19**), the amount of lithiated resin **19** that reacted with the diphenylchloride was only 60%. Presumably, the other 40% was simply polystyrene **17** formed by proton quenching of the lithiated sites. It should be noted, however, that despite the advantages of direct polymerization of functionalized monomers outlined earlier, the approach can be problematic since the functionalized monomers may affect the outcome of the polymerization reaction in a deleterious manner.

The observation that the polymer-supported reagent formed directly from the suspension polymerization reaction had a higher loading as assessed by combustion analysis should be treated with caution since it involved the generation of a reagent upon a lightly crosslinked gel-type resin. If the resin had been a highly crosslinked one, it is likely that the loading results would have been reversed. The reason for this is that the direct polymerization of a functionalized monomer to form a highly crosslinked polymer matrix can result in a significant amount of the functionalized monomer being incorporated into highly crosslinked regions of the polymer matrix. This, in turn, results in a significant number of functional groups being in regions of the bead that are inaccessible to both solvent and other reagents alike. Moreover, this problem is a generic one since the degree to which the functionalized monomer is incorporated into solvent-inaccessible regions of a highly crosslinked resin depends on its partition coefficient between the growing polymer and the solvent/monomer mixture in which the polymerization reaction is conducted, a situation that is always going to be highly monomer and polymerization reaction conditions specific. There is thus no hard and fast rule as to which approach—synthesis upon a preformed polymer or direct access by suspension polymerization reaction involving a functionalized monomer—is the best one to adopt. Regardless of synthetic strategy (bead type, etc.), the best approach is to construct the polymer-supported reagent and to base its loading not upon chemical composition, but upon how many *reactive* functional groups are present per mass of polymer, that is, its "real loading," a number that can be obtained experimentally as outlined earlier in the description of the development of PSP.

With the increasing drive toward more sustainable/"greener" chemistry, recycling is clearly an important criterion when designing either a new polymer-supported reagent or a polymer-assisted synthesis. For reagents covalently linked to the support, two recycling possibilities exist: Either the polymer-supported reagent can be recycled directly back to its original state, or else it may be converted into an alternative polymer-supported reagent. Although the initial reports of the use of polymer-supported triphenyl phosphine did not focus on recycling, it was shown shortly afterward that the reagent **16** may be readily recycled directly from the byproduct,

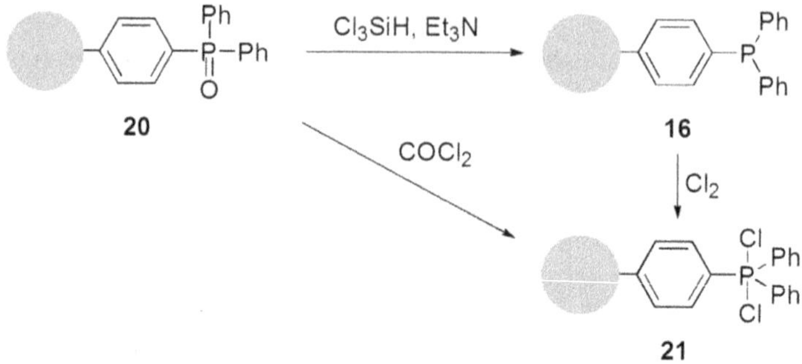

SCHEME 11.7 Recycling/further derivatization of polymer-supported triphenylphosphine oxide **20.**

polymer-supported triphenyl phosphine oxide **20**, by treatment with trichlorosilane and triethylamine [32]. Subsequently it was also shown that phosgene could be employed to convert polymer-supported triphenyl phosphine oxide **20** into the corresponding polymer-supported triphenylphosphine dichloride **21**, which was shown to act as an effective polymer-supported reagent for the conversion of carboxylic acids into their corresponding acid chloride derivatives and alcohols into their corresponding chlorides [33] (Scheme 11.7).

11.1.6 Exemplar Syntheses

The following syntheses have been selected to illustrate some of the techniques that exploit the unique nature of polymer-supported reagents in addition to the standard "add, mix and filter" type reactions described earlier. Specifically, exemplars of each of the following techniques have been chosen to demonstrate the versatility that is possible using solid-supported reagents: scavenging, catch and release, more than one reagent simultaneously, and direct reaction of unstable intermediates.

11.1.6.1 Scavenging. The scavenging strategy was outlined in Fig. 11.1. In this approach, a soluble product is purified by the sequestration of one or more soluble byproducts by the addition of an appropriate scavenger resin. One of the reasons for the necessity of the approach is that often it is desirable to drive a reaction to completion through the addition of an excess of reagent(s). Consider the following theoretical example by which an acid chloride **22** is reacted with an amine **23** to form an amide **24** and the acid chloride is readily available and thus cheap, while the amine is very expensive (Scheme 11.8). Thus, to maximize the yield of the reaction and to ensure that no amine **23** is wasted, acid chloride **22** is added to the reaction in excess. Upon completion of the reaction, the reaction mixture contains solvent, unreacted acid chloride **22**, and the salt triethylamine hydrochloride (the addition of a base, triethylamine, is required for the reaction to proceed). However, the amide product **24** may be obtained in a pure form by addition of a simple amine-based scavenger resin

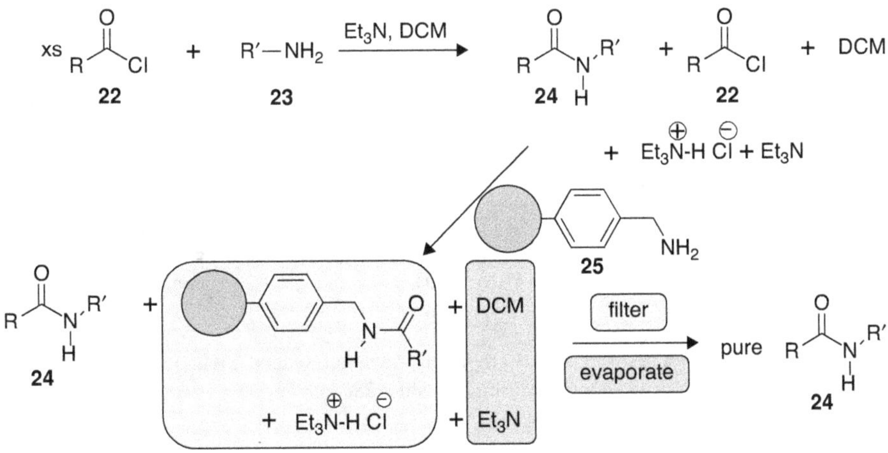

SCHEME 11.8 Exemplar showing how scavenging can be used for the purification of amide **24.**

25 followed by filtration. As the scheme shows, excess acid chloride **22** reacts with the scavenger resin, while the amide, solvent, and triethylamine do not. Following filtration, which removes the triethylamine hydrochloride and the scavenger resin, the solvent and any excess triethylamine is readily removed by evaporation, leaving just the pure amide **24** behind. Such an approach is commonplace in polymer-assisted synthesis, and numerous examples exist [34]. Before using it, however, it is advisable to consider carefully the nature and reactivity of all possible products and byproducts to ensure that only the desired product is left behind following the scavenging steps.

11.1.6.2 Catch and Release. The "catch and release" strategy is very similar to the scavenging strategy, with the exception that the desired product, rather than the byproduct(s), is sequestered reversibly by a polymer-supported reagent, and after filtration the product may be released again by treatment of the isolated product-bearing resin with an appropriate reagent (see Fig. 11.1). The earliest report of such an approach was in 1997 by workers at Eli Lilley [35], who "caught" products from a reductive amination procedure on a cation exchange resin and subsequently "released" them by treatment of the column with ammonia. Scheme 11.9 shows a specific example of this process in which the secondary amine product **28**, formed from the reaction between aldehyde **26** and amine **27**, is caught on a strong silica-based cation exchange material **29** and subsequently released. Moreover, they demonstrated the utility of the approach in combination with an automated synthesizer and showed that a number of varied products from a 3 × 3 reaction array could be purified by "catch and release," although the authors made the point that the "excess of reagents was employed in such a way as to ensure that the desired product was the only ionisable component," emphasizing the importance of a judicious choice of reagents coupled with an intimate understanding of the reaction.

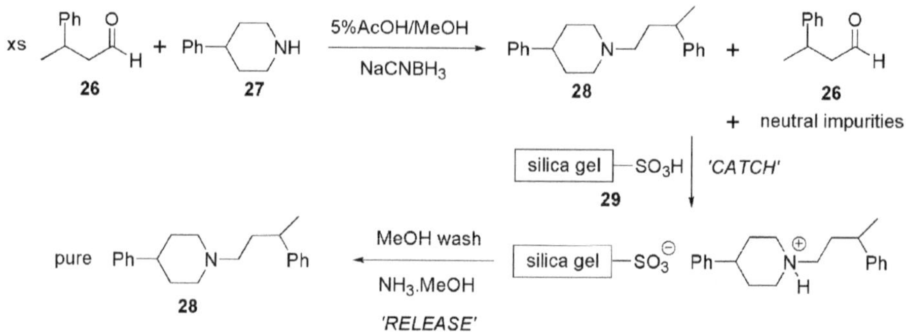

SCHEME 11.9 The first reported example of "catch and release," which employed an ion-exchange process to purify secondary amine product **28**.

11.1.6.3 More Than One Reagent Simultaneously. A unique approach that is only enabled by having individual incompatible reagents/catalysts immobilized on different supports is based on the ability to carry out more than one reaction in a single step in a single reaction vessel. This is possible because the reagents and/or their byproducts are not free in solution and so have no possibility of encountering and thus reacting with each other. An outstanding biological example, well before its time, albeit without the issue of incompatible reactivities, helped pave the way for this approach. In 1971 Mosbach reported [36] the immobilization of both hexokinase and glucose-6-phosphate dehydrogenase onto the same polymeric support. Using these polymer bead-supported enzymes, it proved possible to convert glucose **30** into gluconolactone-6-phosphate **32** via glucose 6-phosphate **31** (Scheme 11.10). The presence of the two enzymes on the same support gave reaction rates that were better than, or at worst the same as, those obtained for the same reactions with the two enzymes free in solution. This observation was rationalized by suggesting that the localized concentration of the intermediate glucose-6-phosphate **31** at the sites of the dehydrogenase enzyme was greater with the on-support reactions, where the intermediate was formed *in situ*, than when the reactions were conducted in solution.

Shortly after this biological example was reported, Pittman and Smith published a similar study [37] that described two metal-containing catalysts being immobilized on

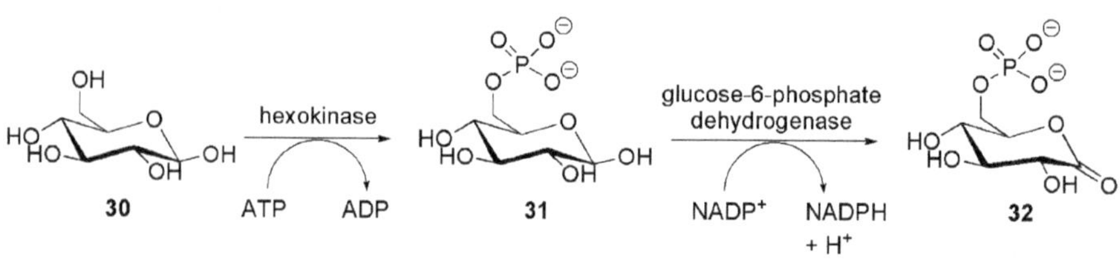

SCHEME 11.10 Enzyme-mediated conversion of glucose **30** into gluconolactone-6-phosphate **32**.

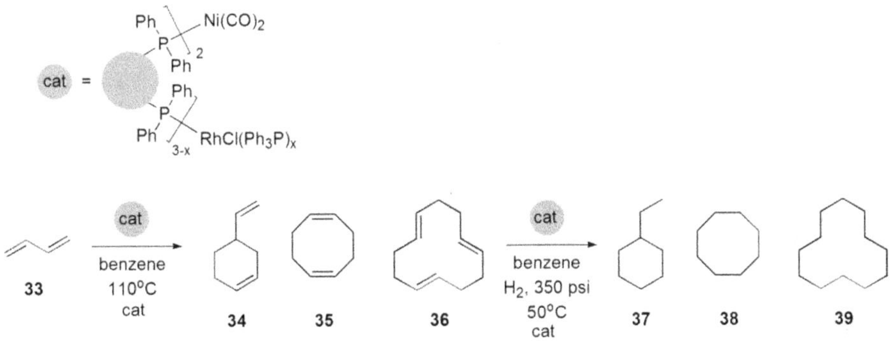

SCHEME 11.11 Sequential cyclooligomerization and reduction of butadiene **33** employing a two-catalyst, one-bead system.

polymer-supported triphenylphosphine **16**. Using the system shown in Scheme 11.11, they were able to catalyze sequential cyclooligomerization and hydrogenation reactions. The two-step process proceeded via cyclooligomerization of butadiene to generate olefins **34–36** and subsequent hydrogenation of these olefins to their fully saturated analogs **37–39**. The two catalysts were both attached to the same support and were present in both steps, but the reaction conditions were changed for the hydrogenation step. Cyclooligomerization and selective hydrogenation and cyclooligomerization and hydroformylation procedures could be conducted similarly.

Another early example [38] involved the simultaneous use of both strong acid and base catalysis in the synthesis of a range of cyclopentenones **41** (Scheme 11.12). Unlike the catalytic examples described previously, this example employed two incompatible reagents: If used together in solution, they would simply react with each other and form a salt, thus precluding the desired reaction from taking place. This very early example is also particularly apposite since only the use of both reagents

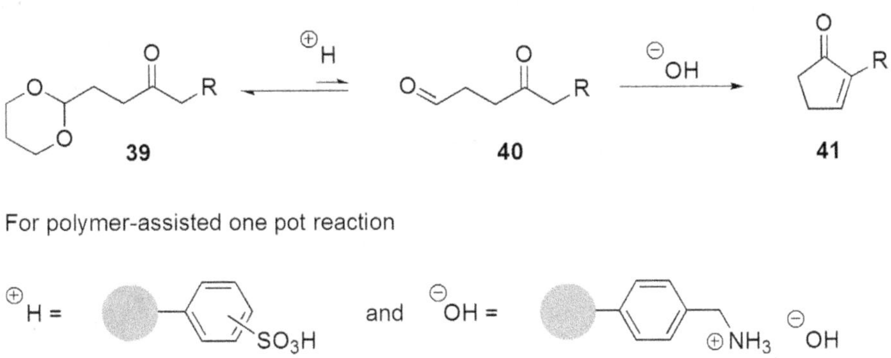

SCHEME 11.12 Synthesis of cyclopentenone **41** by the simultaneous employment of both acid- and base-supported catalysts.

simultaneously enables the product to be formed; their sequential use would result in either none or, at best, trace amounts of the desired product being formed. Specifically, the first reaction is the acid-catalyzed cleavage of acetal **39** into the corresponding aldehyde **40**. This hydrolysis reaction is reversible, and the relative stability of the dioxirane ring is such that only a very small amount of aldehyde **40** is produced even when a very large excess of water is added. Moreover, the base-mediated intramolecular Aldol reaction tends to generate many polymeric byproducts if it is carried out in anything less than very dilute conditions. By using the two reagents in concert, the equilibrium of the first reaction is dragged over in favor of product formation by the product aldehyde **40** being continuously consumed in the second step. At the same time, there is only ever a very low concentration of aldehyde **40** produced, and so the cyclization reaction to generate cyclopentenone **41** is greatly favored over intermolecular polymerization-type processes. Thus, pairing the two steps together enables a reasonable yield (87% crude, 56% distilled, when R = Me) of the desired product to be formed, in much the same way as an enzymatic reaction cascade proceeds.

A more recent but related example [39] that employed multiple reagents simultaneously also generated stereochemical information by including an enantioselective step. This asymmetric example focused on the hydrolysis of the dimethyl acetal of *o*-nitrobenzaldehye using an acid catalyst (Amberlite IR-120 H^+ form) to produce *o*-nitrobenzaldehye. Simultaneous treatment with a basic polymer-supported tripeptide enabled the *o*-nitrobenzaldehyde to react with acetone in an intermolecular Aldol reaction that produced chiral alcohol (Scheme 11.13). While no comment was made as to the product of sequential reactions, the rationale behind this one-pot approach was to mimic an enzyme-mediated cascade process. Overall the approach gave a moderately good enantiomeric excess for the asymmetric Aldol reaction, and the catalyst could be reused a number of times, although the reaction slowed upon recycling. It should be stressed that the two examples just given centering on incompatible reactions focus on using acid and base concomitantly, but the reader should be aware that other incompatible reagent groupings have also been reported. These include oxidant and reductant and hydroformylation catalyst and Wittig reagent.

The examples selected here illustrate one of the likely future directions of this approach, which will be to increase the number of reactions in the cascade and also

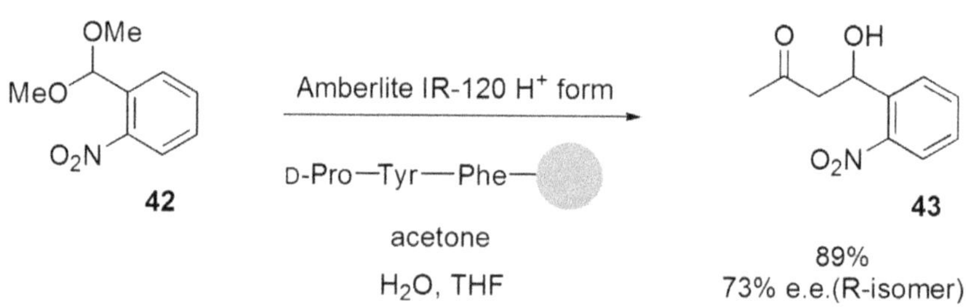

SCHEME 11.13 One-pot enantioselective synthesis employing two polymer-supported reagents simultaneously to generate chiral alcohol **43** with an enantiomeric excess (e.e.) of 73%.

to make the reaction products more complex in terms of both structure and the information that they contain. The ultimate deployment of multiple polymer-supported reagents is in automated synthesizers. However, rather than doing everything in one reaction vessel, compartmentalized flowthrough columns are used to conduct flowthrough synthesis [40].

11.2 POROUS POLYMERS IN ORGANIC SYNTHESIS: RECENT ADVANCES

11.2.1 Introduction

Since the pioneering work of Merrifield and Letsinger, solid-phase organic synthesis has flourished, as exemplified by the applications described earlier. It has become increasingly apparent, however, that it is more appropriately viewed as a synthetic tool to be used in conjunction with classic solution-based synthesis rather than an alternative discipline. There are certainly several issues that can limit the general applicability of the technique; first, analysis of resin-immobilized intermediate products is nontrivial, often necessitating cleavage of the intermediates after each step in the reaction. In addition, particularly when multiple transformations are carried out "on bead", it is important that each transformation reaches quantitative conversion so as to obtain a single, pure product upon cleavage from the resin rather than a complex mixture of deletion products. "On-bead" reactions allow for use of excess reagents in order to drive reactions to completion, which somewhat mitigates the latter issue, but the apolar environment offered by standard polymeric supports (e.g., polystyrene-based supports) is not appropriate for all types of reaction—for example, where ether formation was attempted between a resin-bound alkyl bromide and a metal alkoxide in solution, the apolar nature of the resin caused an increased prevalence of the competing side reaction, β-elimination of the bromide. The apolar nature of Merrifield-type polystyrene supports also makes them incompatible with water and polar solvents (e.g., methanol), which cause the beads to shrink, thus limiting the accessibility of their reactive functional groups to molecules in solution. A number of strategies have been employed to improve the swelling properties of lightly crosslinked polystyrene supports in polar solvents, such as grafting polyethylene glycol onto the polymer backbone (e.g., TentaGel); however, these resins still cannot be viewed as truly aqueous compatible.

An alternative approach to the use of gel-type polymeric supports for organic synthesis that has been increasingly adopted in recent years is the use of highly crosslinked polymers containing porous networks throughout their structure. Due to their highly rigid structure, these polymers do not undergo significant shrinkage or swelling under the influence of the solvent that surrounds them. Instead, molecules in solution can diffuse through their porous network and, in this way, are able to access reactive groups on the surfaces of these pores. This negation of the need to keep the polymer swollen throughout the reaction to allow access to reactive sites leads to enhanced compatibility with a range of solvents. Porous polymers are also less susceptible to their properties being adversely affected by the transformations carried out at the reactive sites (the swelling properties of gel-type resins can be significantly affected by conjugation of highly polar groups to their reactive sites). Achieving a high degree of conversion at all active sites within porous polymers, however, is not a trivial matter. The majority (>99%) of active sites are located within the polymer matrix, and so efficient diffusion of the solvent and reagents throughout the interior of the polymer is essential for achieving high conversions. This has been particularly challenging where monolithic porous polymers (e.g., polyHIPE) have been used as supports, as these structures have a much greater path length through which solvents must diffuse compared to bead-type structures. Despite these issues, both macroporous polymeric beads (e.g., Argopore) and porous monolithic structures have been used to great effect as supports for a range of synthetic applications, which will be highlighted throughout the following section.

11.2.2 Macroporous Beads

As discussed earlier, the limited swelling in all solvents displayed by macroporous polystyrene beads provides enhanced compatibility with reagent/solvent combinations commonly used in classic solution-based synthesis, particularly in water and other polar solvents. The porous network also allows for rapid diffusion of functional groups that are covalently attached to the surface of the pores. These properties allow for non–resin-bound excess reagents to be more easily washed from the beads following a reaction, using smaller volumes of solvent than is required with gel-type resins, making the use of such beads more economical and environmentally friendly. Macroporous beads may also be more rapidly dried in vacuo to remove solvents following a reaction, especially in the case of water. In addition, macroporous beads are more suitable for use in low-temperature reactions than lightly crosslinked, gel-type resins. In terms of comparative loading values of resin-bound functional groups, the range for macroporous resins is generally lower in loading than for similarly functionalized 1% crosslinked resins (e.g., Argopore has typical loading values of 0.6–1.1 mmol g^{-1}, whereas this range is generally 0.6–2.0 mmol g^{-1} for 1% crosslinked polystyrene), but are higher loading than PEG-grafted polystyrene resins (e.g., ArgoGel has a typical loading range of 0.35–0.5 mmol g^{-1}) (Argo gel resins are available from Sigma-Aldrich [41]). Due to the inflexible nature of the resin, steric factors are of greater consideration than when using gel-type resins for solid-phase organic synthesis. For example, macroporous beads would be less suitable for multistep, iterative syntheses (e.g., peptide synthesis), as the final resin-bound product may have considerable steric bulk and thus restrict efficient diffusion of reagents through the porous network. Largely for this reason, it would seem, macroporous resins have generally been of greater value synthetically for use (1) as scavenger resins, removing excess reagents and unwanted byproducts from solution-based reactions, (2) for construction of resin-immobilized catalysts, (3) for two-step catch and release synthetic transformations, and (4) as ion-exchange resins. Macroporous resins are now commercially available with a range of functional groups, including chloride, hydroxyl, amino, sulfonamide, and aldehyde, as well as with more complex linkers such as

Wang alcohol and Rink amide linkers. A range of macroporous ion-exchange type resins is also available in both acidic and basic forms.

11.2.2.1 Macroporous Beads for Use in Ion Exchange. Macroporous-type ion-exchange resins are porous beads that contain either acidic (e.g., sulfonic acid, carboxylic acids) or basic (e.g., tertiary ammonium salts) pendant functionalities. Due to the low cost of these resins and wider range of solvent compatibility, macroporous ion-exchange resins have found use in many industrial applications. For example, weakly basic macroporous resins (e.g., Amberlite FPA 51, styrene–DVB matrix) are routinely utilized in processing to remove so-called "color species"—uncharacterized impurities that are present in a range of pharmaceutical/biopharmaceutical process streams (Amberlite FPA 51, styrene–DVB matrix, Amberlite FPA90 Cl, and Amberlite FPC3500 [methacryl–DVB matrix] are available from the Advanced Biosciences Division of Rohm and Haas [42]). By using these inexpensive resins to remove upstream color impurities, the lifetime of the more expensive chromatography media downstream of the process is increased. Use of anionic ion exchange resins has also minimized, and in some cases replaced, the use of activated charcoal for decolorization. In addition to being used to remove unwanted impurities, strongly basic macroporous ion-exchange resins (e.g., Amberlite FPA90 Cl, styrene–DVB matrix) are also used in the recovery step of certain industrial processes to "fish out" the active target compound from mixtures of products; for example, aminoglycoside, erythromycin, and tylosin antibiotics may all be recovered in this manner [43]. Weakly acidic macroporous ion-exchange resins have also found important industrial applications; for example, Amberlite FPC3500 (methyl methacrylate–DVB matrix) is used for the cationic fixation of enzymes, either as the recovery step of the purification process (e.g., lysozyme) or for enzyme immobilization to enhance the duration of enzyme activity (e.g., cytochrome C, acylase) [43].

11.2.2.2 Use of Cation-Exchange Resins in Organic Synthesis. Since the early reports of the use of sulfonic acid resins as catalysts (e.g., condensation of ketones) discussed in the previous section, macroporous acidic ion-exchange resins (e.g., Dowex-50, Amberlyst 15) have been widely utilized as solid catalysts for a range of synthetic transformations. In this manner, Meshram et al. achieved exclusive α-monohalogenation of both 1,3-ketoesters and cyclic ketones with *N*-halosuccinimides, catalyzed by Amberlyst 15 resin **44** [43]. This macroreticular polystyrene ion-exchange resin, which is functionalized with sulfonic acid, allowed for mild reaction conditions and facilitated trivial catalyst removal through filtration. In addition, the use of solid catalysts of this nature has enabled solvent-free synthesis of certain target compounds. For example, by using Amberlyst 15 as a catalyst, Ko and Yao reported a one-pot condensation of β-naphthol **45** with an aldehyde (e.g., benzaldehyde **46**), under solvent-free conditions, to form a 14-substituted-14*H*-dibenzo[*a,j*]xanthene **47** (Scheme 11.14) [44]. Where benzaldehyde **46** was used in this reaction, the presence of Amberlyst 15 resin facilitated a 94% yield of the desired product **47**, compared with a 36% yield in the absence of the solid catalyst.

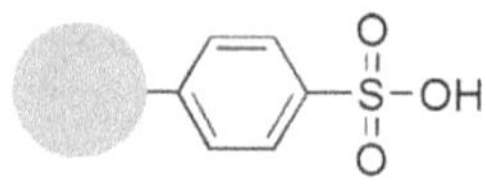

44

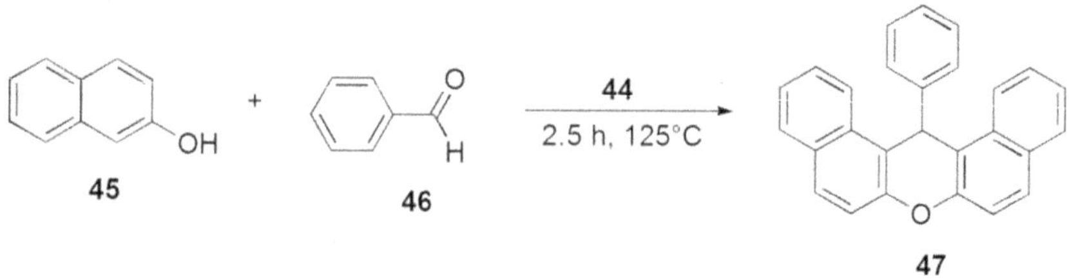

SCHEME 11.14 Amberlyst 15 **44**–catalyzed condensation of β-naphthol **45** with benzaldehyde **46.**

Another recent example of a reaction that has been catalyzed with Amberlyst 15 resin **44** is the condensation of aromatic aldehydes (e.g., **46**) with homopropargyl alcohol **48** to form β-butynyloxy enones [e.g., 5-(3-butynyloxy)-1-phenyl-(E)-1-penten-3-one **49**] (Scheme 11.15) [45]. Here, the use of the macroporous polystyrene was key to the success of the synthetic strategy, with both alternative solid acid catalysts (e.g., montmorillonite KSF, heteropolyacid $H_3PW_{12}O_{40}$, and H-ZSM) and more commonly utilized Lewis acids [e.g., $Sc(OTf)_3$, $Bi(OTf)_3$, and $In(OTf)_3$] giving low conversion yields. The researchers also demonstrated the somewhat recyclable nature of the Amberlyst 15 resin **44**; in the example of the reaction between **46** and **48**, the conversion yield achieved was 85%, 82%, 76%, and 71% over four repeated cycles [45]. Amberlyst 15 has also been demonstrated to be an efficient catalyst for the coupling of alkynes (e.g., phenyl acetylene **50**) with aldehydes (e.g., formaldehyde **51**) to form α,β-unsaturated ketones (e.g., 1-phenylprop-2-en-1-one **52**), giving the desired products in high yields (Scheme 11.16) [46]. As with the previous example, alternative solid acids (e.g., montmorillonite KSF and $H_3PW_{12}O_{40}$) were shown to be ineffective at catalyzing this transformation.

For certain chemical transformations, it has been favorable to catalyze reactions over more strongly acidic macroporous ion-exchange resins. For example, the over-sulfonated macroporous polystyrene resin Amberlyst 35 was found to be more active than Amberlyst 15 **44** for the reaction between 2-propanol **53** and *iso*-butene **54** in

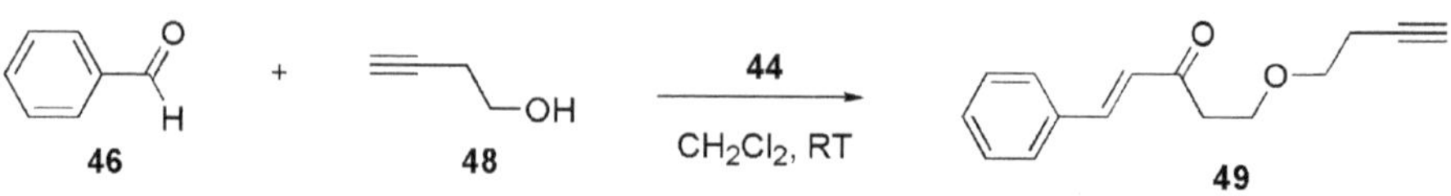

SCHEME 11.15 Amberlyst 15 **44**–catalyzed condensation of benzaldehyde **46** with homopropargyl alcohol **48.**

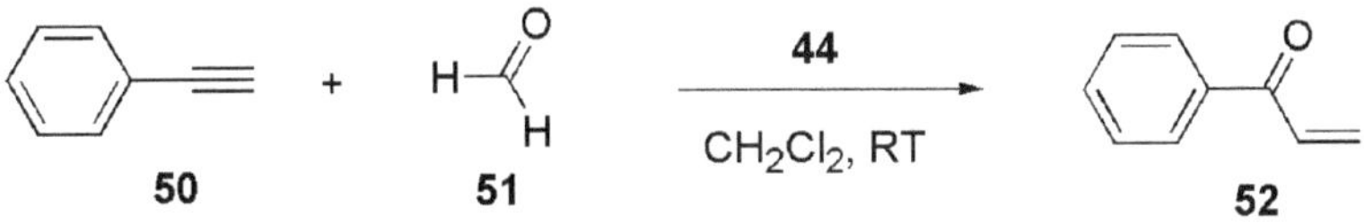

SCHEME 11.16 Amberlyst 15 **44**–catalyzed coupling of phenyl acetylene **50** with formaldehyde **51**.

the solution phase to form *iso*-propyl *tert*-butyl ether (IPTBE) **55** (Scheme 11.17) [47]. Amberlyst 35 was also shown to be a considerably more active catalyst than high-silica H-exchanged zeolites, which were previously used in the reaction (e.g., H-beta [48]), while maintaining high selectivity toward the formation of IPTBE (up to 97%). In general, H-exchanged zeolites have a lower number of acidic active sites than sulfonated macroporous polystyrene resins, and their catalytic activity is only thought to arise from macropores in the borders between particles [48]. Conversely, molecules in solution are able to diffuse within macroporous polystyrene beads and in this way may react with acidic groups throughout their interior, greatly enhancing their relative functional loading. The aforementioned examples highlight the versatility of macroporous cation-exchange resins as replacements for more conventional solution-phase acid catalysts such as *p*-toluene sulfonic acid.

Macroreticular cation-exchange resins of the type described may be further functionalized in order to further expand the scope of reactions that can be catalyzed within their porous matrix. Functionalization of the resin is generally achieved by immobilization of the desired catalytic species by utilization of the electron-donating abilities of the sulfonic acid groups, either through ionic or ligand-type interactions. For example, palladium-doped resins were produced by the addition of Amberlyst 15 resin **44** to a solution of the palladium precursor $[Pd(NH_3)_4](NO_3)_2$, followed by reduction with an H_2/argon mixture [49]. Association between the porous surface of the resin and the metallic surface of the Pd is facilitated by the sulfonic acid groups acting as ligands. The 1% Pd-doped Amberlyst resin **56** thus prepared was used as a catalyst for the preparation of the potential analgesic (−)-(2*R*,3*R*)-[3-(3-methoxy-phenyl)-2-methyl-pentyl]-dimethylamine **58** from (−)-(2*S*,3*S*)-1-dimethylamino-3-(3-methoxy-phenyl)-2-methyl-pentan-3-ol **57** (Scheme 11.18). The bifunctional nature of this support (i.e., its dual metallic and acidic properties) enabled dehydration of the tertiary alcohol and hydrogenation of the olefin to be carried out in a "one-pot" manner. In a more recent report, 1% Pd-doped Amberlyst resin **56** was demonstrated

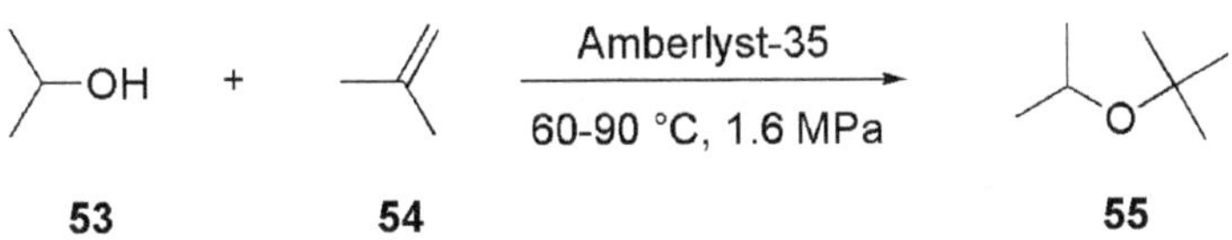

SCHEME 11.17 Amberlyst 35–catalyzed reaction between 2-propanol **53** and *iso*-butene **54.**

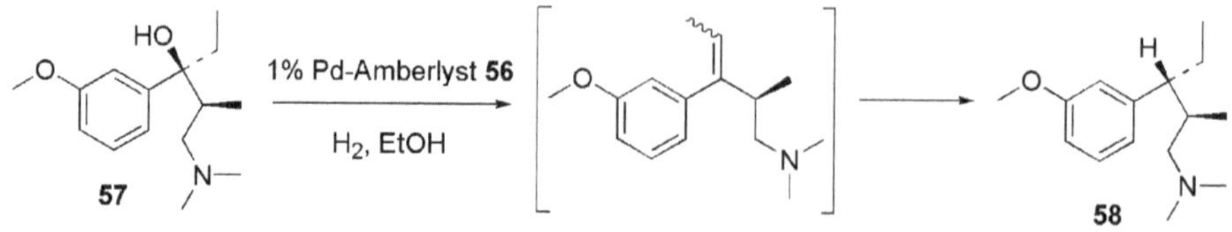

SCHEME 11.18 The 1% Pd-doped Amberlyst resin **56**–catalyzed preparation of the potential analgesic (−)-(2*R*,3*R*)-[3-(3-methoxy-phenyl)-2-methyl-pentyl]-dimethylamine **58.**

to promote the formation of 2-ethylhexanal **60** from crotonaldehyde **59** through a "one-pot" catalytic hydrogenation and Aldol condensation in supercritical carbon dioxide ($scCO_2$) (Scheme 11.19) [50]. Under optimized conditions, this catalyst gave the desired product in much higher yields than the other hydrogenation catalysts used for comparison (Pd/Al_2O_3, Pd/C) and with higher selectivity.

An example of an Amberlyst 15–supported reagent formed by ionic interactions is that of a resin-supported nitrosonium ion formed *in situ* by reacting the ion-exchange resin with sodium nitrite. Thus, a range of protected aldehydes and ketones (oximes, phenylhydrazones, *p*-nitrophenylhydrazones, semicarbazones, and tosylhydrazones) were deprotected efficiently using sodium nitrite in the presence of Amberlyst 15 **44** [51]. When either sodium nitrite or Amberlyst 15 **44** was used alone, no deprotection reaction was observed under analogous reaction conditions. The mechanism for the formation of the resin-bound active species **61** and conversion of benzaldehyde oxime **19** to its parent ketone **46** by this *in situ*–generated reagent is illustrated in Scheme 11.20.

11.2.2.3 Use of Anion-Exchange Resins in Organic Synthesis. As mentioned previously, in addition to the acidic macroporous polystyrene resins described earlier, a variety of basic anion-exchange resins with analogous structures are commercially available. These resins most commonly comprise tertiary ammonium groups throughout the porous network grafted to the polystyrene backbone. These moieties carry a highly stabilized positive charge and are, therefore, able to immobilize a series of desirable negatively charged counter-ions. The most common anion-exchange resin of this type is Amberlyst A-26 **63**, which has tertiary amines bonded to one benzylic group, through which the amine is attached to the polystyrene support, and three methyl groups. Amberlyst A-26 **63** is available commercially in a

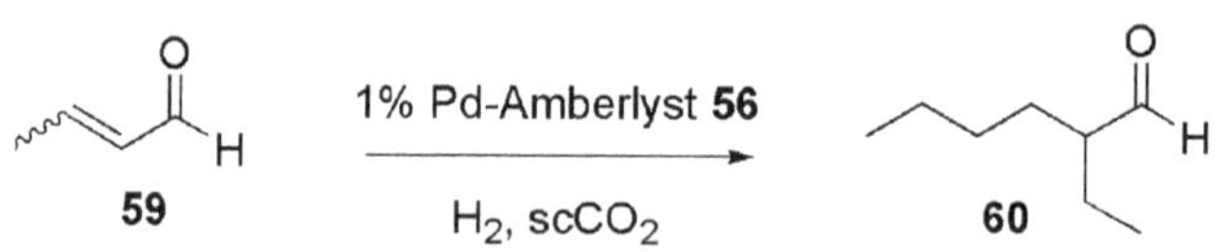

SCHEME 11.19 The 1% Pd-doped Amberlyst resin **56**–mediated conversion of crotonaldehyde **59** into 2-ethylhexanal **60** using supercritical carbon dioxide ($scCO_2$) as the reaction solvent.

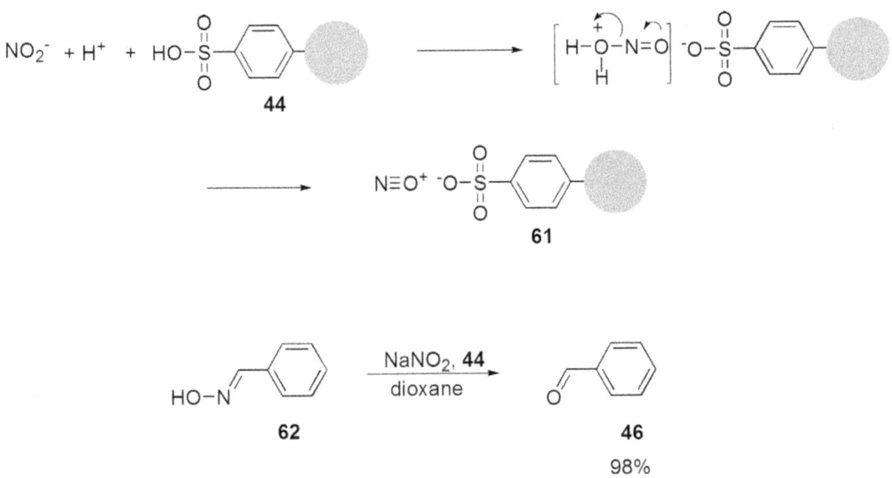

SCHEME 11.20 *In situ* generation of a resin-supported nitrosonium ion **61** and use of this reagent in converting oxime **62** into benzaldehyde **46.**

variety of different forms that allow for its use in a diverse range of synthetic transformations based on alternative counter anions immobilized by the tertiary amine groups (e.g., X = bromide, hydroxide, fluoride, borohydride, etc.). The type of reaction effected by each form of the resin is entirely dependent on the nature of the counter ion. Alternatively, the bromide form ($X^- = Br^-$) can be used for the bromination and chlorobromination of organic compounds through nucleophilic attack of the bromide counter ion [52].

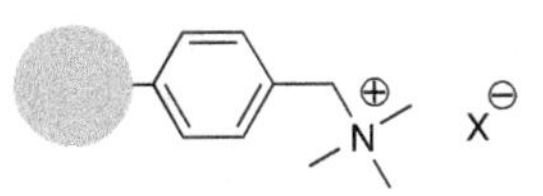

63

The diversity of Amberlyst A-26 **63** in organic synthesis is further highlighted by the fact that it has been utilized as a reagent for both oxidation and reduction. The borohydride form of the resin ($X^- = BH_4^-$) has previously been utilized for the mild reduction of conjugated olefinic bonds, for example, the reduction of α,β-unsaturated nitroalkenes to nitroalkanes [53, 54]. Conversely, the chromic acid form of Amberlyst A-26 ($X^- = CrO_4H^-$) has been used for the oxidation of alcohols to aldehydes and ketones [55]. This reaction was generally applicable for allylic, benzylic, and saturated primary and secondary alcohols, giving exclusively the corresponding aldehydes and ketones with no trace of carboxylic acids or other byproducts. However, it should be noted that this methodology has largely been superseded by the development of

Ley's polymer-supported perruthenate (PSP) whose development is described in the first part of this chapter.

11.2.3 PolyHIPEs

High-internal-phase emulsions (HIPEs) are defined as those in which the dispersed phase emulsion droplets occupy a volume greater than 74% of the total emulsion volume. When the continuous phase of a HIPE comprises one or more monomers (most commonly vinyl), polymerization of the continuous phase leads to the formation of porous monoliths with open-cell structures, commonly termed polyHIPEs, a generic term coined by Unilever [56], although the polymerization of a HIPE was first described by Bartl and von Bonin [57,58]. As discussed previously, while traditional porous monoliths have been utilized as polymeric supports for organic synthesis, a compromise must be reached between the opposing needs for sufficiently large pores to achieve the required level of permeability and sufficiently small pores to supply a large enough surface area that provides access to the polymer-bound reactive groups by substrates in solution. Uniquely, however, polyHIPEs circumvent this issue, as their structure comprises a network of large pores (or voids) connecting a network of smaller pores, thus providing these porous monolithic structures with both a high surface area and efficient permeability at low flow pressures. Due to these desirable properties, polyHIPEs continue to receive considerable interest, particularly as candidates for use as solid supports in continuous-flow reaction vessels. The patent protecting the concept embodied by polyHIPEs, filed by Unilever researchers Barby et al. in 1982 [59], includes examples of polyHIPEs formed from monomer mixtures including styrene–divinylbenzene (DVB), styrene–DVB–ethylvinylbenzene, and styrene–butyl methacrylate–allyl methacrylate. In order to facilitate their use as solid supports for organic synthesis, there has been considerable focus upon chemical modification of polyHIPEs for the introduction of reactive functional groups. While they have not yet come close to achieving the ubiquity of Merrifield resin in the field of solid-phase organic synthesis, recent developments are likely to further increase the popularity of suitably functionalized polyHIPEs for use as supported reagents, catalysts, and scavengers and in various bioapplications. The following subsection focuses on the various strategies by which both styrenic- and acrylic-based polyHIPEs have been rendered chemically reactive, and provides examples of the range of use of the supports thus functionalized within the field of organic synthesis.

11.2.3.1 Chemical Functionalization of PolyHIPE Supports. By far, the most common polyHIPE reported in the literature is composed of poly(styrene–DVB). Consequently, the most prevalent functionalizable groups within these polymeric monoliths are aromatic rings, and thus aromatic substitution provides a direct route to chemical functionalization of these supports. To this end, Sherrington and coworkers studied a series of electrophilic substitution reactions—sulfonation, nitration, and bromination—with a view to developing mild and support-compatible reaction conditions for each [60]. The authors suggested three criteria that must be met to ensure that poly(styrene–DVB) polyHIPEs are chemically modified uniformly throughout their porous interior: (1) the reaction solvent should swell the polymer and maintain this swollen state throughout the reaction, (2) reagents should be sufficiently hydrophobic to provide compatibility with the hydrophobic polymer matrix, and (3) the rate of reaction should be low at or below room temperature, thus allowing equilibrium of the reagents throughout the monolith prior to the reaction, but moderate heating should provide a sufficiently high rate of chemical modification. The second of these criteria provides an explanation as to why the use of acetyl sulfate in 1,2-dichloroethane, while giving almost uniform sulfonation throughout the polyHIPE support, was reported to only produce a modest yield (35%). Conversely, the use of the more hydrophobic lauryl sulfate in cyclohexane gave a much improved reaction yield (70%, 2.4 mmol g^{-1}), but sulfonation was nonuniform throughout the structure due to cyclohexane being a poorer swelling solvent for the support. However, by evacuation of the monolithic polyHIPE for 1 h prior to the reaction, the rapid equilibrium afforded led to much more uniformity (variation of ~1 mmol g^{-1} from the surface to the center) and a conversion yield of 42% using the latter reagent/solvent system. Again, where an insufficiently hydrophobic system was utilized for nitration of the polyHIPE, nitration was of low yield and nonuniform, with very low interior nitration. The use of an alternative, more hydrophobic nitrating system—tetrabutylammonium nitrate–trifluoroacetic anhydride—was required to provide improved homogeneity between external and internal functionalization, although again only a low nitration yield was obtained. The authors reported that the greatest success with this type of aromatic substitution reaction was through functionalization of poly(styrene–DVB) polyHIPEs by bromination. Under optimized conditions, the use of bromine–stannic chloride in dichloromethane gave a high yield and excellent homogeneity (3.6 mmol g^{-1} of bromine in the center of the monolith and 3.7 mmol g^{-1} on the surface).

In addition to prevalent aromatic groups, polyHIPEs that incorporate DVB contain a synthetically useful level of residual vinyl groups. For example, polyHIPEs synthesized from a HIPE with a continuous phase composed of 80 wt% DVB and 20 wt% ethylvinylbenzene were found to contain around 3 mmol g^{-1} of residual vinyl groups [61]. Furthermore, the authors reported chemical modification of vinyl-containing polyHIPE supports through the use of various electrophilic addition reactions [61, 62], using both batch (small blocks of polyHIPE suspended in the reaction solution) and flow (a continuous flow of the reaction solution through the monolith) methods. For example, bromination of the pendant vinyl groups was explored, as this would provide convenient sites for subsequent chemical elaboration of the support via nucleophilic substitution of the bromine atom [61]. Under anhydrous conditions and in the presence of a free radical initiator, β-addition (also termed anti-Markovnikov addition) at the least-substituted, terminal carbon of the vinyl double bond would be expected (Scheme 11.21). For purposes of providing sites for nucleophilic addition, bromination at this position was favorable to α-addition, which would be expected under ionic conditions. Under batch conditions, the authors reported that the polyHIPE was exclusively brominated at the desired β-position of its pendant vinyl groups by reaction with HBr (generated *in situ* by slow addition of an equimolar quantity of water to chloromethylsilane and anhydrous lithium bromide) in the presence of azo-bis-*iso*-butyronitrile (AIBN). However, the same reaction conditions were found

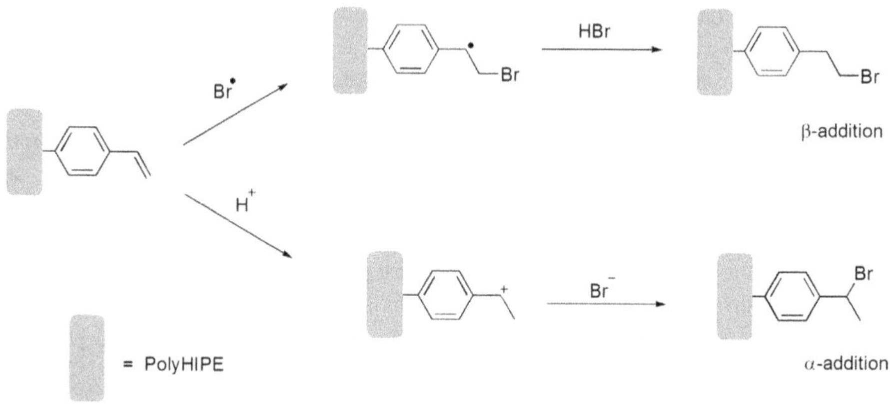

SCHEME 11.21 The two potential modes for the bromination of pendant vinyl groups on poly(styrene–DVB) polyHIPE supports: (top) free radical addition and (bottom) ionic addition.

to give a mixture of α- and β-addition when used to brominate the same type of polyHIPE support under continuous-flow methods. This anomaly was ascribed to the difficulty of completely drying a large piece of molded monolith, with the excess water providing a degree of ionic character to the reaction conditions. It should be noted that, unlike the previously discussed report by Sherrington and coworkers [60], the authors did not provide information as to the homogeneity of the bromination reaction throughout the monolithic structure, which makes assessment of the true success of the methodology more difficult to ascertain.

In order to furnish these vinyl-polyHIPEs with hydroxyl groups, Mercier et al. also explored hydroboration across the vinyl double bonds and subsequent hydrolysis [61]. The hydroboration step was achieved using BH_3-THF for both batch and continuous-flow methods, and the hydrolysis step was subsequently achieved by treatment with sodium hydroxide and hydrogen peroxide. Based on the residual vinyl content of the hydroxyl-functionalized polyHIPE supports thus synthesized, the authors estimated a hydroxyl loading of 1.7 mmol g^{-1} for the batch method and 0.7 mmol g^{-1} for the continuous-flow method. IR spectroscopic analysis of both supports revealed that, under these conditions, both α- and β-addition products were formed, although the latter was primarily favored. Free radical–mediated thiol addition, a highly useful vinyl addition reaction, was utilized by Mercier et al. to functionalize vinyl-polyHIPEs with a variety of useful functionalities [61, 62]. Primary thiols were attached to pendant vinyl groups on the polyHIPE via robust thioether linkages, using AIBN as a free radical initiator (Scheme 11.22). Furthermore, a range of synthetically useful functional groups, including hydroxyls, siloxanes, amines, chloride, and carboxylic acids, are tolerant of the thiol addition reaction conditions, and so small organic compounds containing both primary thiols and these functional groups (e.g., 2-mercatoethanol) may be attached to vinyl-polyHIPEs by this method without the need for protecting-group strategies. The thiol addition reaction proceeds

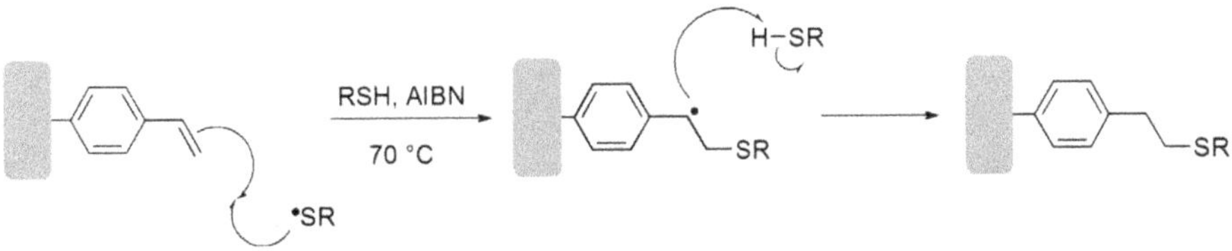

SCHEME 11.22 Radical-mediated thiol (RSH) addition to pendant vinyl groups on poly(styrene–DVB) polyHIPEs.

exclusively in the anti-Markovnikov, terminal position of vinyl groups, and provides functional loadings of up to 1.90 mmol g^{-1} for the batch method and 1.50 mmol g^{-1} for the continuous-flow method [61].

Through addition of thioacetic acid to vinyl-polyHIPE using this free radical addition methodology and subsequent aminolysis (*n*-butylamine and sodium borohydride) of the labile acetate group, it was also possible to furnish the polyHIPE support with primary thiol groups (thiol loading = 0.3 mmol g^{-1}) (Scheme 11.23) [62]. Alternatively, cleavage of the acetate group under acidic, oxidizing conditions (slow addition of 39% acetic acid) led to the formation of a polyHIPE-supported sulfonic acid (sulfonic acid loading = 0.7 mmol g^{-1}). These transformations further demonstrate the versatility of this thiol addition reaction.

Recently, vinyl-polyHIPEs have been utilized for the immobilization of a Grubbs initiator [$Ru(=CHPh)(PCy_3)_2Cl_2$)] to form a polymer-supported catalyst intended for use in ring-opening metathesis polymerization (ROMP) and ring-closing metathesis (RCM) [63]. In order to analyze the content of residual vinyl groups in their DVB-derived polyHIPE support, the authors carried out a quantitative bromination reaction (as indicated by the disappearance of peaks corresponding to the vinyl groups in the IR spectrum of the polymer) (Scheme 11.24) and used elemental analysis to determine the bromine content. Through backcalculation, this indicated that the polyHIPEs formed had vinyl loadings of between 2.17 and 4.36 mmol g^{-1}. Immobilization of the ruthenium initiator was achieved via an alkylidene exchange reaction simply

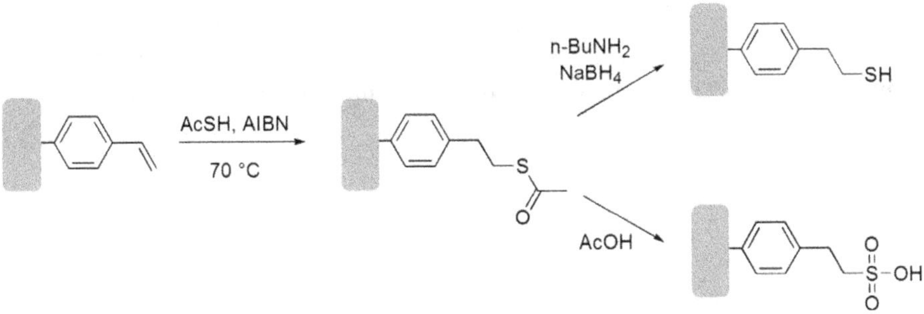

SCHEME 11.23 Addition of thioacetic acid to a poly(styrene–DVB) polyHIPE and conversion to a polyHIPE-supported (top) primary thiol or (bottom) sulfonic acid.

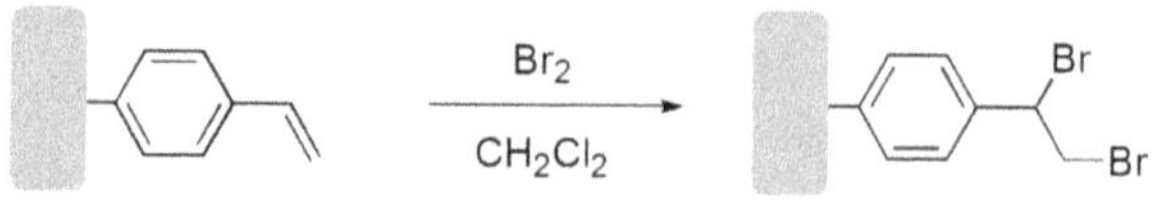

SCHEME 11.24 Bromination of a poly(styrene–DVB) polyHIPE.

by adding the vinyl-polyHIPE (in powder form) to a solution of the initiator in chloroform and incubating at room temperature for 1 h (Scheme 11.25) in a glove box. The loading level of the Grubbs initiator was calculated as 0.053–0.097 mmol g^{-1} by use of an ion beam analysis technique (Rutherford backscattering). The authors have yet to publish a follow-up paper in which the practical applications of the polyHIPE-supported initiator are described; such a publication would represent a notable development in the use of polyHIPE supports in organic synthesis and is eagerly anticipated.

The examples of functionalized polyHIPEs described so far have been prepared by postfunctionalization of what could be described as "nude" polyHIPE supports, that is, those that, while reactive under certain conditions, have not been formed using monomers specifically selected for the incorporation of pendant reactive functional groups. An alternative strategy, however, is to incorporate custom-made, functional comonomers directly into the polymerized continuous phase of the HIPE in order to provide a controlled level of reactive moieties as sites for subsequent derivatization of the polyHIPE formed. The most commonly utilized functional comonomer for these purposes is 4-vinylbenzyl chloride (VBC), which affords the polymer formed with chloromethyl groups that readily undergo nucleophilic substitution reactions. However, care must be taken when incorporating functionality via this method because the presence of such comonomers can have significant effects upon the structure and physical properties of the polyHIPE produced. For example, Barbetta et al. reported that the cell size (i.e., the size of the larger pores) of poly(DVB-VBC) polyHIPEs decreases with increasing VBC content [64]. The authors believe that this phenomenon occurs due to adsorption of VBC at the emulsion interface, which causes a lower interfacial tension and a smaller droplet size in the dispersed phase. This is clearly an important consideration, as too great a reduction in cell size of the polymeric foam would lead to impeded permeability through the support. Fortunately, in practice, the use of VBC at the levels required for a synthetically useful loading level has not proven to be significantly detrimental to the cell size. For example, polyHIPEs with a pore volume of 90% can be synthesized from a monomer mixture containing VBC

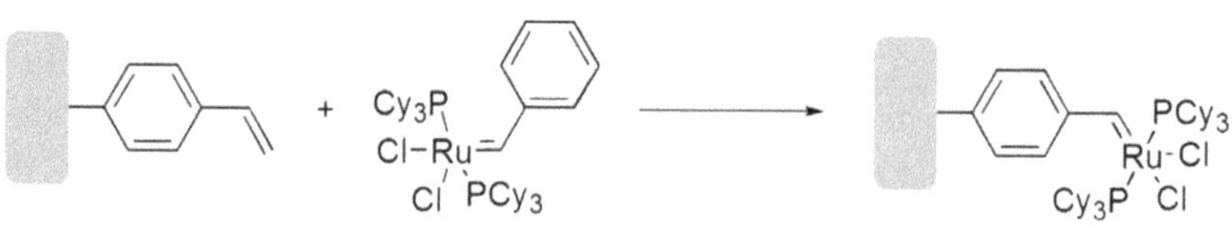

SCHEME 11.25 Immobilization of a ruthenium initiator (Grubbs catalyst) on a poly(styrene–DVB) polyHIPE.

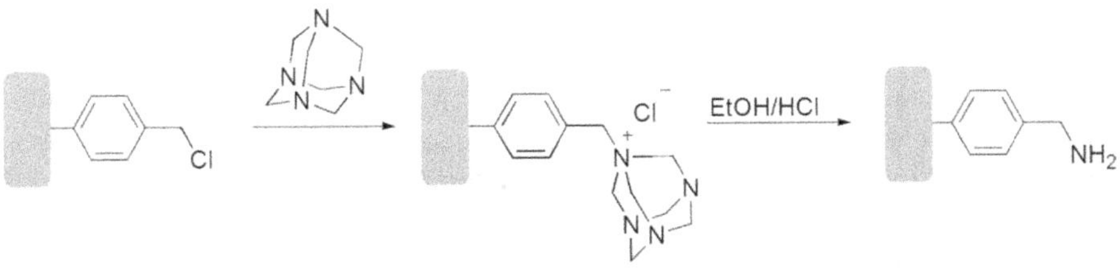

SCHEME 11.26 Synthesis of a polyHIPE-supported benzylamine from a poly(VBC-DVB) polyHIPE.

and DVB in a molar ratio of 6:4 [65]. The authors demonstrated the postfunctionalization potential of these chloromethylated polyHIPE supports through reactions with tris(2-aminoethyl)amine, 4-aminobutanol, tris(hydroxymethyl)-aminomethane, morpholine, and hexamethylenetetramine for the introduction of both amino and hydroxyl groups onto the support. In addition, following the reaction of hexamethylenetetramine with the polyHIPE support, treatment with hydrochloric acid afforded a benzylamine-functionalized polyHIPE (Scheme 11.26).

The introduction of hydroxyl groups into polyHIPE supports is desirable for applications in both organic synthesis (as a functional linker) or for scavenging applications, which will be discussed further later. With a view to creating porous monolithic supports for both types of application, Cameron et al. developed various strategies for the chemical modification of poly(VBC-DVB) polyHIPEs to introduce hydroxyls. Their initial focus was upon the introduction of a 4-hydroxymethyl phenyl group (commonly known as a Wang linker), which is commonly utilized in solid-phase organic synthesis as a carboxylic acid reactive linker, for example, for the immobilization of amino acids in polymer-supported peptide synthesis [66]. The authors reported the introduction of this linker by a two-step process; in the first step the polyHIPE support, in granular form, was heated under reflux with 4-hydroxybenzaldehyde, potassium carbonate, and sodium iodide in acetonitrile (Scheme 11.27). Quantification of the residual chloride content of the polyHIPE following this transformation

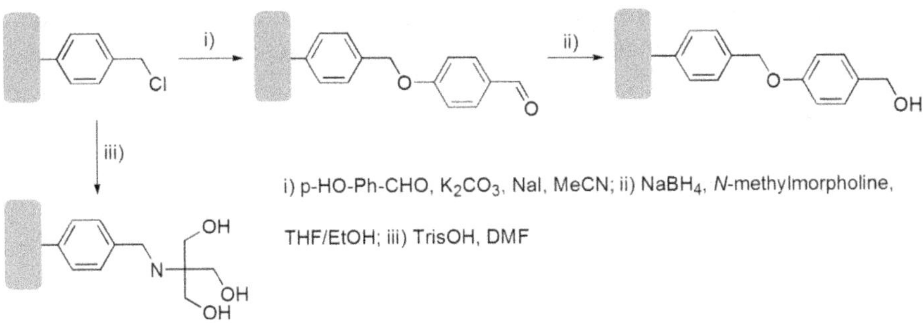

SCHEME 11.27 Synthesis of hydroxyl-functionalized polyHIPE supports from a poly(VBC-DVB) polyHIPE.

suggested that the reaction proceeded with a 97% conversion yield, giving an aldehyde loading of 3.14 mmol g^{-1}. The majority of aldehyde groups were subsequently reduced by treatment with sodium borohydride in tetrahydrofuran/ethanol in order to produce the desired Wang alcohol linker. In order to produce a polyHIPE support with a higher hydroxyl loading, which would be particularly desirable for scavenging applications, the authors also reacted a poly(VBC-DVB)–based support with an amino-functionalized triol, tris(hydroxymethyl)aminomethane (Scheme 11.27), with around a 90% conversion yield.

In an effort to produce polyHIPE supports with alternative functionalities to chloromethyl incorporated into the polymer matrix, the use of several different activated acrylates as monomers for polyHIPE synthesis has been studied [68, 69]. For example, 4-nitrophenyl acrylate (NPA) has been incorporated as a comonomer with styrene and DVB, although this was found to result in a polyHIPE material with a nonhomogeneous void size [68]. This observed change in morphology is thought to result from the more hydrophilic nature of the acrylate causing a reduction in the emulsion stability and, therefore, coalescence of the water droplets during polymerization. Greater success has been achieved by use of an alternative activated acrylate, (2,4,6-trichlorophenyl)acrylate (TCPA), which was more hydrophobic and, consequently, when used as a comonomer with styrene and DVB, produced polyHIPEs with a more homogeneous void size. PolyHIPEs containing both types of functional comonomer (either NPA or TCPA) have been subsequently transformed by several chemical reactions to produce polyHIPEs with desirable functional groups [67]. Conversion to an acid chloride was achieved in two steps: (1) treatment with aqueous sodium hydroxide to convert the activated esters to carboxylic acids, and (2) treatment with thionyl chloride in acetonitrile, giving the acid chloride in greater than 95% conversion (Scheme 11.28). Both types of active ester polyHIPE also readily underwent nucleophilic substitution reactions with amines, with both tris(hydroxymethyl)aminomethane (hydroxyl loading of the resultant polyHIPE = 10.9 mmol g^{-1}) and tris(2-aminoethyl) amine (amine loading = 5.4 mmol g^{-1}) being efficiently loaded onto the support (Scheme 11.29).

For the production of polyHIPE supports that contained no styrene groups, ethylene glycol dimethacrylate (EGDMA) was introduced as an alternative crosslinker to DVB, and the synthesis of poly(TCPA-EGDMA) polyHIPEs was explored [68]. Using standard surfactant combinations, it was not possible to form stable emulsions using this more hydrophilic crosslinker; however, an increase in salt concentration

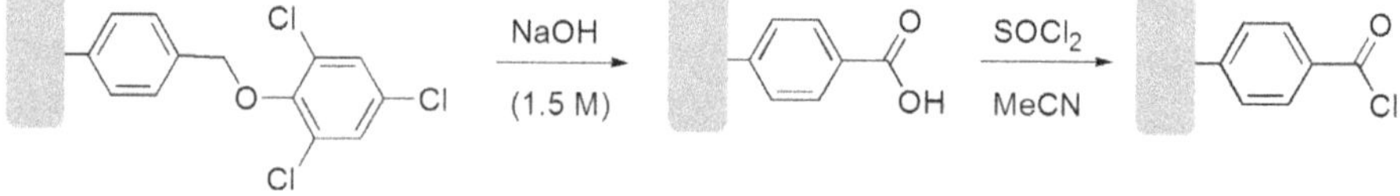

SCHEME 11.28 Synthesis of a polyHIPE-supported acid chloride from a TCPA-functionalized polyHIPE.

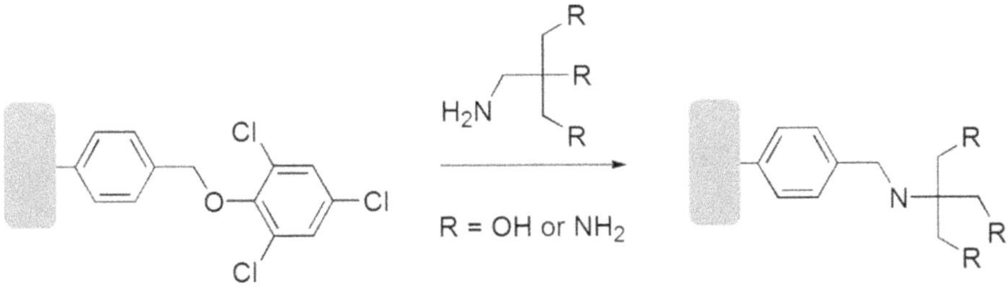

SCHEME 11.29 Immobilization of tris-OH and tris-NH_2 using a TCPA-functionalized polyHIPE.

in the dispersed phase [from 2% to 5% (w/w) of calcium chloride] resolved this issue and allowed the formation of emulsions with up to 90% aqueous phase. The utility of the activated trichlorophenyl esters in the resultant polyHIPEs was demonstrated via several chemical transformations. Ester hydrolysis, to give a carboxylic acid functional polyHIPE, was again achieved by immersion of the support, either as monolithic cubes or in powdered form, in an aqueous solution of sodium hydroxide. Surprisingly, more efficient conversion yields were obtained for the polyHIPEs when reacted in monolithic/cube form (64%–93%) than in powdered form (58%–88%). The activated trichlorophenyl esters were also reacted with tris(2-aminoethyl)amine, as with their styrene-containing analogs (described earlier) attaching the tetramine to the polyHIPE support via an amide bond. Again, high conversion yields were achieved using the EGDMA-crosslinked polyHIPE support (90%–94%), although the trifunctional tris(2-aminoethyl)amine additionally caused a significant degree of crosslinking between adjacent reactive sites on the polymer support.

11.2.3.2 Secondary Functionalization of PolyHIPE Supports. In order to open up greater possibilities for surface functionalization using "nonfunctionalized" poly(styrene–DVB) polyHIPEs as precursors, various researchers have explored carrying out secondary polymerization reactions that coat the porous polyHIPE void surfaces with a thin layer of a functionalized polymer. Sherrington and Small employed such an approach for the preparation of polyHIPE supports for use in peptide synthesis (discussed further later) [69, 70]. Specifically, they include an example in which a 10% crosslinked poly(styrene–DVB) polyHIPE support (milled and sieved to a particle size of 850–1200 μm) was reacted with *N*-(2-(4-acetoxyphenyl)ethyl) acrylamide and crosslinker ethylene *bis*(acrylamide), with AIBN as initiator. Following formation of the acrylamide polymer layer within the polyHIPE, a deprotection step with hydrazine hydrate in *N*,*N*-dimethylformamide (DMF) provided free phenol groups on the support (Scheme 11.30).

In the previously described protocol, residual vinyl groups within the polymer were utilized for anchoring the polymer chains that formed the secondary polymeric coating. However, Sherrington and Small additionally described a method of introducing pendant acrylic groups for subsequent anchoring of a secondary polymer

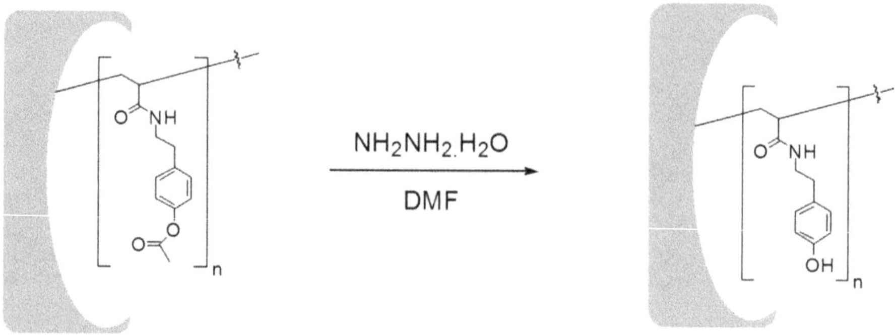

SCHEME 11.30 Deprotection of acetoxyphenol groups on a poly(N-(2-(4-acetoxyphenyl) ethyl) acrylamide)–coated polyHIPE.

layer [69, 70]. The first step of this process was the formation of an aminomethyl-functionalized polyHIPE using a poly(styrene–DVB) polyHIPE as a precursor. This was achieved by reacting the support with *N*-hydroxymethyl phthalimide, using trifluoromethanesulfonic acid as a catalyst, followed by deprotection of the phthalimido group by hydrazinolyisis (Scheme 11.31). The resin-bound amino methyl groups were then reacted with acryloyl chloride in order to introduce the desired acrylic groups. This support was derivatized by reaction with acryloyl tyramine acetate, with *N′N′*-ethylene *bis*-acrylamide as crosslinker and AIBN as initiator. This provided a polyHIPE support with terminal acetate-protected phenolic functional groups. Alternatively, the acrylate-functionalized polyHIPE scaffold could be reacted with acryloyl sarcosine to provide a methyl ester as the terminal functional group in the resultant support.

Cummins et al. reported an elegant strategy for the functionalization of polyHIPEs with functionalized polymer brushes via atom transfer radical polymerization (ATRP) [57]. The polyHIPE support utilized as a scaffold for subsequent ATRP was acrylate based, formed primarily from a monomer mixture of 2-ethyl hexyl acrylate, isobornyl acrylate, and trimethylolpropane triacrylate (crosslinker). In addition, the authors added 9 mol% (with respect to the constituent acrylates) of a polymerizable

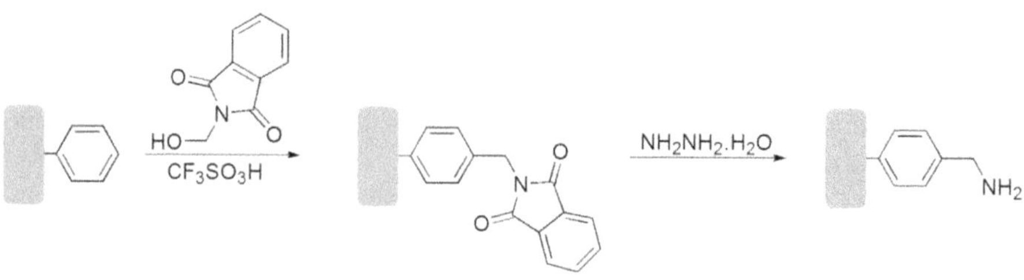

SCHEME 11.31 Amine functionalization of a poly(styrene–DVB) polyHIPE via aromatic substitution.

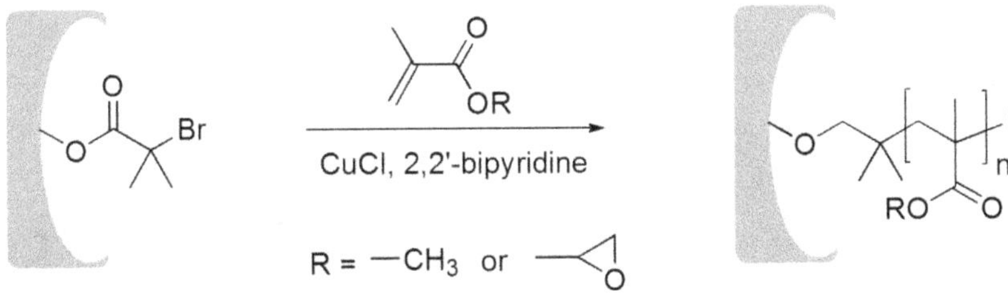

SCHEME 11.32 Controlled polymerization from the surface of an ATRP initiator–functionalized polyHIPE.

ATRP initiator, 2-acryloxyethyl-2′-bromoisobutyrate, in order to provide sites for the growth of secondary polymer chains. Following the formation of a stable HIPE with this monomer mixture and photopolymerization to form a polyHIPE, ATRP from the polyHIPE surfaces was initially tested using methyl methacrylate (MMA). The ATRP reaction was carried out in the presence of a catalyst (CuCl) and ligand (2,2′-bipyridine) in methanol/water (Scheme 11.32). Changes in the surface morphology, as seen using SEM, and increased brittleness of the resultant polyHIPE were indicative of successful formation of the PMMA-grafted polyHIPE, termed poly(HIPE-*g*-MMA). Due to the "living" nature of ATRP, the grafted polymer chains retained the ATRP initiator at their terminus, and this feature was utilized to incorporate reactive functional groups through the formation of a block copolymer. Accordingly, poly(HIPE-*g*-MMA) was reacted with 2-hydroxylethyl methacrylate (HEMA) under the previous ATRP reaction conditions to produce the hydroxyl-functionalized support poly(HIPE-*g*-MMA-*b*-HEMA).

In order to incorporate a higher level of functionality using their ATRP polymer grafting protocol, Cummins et al. also reacted the initial ATRP initiator-functionalized polyHIPE with the epoxide-functionalized monomer glycidyl methacrylate (GMA). Under the ATRP reaction conditions described previously, the authors found that the polymerization proceeded without causing ring opening of the epoxide groups, leaving these moieties available for subsequent chemical modification. As a demonstration of this potential for postfunctionalization, poly(HIPE-*g*-GMA) was treated with concentrated HCl in THF to hydrolyze the epoxide groups to glycols. The primary alcohols of these glycols were then esterified by treatment with pentafluorobenzoyl chloride in dichloromethane in the presence of triethylamine and dimethylamino pyridine (Scheme 11.33). The change in the hydrophobic character of the surfaces of this modified polyHIPE caused by the presence of highly hydrophobic pentafluorobenzyl groups was demonstrated by water contact angle measurements.

11.2.3.3 Applications of Functionalized PolyHIPE Supports in Solid-Phase Organic Chemistry. The examples of the postfunctionalization of polyHIPEs described here highlight various chemistries that are compatible with these porous monoliths; however, they describe the construction of synthetic tools rather than practical applications of these tools. Despite an initial surge of interest in the

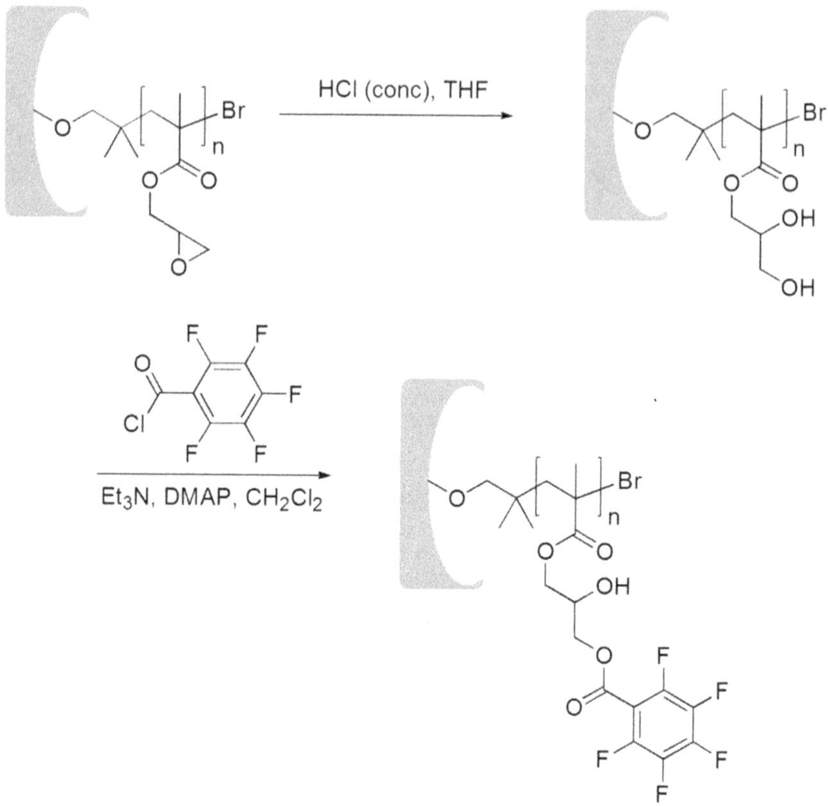

SCHEME 11.33 Ring opening of epoxide groups on an ATRP-grafted polyHIPE and subsequent esterification.

1980s and early 1990s, when functionalized polyHIPEs were demonstrated to have practical utility (e.g., peptide synthesis [71]), the literature contains relatively few examples of the practical application of polyHIPE materials to synthetic challenges. However, in recent years polyHIPEs appear to be experiencing somewhat of a renaissance within the field of solid-phase organic synthesis, particularly for the formation of polymer-supported catalysts and biocatalysts. Combined with their suitability for use in continuous-flow reactor vessels, in which polyHIPEs are not susceptible to deleterious "channeling" issues associated with the use of flow columns packed with functionalized polymer beads, polyHIPE-supported reagents may well have a bright future in high-throughput synthesis. This section reviews earlier reports of polyHIPE applications in organic synthesis, as well as highlights some pertinent recent examples that provide an indication of the areas in which polyHIPE research is likely to progress in the coming years.

In the early 1990s, several U.S. patents were filed (inventors: Sherrington and Small; assignees: Unilever) that provided examples of the utility of polyHIPEs as

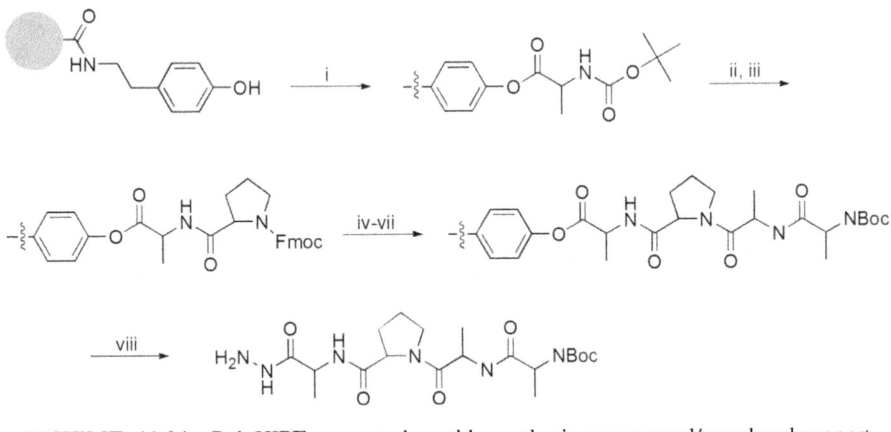

SCHEME 11.34 PolyHIPE-supported peptide synthesis on a ground/powdered support.

a support for peptide synthesis [69, 70]. As described earlier, the functional supports utilized for this application were derived from a poly(styrene–DVB) polyHIPE, which was subsequently coated with a functionalized acrylate polymer in order to introduce phenolic groups (Scheme 11.30). In the first example of peptide synthesis provided, the starting amino acid, *N*-(*t*-butoxycarbonyl)alanine, was activated using dicyclohexylcarbodiimide (DCC) in DMF and immobilized onto the functionalized polyHIPE support via the formation of phenolic ester linkages (Scheme 11.34). The *t*-butoxycarbonyl (Boc) protecting group was then removed by treatment of the polyHIPE support with boron trifluoride etherate in benzyl alcohol. Prior to further elongation of the peptide chain, the polyHIPE support was transferred to the column of a peptide synthesizer in order that subsequent iterations could be carried out by a semi-automatic, continuous-flow method. Loading of a second amino acid was achieved by a reaction with the symmetric anhydride of *N*-fluorenylmethyloxycarbonyl (Fmoc) proline, and the Fmoc group was removed by treatment with 20% diethylamine in DMF. This was followed by two further coupling steps [coupling Fmoc–alanine and *N*-*tert*-butoxycarbonyl (Boc)–alanine, respectively], and the final peptide was cleaved from the polyHIPE support (100 mg of the sample) by treatment with hydrazine hydrate in DMF. For the 100 mg sample cleaved, 61 mg of the tetrameric peptide Boc-Ala-Ala-Pro-Ala-N_2H_3 was isolated. It should be noted that in this example, the polyHIPE support was ground to a fine powder prior to peptide synthesis, so it does not represent an example of peptide synthesis within a porous monolith. It does, however, indicate that this type of polyHIPE support is compatible with the various reaction conditions necessary for peptide synthesis. In a further example of peptide synthesis, the authors coupled (Fmoc-Leu)$_2$O (the symmetric anhydride of Fmoc–leucine) to an amino-functionalized poly(styrene–DVB) polyHIPE support (again in powder form) and removed the Fmoc group with 20% piperidine, calculating a functional loading of 0.12 mmol g^{-1}. An acid-labile linker molecule **64** was coupled to the polyHIPE-supported amino acid via an amide bond, and the terminal benzyl alcohol was then

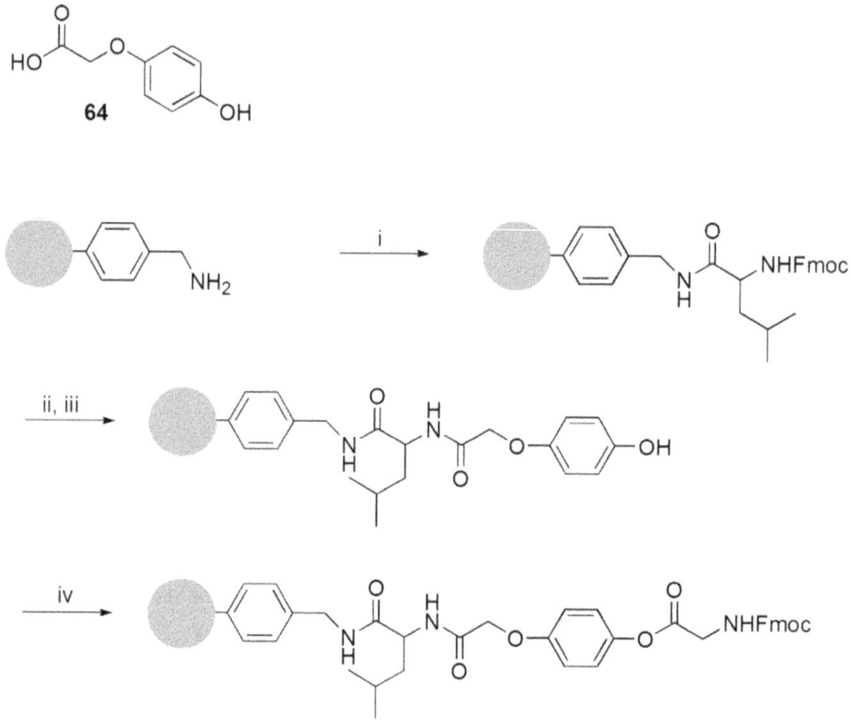

i) $(Fmoc\text{-}Leu)_2O$, DMF; ii) piperidine (20% in DMF); iii) **64**, EDC, DMF; iv) $(Fmoc\text{-}Gly)_2O$

SCHEME 11.35 Acid-labile linker molecule **64** and a polyHIPE-supported peptide synthesis that incorporates this linker.

esterified with $(Fmoc\text{-}Gly)_2O$ (Scheme 11.35). The observed 1:1 ratio of Gly:Ala indicated that these two coupling steps both proceeded with quantitative conversion.

As discussed previously, macroporous beads have found widespread industrial use as scavengers, in particular for removing impurities from process streams. Although they have not yet been similarly adopted for industrial applications, there are various reports in the literature on the use of polyHIPEs for scavenging applications. Amino-functionalized polyHIPEs have been shown to be efficient scavengers of acid chlorides [65]. For example, a tris(2-aminoethyl)amine polyHIPE derivative was used to sequester 4-chlorobenzoyl chloride from solution (6.7-fold excess of amine to acid chloride), leaving no detectable trace of acid chloride after 1 h. Furthermore, these trisamine–polyHIPE supports were found to be more efficient than a commercial trisamine support (Argonaut PS-trisamine, 3.2 mmol g^{-1} of NH/NH_2), with the former scavenging up to 3.39 mmol g^{-1} of the acid chloride compared to 1.92 mmol g^{-1} for the commercial resin. Hydroxyl-functionalized polyHIPEs have also been shown to be efficient scavengers for acid chlorides [72]. A tris-(hydroxymethyl)aminomethane–functionalized polyHIPE added to a solution

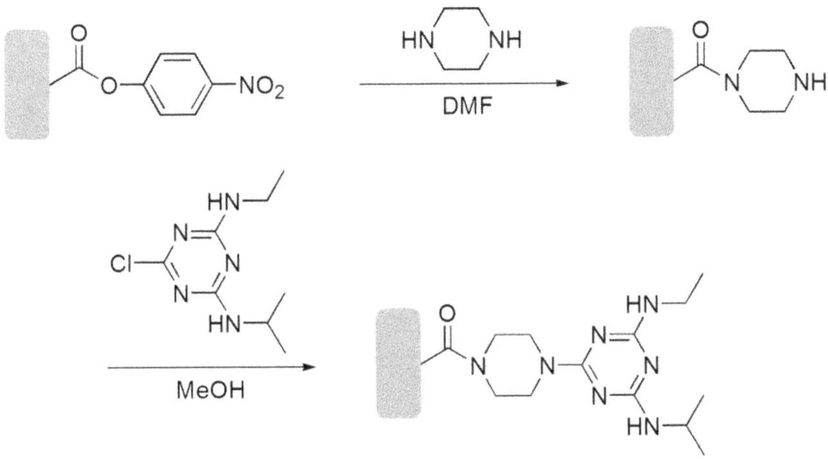

SCHEME 11.36 Synthesis of piperazine-functionalized polyHIPE and scavenging of atrazine.

containing excess acetyl chloride was shown to scavenge 5.5 mmol g^{-1} of the acid chloride. Recently, piperazine-functionalized polyHIPEs have been utilized for scavenging atrazine from aqueous solutions [73]. Atrazine is used agriculturally as a herbicide, but has been shown to be toxic toward animals, including humans, so its removal from drinking water is a desirable practical application. The scavenger was prepared by loading piperazine onto a nitrophenyl-functionalized poly(styrene–DVB) polyHIPE support (Scheme 11.36). With the use of a 3200-fold excess of immobilized piperazine to atrazine (33 ppb), the polyHIPE support sequestered almost 100% of the atrazine from water after 48 h. With a smaller excess of piperazine (320-fold, 330 ppb atrazine), the reaction kinetics were slightly slower, but near-complete removal of atrazine was still achieved after 72 h (Scheme 11.36). Poly(styrene–DVB) polyHIPEs coated (both internally and externally) with iron oxides as adsorbing agents have been shown to be very efficient at removing inorganic arsenic anions from water [74]. The iron oxide coating process was achieved by treating the polyHIPE support (cut into cubes of <5 mm) with ferric nitrate in a dilute sodium hydroxide solution (1 N, pH 5), giving an iron oxide loading of up to 75 mg. Treatment of aqueous solutions containing 50 μg L^{-1} of arsenic(V) ions led to a reduction of these toxic metal ions to levels well below the 10 μg L^{-1} safe limit set by the European Commission. PolyHIPEs with poly(4-vinyl pyridine) grafted to their surfaces have also been shown to be efficient at removing heavy metal ions from solution [75]. To prepare the active support, a poly(styrene-VBC-DVB) polyHIPE was treated with sodium thiosulfate to introduce free radical initiation sites, from which 4-vinyl pyridine was readily photopolymerized. This modified polyHIPE was then used to scavenge both iron from solution in 10 N HCl and plutonium from solution in 7 N nitric acid.

The scope for the utility of polyHIPEs in modern organic synthesis has been extended in the last 10 years by various notable reports of synthetically powerful

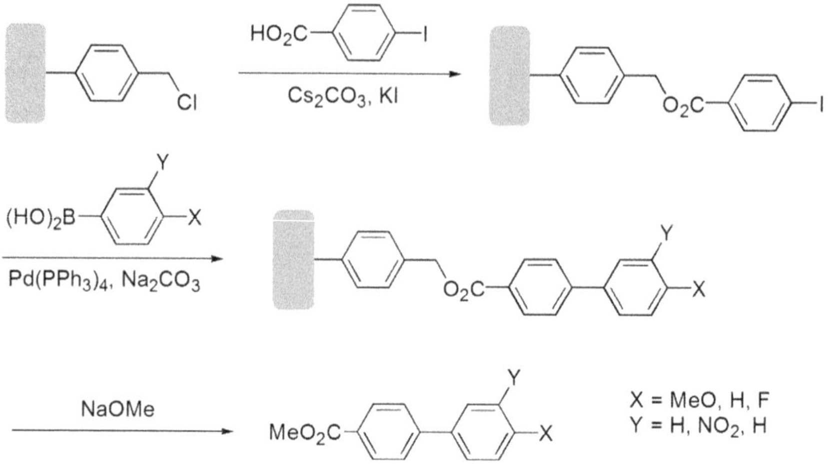

SCHEME 11.37 Immobilization of 4-iodobenzoic acid onto a polyHIPE support and Suzuki cross coupling.

transformations being carried out on these supports. Cameron et al. reported the use of polyHIPEs as a support for Suzuki cross-coupling reactions under both batch and flowthrough conditions [76]. The reactions were carried out on a poly(VBC-DVB) polyHIPE support that was cut into pieces or granulated (for batch reactions) or synthesized in PTFE reactor tubes for flowthrough reactions. The substrate, 4-iodobenzoic acid, was immobilized on this support through reaction of its carboxylic acid moiety with chloromethyl groups on the polyHIPE (Scheme 11.37). This reaction proceeded in high conversion yields for both batch (69% conversion) and flowthrough (74% conversion) methods. Suzuki cross-coupling reactions with various monosubstituted and disubstituted phenyl boronic acids were then carried out on this polyHIPE-supported substrate in the presence of a palladium catalyst [$Pd(PPh_3)$] (Scheme 11.37). These reactions were similarly efficient, with conversion yields of up to 96% obtained for the batch method and up to 86% for the flowthrough method. The resultant biaryl products were subsequently cleaved from the polyHIPE support by treatment with a 2 M solution of sodium ethoxide.

Cummins et al. utilized poly(HIPE-*g*-GMA) (see prior discussion) for the formation of a polymer-supported azide group, allowing click chemistry to be performed upon the polyHIPE support [77]. The polyHIPE-supported azide was formed by treatment of poly(HIPE-*g*-GMA) with sodium azide in DMF (Scheme 11.38). A complete disappearance of the IR signal corresponding to the epoxy groups suggested that this transformation proceeded with an almost quantitative yield. The polyHIPE-supported azide was then reacted with a series of molecules containing alkyne moieties, using Cu(I)Br as a catalyst and THF as the reaction solvent (Scheme 11.38). Under optimized conditions, conversion yields of up to 80% were obtained. Furthermore, this reaction gave an almost equal level of conversion on the interior and exterior of the polyHIPE support. To provide a visual representation of the homogeneous nature of

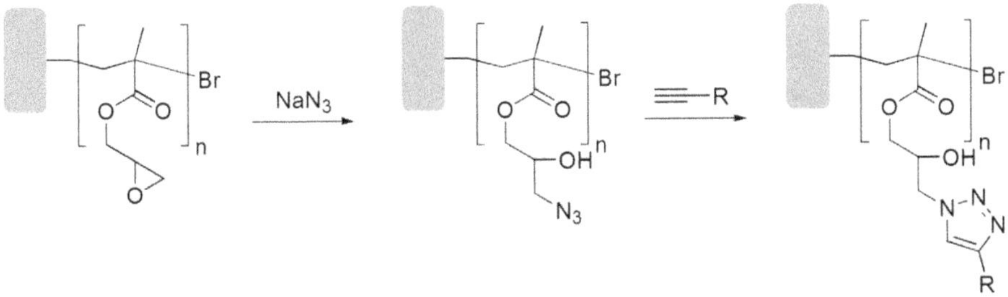

SCHEME 11.38 Synthesis of an azide-functionalized polyHIPE and "click-chemistry" reaction with an alkyne.

this functionalization, the authors additionally utilized the click reaction to couple a fluorescent dye (3,4-difluorophenyl acetylene) to the polyHIPE support.

Over the last 10 years, polyHIPEs have proven to be versatile supports for immobilized reagents and catalysis, including biocatalysts. An early example of the use of a sulfonated polyHIPE as a polymer-supported catalyst was reported by Ottens et al. in 2000 [78]. Specifically, a sulfonated poly(styrene–DVB) polyHIPE support was prepared by aromatic electrophilic substitution (see earlier discussion) and was utilized as a catalyst for the hydration of cyclohexene (Scheme 11.39). The efficiency of the polyHIPE-supported catalyst for this reaction was found to be comparable to that obtained with more commonly used ion-exchange resin-supported catalysts. Indeed, when the reaction was carried out in a two-phase system (cyclohexene in a 90 mol% aqueous solution of sulfolane), the sulfonated polyHIPE was found to be a more efficient support than the ion-exchange bed.

Mercier et al. developed a polyHIPE-supported organotin catalyst, which has been used as a dehalogenation reagent [79]. The organotin moiety was introduced during the initial synthesis of the polyHIPE support, in which the functional monomer 4-ethyl-(2-dibutyl chlorostannyl)styrene was added to the continuous phase of the HIPE as a comonomer with styrene and DVB. The resultant support had an SnCl loading of 0.6 mmol g^{-1}, and these moieties were reduced to SnH using sodium borohydride to produce the active dehalogenation reagent (50% conversion yield). To demonstrate the utility of this reagent, the polyHIPE-supported organotin chloride reagent was employed for the reduction of bromoadamantane in the presence of sodium borohydride to provide *in situ* reduction of the supported reagent to its active

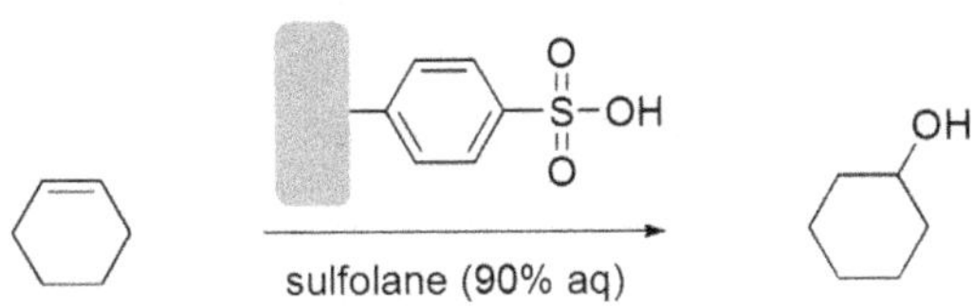

SCHEME 11.39 Hydration of cyclohexene using a sulfonated polyHIPE as a catalyst.

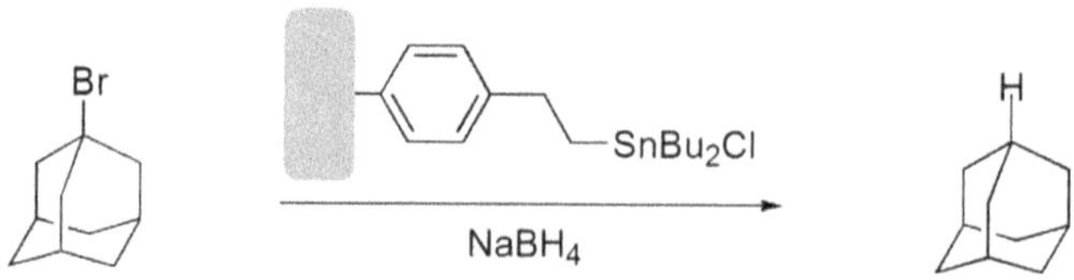

SCHEME 11.40 Reduction of bromoadamantane using a polyHIPE-supported organotin catalyst.

tin hydride form (Scheme 11.40). This reaction was highly efficient, proceeding in quantitative conversion for five successive runs with no loss of activity of the supported reagent. The supported organotin reagent was also successfully used for the radical cyclization of 1-bromo-2-(prop-2-enyloxy)benzene, which proceeded in an 89% yield and with a 9:1 selectivity in favor of the desired cyclic product **65** over a dehalogenated side product **66** (Scheme 11.41).

In recent years, suitably functionalized polyHIPEs have additionally been used for the immobilization of biomolecules, primarily enzymes for the formation of bioactive supports. For example, acrylate-based polyHIPEs with *N*-hydroxysuccinimide functional groups have been used to immobilize several proteins, including the lipase enzyme CAL-B, through covalent coupling to their lysine residues [80]. Due to the limited thermal stability of the functional comonomer used to synthesize this activated polyHIPE (*N*-acryloxysuccinimide), a photopolymerization protocol was utilized for synthesis of the support. Initially, a green fluorescent protein (rAceGFP) was immobilized onto the support for ease of monitoring the reaction via the protein's fluorescent signal. Following protein conjugation, the fluorescence intensity of the immobilized GFP was found to be directly proportional to the succinimide loading level of the polyHIPE support utilized. In addition, confocal microscopy suggested that bioconjugation was uniform throughout the polyHIPE support. In the case of CAL-B immobilization, an enzyme loading of up to 0.8 wt% was achieved using *N*-hydroxysuccinimide–functionalized polyHIPE. Conversely, no CAL-B was immobilized on an analogous polyacrylate polyHIPE containing no succinimide groups, demonstrating that noncovalent, physical adsorption did not contribute to the enzyme loading level observed previously. Subsequently, a 0.8 wt% CAL-B polyHIPE was utilized for the biocatalytic ester hydrolysis of *para*-nitrophenol acetate

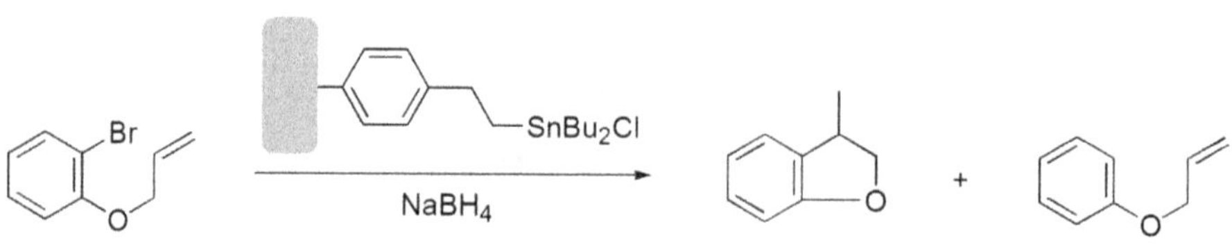

65 66

SCHEME 11.41 Radical cyclization using an organotin-supported polyHIPE catalyst.

(PNPA), giving a conversion rate of 460 μmol PNPA $min^{-1}g^{-1}$. This represented a more efficient conversion rate than that of CAL-B immobilized on more traditional polymer beads. Polystyrene-based polyHIPEs have also been used to immobilize enzymes for the formation of bioactive supports. Recently, a polyglutaraldehyde-activated poly(styrene-DVB) polyHIPE has been used to immobilized the microbial lipase *Thermomyces lanuginosus* through reaction of aldehyde groups with pendant functional groups on the enzyme (e.g., amines, thiols) [81]. Aldehyde groups were incorporated onto the porous surfaces of the polyHIPE support through polymerization of a styrene–DVB HIPE that contained polyglutaraldehyde in the aqueous phase. Using this support for lipase immobilization, the authors reported a maximum protein loading of 11.4 mg protein/g polyHIPE [81]. In addition to retaining antibacterial activity, the polyHIPE-supported lipase was shown to be more resistant to extreme temperatures than the free enzyme. For example, at 80°C, a temperature that caused the free enzyme to lose all activity, the immobilized lipase retained 35% of its initial activity. The immobilized enzyme also retained full activity for 30 days when stored at 4°C, whereas the free enzyme lost 55% of its initial activity in this time frame under the same storage conditions.

11.3 CONCLUSIONS

In this chapter we hope that we have been able to convey the excitement of those who work with catalysts and reagents immobilized on crosslinked polymer supports. We have attempted to illustrate the length and breadth of this vibrant area through the historical description of these developments and by pertinent examples from the more recent past. In terms of the future development of this field, first, there is still a need for new and more efficient supported reagents and scavengers to be developed. Work to this end will no doubt continue, although, as many supported reagents and catalysts now exist, it is likely that progress will perhaps proceed at a slower rate than in the past. Similarly, new supports will continue to be developed to circumvent problems such as physical fragility, chemical instability, and reaction/solvent system incompatibilities. Environmental concerns, which are already influencing synthesis, will, perhaps, also come to bear on the types of support used in polymer-assisted synthesis. This, in turn, could see a drive toward more environmentally friendly supports being developed, especially for larger-scale polymer-assisted synthesis activities.

Perhaps, more significantly, there will be a greater shift toward automation. Flowthrough synthesis using supported reagents and catalysts is by no means a new area; as far as we are aware, the first example of flowthrough synthesis employing a non–ion-exchange resin system was described by Harrison and Hodge [82], but, until recently, progress in this area has been relatively slow. In Section 11.1.6.3 we alluded briefly to flowthrough synthesis and made reference to a couple of excellent recent examples by the Ley group [40]. Other modern-day pioneers of flowthrough synthesis include Luis [83] and Kirschning [84], but the area is expanding rapidly, and we refer the interested reader to a recently published series of open-access papers that describe the work of a large number of groups currently leading the area [85]. One of the current trends in this area is the move toward using monolithic supports in these flowthrough systems rather than beaded supports [86]. To this end, polyHIPE monoliths, with their excellent flowthrough properties, may well find increasing application in flowthrough systems in the future.

ABBREVIATIONS

ADP	Adenosine diphosphate
AIBN	Azo-bis-*iso*-butyronitrile
ATP	Adenosine triphosphate
ATRP	Atom transfer radical polymerization
Boc	*tert*-Butoxycarbonyl
cat	Catalyst
conc	Concentrated
DCM	Dichloromethane
DMAP	4-Dimethylaminopyridine
DMF	*N,N*-Dimethylformamide
DMSO	Dimethyl sulfoxide
DVB	Divinylbenzene
EDC	1-Ethyl-3(3-dimethylaminopropyl) carbodiimide
EGDMA	Ethylene glycol dimethacrylate
Fmoc	Fluorenylmethoxycarbonyl
GFP	Green fluorescent protein
GMA	Glycidyl methacrylate
HCl	Hydrochloric acid
HIPE	High internal phase emulsion
IPTBE	*iso*-Propyl *tert*-butyl ether
IR	Infrared
MC	Microencapsulated
MMA	Methyl methacrylate
$NADP^+$	Nicotinamide adenine dinucleotide phosphate
NADPH	Reduced form of nicotinamide adenine dinucleotide phosphate
NMO	*N*-Methyl morpholine *N*-oxide
NPA	4-Nitrophenyl acetate
PdEnCat	Encapsulated palladium catalyst
PEG	Poly(ethylene glycol)
PNPA	*para*-Nitrophenol acetate
PS	Polystyrene
PSP	Polymer-supported perruthenate
RCM	Ring-closing metathesis
ROMP	Ring-opening metathesis polymerization
RT	Room temperature
$scCO_2$	Supercritical carbon dioxide

TPAP	Tetra-*n*-propylammonium perruthenate
TCPA	(2,4,6-Trichlorophenyl)acrylate
Tf	Triflate
THF	Tetrahydrofuran
Tris	Tris(hydroxymethyl)aminomethane
VBC	4-Vinylbenzyl chloride

REFERENCES

1. Miao, W.; Chan, T. H. *Acc Chem Res* 2006, **39**, 897–908. Chen, J.; Yang, G.; Zhang, H.; Chen, Z. *React Funct Polym* 2006, **66**, 1434–1451. Toy, P. H.; Janda, K. D. *Acc Chem Res* 2000, **33**, 546– 554.
2. Merrifield, R. B. *J Am Chem Soc* 1963, **85**, 2149–2154.
3. Letsinger, R. L.; Kornet, M. J. *J Am Chem Soc* 1963, **85**, 3045–3046.
4. Letsinger, R. L.; Mahadevan, V. *J Am Chem Soc* 1965, **87**, 3526–3527.
5. Gutte, B.; Merrifield, R. B. *J Biol Chem* 1971, **246**, 1922–1941.
6. Sherrington, D. C. *J Polym Sci A Polym Chem* 2001, **39**, 2364–2377.
7. Merrifield, R. B. *Angew Chem Int Ed Engl* 1985, **24**, 799–810 [Merrifield's Nobel lecture].
8. Furka, Á.; Sebestyén, F.; Asgedom, M.; Dibó, G. In *Abstracts 14th International Congress of Biochemistry*, Prague, 1988, p. 47; Furka, Á.; Sebestyén, F.; Asgedom, M.; Dibó, G. In *Abstracts 10th International Symposium on Medicinal Chemistry*, Budapest, 1988, p. 288; Furka, Á.; Sebestyén, F.; Asgedom, M.; Dibó, G. *Int J Peptide Protein Res* 1991, **37**, 487–493; Furka, Á. *Drug Discovery Today* 2002, **7**, 1–4.
9. Ley, S. V.; Schucht, O.; Thomas, A. W.; Murray, P. J. *J Chem Soc Perkin Trans 1* 1999, 1251–1252; Caldarelli, M.; Habermann, J.; Ley, S. V. *Bioorg Med Chem Lett* 1999, **9**, 2049–2052. Baxendale, I. R.; Ley, S. V. *Bioorg Med Chem Lett* 1999, **10**, 1983–1986.
10. Ley, S. V.; Baxendale, I. R.; Bream, R. N.; Jackson, P. S.; Leach, A. G.; Longbottom, D. A.; Nesi, M.; Scott, J. S.; Storer, R. I.; Taylor, S. J. *J Chem Soc Perkin Trans 1* 2000, 3815–4195.
11. Udenfried, S.; Gerber, L. D.; Brink, L.; Spector, S. *Proc Natl Acad Sci* USA 1985, **82**, 8672–8676; Hart, H. E.; Greenwald, E. B. *Mol Immunol* 1979, **16**, 265–267.
12. Sukata, K. *Bull Chem Soc Jpn* 1987, **60**, 1085–1089; Sukata, K. *J Org Chem* 1985, **50**, 4388–4390.
13. Regen, S. L.; Quici, S.; Liaw, S.-J. *J Org Chem* 1979, **44**, 2029–2030. Bram, G.; Filleben-Khan, T.; Geraghty, N. *Synth Commun* 1980, **10**, 279–289; Ando, T.; Kawate, T.; Yamawaki, J.; Hanafusa, T. *Chem Lett* 1982, **11**, 935–938.
14. Akiyama, R.; Kobayashi, S. *Chem Rev* 2009, **109**, 594–642.
15. Kobayashi, S.; Nagayama, S. *J Am Chem Soc* 1998, **120**, 2985–2986.
16. Nagayama, S.; Endo, M.; Kobayashi, S. *J Org Chem* 1998, **63**, 6094–6095.
17. Lattanzi, A.; Leadbeater, N. E. *Org Lett* 2002, **4**, 1519–1521.
18. Akiyama, R.; Kobayashi, S. *Angew Chem Int Ed* 2001, **40**, 3469–3471.
19. Ramarao, C.; Ley, S. V.; Smith S. C.; Shirley I. M.; DeAlmeida N. *Chem Commun* 2002, 1132–1133.

20. Ishida, T.; Akiyama, R.; Kobayashi, S. *Adv Synth Catal* 2005, **347**, 1189–1192.
21. Harland, C. E. *Ion Exchange: Theory and Practice*, 2nd ed., Royal Society of Chemistry, Cambridge, 1994.
22. Griffiths, W. P.; Ley, S. V.; Whitcombe, G. P.; White, A. P. *Chem Commun* 1987, 1625–1627.
23. Hinzen, B.; Ley, S. V. *Perkin Trans 1* 1997, 1907–1908.
24. Hinzen, B.; Lenz, R.; Ley, S. V. *Synthesis* 1998, 977–979.
25. Hinzen, B.; Ley, S. V. *Perkin Trans 1* 1998, 1–2.
26. Bleloch, A.; Johnson, B. F. G.; Ley, S. V.; Price, A. J.; Shephard, D. S.; Thomas, A. W. *Chem Commun* 1999, 1907–1908.
27. Leebrick, J. R.; Ramsden, H. E. *J Org Chem* 1958, **23**, 935–936.
28. Rabinowitz, R.; Marcus, R. *J Org Chem* 1961, **26**, 4157–4158.
29. Camps, F.; Castells, J.; Font, J.; Vela, F. *Tetrahedron Lett* 1971, **12**, 1715–1716.
30. Heitz, W.; Michels, R. *Angew Chem Int Ed Engl* 1972, **11**, 298–299.
31. McKinley, S. V.; Rakshys, J. W. *J Chem Soc Chem Commun* 1972, 134–135.
32. Heitz, W.; Michels, R. *Justus Liebigs Ann Chem* 1973, **764**, 227–230.
33. Relles, H. M.; Schluenz, R. W. *J Am Chem Soc* 1974, **96**, 6469–6475.
34. See for example: Solinas, A.; Taddei, M. *Synthesis* 2007, 2409–2453: Guino, M.; Hii, K. K. M. *Chem Soc Rev* 2007, **36**, 608–617.
35. Siegel, M. G.; Hahn, P. J.; Dressman, B. A.; Fritz, J. E.; Grunwell, J. R.; Kaldor, S. W. *Tetrahedron Lett* 1997, **38**, 3357–3360.
36. Mosbach, K. *Sci Am* 1971, **224**, 26–33.
37. Pittman, C. U.; Smith, L. R. *J Am Chem Soc* 1974, **97**, 1749–1754.
38. Stowell, J. C.; Hauck, Jr., H. F. *J Org Chem* 1981, **46**, 2429–2431.
39. Akagawa, K.; Sakamoto, S.; Kudo, K. *Tetrahedron Lett* 2007, **48**, 985–987.
40. Baxendale, I. R.; Ley, S. V.; Mansfield, A. C.; Smith, C. D. *Angew Chem Int Ed* 2009, **48**, 4017–4021. Palmieri, A.; Ley, S. V.; Hammond, K.; Polyzos, A.; Baxendale, I. R. *Tetrahedron Lett* 2009, **50**, 3287–3289
41. Sigma-Aldrich. Available at: http://www.sigmaaldrich.com.
42. Advanced Biosciences, Rohm and Haas. Available at: http://www.advancedbiosciences.com.
43. Meshram, H. M.; Reddy, P. N.; Sadashiv K.; Yadav, J. S. *Tetrahedron Lett* 2005, **46**, 623–626.
44. Ko S.; Yao, C.-F. *Tetrahedron Lett* 2006, **47**, 8827–8829.
45. Yadav, J. S.; Subba Reddy B. V.; Vishnumurthy, P. *Tetrahedron Lett* 2005, **46**, 1311–1313.
46. Yadav, J. S.; Subba, B. V.; Vishnumurthy, P. *Tetrahedron Lett* 2008, **49**, 4498–4500.
47. Pera-Titus, M.; Bausach, M.; Tejero, J.; Iborra, M.; Fité, C.; Cunill, F.; Izquierdo, J. P. *Appl Catal A* 2007, **323**, 38–50.
48. Iborra, M.; Tejero, J.; Fité, C.; Cunill, F.; Izquierdo, J. F. *J Catal* 2005, **231**, 77–91.
49. Wissler, M. C.; Jagusch, U.-T.; Sundermann, B.; Hoelderich, W. F. *Catal Today* 2007, **121**, 6–12.
50. Seki, T.; Grunwaldt, J. D.; van Vegten, N.; Baiker, A. *Adv Synth Catal* 2008, **350**, 691–705.
51. Lakouraj, M. M.; Noorian, M.; Mokhtary, M. *React Funct Polym* 2006, **66**, 910–915.

52. Bongini, A.; Cainelli, G.; Contento, M.; Manescalchi, F. *Synthesis* 1980, 143–146.
53. Nag, A.; Sarkar, A.; Sarkar, S. K.; Palit, S. K. *Synth Commun* 1987, **17**, 1007–1013.
54. Goudgaon, N. M.; Wadgaonkar, P. P.; Kabalka, G. W. *Synth Commun* 1989, **19**, 805–811.
55. Cainelli, G.; Cardillo, G.; Orena, M.; Sandri, S. *J Am Chem Soc* 1976, **98**, 6737–6738.
56. Deleuze, H.; Maillard, B.; Mondain-Monval, O. *Bioorg Med* Chem Lett 2002, **12**, 1877–1880.
57. Cummins, D.; Wyman, P.; Duxbury, C. J.; Thies, J.; Koning, C. E.; Heise, A. *Chem Mater* 2007, **19**, 5285–5292.
58. Bartl, H.; von Bonin, W. *Makromol Chem* 1962, **57**, 74–95; Bartl, H.; von Bonin, W. *Makromol Chem* 1963, **66**, 151–156.
59. Barby, D.; Haq, Z. European Patent 060138, September 1982.
60. Cameron, N. R.; Sherrington, D. C.; Ando, N.; Kurosu, H. *J Mater Chem* 1996, **6**, 719–726.
61. Mercier, A.; Deleuze, H.; Mondain-Monval, O. *React Funct Polym* 2000, **46**, 67–79.
62. Mercier, A.; Deleuze, H.; Mondain-Monval, O. *Macromol Chem Phys* 2001, **202**, 2672–2680.
63. Cetinkaya, S.; Khosravi, E.; Thompson, R. *J Mol Cat A* 2006, **254**, 138–144.
64. Barbetta, A.; Cameron, N. R.; Cooper, S. J. *Chem Commun* 2000, 221–222.
65. Krajnc, P.; Brown, J. F.; Cameron, N. R. *Org Lett* 2002, **4**, 2497–2500.
66. Wang, S. S. *J Am Chem Soc* 1973, **95**, 1328–1333.
67. Kranjc, P.; Štefanec, D.; Brown, J. F.; Cameron, N. R. *J Polym Sci A Polym Chem* 2005, **43**, 296–303.
68. Leber, N.; Fay, J. D. B.; Cameron, N. R.; Krajnc, P. *J Polym Sci A Polym Chem* 2007, **45**, 4043–4053.
69. Sherrington, D. C.; Small, P. W. U.S. Patent 4,965,289, October 23, 1990.
70. Sherrington, D. C.; Small, P. W. U.S. Patent 5,066,784, November 19, 1991.
71. Bycroft, B. W.; Chan, W. C.; Ram Chhabra S.; Hone, N. D. *Chem Commun* 1993, 778–779.
72. Krajnc, P.; Leber, N.; Brown, J. F.; Cameron, N. R. *React Funct Polym* 2006, **66**, 81–91.
73. Pulko, I.; Kolar, M.; Krajnc, P. *Sci Total Environ* 2007, **386**, 114–123.
74. Katsoyiannis, I. A.; Zouboulis, A. I. *Water Res* 2002, **36**, 5141–5155.
75. Benicewicz, B. C.; Jarvinen, G. D.; Kathios, D. J.; Jorgenson, B. S. *J Radioanal Nucl Chem* 1998, **235**, 31–35.
76. Brown, J. F.; Krajnc, P.; Cameron, N. R. *Ind Eng Chem Res* 2005, **44**, 8565–8572.
77. Cummins, D.; Duxbury, C. J.; Quaedflieg, P. J. L. M.; Magusin, P. C. M. M.; Koning, C. E.; Heise, A. *Soft Matter* 2009, **5**, 804–811.
78. Ottens, M.; Leene, G.; Beenackers, A. A. C. M.; Cameron, N.; Sherrington, D. C. *Ind Eng Chem Res* 2000, **39**, 259–266.
79. Mercier, A.; Deleuze, H.; Maillard, B.; Mondain-Monval, O. *Adv Synth Catal* 2002, **344**, 33–36.
80. Pierre, S. J.; Thies, J. C.; Dureault, A.; Cameron, N. R.; van Hest, J. C. M.; Carette, N.; Michon, T.; Weberskirch, R. *Adv Mater* 2006, **18**, 1822–1826.
81. Dizge, N.; Keskinler, B.; Tanriseven, A. *Coll Surf B Biointerf* 2008, **66**, 34–38.
82. Harrison, C. R.; Hodge, P J Chem Soc Perkin Trans 1 1976, 2252–2254.

83. Burguete, M. I.; Cornejo, A.; García-Verdugo, E.; Gil, M. J.; Luis, S. V.; Mayoral, J. A.; Martinez-Meríno, V.; Sokolova, M. *J Org Chem* 2007, **72**, 4344–4350.

84. Kirschning, A.; Altwicker, C.; Drager, G.; Harders, J.; Hoffmann, N.; Hoffmann, U.; Schonfeld, H.; Solodenko, W.; Kunz, U. *Angew Chem Int Ed Engl* 2001, **40**, 3995–3998.

85. Kirschning, A., Ed. *Beilstein J Org Chem* 2009(Thematic Series 4), articles 15–19, 21, 23, 27, 29, 30, 34, 35, 40, 56, 70.

86. Kovačič, S.; Krajnc, P. *J Polym Sci A* 2009, **47**, 6726–6734.

CHAPTER 12

Templates for Porous Inorganics

ARNE THOMAS, JENS WEBER, and MARKUS ANTONIETTI
Max-Planck-Institute of Colloids and Interfaces, Potsdam, Germany

12.1 INTRODUCTION

Polymer gels can be classified into microporous, mesoporous, and macroporous, depending on the pore size (according to the International Union of Pure and Applied Chemistry, microporous, diameter $D < 2$ nm; mesoporous, $2 < D < 50$ nm; macroporous, $D < 50$ nm). Gels with mesoporosity turn out to be especially useful for ion exchange, as catalyst supports, or for sorption, while porous polymer scaffolds with macropores are especially useful for transporting fluids though their structure. The combination of mesoporosity and macroporosity is most advantageous and is usually described as "hierarchical pore system." Most of the systems dealt with in this chapter are designed to solve complex problems and are therefore mostly hierarchical.

The procedures for making such gels have greatly improved, and the range of obtainable pore sizes, pore connectivities or morphologies, surface functionalities, and overall polarities has been significantly extended. With this growing versatility, new, promising fields of application have been developed. Starting from the first applications of porous polymers as insulators and ion-exchange resins, their outstanding performance in different types of column chromatographies and as gel separation media was rapidly recognized. Meanwhile, porous polymers found application as high-tech materials for sensors, artificial antibodies, and enantioselective catalysts, and their use for electrooptical and (micro)electronic devices, implant materials, and supports for flowthrough reactors is being explored [1]. This includes some of the hybridization techniques discussed in this chapter.

Whereas synthesis and direct applications are discussed in other chapters, we delineate how the as-fabricated polymer gels can be employed as structured templates. Porous polymer gels or materials can possess outstanding and unusual properties, but sometimes complementary properties such as magnetism, metal-like behavior, or ceramic-like high-temperature stability are demanded. Then, the useful pore

Porous Polymers, Edited by Michael S. Silverstein, Neil R. Cameron, and Marc A. Hillmyer

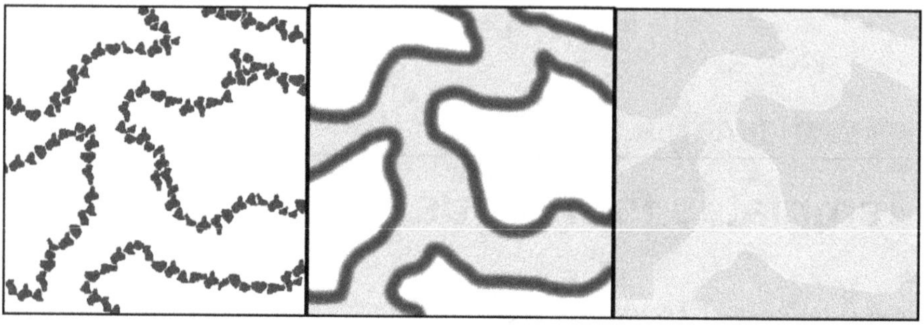

FIGURE 12.1 Schematic presentation of different template processes. In "nano-Pickering," only nanoparticles are deposited, the mutual arrangement of which, however, depends on the gel template. "Nanocoating" gives a thin layer of the second, templated material, while nanocasting results in the structural inversion of the template.

and scaffold structure can be replicated into the other material using the polymer as a structure-donating template. This process is also sometimes called "structural transcription," in analogy to the biological case of DNA and the coupled resulting peptide structure.

If, after replication, the polymer is removed, for example, by solvent treatment or calcination, a hollow replica of the primary structure is obtained. In many cases, however, this removal is not necessary, and a hybrid material that combines the functional properties of the templated material with the typical properties of a porous organic structure (ductility, elasticity, lightweight design) is obtained. Such hybrids with porous polymers sometimes have extreme mechanical properties, and bone or seashell are examples from nature in which a porous polymer structure is used as a template for natural mineralization, creating extraordinary properties [2, 3].

Figure 12.1 presents the principal differences among "nano-Pickering," "nanocoating," and "nanocasting," where the gel structure defines the principal morphology of the second phase to be templated. If the gel is kept, we speak of a hybrid material, while we speak of transcription when the gel is removed after structure donation.

12.2 SYNTHESIS OF POROUS POLYMER NETWORKS SUITABLE FOR TEMPLATING

Most properties of porous polymers, such as permeability, mechanical strength, transparency, size exclusion/filtering, and adsorption power, mainly depend on the topology or connectivity of the polymer strands or the pore system they constitute. Each application is optimally served with a special architecture to be generated.

Foaming techniques make use of gaseous porogens. These might be generated by evaporation of solvents during temperature increase or pressure drop (e.g., polystyrene foam), or are produced by chemical reactions during polymerization (e.g., polyaddition of urethanes). Because of the comparably high interface tension of gas bubbles in polymers, they undergo rapid Ostwald ripening, and the resulting pore structure can be open or closed, but both absolute porosity and pore size are usually very large, that is, macroporous polymer gels are formed. Formation of finer pore systems relies on the addition of surfactants, for example, silicone surfactants.

Porogenesis can also be provided by polymer–solvent phase separation processes. The phase separation can be induced throughout polymerization and crosslinking in different ways: (1) chemically induced (e.g., the polymerization is performed in a monomer/nonsolvent mixture, the polymerization itself depletes the monomer, and insolubility is induced), or (2) by thermally induced phase separation (TIPS). These are the standard systems used in templating.

Pores can also be produced by templates, using also concepts of supramolecular and colloid chemistry. Syntheses within self-organized media are convenient one-step reactions toward porous polymers and open manifold opportunities to control the pore architecture in a much broader range than possible by other techniques. The highest goal of a template synthesis, namely the direct replication of a self-organized structure into a permanent (polymeric) material, however, is complicated. This is due to the fact that changes of the free energy of mixing throughout polymerization usually destabilize the well-chosen starting situation. In most cases only kinetic control of the pore morphology and therefore indirect structure control are obtained. Polymer gels based on supramolecular assembly, however, are highly significant for templating applications.

Besides the pore geometry, the hydrophilicity/hydrophobicity and the chemical surface functionalization also have to be adjusted for gel templates. The process of templating relies on an intense coupling of template and the material to be shaped, cast, or molded, and the net interface energy should be as small as possible, in the best case even negative. In ternary systems when a more unfortunate surface contact is resolved, say between the primary gel and the solvent, the presence of the third component can significantly lower the free energy of the system. For liquids, this phenomenon is called spreading and is characterized by a spreading coefficient, but the phenomenon also exists for solids and is the basis of powder coating.

For successful gel templating, the surface of the gel should be chosen to interact with the material used for the coating to generate this negative free energy, for example, for minerals potential ligands or carboxylic acids are favorable on the gel surface, for protein coats hydrogen bridge systems are advised, and gold layers can be bound by thiols. Therefore, addition of appropriate binder monomers may be considered throughout polymer gel synthesis.

12.3 POLYMERS GELS AS TEMPLATES: SOL–GEL NANOCOATING AND NANOCASTING

Such porous polymers, when used as templates, represent a mesostructured reaction medium. Beside other effects, this means that the outcome of the reaction depends on the polymer network structure and might be seriously different from the corresponding bulk reactions. This is due not only to altered transport properties, but also to chemical size quantization effects and to the local presence of large areas of functional polymer surface. Most especially, nucleation and crystallization can occur in completely altered fashion. Homogeneous nucleation is usually suppressed in small volumes, and, assuming a materials-friendly surface functionalization, heterogeneous nucleation can dominate.

A usually neglected parameter is solvent quality. To allow spreading or sufficient material binding, the solvent should be as poor as possible for both the porous polymer gel and the material to be templated. Even if the reaction medium has to be water, solvent quality and spreading can be adjusted by temperature, salt addition, or consideration of a mixed solvent situation.

The use of porous polymer gels made from bicontinuous microemulsions for the synthesis of organic–inorganic nanohybrids was demonstrated by polymer-directed mineralization of iron oxide within a bicontinuous polystyrene-*co*-polyacrylic acid gel [4] where the acrylic acid hydrophilizes the surface and acts as a mineral binder. Here, a nanocomposite consisting of a porous polymer gel functionalized by homogeneously distributed, superparamagnetic magnetite particles of about 15 nm diameter was obtained. The overall hybrid structure represents an "elastic magnet" that couples electromagnetic and mechanic behavior. Of interest, the observed morphology is reminiscent of the structure of a mollusk tooth, which consists of magnetite colloids of comparable sizes within a protein–polysaccharide matrix. The mollusk uses the combination of elastic gel and mineral nanoparticles as an abrasive tool.

If the amount of material to deposit is increased, we move from a dispersion to a coating situation [5]. Highly porous oxidic semiconductor networks that could be used for photocatalytic and solar cell applications were generated in this way by using porous polymers as templates [6]. Titanium dioxide networks were produced, resulting in a coral-like structure with a wall thickness of about 100–150 nm and transport pore sizes ranging from 100 nm to micrometers in diameter. Calcination resulted in the production of either the anatase or rutile crystal phase without carbon contamination from the starting polymer materials, with the individual titanium dioxide particles contacting neighbors, thereby forming a continuous network. The results obtained are of importance to various fields of research, as the open "coral-like" network allows fast and direct access of the titanium dioxide surface to the reaction medium.

This approach was later repeated with preformed commercial polyamide and cellulose acetate filter membranes with hierarchical porosity, and correct structural transcription was obtained, besides a slight shrinking, on all lengths scales up to the preservation of the original membrane geometry [7]. The resulting structures are depicted in the scanning electron microscopy (SEM) and transmission electron microscopy (TEM) images in Fig. 12.2.

Polymer gel templating has the big advantage that the scaffold supports the reaction center as long as demanded, while it also can be removed whenever needed. This results in multifunctional, structured replicas with astonishing precision and layout—for example, functional gradient structures and layered scaffolds. This was demonstrated in Reference 8. Here, a microemulsion-based globular gel with

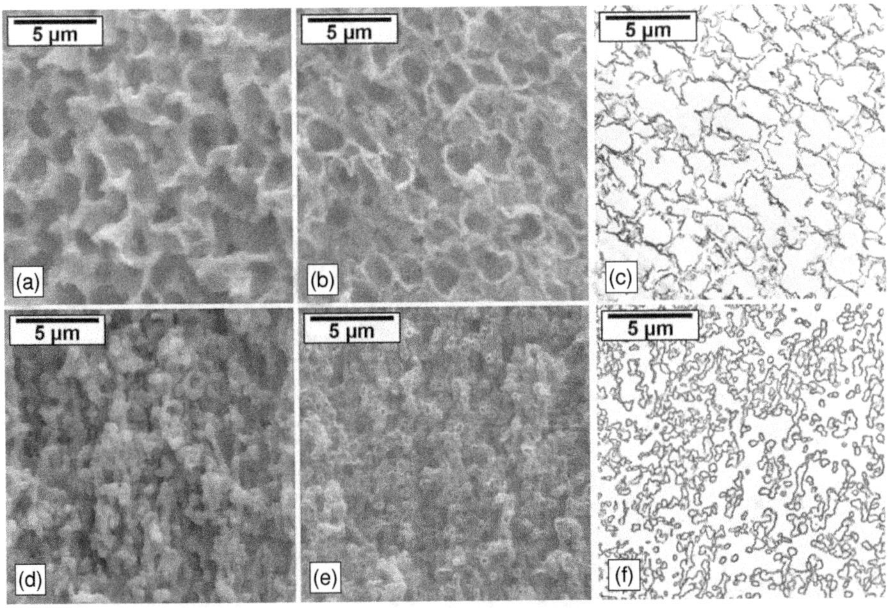

FIGURE 12.2 **(a)** SEM images of the polyamide membrane, **(b)** the titanium dioxide structure obtained by coating this membrane three times with titanium (IV) isopropoxide (17 wt%), and **(c)** a TEM image of the ultramicrotomed slice of the inorganic structure. SEM images of **(d)** cellulose acetate membrane and **(e)** the titanium dioxide structure obtained by coating this template three times with titanium (IV) isopropoxide (17 wt%), and **(f)** TEM image of the ultramicrotomed slice of the inorganic structures. (From Reference 7.)

dimension of 1 cm × 1 cm × 1 cm was infiltrated/nanocoated in a consecutive fashion with titania- and zirconia-isopropoxide solution.

The high quality of structural replication between the original polymer gel and the (hollow) nanocoats of both materials is shown in Fig. 12.3. Note that, similar to atomic layer deposition [9] but on the colloidal scale, the thickness of the coats can be controlled in the 10- to 200-nm range by the number of repetitions of gel impregnation and the concentration of precursor solution.

Dipping the gels in different solutions (enabling cycles of gel impregnation and fixation) results in multilayer structures of flexible geometry, the thickness and composition of which depend on the demands of application. The whole construction is mechanically stabilized by the porous gel against fracture and collapse. In this case, the structure is mineralized at the end, thus allowing a comparison of the quality of the replica with that of its original polymer gel template

Both systems were, for instance, characterized by SEM, but also by mercury porosimetry (Fig. 12.4). We can find preservation of all essential structural features of the pore system, while throughout calcination a linear shrinkage by a factor 2 was observed. This is remarkable, as the final inorganic scaffolds still have an absolute porosity of 98%, while the crystal hollow foam structure is comparably stable.

FIGURE 12.3 SEM images of (a) the initial polymer gel template, (b) the pure ZrO_2 network, and (c, d) the pure TiO_2 network (different magnifications). Scale bars are 5 μm, 5 μm, 5 μm, and 2 μm.

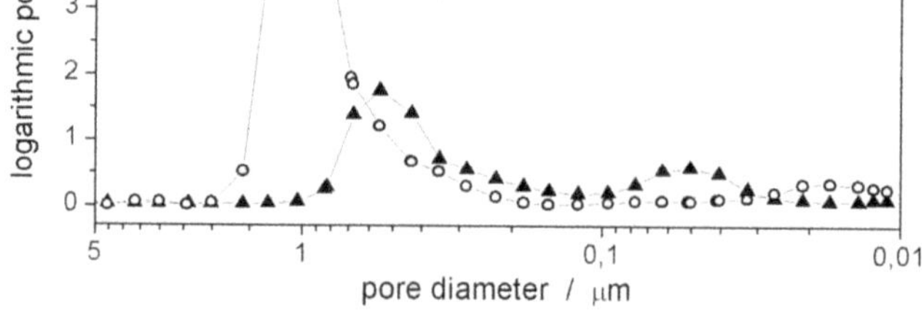

FIGURE 12.4 Mercury intrusion results for the polymer gel template and the titania/zirconia material.

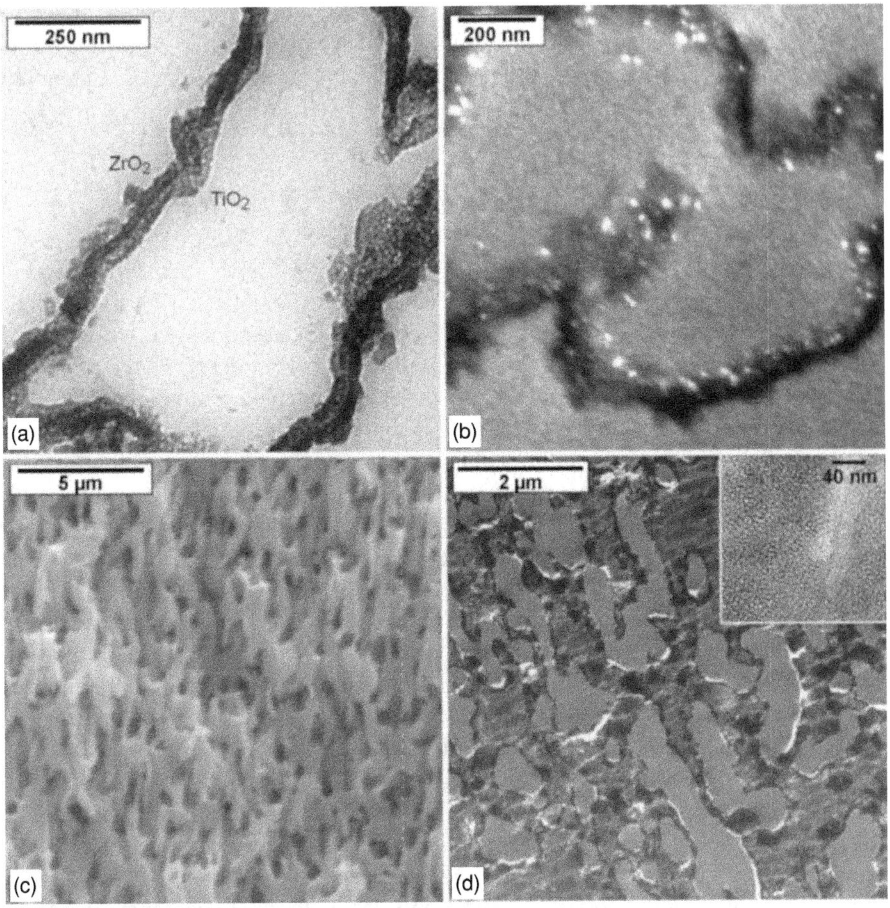

FIGURE 12.5 The sequential coating approach was used to obtain complex materials. **(a)** TEM and **(b)** dark-field TEM images of the two layers of titanium dioxide followed by two layers of zirconium dioxide. **(c)** SEM image of the TiO_2/SiO_2 material (two layers TiO_2 followed by cast with silica in the presence of C16E10). **(d)** TEM image of an ultramicrotome of the TiO_2/SiO_2. Insert shows the mesoporous character of the silica as observed by TEM.

The local structure of the titania/zirconia coating is best resolved in TEM on microtomed specimens (Fig. 12.5). Due to the higher electron density of zirconia, the layered setup becomes visible as a binary layout in two different shades of gray.

Such systems are useful for improving photoinduced charge separation by distribution among two different materials (a colloidal "diode" construction). Indeed, such systems showed a remarkably improved efficiency on the light-induced deterioration of water-soluble organic compounds [8]. In the present context, it is worth mentioning that the structure can be stabilized for the real application by a final filling of the coated gel with silica, thus going from nanocoating to nanocasting. Then the final material is a structural negative of the primary gel, as depicted in Fig. 12.5c, d. Of interest, the active diode structure is still on the surface of the silica monolith, although now in an inverted order.

12.4 MESOPOROUS AND MICROPOROUS POLYMERS AS TEMPLATES: DOWN TO THE MOLECULAR SCALE

The previous examples dealt mainly with macroporous polymers. This section highlights some examples of using mesoporous or microporous polymers as templates. Generally speaking, one finds far less use of mesoporous or microporous polymers as templates. This is in accordance with the more sophisticated synthetic requirements that have to be met in order to achieve stable, homogeneous porosity and a low-pore-size polydispersity. In addition, the pore size now becomes compatible with the size of molecules, thus hindering structural replication both by transport and by the "pixel-size problem" of chemical pictures (the "pixel problem" in chemistry is that the minimal structural features of an object or a curve are given by the elemental unit of the network, here, e.g., a silica tetrahedron).

In the case of mesostructured polymeric materials, one finds replication of both bulk systems and thin films, but the procedures used are comparable. Filling the mesopores of a well-defined porous film with a precursor solution of the targeted material followed by reaction of these precursors yields, depending on the pore morphology, nanostructures like nanowires or nanoparticles.

For example, it is possible to synthesize metal nitride nanoparticles within the mesopores of a graphitic carbon nitride–type polymeric material [10]. After filling of the pores, which had a diameter of 12 nm, with solutions of metal chlorides in ethanol, the precursors reacted in a sol–gel process and formed metal oxide nanoparticles. These were transformed into metal nitrides by thermal decomposition of the polymeric matrix, leaving the pure nanoparticles behind. Cobalt nanowires having a diameter of 14 nm and a length of ~500 nm were grown in a mesoporous polymer film [11]. A PS-*b*-PMMA polymer was spin coated on an electrically conducting surface, and after annealing under the influence of an electric field, degradation of the PMMA block, and crosslinking of the PS block, cylindrical pores that were aligned perpendicular to the surface were obtained. These pores could be infiltrated with a cobalt salt, and the cobalt wires were grown electrochemically. If an ultrathin film of a block copolymer is deposited, it is possible to use the microphase-separated structure to generate the porous polymer structure and then to use the regular array of holes for lithography [12]. Regular arrays of silicon nitride holes or dots with sizes down to 15 nm and periods of 30 nm are accessible using this pathway.

As delineated at the beginning, it is not always necessary to remove the template, as the resulting hybrid material might profit from the existing polymer scaffold. Such a case is the use of mesoporous poly(benzimidazole) (PBI) as a host for polyphosphoric acid. Adducts of PBI and phosphoric acid are good proton conductors, but usually the proton transport at zero humidity suffers from the homogeneous character of the material. To overcome this problem, a mesoporous PBI network with pore diameters of 12 nm was synthesized by a hard-templating approach using silica nanosols, and the pores were subsequently filled with phosphoric acid [13]. In that case, the porous polymer can be regarded as a template and scaffold for microstructured domains of phosphoric acid. It was shown that such a nanostructural control of the domain size was beneficial for the proton conductivity properties.

There are a variety of other microporous and mesoporous polymers that have potential to be used in templating applications—for example, microporous and mesoporous conjugated polymers that probably allow the formation of interpenetrating networks with charge transport materials for hybrid solar cells [14, 15]—but generally this field is just at its beginning, and more work is still to be done.

12.5 POROUS POLYMER NETWORKS FOR MINERALIZATION EXPERIMENTS

The templating of porous polymer gels is not only a novel synthetic process for controlling the structure and properties of structured matter, it is also a biological process employed for ca. 500 million years for the benefit of living species. It was already mentioned that bone and seashells ("nacre") are essentially highly mineralized biogels, usually made from collagen, with some secondary grafted polymers used to modify wetting behavior and binding to the mineral. It was therefore of interest to analyze how nacre is in reality set up and to see whether nacre formation can be copied by simple chemical means [16]. Nacre mineralization essentially starts from a soft collagen structure that is a stable porous polymer hydrogel. The structures of these gels can be analyzed by demineralizing a seashell, and they were found to be lamellar with slit pores 100–300 nm thick and 1–30 μm long (see Fig. 12.6a). Of interest, the gel layers contain a second type of "perforating mesopore" with a size of 30–50 nm; these are clearly seen in the TEM pictures within the gel layers and presumably function as the transport entries to perform the mineralization within the larger macropores. Biological processes usually deposit $CaCO_3$ into those gels, which then recrystallizes as aragonite.

"Synthetic" chemical nanocasting (see Fig. 12.1) can be performed in a similar fashion. Due to the mesopore entries, $CaCO_3$ was added as amorphous nanoparticles that deposited onto the collagen gels. Recrystallization and continued nanoparticle transport slowly lead to a controlled "nanocasting" of the gel (Figs. 12.6b–d), leading finally to a densely mineralized gel structure in which the gel influences not only the shape and texture of the crystals, but also their joint crystal orientation.

The final synthetic material seems very similar to the natural material, exhibiting similar mechanical behavior and fracture surfaces. The practically equivalent microscale textures of synthetic and natural nacre are shown in Fig. 12.7a, b, respectively, indicating that the gel template encoded practically complete structural information into the hybrid material. It is noteworthy that the synthetic nacre was made up from calcite (presumably due to the absence of appropriate polymorph-controlling proteins) instead of aragonite.

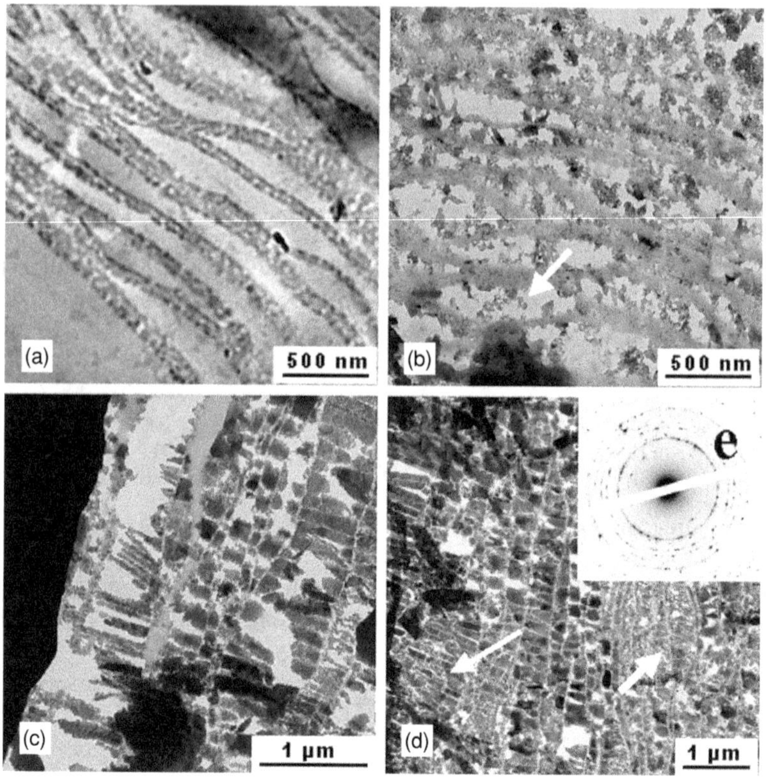

FIGURE 12.6 TEM pictures of **(a)** a completely demineralized collagen gel matrix. **(b–d)** Matrix after 24 h of remineralization; **(b)** area with low mineralization, containing nanoparticles (ca. 30 nm) starting to form aggregates; **(c)** area with higher mineralization, showing aggregates; **(d)** highly mineralized area, with platelets (see Fig. 121a, b) as well as aggregates observable. **(e)** Electron diffraction pattern of the aggregates in panel (d). The sample in panel (a) was critical point-dried, and the samples in panels (b)–(d) were prepared by microtoming only. (From Reference 16 with permission.)

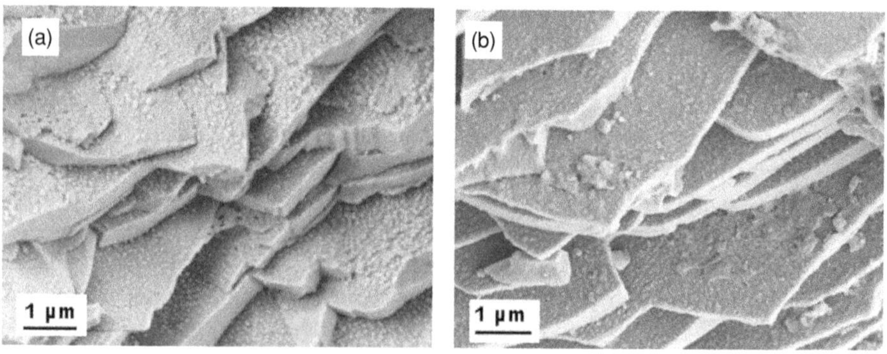

FIGURE 12.7 SEM micrographs of fracture surfaces of **(a)** synthetic nacre made by biogel templating and **(b)** original nacre from *Haliotis laevigata*.

There are many aspects to be learned from such a retro-synthetic approach. In the present context, it is only important to repeat that (1) it was the polymer template gel that controlled the structure, (2) templating is not restricted to sol–gel chemistry or reductions but can be extended to mineralization, and (3) nature uses porous polymer gel templating.

12.6 SUMMARY AND OUTLOOK

This chapter indicated that porous polymer gels are highly valuable as templates for the "morphosynthesis" (the controlled chemical generation of shape and form) of material and material systems, such as binary hybrids. The very special structure of polymer gels is transcribed by templating into a secondary material that is otherwise not easily shaped and structured. This can occur via deposition of nanoparticles onto the gel structure ("nano-Pickering"), the coating of the gel structure ("nanocoating"), or complete filling and structural replication of the gel ("nanocasting"). Any progress in structuring and functionalizing porous polymer gels can thereby directly be transferred into structuring materials in general. Gel templates for functional design are omnipresent in the natural world, yielding a rich harvest in properties and functions that are unobtainable by synthetic chemistry. In other words, the most significant advances are still to come.

The potential applications of the resulting structures are manifold: The cases delineated in this chapter already range from elastic magnets and switchable soft materials, to photocatalyst and catalyst support materials, to nanostructured layers for solar cells and fuel cell membranes, as well as to new sustainable hybrid materials for mechanical purposes.

REFERENCES

1. Hentze, H.-P.; Antonietti, M. *Curr Opin Solid State Mater Sci* 2001, **5**, 343–353.
2. Fratzl, P.; Gupta, H. S.; Paschalis, E. P.; Roschger, P. *J Mater Chem* 2004, **14**, 2115–2123.
3. Jackson, A. P.; Vincent, J. F. V.; Turner, R. M. *Proc R Soc Lond B Biol Sci* 1988, **234**, 415–440.
4. Antonietti, M.; Hentze, H.-P. *Coll Polym Sci* 1996, **274**, 696–702.
5. Caruso, R. A.; Antonietti, M. *Chem Mater* 2001, **13**, 3272–3282.
6. Caruso, R. A.; Giersig, M.; Willig, F.; Antonietti, M. *Langmuir* 1998, **14**, 6333–6336.
7. Caruso, R. A.; Schattka, J. H. *Adv Mater* 2000, **12**, 1921–1923.
8. Schattka, J. H.; Shchukin, D. G.; Jia, J.-G.; Antonietti, M.; Caruso, R. A. *Chem Mater* 2002, **14**, 5103–5108.
9. Leskela, M.; Ritala, M. *Angew Chem Int Ed* 2003, **42**, 5548–5554.
10. Fischer, A.; Antonietti, M.; Thomas, A. *Adv Mater* 2007, **19**, 264–267.
11. Thurn-Albrecht, T.; Schotter, J.; Kastle, G. A.; Emley, N.; Shibauchi, T.; Krusin-Elbaum, L.; Guarini, K.; Black, C. T.; Tuominen, M. T.; Russell, T. P. *Science* 2000, **290**, 2126–2129.

12. Park, M.; Harrison, C.; Chaikin, P. M.; Register, R. A.; Adamson, D. H. *Science* 1997, **276**, 1401–1404.
13. Weber, J.; Antonietti, M.; Thomas, A. *Macromolecules* 2007, **40**, 1299–1304.
14. Weber, J.; Thomas, A. *J Am Chem Soc* 2008, **130**, 6334–6335.
15. Jiang, J. X.; Su, F.; Trewin, A.; Wood, C. D.; Campbell, N. L.; Niu, H.; Dickinson, C.; Ganin, A. Y.; Rosseinsky, M. J.; Khimyak, Y. Z.; Cooper, A. I. *Angew Chem Int Ed* 2007, **46**, 8574–8578.
16. Gehrke, N.; Nassif, N.; Pinna, N.; Antonietti, M.; Gupta, H. S.; Cölfen, H. *Chem Mater* 2005, **17**, 6514–6516.

INDEX

Porous Polymers, Edited by Michael S. Silverstein, Neil R. Cameron, and Marc A. Hillmyer

www.ingramcontent.com/pod-product-compliance
Lightning Source LLC
LaVergne TN
LVHW011901200926
848484LV00002B/499